Ethology and Behavioral Ecology of Marine Mammals

Series Editor
Bernd Würsig, Department of Marine Biology, Texas A&M University at Galveston, Galveston, TX, USA

The aim of this series is to provide the latest ethological information on the major groupings of marine mammals, in six separate books roughly organized in similar manner. These groupings are the 1) toothed whales and dolphins, 2) baleen whales, 3) eared seals and walrus, 4) true seals, 5) sea otter, marine otter and polar bear, and 6) manatees and dugong, the sirens. The scope shall present 1) general patterns of ethological ways of animals in their natural environments, with a strong bent towards modern behavioral ecology; and 2) examples of particularly well-studied species and species groups for which we have enough data. The scope shall be in the form of general and specific reviews for concepts and species, with an emphasis especially on data gathered in the past 15 years or so. The editors and authors are all established scientists in their fields, even though some of them are quite young.

More information about this series at http://www.springer.com/series/15983

Bernd Würsig
Editor

Ethology and Behavioral Ecology of Odontocetes

Editor
Bernd Würsig
Texas A&M University at Galveston
Galveston, TX, USA

ISSN 2523-7500 ISSN 2523-7519 (electronic)
Ethology and Behavioral Ecology of Marine Mammals
ISBN 978-3-030-16665-6 ISBN 978-3-030-16663-2 (eBook)
https://doi.org/10.1007/978-3-030-16663-2

© Springer Nature Switzerland AG 2019, corrected publication 2020
This work is subject to copyright. All rights are reserved by the Publisher, whether the whole or part of the material is concerned, specifically the rights of translation, reprinting, reuse of illustrations, recitation, broadcasting, reproduction on microfilms or in any other physical way, and transmission or information storage and retrieval, electronic adaptation, computer software, or by similar or dissimilar methodology now known or hereafter developed.
The use of general descriptive names, registered names, trademarks, service marks, etc. in this publication does not imply, even in the absence of a specific statement, that such names are exempt from the relevant protective laws and regulations and therefore free for general use.
The publisher, the authors, and the editors are safe to assume that the advice and information in this book are believed to be true and accurate at the date of publication. Neither the publisher nor the authors or the editors give a warranty, express or implied, with respect to the material contained herein or for any errors or omissions that may have been made. The publisher remains neutral with regard to jurisdictional claims in published maps and institutional affiliations.

Cover illustration: On the left: Part of a subgroup of rough-toothed dolphins, *Steno bredanensis*, off Kona, Hawai'i. They have very large brain to body size ratios, and sophisticated rapid learning. Photo by Deron S. Verbeck. On the right: Short-finned pilot whales, *Globicephala macrorhynchus*, socialize in the waters off Tenerife, Canary Islands. Photo by Teo Lucas.

This Springer imprint is published by the registered company Springer Nature Switzerland AG.
The registered company address is: Gewerbestrasse 11, 6330 Cham, Switzerland

Dusky dolphin (*Lagenorhynchus obscurus*) males "scramble competing" for a female below and in front of them, with males leaping high to home onto the female, and very rapidly to out-leap/out-distance other males. These races and leaps take much energy, can go on for more than one hour, and are part of the multi-male/multi-female system of sexual strategy quite common in delphinid society. While several males inseminate a female during her estrus, they also have large testes and compete not only by behavior as shown here but by sperm competition of volume of sperm in one female at one time of estrus. Researchers are just now beginning to unravel aspects of behavioral and possible physical/physiological female choice that must be very important in scramble competition strategy (Photo by B. Würsig, reprinted from cover of *Aquatic Mammals* 2018, 44(6), with permission)

Dolphin, leaping high in the sky
Diving deeply, far past our sight.
You seem quite joyful, and adventurous...
And yet, we hardly know you.

Decades of research
Following in small boats.
Documenting behaviors from oceangoing
ships
Tracking and observing from high cliffs.

We immerse ourselves in water
To chance upon an intimate glimpse of life.
What is your fabric of society?
How do you relate to others; what do they
mean, to you?

What clues of life can we learn
To study foraging, social, sexual, calf rearing,
predator avoidance behaviors?
These are, in so many ways, mysteries...
And so, in we dive.

Robin Vaughn-Hirshorn
Mt. Airy, MD, USA

Introduction to the Series

We are pleased to provide the reader with a series on ethology and behavioral ecology of marine mammals. We define ethology as "the science of animal behavior" and behavioral ecology as "the science of the evolutionary basis for animal behavior due to ecological pressures." In our assessment, those ecological pressures include us, the humans. We determine, somewhat arbitrarily but with some background, that "marine mammals" are those mammals that habitually feed in the sea, but also include several mammals that went from saltwater oceans back into rivers, as seen in the chapter by Sutaria et al. in the first book of the series (Odontocetes).

Thus, we include toothed whales (the odontocetes) as Book 1; baleen whales (the mysticetes) as Book 2; sea lions and fur seals (the otariids) as well as the walrus as Book 3; true seals (the phocids) as Book 4; the special cases of sea otters and polar bears as Book 5; and manatees and the dugong (the sirens) as Book 6. Now, each of our chosen editors and their chapter authors have their own schedules, so the series will likely not arrive in the order given above, but we have full faith that within the next 5 years, i.e., 2019–2023, all six marine mammal books on "Ethology and Behavioral Ecology of Marine Mammals" will have seen the light of day, and you, the readers, will be able to ascertain their worth and their promise, as to present knowledge and to accumulating data while our fields of science advance.

For those of you paying close attention, we admit that not all mammals that occur in marine waters are represented, nor all who have gone back to freshwater. Thus, there is nary a mention of marine-feeding bats, marine-feeding river otters, those aspects of beluga whales who foray way up into major rivers, seals living in land-locked lakes at times thousands of kilometers from the ocean, and other species that occasionally make the marine environment or—as generally accepted marine mammals—adjacent freshwater systems their home. Such are the ways of a summary, and we apologize that we will not be able to please all.

As the series editor, I confess to having been a science party to all of the major taxonomic entities of this series, but this is only because I have been in the marine mammal field for about 50 years now, with over 65 graduate students who—in aggregate—have conducted research on all continents, and in no manner do I pretend to have kept up with all aspects of diverse fields of modern enquiry. It is a special privilege (and delight) to have multiple up-to-date editors and their fine authors involved in this modern compilation, and I am extremely grateful for this. I am learning, still and ever.

I confess to lending only a minor hand to all but the first book on odontocete cetaceans. At first, it was envisioned that each book be similar to the structure of that first book, with a general section on patterns of ethology, and several in-depth species-specific chapters after that. But chosen coeditors for the other books soon told me that this would not work—the knowledge base of other-than-odontocetes is simply not the same for those other taxa, and the books need to represent this. I (at first reluctantly) agreed. We are here to provide the reader with up-to-date excellent new summaries of the state of the art of these taxa, and it would be inefficient to stick to a particular formula of presentation for all. Each chapter is excellently reviewed by the book editors, peer reviewed by other scientists as chosen by the editors, and perused and commented on by me. If you learned something new and imparted that to your colleagues and students, then the series and sections of it shall have been worthwhile.

With respect and best wishes

Tortolita Desert, AZ, USA   Bernd Würsig
December 2018

Preface

It is tremendously satisfying to help put together a book of this nature, an up-to-date summary of data and thoughts about ethology and behavioral strategies of odontocetes, the toothed whales. Choosing the authors (who chose their own coauthors, if any) was a challenge—each and every subdiscipline of our wonderful field could have had other talented authors, from overall topics such as mating strategies, mom-calf rearing, and foraging to specifics of a chosen few species with goodly amounts of data (and many more would have been possible to represent), and could have gone in many different ways. But the authors who said "yes" rose to their task in an admirable and timely fashion and produced what I (in an admittedly biased view) consider to be a wonderful compendium, suitable as a "kick-off" (chose either traditional football or the US version as analogy) to other editions to come forth, on the same topic with 2, mysticetes; 3, otariids and walrus; 4, phocids; 5, sea otter and polar bear; and 6, sirens. I am pleased by the results from this first volume on odontocetes and hope that you may be as well.

The layout of this book is in two parts: (1) general patterns and (2) species examples. Chapter 1 (Gowans) discusses grouping patterns and potential reasons for these, and somewhat laments the present state of our science, where we are learning much about overall patterns but are still in early steps of delving into individuality of dolphin societies, i.e., how does each dolphin—with its own uniqueness—adapt and survive within the society. I will come back to this point in the final short chapter of this volume, as I believe it important in addressing our concept of "dolphin" as an individual interacting with others in a complex society. Chapter 1 lays the groundwork for all chapters to follow, and there are references of all other chapters in this one. I regard it seminal to the flow of the book.

Chapter 2 (Tyack) provides us with a primer to modalities of communication, with the understanding that communication is of vital importance in the flow of information of societal creatures. Chapter 3 (Vaughn-Hirshorn) takes us into a discussion of one of the most important requisites to sustain life—feeding, and Chapters 4 (sexual strategies, Orbach) and 5 (maternal care, Mann) discuss what comes out of successful feeding—the capability and promise of making new life.

xi

Chapter 6 (Bräger and Bräger) discusses how odontocetes move relative to requisites of food, predator avoidance, and social-sexual rearing of young, and Chap. 7 (Srinivasan) reminds us that neither feeding nor sex nor calf rearing will work successfully if a dolphin is maimed or killed by a predator. "The ecology of fear" is indeed a major guidance for sociality of odontocete cetaceans.

The final three chapters of this first section illustrate diverse parts of odontocete cetacean life. Chapter 8 (McHugh) summarizes the excellent knowledge we now have of social strategies and tactics of odontocetes close to shores, and Chap. 9 (Mesnick et al.) discusses the amazing "other" needs of dolphins and multi-species societies of delphinids in deep waters, far from shores, with the eastern tropical Pacific as a major rather well-studied paradigm. Both authors highlight that dolphins near shore and in deep oceanic waters have been and are subject to intensive (and very different) human-caused influences. This topic is admirably taken up in the final chapter of this session, Chap. 10 (Bearzi et al.), with a cogent and up-to-date discussion of odontocetes adapting—as much as may be possible—to human activities, and we humans attempting to do likewise.

As in any compendium of this sort, there are editorial needs of regret, and I will not hide mine. I am sorry that we did not get into the lovely topic of dolphins and other toothed whales diving (they spend most of their lives underwater, after all). We are learning about social communication, potential coordination, and affiliation needs of odontocetes somewhat and way below the surface of the water, and we have inadequately represented this topic here. As well, there are details of certain societies, such as those of isolated islands and atolls, which show us strong indication of operating differently simply because of such isolation. While there are references to such excellent works, it would have been good to have a separate chapter on "insularity," but this did not happen due to author constraints on time. Similarly, we are missing a chapter on intelligence and cognition, including the topic of grieving. Much ink has been spilled elsewhere on topics related to potential mental experiences, and we do attempt to lead the readers to that vast literature, especially in Chap. 23 (Würsig). Finally, the concept of culture—of passing on learned traits from generation to generation (and, at times introducing new behaviors, "fads," rapidly within one generation)—could have had its own chapter. Much has also been written about it, and I hope that references to culture throughout the book will help give the reader a flavor of the importance of this topic.

The second part of this volume, on examples of odontocete knowledge, largely follows a species by species plan. We move from matriarchal societies (Chaps. 11, 12, and 13, with authors Ford, Cantor et al., and Boran and Heimlich, respectively) to two examples of beaked whales (Chap. 14, Baird, of animals believed to be not so social), to our best examples of long-term research on dolphins, the common bottlenose (Chap. 15, Wells) and Indo-Pacific bottlenose (Chap. 16, Connor et al.) dolphins of generally near-shore waters. These two chapters provide much of the informational depth to discussions of sociality for the rest of the book, represented well by Chap. 1 (Gowans). Again, other species and species groups could have been chosen, such as on our quite large knowledge of humpback dolphins (*Sousa* sp.), common dolphins (*Delphinus* sp.), Risso's dolphins (*Grampus griseus*), bottlenose

Preface

whales (*Hyperoodon* sp.), Atlantic spotted dolphins (*Stenella frontalis*), and narwhals and beluga (Family Monodontidae). Each of these species discussions would have further enriched the book, but then there is a time and need to say "enough," and we have here a summary of important species, never mind what could have been.

Chap. 17 (Lammers) on spinner dolphins and Chap. 18 (Pearson) on dusky dolphins have similar trajectories to each other, with data and discussions about near-shore small and highly gregarious cetaceans that live relative to feeding in an efficient manner while needing to be ever vigilant about predators that can wipe them out "at any moment" (see also Chap. 7, Srinivasan). Chap. 19 (Sutaria et al.) discusses the special case of riverine dolphins and porpoises, with a developing model of potential sociality as related to evolution, and possibilities and constraints of habitats. Chapter 20 (Constantine) discusses the New Zealand Hectors dolphin that lives close to shore in small societies and is beleaguered by human attention wherever it goes. Chapter 21 (Teilmann and Sveegard) does similarly with harbor and other porpoises the world over, excellently weaving through diverse aspects of their behavior and ecology. Chapter 22 (Jefferson) reminds us that no matter what knowledge we have or are gaining, this will not do the animals (or us!) much good in the face of multiple examples of known risks of population and species demise due to human actions. Chapter 10 (Bearzi et al.) and this one might be taken as the most important ones of this book—present and rational discussions of dangers and needed conservation actions of social animals in the seas. The final short Chap. 23 (Würsig) is an attempt at a synthesis of the excellent information of this volume, with personal insights as to what may be important to study as we advance our knowledge of odontocetes, and of other marine mammals of oceans, bays, and rivers.

It is always a pleasure to thank the numerous people who have made such a compilation possible. First and foremost, hearty thanks go to the authors, from 15 different countries and all major continents. You have provided us with gifts of fine science and thought, and I am grateful. I also thank the numerous authors and others who are eagerly (and for free!) sharing your beautiful photos and illustrations of animals in nature. I thank over 50 anonymous reviewers who helped to make the present contributions better by your wise comments; and the editors and advisors of Springer International, with special commendations to Éva Loerinczi, Uma Periasamy, and Bibhuti Sharma. As always when I help to put a book together, Melany Würsig has been a stalwart companion to critique, advise, and simply listen. Thank you.

Submitted with respect

Atawhai, Kaikoura, New Zealand   Bernd Würsig
January 2019

Hector's dolphin, *Cephalorhynchus hectori*, mother and calf surfacing off Kaikoura, New Zealand. These little nearshore dolphins are endangered in much of their range around the south island of New Zealand, and critically endangered for their subspecies relative, the Maui's dolphin, *C. h. maui*. Photo by Bernd Würsig

Contents

Part I Patterns of Odontocete Ethology and Behavioral Ecology

1 Grouping Behaviors of Dolphins and Other Toothed Whales 3
Shannon Gowans

**2 Communication by Sound and by Visual, Tactile, and Chemical
Sensing** 25
Peter Tyack

3 Social Ecology of Feeding in an Open Ocean 51
Robin Vaughn-Hirshorn

4 Sexual Strategies: Male and Female Mating Tactics 75
Dara N. Orbach

5 Maternal Care and Offspring Development in Odontocetes 95
Janet Mann

6 Movement Patterns of Odontocetes Through Space and Time 117
Stefan Bräger and Zsuzsanna Bräger

7 Predator/Prey Decisions and the Ecology of Fear 145
Mridula Srinivasan

8 Odontocete Social Strategies and Tactics Along and Inshore 165
Katherine McHugh

**9 Oceanic Dolphin Societies: Diversity, Complexity,
and Conservation** 183
Sarah L. Mesnick, Lisa T. Ballance, Paul R. Wade, Karen Pryor,
and Randall R. Reeves

10 Odontocete Adaptations to Human Impact and Vice Versa 211
Giovanni Bearzi, Sarah Piwetz, and Randall R. Reeves

xvii

Part II Examples of Odontocete Ethology and Behavioral Ecology: Present Knowledge and Ways Forward

11 Killer Whales: Behavior, Social Organization, and Ecology of the Oceans' Apex Predators 239
John K. B. Ford

12 Sperm Whale: The Largest Toothed Creature on Earth 261
Mauricio Cantor, Shane Gero, Hal Whitehead, and Luke Rendell

13 Pilot Whales: Delphinid Matriarchies in Deep Seas 281
Jim Boran and Sara Heimlich

14 Behavior and Ecology of Not-So-Social Odontocetes: Cuvier's and Blainville's Beaked Whales 305
Robin W. Baird

15 Common Bottlenose Dolphin Foraging: Behavioral Solutions that Incorporate Habitat Features and Social Associates 331
Randall S. Wells

16 The Indo-Pacific Bottlenose Dolphin (*Tursiops aduncus*) 345
Richard C. Connor, Mai Sakai, Tadamichi Morisaka, and Simon J. Allen

17 Spinner Dolphins of Islands and Atolls 369
Marc O. Lammers

18 Dusky Dolphins of Continental Shelves and Deep Canyons 387
Heidi C. Pearson

19 Cetacean Sociality in Rivers, Lagoons, and Estuaries 413
Dipani Sutaria, Nachiket Kelkar, Claryana Araújo-Wang, and Marcos Santos

20 Hector's and Māui Dolphins: Small Shore-Living Delphinids with Disparate Social Structures 435
Rochelle Constantine

21 Porpoises the World Over: Diversity in Behavior and Ecology 449
Jonas Teilmann and Signe Sveegaard

22 Endangered Odontocetes and the Social Connection: Selected Examples of Species at Risk 465
Thomas A. Jefferson

23 Ethology and Behavioral Ecology of Odontocetes: Concluding Remarks 483
Bernd Würsig

Contents

xix

Correction to: Ethology and Behavioral Ecology of Odontocetes C1
Bernd Würsig

Index 495

Part I
Patterns of Odontocete Ethology and Behavioral Ecology

Chapter 1
Grouping Behaviors of Dolphins and Other Toothed Whales

Shannon Gowans

Abstract Social structures and grouping behaviors of odontocetes are under strong evolutionary pressures, as they influence mating opportunities, calf survival, predation, and foraging efficiency. The fission-fusion nature of most odontocete groups facilitates fine-scale spatial and temporal variability and establishes relationship complexity at the core of odontocete social structure. Different communities of the same species vary in social structure, as do different individuals within the same community. The same individuals can also display different social structures in different conditions. Advances in field techniques and data analyses have improved our ability to tease out nuances of social structure. Future work is essential for understanding evolutionary bases of odontocete social grouping, to focus more on interactions as well as associations, incorporate long-range acoustic communication and multivariate analyses of social and ecological variables, and move to agent-based analysis, where each individual (or collection of individuals) can act autonomously and independently. Social structural plasticity is key to the success of odontocetes and may also play an important role in speciation.

Keywords Social structure · Behavior · Social organization · Grouping · Odontocete · Toothed whale · Dolphin · Social organization · Relationship complexity · Network

We have been fascinated by whales and dolphins throughout the ages, especially by their behavior and social structure. Why do dolphins at times appear to help injured sailors? Why do some dolphins and whales mass strand? Why do some dolphins carry their dead calves around for days? It is clear that social structure of odontocetes is key to their survival and plays an important part in their success in the ocean environment.

S. Gowans (✉)
Eckerd College, St. Petersburg, FL, USA
e-mail: gowanss@eckerd.edu

© Springer Nature Switzerland AG 2019
B. Würsig (ed.), *Etthology and Behavioral Ecology of Odontocetes*, Ethology and Behavioral Ecology of Marine Mammals,
https://doi.org/10.1007/978-3-030-16663-2_1

Early work on odontocete social structure involved categorization of group size (solitary, small herds, bachelor pods, and the like), but increasingly we recognize that for most odontocetes, social lives are richer and more varied than a simple one- or two-word description. In fact, relationship complexity is key to understanding the social structure of whales and dolphins, and I hope that throughout these chapters on odontocetes, and volumes on the topic of ethology and behavioral ecology of other marine mammals, we will come to appreciate just how flexible are their social lives. This chapter is not meant to be an exhaustive review of odontocete social structure but instead focuses on how behavioral plasticity in grouping patterns has led to the evolutionary success of odontocetes.

1.1 Why Live in Groups?

Behavioral ecologists have documented the pros and cons of living in groups in a wide variety of species. In general, group living is favored when it reduces predation risk and increases access to resources or when resources are clustered such that individuals must share the same location. However, living in groups can also increase predation risk, enhance competition for resources, and increase the risk of disease and parasite transmission (Alexander 1974; Bertram 1978).

For odontocetes, the benefits of grouping appear to outweigh the costs, as most species are generally described as social (Whitehead and Mann 2000; Gowans et al. 2008). Predation risk strongly influences cetacean grouping behavior, especially in the open ocean where there are few other options to reduce predation (Norris and Dohl 1980; Norris and Schilt 1988; Gowans et al. 2008), while increased competition for resources often appears to limit group size (Gowans et al. 2008; Möller 2012). Once in groups, more complex behavior can evolve (Alexander 1974).

In mammals, female reproductive success is often driven by the survival of offspring; therefore, female distribution and social structure is often driven by availability of food and avoidance of predators (Fig. 1.1). However, for males, reproductive success is often maximized by mating with the largest possible number of females; therefore, male distribution and overall social organization are often driven by the distribution of females (Wrangham 1980). Thus, males and females likely respond differently to similar ecological conditions.

Primates, odontocetes, elephants (*Loxodonta africana*), and spotted hyenas (*Crocuta crocuta*) have evolved a flexible grouping pattern generally called fission-fusion where group membership and size can change over time. Individuals (or groups of individuals) join or leave groups depending on the current benefits or constraints to certain group sizes (Aureli et al. 2008). The term fission-fusion encompasses a wide variety of social structures. For sperm whales (*Physeter macrocephalus*; Christal and Whitehead 2001; Chap. 12) and elephants (Wittemyer et al. 2005), subsets of individuals remain together for years while still interacting with other individuals over shorter time spans, while in Indo-Pacific bottlenose dolphins (*Tursiops aduncus*) in Shark Bay, Australia, group membership changes

Fig. 1.1 Long-term maternal bonds are at the heart of female relationships, as female odontocetes typically spend years in close association with dependent calf. This young common bottlenose dolphin, *Tursiops truncatus*, is only a few weeks old, still with its fetal folds from being in the womb, and will remain near its mother for at least the next 3 years. (Photo credit: Shannon Gowans, photo taken under the authority of NMFS LOC. No. 15512)

on the order of every 10–20 min (Connor and Krützen 2015; Galezo et al. 2018; Chap. 16). Referring to all of these variations under the same terminology of fission-fusion overlooks much of the social complexity found in these species. Aureli et al. (2008) suggest that temporal variation in spatial cohesion, group size, and group composition are key to understanding the variability in social structure. They map the social landscape with cohesiveness on the x axis and social complexity on the y axis to create four quadrats ranging from low cohesiveness and low complexity (typically solitary individuals that only interact to mate and fight over resources) to highly cohesive groups with high levels of differentiation in relationships between different individuals. While this framework has not been systematically applied to odontocetes, numerous studies have been investigating the nuances of odontocete grouping patterns, going far beyond simple descriptions of group size and age/sex classification. Most odontocetes live in relatively complex societies; however, there is a great deal of variation in how long individuals and groups stay together.

Complex social structure can exist in many different forms. Eusocial organisms such as ants (family Formicidae) and naked mole rats (*Heterocephalus glaber*) form castes of specialized non-breeding individuals who assist in the reproduction of others (Bennett and Faulkes 2000; Wilson 1971). In other species, social structure is characterized by complex relationships between different individuals as individuals compete to increase their own reproductive success. Kinship plays a large role in establishing the type of social complexity. Organizational complexity is favored

Fig. 1.2 Surface observations of behaviors are often very limited, and it can be difficult to determine how dolphins are interacting with each other. Synchrony between these pairs of common bottlenose dolphins, *Tursiops truncatus*, suggests that they are interacting with each other, but this is not at all clear. It can also be challenging to determine behavioral context from surface observations alone. (Photo credit: Shannon Gowans, photo taken under the authority of NMFS LOC. No. 15512)

with high female relatedness and revolves around cooperation between individuals and reproductive division of labor. It includes, but is not restricted to, eusocial organisms but has not been documented in odontocetes. In contrast, relationship complexity occurs when there are conflicts between individuals and complexity is favored when average female relatedness is relatively low. In these societies, dominance hierarchies or competitive alliances can form (Lukas and Clutton-Brock 2018). Indo-Pacific bottlenose dolphins form some of the most complex competitive alliances, which rival the complexity of human alliances (Connor and Krützen 2015; Chap. 16).

The cryptic nature of odontocete behavior has been an important constraint in investigating their social structure and grouping patterns. At the surface, we have only brief glimpses of what these animals are doing. Most of their behavior occurs at depths out of sight (Fig. 1.2). In only a few locations is water clarity sufficient for us to see details of their lives (e.g., Bahamas: Herzing and Johnson 1997; Rossbach and Herzing 1997). Defining a "group" of whales or dolphins has been particularly problematic when we rarely see all individuals at the surface at the same time. Most research studies investigating odontocete social structure use a spatial definition (e.g., 10 body lengths; chain rule) and may incorporate a behavioral component (e.g., behaving similarly); see Mann (1999) for a review.

However, these proximity-based definitions of group typically fail to consider the impact that long-range acoustic communication may have on behavior. For example, sperm whale echolocation clicks can be detected by hydrophones located 10–16 km

1 Grouping Behaviors of Dolphins and Other Toothed Whales

from the echolocating whale. Slow clicks, which are typically produced only by males, may be detected at four to six times that distance (Madsen et al. 2002). Coda exchanges between at least two sperm whales are thought to have a social communication role and have been recorded between individuals well over 300 m apart (Schulz et al. 2008; Chap. 12). Many delphinids use tonal sounds such as whistles for social communication. These types of social communication sounds likely do not transmit as far as echolocation clicks, but they are likely still detectable at much greater distances than used for proximity-based group definitions. For example, whistles recorded from baiji (Yangtze River dolphin, *Lipotes vexillifer*) were detected over 6 km from the sound-producing dolphin in a relatively quiet seminatural reserve, although under typical noisy conditions in the Yangtze River, these detection distances would be less (Wang et al. 2006). Dolphin whistles likely serve as a communication signal between dolphins well over 1 km apart (Jensen et al. 2012; Frankel et al. 2014; Chap. 2). Even species typically considered to be relatively asocial may be much more social than previously thought, when acoustic communication is considered, as it should be. During acoustic tag deployment on six wild harbor porpoises (*Phocoena phocoena*), social communication sounds were routinely detected not only from the tagged individual but also from nearby conspecifics, suggesting that these porpoises are actually much more social than indicated by surface activity alone (Sørensen et al. 2018; Chap. 21). Therefore, describing the social structure of odontocetes based predominantly on spatial proximity at the surface is likely missing important components of social behavior among individuals spaced (at times much) further apart (see also discussion of riverine dolphins and porpoises, by Chap. 19).

1.2 Evolution of Social Structure in Odontocetes

Three different but interconnected frameworks have been proposed to describe the evolution of social structure in odontocetes. Bräger (1999) suggested that body size correlates with the stability of groups, such that smaller odontocetes like Hector's dolphins (*Cephalorhynchus hectori*) tend to live in fluid groups that frequently change while larger odontocetes like killer (*Orcinus orca*), pilot (*Globicephala* spp.), and sperm whales live in groups that change group membership less frequently. Body size is unlikely to directly influence social structure: instead, body size influences key life-history traits such as longevity and lactation. Long lactation periods and the subsequent long-term interactions between mother and calf are likely at the core of social stability. Bräger's hypothesis does help describe the social structure of odontocetes, but it does not adequately explain the variability in social structure observed within many odontocete species. For example, why are there differences in social structure between resident and transient killer whales off the Pacific West Coast of North America when they are similar in size (Riesch et al. 2012; Chap. 11)?

Gowans et al. (2008) suggested that ranging behavior, which is largely determined by food availability and predation risk, can drive the evolution of group size and social behavior. When food is predictably available year-round and relatively abundant, it pays for individuals to remain in the same location. In doing so, individuals can recognize areas of higher and lower predation risk, and therefore the advantages of forming groups to avoid predation may be reduced. Group size may be predominantly set by competition between individuals for food resources, typically limiting individuals to relatively small groups while foraging (Chap. 3). However, it may be advantageous to form larger groups when socializing or when predation risk is high. This pattern of association is common in inshore waters where group size may change frequently (Chaps. 8 and 15). In contrast, when food is only available in rare but large patches, such as mobile schools of fish, individuals must range widely to locate sufficient food (Chap. 9). Predator distribution is also likely to be patchy; therefore forming large groups may be essential to reduce predation risk, and competition for food resources may be reduced if individuals cooperate to find rare but profitable schools of fish (Chap. 7). Thus, offshore odontocetes tend to be found in large and relatively unstructured groups, such as dolphin groups in the Eastern Tropical Pacific. However, if cooperation between females is required to successfully rear offspring (as is found in sperm whales), long-term bonds between females can form. This framework helps to elucidate the role of ecological conditions in determining grouping patterns and social structure and allows for intraspecific variation. But, life-history traits and philopatry have also been shown to be important in the evolution of social structure in many other species including birds (Hatchwell 2009), canids (Bekoff and Daniels 1984), primates (Sterck et al. 1997), and ungulates (Bro-Jørgensen 2011).

Möller (2012) proposed a framework that incorporates philopatry, dispersal, and life-history traits in the development of social structure in delphinids. In this framework, female bonds are largely determined by philopatry such that in inshore waters where female philopatry is common, females form medium to strong bonds, but in offshore waters, females are less likely to be related to each other and therefore have weaker bonds. In contrast, male social structure is tied to life-history patterns; male bonds are likely to form in species with slight sexual dimorphism and a male-biased operational sex ratio (caused by long interbirth intervals and calf dependency when most females are unavailable for mating). Long-term bisexual bonds in both killer and pilot whales may have evolved in sexually dimorphic species so as to reduce the effectiveness of male coercion. Environmental conditions impact dispersal and life-history patterns, which in turn influence social structure (Chaps. 4, 5, and 15).

Each of these frameworks attempts to explain the evolution of social structures in odontocetes. However, none perfectly predicts all odontocete social structures, nor do they adequately explain the large variability in odontocete social systems.

1.3 Evidence for Plasticity in Social Structure in Odontocetes

One of the complications in developing a framework to describe evolution of social structure in odontocetes is variability in social structure. Each additional species described adds a new nuance to the overall pattern, and differences among populations of the same species abound. We are increasingly documenting individual level differences within the same community. It is clear that odontocete social structure is variable and likely finely tuned to environmental conditions that can change rapidly. However not all odontocete species display high social plasticity. Matriarchal odontocetes such as sperm whales, resident killer whales, and pilot whales appear to display much less diversity in social behavior, and instead relationships are centered around maternal bonds that can persist for many decades (i.e., Bigg et al. 1987; Christal and Whitehead 2001; de Stephanis et al. 2008; Gero et al. 2008; Riesch et al. 2012; Chaps. 11, 12, 13). Further work on these and other matrilineal species will elucidate the influence of a matriarchal social structure on social plasticity.

1.3.1 Variability Between Communities: Tursiops spp.

One of the complications in trying to develop a framework to describe evolution of social structure in odontocetes is in adequately capturing variability in social structure within a single species or genus. Social structure and grouping patterns have been studied in many different species, but the best-studied taxon is the bottlenose dolphin (*Tursiops* spp.). While the number of species and relationships between species within the *Tursiops* genus has yet to be resolved, most studies recognize the common bottlenose dolphin (*T. truncatus*) as separate but related to the Indo-Pacific bottlenose dolphin (*T. aduncus*). The common bottlenose dolphin is widely distributed and the majority of studies have been carried out on it (Chap. 15), but a long-term study has been conducted on Indo-Pacific bottlenose dolphins of Shark Bay, Australia, and additional studies on other communities are ongoing (Chaps. 5, 16).

While all *Tursiops* spp. societies described to date can be called fission-fusion, they do not behave the same. Some communities, particularly those in isolated habitats, and where cooperation is favored, tend to have longer-term bonds and stronger association levels (e.g., Doubtful Sound New Zealand; Lusseau et al. 2003 and in the Laguna Estuary in Southern Brazil; Daura-Jorge et al. 2012).

In contrast, associations in Shark Bay, Australia, and Sarasota, Florida, are typically described as a loose network of associations where females with dependent calves have a tendency to associate with other females with dependent calves (Whitehead and Mann 2000; Frère et al. 2010a; Connor et al. 2000; Wells 2003). Males, on the other hand, tend to have longer-term bonds with other males, apparently to facilitate mating with females. Numbers of affiliates and strengths and longevities of bonds vary among communities and individuals (Owen et al. 2002; Connor and Krützen 2015).

The social structure of common bottlenose dolphins is highly variable. In the Moray Firth of Scotland and in the Sado Estuary of Portugal, no sex-specific pattern of associations was detected, and in the Portugal population no pattern by age class was detected (Scotland; Wilson 1995, Portugal; Augusto et al. 2012). In the Shannon Estuary, Ireland, the strongest bonds were between juveniles, and the population was split into inner and outer estuary communities with lower levels of interactions between the two communities (Baker et al. 2018). In the Indian River Lagoon of eastern Florida, different communities occur along the long narrow estuary with limited mixing between communities. Along this estuary, habitat characteristics were also key drivers to the observed social structure (Titcomb et al. 2015). Similarly, in the open coastal waters off the Normandy coast, the social structure of common bottlenose dolphins appears to be predominantly driven by foraging differences among community members (Louis et al. 2018).

Additionally, there are no consistent patterns differentiating social structure between *Tursiops* spp. Common bottlenose dolphins in Sarasota, Florida, share more similarities in social structure with Indo-Pacific bottlenose dolphins in Shark Bay, Australia, than they do with other communities of common bottlenose dolphins in Scotland, Portugal, or Ireland. Moreover, Indo-Pacific bottlenose dolphins in Port Stephens, Australia, displayed similarities in social structure with the bisexually bonded common bottlenose dolphins in Scotland and New Zealand (Wiszniewski et al. 2010; Chap. 8).

1.3.2 Variability Within Communities: Indo-Pacific Bottlenose Dolphins in Shark Bay

Not only do communities of bottlenose dolphins differ in social structure, individual bottlenose dolphins in each of these communities do not behave the same. While age and sex differences in social structure were predicted, the degree of variability between different individuals in Shark Bay, Australia, was surprising. Access to females appears to be at the core of male social structure in bottlenose dolphins in Shark Bay. Males form "first-order alliances," consisting of two to three males that cooperate to coerce estrus females, both to guard the females from leaving and to defend them against other males that attempt to coerce the females themselves. The stability of male partnerships within these first-order alliances varies between individuals, with some individuals maintaining long-term bonds with two or three other males, and others routinely switching partners, selecting from a set of up to 14 other adult males to form brief partnerships. Males also form "second-order alliances," which consist of 4–14 males who cooperate to steal females away from other first-order alliances. Additionally, sets of these second-order alliances also display low levels of association with each other and are likely involved in female acquisition (Connor and Krützen 2015). In addition to the complex set of nested alliances, male social structure systematically varies from north to south in Shark Bay, with an increase in number of trios forming first-order alliances and the resulting rate that males consort females in the more northern parts of the study area. This variability within an open network where males associate with many other males suggests an ecological basis that influences the economics of mate acquisition and defense. The lower consortship rate and proportion of trios in the south may reflect lower densities of females based upon the lower quality of the habitat in the south (Connor et al. 2017; Chap. 16).

1 Grouping Behaviors of Dolphins and Other Toothed Whales

In Shark Bay, female bottlenose dolphins tend to be more solitary than males, although some females are highly social (Galezo et al. 2018). Some aspects of female social structure are correlated with foraging style, with dolphins that share similar foraging style tending to preferentially associate with each other even when controlling for sex, location, and maternal relatedness (Mann et al. 2012). Additionally, reproductive success influences social structure as females with high calving success preferentially associate with other females with high calving success (Frère et al. 2010b). Finally, female association patterns are influenced by home range overlap, as well as by maternal and biparental kinship (Frère et al. 2010a; Chap. 5).

1.3.3 Variability Over Time by Individual

Not only does cetacean social structure vary within the same species and community, individuals can change social structure over time. While changes in social structure are expected as individuals age (e.g., Mann and Sargeant 2003; Mann and Smuts 1998, 1999; McHugh et al. 2011), other systematic changes in social structure are less easy to explain.

Most well-studied dolphin communities are relatively resident (facilitating their long-term study), but annual migrations occur in several species (Chap. 6). Dusky dolphins (*Lagenorhynchus obscurus*) off the east coast of New Zealand display high social fluidity with many weak bonds, but some dyads form strong long-term bonds. These individuals display some of the most dramatic examples of variability in social structure in cetaceans as these same individuals are found in the summer in large open societies in coastal waters near Kaikoura, New Zealand, during the day and then feeding at night on the deep-scattering layer in the nearby deepwater canyon (Benoit-Bird et al. 2004). During the winter some of these same dolphins migrate to the shallow waters of Admiralty Bay where they live in small fission-fusion groups feeding during the day on schooling surface fishes (Whitehead et al. 2004, Würsig and Pearson 2014; Chap. 18). Group sizes are very different in these two locations (~50–1000 in Kaikoura and ~7 in Admiralty Bay), reflecting different ecological conditions for both foraging opportunities and predation risk, as deepwater sharks and killer whales are found in the deep water near Kaikoura but not in the shallow waters of Admiralty Bay. Social bonds persist even when individuals switch locations, alternating between Kaikoura and Admiralty Bay. Cooperation appears to be the basis for these strong bonds, as dolphins cooperate in Admiralty Bay to feed upon schooling "bait balls" and cooperate in Kaikoura for mating, calf rearing, and predation avoidance. The persistence of these bonds in both locations suggests that ecological pressures may not be as important for bond formation, although group size responds intensely to these ecological differences (Pearson et al. 2017).

Grouping and bonding patterns can change rapidly as individuals change behavioral state. In Cedar Key, Florida, common bottlenose dolphins formed stronger repeated connections with other individuals when socializing, weaker connections during travel, and the weakest connections while foraging. This suggests that competition may limit interactions during foraging, but individuals within the community preferentially interact with certain other individuals during socializing (Gazda et al. 2015). In contrast, Gero et al. (2005) indicated that preferred associations between Indo-Pacific bottlenose dolphins in Shark Bay, Australia, occurred during both foraging and socializing, although connections during foraging were weaker than for socializing. Baker et al. (2018) found that common bottlenose dolphins in the Shannon Estuary, Ireland, had similar behaviorally based association patterns as those found in Shark Bay. This suggests that dolphins have high behavioral flexibility and adjust their group size and structure to optimize conditions in different behavioral states (see also Würsig and Würsig 1980, for dusky dolphins in different behavioral states of resting, traveling, herding food, and socializing).

Social structure of a community can also alter over time in response to demographic or ecological changes. In the Bahamas, two hurricanes within less than 1 month in 2004 altered the structure of the resident common bottlenose and Atlantic spotted dolphin (*Stenella frontalis*) communities, after approximately 1/3 of the dolphins in both communities were lost. The spotted dolphin community retained its reduced size as no new dolphins immigrated into the area. The social structure of the community remained relatively consistent post-hurricane, but there was increased cohesion between community members and a reduction in male alliance complexity (Elliser and Herzing 2014). In contrast, the common bottlenose dolphin community was replenished with approximately the same number of new individuals immigrating into the community after the hurricanes. Post-hurricane, the community divided with few interactions between these new clusters. Both new clusters showed social cohesion, although one cluster became more cohesive than the other. Immigrants integrated into both of the new clusters, although interaction rates were higher between residents and immigrant males than with immigrant females (Elliser and Herzing 2011). In Moreton Bay, Australia, changes in food availability led to the cohesion of two previously distinct communities of Indo-Pacific bottlenose dolphins. In the 1990s, commercial trawling for prawns was common, and two behaviorally distinct communities of dolphins formed in the bay, differentiated by whether or not they interacted with commercial fisheries (trawl associated dolphins were frequently observed foraging on bycatch). These two communities overlapped spatially but did not interact socially. Regulations have dramatically reduced trawling in the bay, and trawling has been banned from most of dolphin core habitat. With the loss of food from the trawling industry, these two communities of dolphins have converged, now forming one cohesive community, composed of many of the same individuals that previously avoided each other (Ansmann et al. 2012).

1.4 Social Plasticity and Speciation

As a species, killer whales show a great deal of social plasticity (quite unlike the within community lack of plasticity mentioned above), and the species has been divided into a number of different ecotypes that have foraging specializations.

Ecotypes are also differentiated by group size, stability, and strengths of social bonds. Sympatric ecotypes occur in several highly productive locations, but individuals from different ecotypes do not interact with each other, leading to behaviorally mediated reproductive isolation. Cultural differences appear to be at the heart of ecotype formation and reproductive isolation. While all ecotypes are considered part of the same species, some researchers have argued that incipient speciation is ongoing (Riesch et al. 2012; Chap. 11).

The argument for ecotype formation in common bottlenose dolphins is not as well advanced, but along the Normandy coast of France, three separate foraging types have been documented (based on stable isotope analysis). Individuals tend to associate only with other individuals with similar foraging type (Louis et al. 2018), which may lead to social and genetic isolation, as has previously been observed between inshore and offshore bottlenose dolphins (e.g., Louis et al. 2014; Sellas et al. 2005).

Whitehead and Ford (2018) modeled the consequences of cultural speciation, similar to observations in killer whales. In this model, cultural specialization typically leads to narrow niche breadth as individuals focus on only a portion of the entire niche. The models indicate that this is adaptive in the short term, as individuals efficiently exploit food resources, but in the long term, it can lead to an increase in rate of group extirpation. In comparisons to models of natural selection for niche breadth, cultural specialization was faster than natural selection, but due to the high rates of extirpation, it was less stable than natural selection. The authors argue that the high rate of group extirpation may explain why killer whales have not yet split into multiple species, even though most other delphinid genera are represented by multiple species (Whitehead and Ford 2018). However, it is also possible that if niche size changed rapidly, then rapid speciation could occur through social differences leading to reproductive isolation.

Social plasticity is a key feature leading to the ecological success of odontocetes, and it may be responsible for the rapid speciation of a group, especially of delphinids. Mesoplodons (of the beaked whale group, family Ziphiidae) are another taxon of cetaceans with rapid speciation. Unfortunately, relatively little is known about the biology of most beaked whales, and even less is known about their social structure, although they are often described as relatively nonsocial (Chap. 14). There is some evidence that at least some beaked whales, and even some mesoplodons, live in resident communities, often associated with submarine canyons or other dramatic bathymetric features (Hooker et al. 2002; McSweeney et al. 2007; Claridge 2013; Baird et al. 2016). In these types of communities, long-term relationships may develop, similar to those shown in northern bottlenose whales (*Hyperoodon ampullatus*; Gowans et al. 2001). Social plasticity is also key to the current survival of many species and communities of odontocetes, and incipient speciation may be ongoing in several different odontocetes, based primarily on social isolation. Future work on odontocete social structure will help reveal these potential patterns (Chaps. 17 and 20).

1.5 Advances in Understanding Odontocete Social Behavior

1.5.1 Photo-Identification

The heart of most cetacean social organization work relies on photo-identification of natural markings to document which individual animals are present when, where, and with which other individuals (Würsig and Würsig 1977; Bigg et al. 1987; Ballance 2018). Photo-identification has vastly improved over the years as cameras have become more reliable, durable, and affordable. The switch from analog film or slide to digital has facilitated the greatest increase in the ability to use photo-identification for cetaceans (Markowitz et al. 2003). In the field we are no longer limited by the cost of film or having to pause and change rolls of film. We do not have to wait weeks to process film, but instead we can check in the field if we have a high-quality photograph of a certain individual (Markowitz et al. 2003). Initially the resolution of digital images was limited, but high-resolution digital cameras are now readily available and relatively inexpensive. It is now much easier to ensure the long-term storage of images, as backups can be readily created and stored in multiple locations, including cloud-based storage options. Digital images can also be easily shared between researchers, facilitating collaborations between studies and the development of several large-scale regional catalogs (e.g., Gulf of Mexico Dolphin Identification System, Balmer et al. 2016). Some of these catalogs are available on the web and incorporate elements of citizen science to add new images or make matches between individuals (e.g., www.flukebook.org).

Digital images have also facilitated the development of computer-assisted matching programs. Several different programs are freely available and are widely used to reduce the time burden of identifying individuals especially in large catalogs (e.g., Digital Analysis and Recognition of Whale Images on a Network; http://darwin.eckerd.edu). While the underlying approaches differ between these programs, none of these programs completely automate the match process, and human eyes are still required to confirm each match. Additionally, most of the programs struggle to match individuals that have experienced major mark-change.

The use of digital images has also facilitated many aspects of database management. Associated with each digital image is an EXIF file that contains information about the time the photograph was taken, camera settings, and additional information such as photographer and GPS position.

1.5.2 Genetic and Other Tissue Sampling

In addition to photo-identification, studies of genetic sampling of individuals have been instrumental in revealing a deeper understanding of social relationships. Initial application of genetic analysis focused on identifying the sex of the individual and allowed researchers to investigate differences in grouping patterns and social structure based on sex (e.g., Baker et al. 2018). However, as genetic tools improved and researchers were able to sample larger proportions of their study populations, they have been able to investigate maternal and paternal relationships between individuals in relation to association patterns (e.g., Frère et al. 2010a). Sex-based philopatry has been investigated, and both female philopatry (Charlton-Robb et al. 2015) and male philopatry (Ball et al. 2017) have been found, as well as kinship-based relationships when neither sex disperses (Wallen et al. 2017). Genetic analysis of paternity has also been key to understanding the potential reproductive success of alternative male strategies (Connor and Krützen 2015; Wells 2014).

Stable isotope and fatty acid analysis for dietary preferences and differences can also be conducted on biopsy samples obtained for genetic analysis. Incorporating multiple variables such as kinship, sex, and diet in social structure analysis can reveal important differences between locations and individuals (e.g., Louis et al. 2018).

1.5.3 Data-Logging Tags

The application of data-logging tags to individual animals has permitted glimpses into what whales and dolphins do when they are not at the surface and when researchers are not around. Early tags emitted a VHF signal, allowing researchers to track individual movements in real time (Hooker et al. 2002; Martin and Da Silva 2004; Würsig et al. 1991). More modern tags can remotely collect a wealth of information including location, dive depth and duration, acceleration, body orientation, water characteristics, as well as sounds produced by the animal and background noises (Laplanche et al. 2015; Nowacek et al. 2016). These types of tags permit researchers to combine ranging and movement behavior with association patterns.

Data-logging tags can also reveal important aspects of foraging behavior. Subtle differences in foraging behavior can also influence social behavior and grouping (Kovacs et al. 2017; Mann et al. 2012) or potentially even be a key driver of population structure (Cantor and Farine 2018). While most work to date relating foraging behavior and social structure have relied on visual observations at the surface of foraging (Gero et al. 2005; Gazda et al. 2015), in the future, data loggers will likely be important in investigating social structure in species which tend to forage at depth, especially deep diving cetaceans like sperm and beaked whales.

Acoustic tags will likely become increasingly important in understanding social structure as we seek to incorporate social communication (Nowacek et al. 2016). Common bottlenose dolphins are well known for their signature whistles that convey individual identity, and it is likely that several other dolphin species also produce signature whistles (Janik and Sayigh 2013). Therefore, it may be possible to investigate social structure based on detection of individually identifiable whistles in a similar manner as photo-identification is currently used.

1.5.4 Analytical Advances

All of these advances in data collection have led to challenges in data analysis. In 1997, Hal Whitehead developed and freely distributed a set of programs called SOCPROG designed to facilitate the analysis of social structure based on individual identifications (Whitehead 2009). The program has gone through several iterations that included major advances and incorporated additional analyses, and it is currently the predominant software package used to analyze these types of data, although new techniques continue to be developed. Network analysis (e.g., Baker et al. 2018; Louis et al. 2018) and multivariate analysis are becoming increasingly common (Gero et al. 2005; Baker et al. 2018; Gazda et al. 2015; Louis et al. 2018) and will reveal additional aspects of social structure.

1.6 Current Constraints to Understanding Odontocete Social Structure

Many studies of odontocete social structure represent brief snapshots into the lives of dolphins. While several studies have been investigating odontocete social structure for decades (e.g., work on common bottlenose dolphins in Sarasota Bay, Florida, killer whales off British Columbia, and Indo-Pacific bottlenose dolphins in Shark Bay, Australia), most projects are still shorter than the life-span of many of their subjects. While some animals have been tracked from birth to death, we are just beginning to track the entire life-span of some of the most successful individuals (Chap. 15). Most other studies are much shorter in duration, and even with long-term studies, papers are typically published only on a small subset of data. For example, McHugh et al. (2011) investigated juvenile behavior in Sarasota Bay, Florida, between 2005 and 2007, rather than on the entire dataset from the 1970s. These truncations in data inclusion are often logistical; the appropriate data were not available for the entire dataset, methods and protocols have changed over time, and therefore the entire dataset is not amenable to the proposed analysis, or the analytical tools to incorporate the full dataset are not available.

Descriptions of cetacean social organization predominately rely on a framework developed by Robert Hinde (1976), where repeated interactions between individuals build up relationships that can then be summarized to describe overall social structure. At the heart of this analysis are observations of interacting individuals. Ideally, observations represent meaningful interactions between individuals, such as animal "A" groomed animal "B." However, we rarely observe these types of detailed interactions in cetaceans. Instead, most researchers rely on "the gambit of the group" which is the idea that individuals found in the same group together have the potential to interact, and if individuals are found in the same group repeatedly, then interactions likely occur. Thus, the frequency of being found in the same group is a measure of social affiliation among individuals (Whitehead and Dufault 1999). However, it is also possible that associations between individuals do not always represent interactions but are instead simple aggregations for resources (such as food or mates, or the result of demographic pressures, or just chance). While techniques have been developed to attempt to tease out how meaningful these associations are (e.g., Bejder et al. 1998; Whitehead 2008), analysis based on "the gambit of the group" limits the ability to make more detailed inferences about the meaning of these associations. Temporal or spatial methods of defining group membership, and thus associations, show some promise (e.g., nearest neighbor Scott 1991; Allen et al. 2001 or photographed within a certain time frame, e.g., Tavares et al. 2017; Johnston et al. 2017) but still cannot differentiate who did what to whom (Chap. 23). An added complication to "the gambit of the group" approach to defining associations is that the majority of group definitions are based upon surface observations that do not necessarily include underwater behaviors or acoustic interactions among individuals that are spatially relatively far apart, i.e., on the order of 0.5–2 km or more apart.

Most cetacean research has focused on inshore and coastal species. We have only a few insights on offshore dolphin social structure from research associated with the Eastern Tropical Pacific tuna fishery (Pryor and Kang Shallenberger 1991; Norris and Dohl 1980; Chap. 9). However, these studies can reveal important differences between inshore and offshore communities. In the open ocean, spinner dolphins (*Stenella longirostris*) live in large open communities of several hundred to several thousand individuals, often intermingling with pantropical spotted (*S. attenuata*), striped (*S. coeruleoalba*), and common dolphins (*Delphinus* sp.). These groups range widely and appear to remain together for extended time periods. It is likely that smaller subgroups form within these larger groups, which may be based on age or sex classes (Pryor and Kang Shallenberger 1991; Norris and Dohl 1980; Chap. 9); however little is known about the dynamics of these large groupings. In contrast, spinner dolphins have also been studied off the Big Island of Hawaii, where dolphins feed offshore at night in the deep-scattering layer and rest during the daytime in protected shallow bays (Norris and Dohl 1980; Würsig et al. 1994). Although individuals remain resident along a stretch of coastline, group composition changes frequently within the bays (Würsig et al. 1994), while at night pairs of dolphins coordinate to forage at depth (Benoit-Bird 2004; Benoit-Bird and Au 2003; Chap. 17). In the remote Hawaiian atolls, small isolated communities have developed which remain separate from other atoll-associated communities. Within each atoll, individuals rest during the daytime and associate with all other community members. At night, dolphins forage in the deep waters but return to the same atoll, likely restricted by geographic distance from frequently switching atoll location or affiliation (Karczmarski et al. 2005). Thus we see three different patterns of association in spinner dolphins related to habitat and isolation from other communities, exemplifying the adaptability of social behavior to varying ecological conditions.

Small resident communities of beaked whales may also be amenable to studies of their social structure (e.g., Gowans et al. 2001; McSweeney et al. 2007; Falcone et al. 2009; Chap. 14) as well as several communities of sperm whales (e.g., Gero et al. 2008; Christal and Whitehead 2001); however, most offshore communities remain unstudied. Similarly, the social structure of most species or communities that range widely has not been well investigated (e.g., Defran and Weller 1999, for common bottlenose dolphins), as most social structure analysis relies heavily on multiple identifications of the same individual over time. Large-scale cooperative catalogs will help to facilitate the understanding of the behavior and social structure of these individuals.

As most studies have been conducted in coastal waters, they also represent communities of dolphins that have potentially been exposed to a greater amount of anthropogenic disturbance (Chaps. 10 and 22). For example, the Shark Bay Dolphin Project in Australia began as a study on a group of provisioned dolphins (Gawain 1981), a practice of hand-feeding that continues today. This intentional feeding has influenced the survival, reproduction, and social structure of some members of the local dolphin community (Foroughirad and Mann 2013). There is also an active dolphin watching industry in Shark Bay, which also influences the survival of individuals in this community (Bejder et al. 2006). Similarly, the dolphins of Sarasota Bay, Florida, are routinely exposed to vessel traffic, predominantly recreational boats (Buckstaff 2004), and can also be exposed to food provisioning (Christiansen et al. 2016) that may influence social structure and behavior. Undisturbed communities of coastal dolphins are difficult to locate, and even if they were located, operation of the research vessel itself has the potential to influence dolphin behavior (Chap. 10).

1.7 Future Outlook

Multifaceted studies that incorporate ecological and life-history variables, cultural components, kinship measures, and detailed behavioral aspects will be key to understanding the nuanced details of the evolutionary bases of odontocete grouping patterns. Examples of these types of studies include work in Shark Bay, Australia, investigating the influence culture has on foraging techniques (such as tool use) and social interactions, including elements of prolonged maternal care (Mann et al. 2012; Chap. 5); influence of low food availability, linked to a severe harmful algal bloom, has on the development of social networks in juvenile dolphins in Sarasota, Florida (McHugh et al. 2011); and influences of dietary differences and kinship on social structure in dolphins in open bays of the English Channel (Louis et al. 2018).

The use of unmanned aerial systems (UAS) will provide much richer behavioral observations of whales and dolphins at the surface (and sometimes up to several meters below) and will allow researchers to incorporate interactions between individuals into social structure analysis (Nowacek et al. 2016). While only limited behavioral data of odontocetes have been published to date using UAS (e.g., Ramos et al. 2018; Fiori et al. 2017; Weir et al. 2018), it is clear that these tools will be increasingly used to document cetacean behavior.

The further development of long-term studies that follow individuals over entire life-spans will enhance our understanding of cetacean social structure and behavior (e.g., Connor and Krützen 2015; Wells 2014). Improved analytical tools that focus on the individual, rather than categorical groups of individuals, will also assist our understanding. The use of agent-based modeling is developing rapidly in this field (e.g., Pirotta et al. 2014; Whitehead and Ford 2018), and traditional behavioral gathering techniques combined with newer sophisticated statistical and modeling studies indicate that we are reaching a better understanding of the grouping patterns of these very complex animals.

1.8 Conclusion

The work in this chapter and throughout these volumes clearly indicates that many of the facets of cetacean grouping and social organization are beginning to be understood. However, it is also clear that much remains to be done. It is imperative that this work be conducted quickly as we are increasingly recognizing the importance of understanding social structure to help with effective management and conservation of cetaceans—animals all too often beleaguered due to human influence (Chaps. 10 and 22).

References

Alexander RD (1974) The evolution of social behaviour. Annu Rev Ecol Syst 5:325–383
Allen MC, Read AJ, Gaudet J, Sayigh LS (2001) Fine-scale habitat selection of foraging bottlenose dolphins *Tursiops truncatus* near Clearwater, Florida. Mar Ecol Prog Ser 222:253–264
Ansmann IC, Parra GJ, Chilvers BL, Lanyon JM (2012) Dolphins restructure social system after reduction of commercial fisheries. Anim Behav 84:575–581
Augusto JF, Rachinas-Lopes P, Dos Santos ME (2012) Social structure of the declining resident community of common bottlenose dolphins in the Sado Estuary, Portugal. J Mar Biol Assoc U K 92:1773–1782
Aureli F, Schaffner CM, Boesch C, Bearder SK, Call J, Chapman CA, Connor R, Fiore AD, Dunbar RI, Henzi SP, Holekamp K (2008) Fission-fusion dynamics: new research frameworks. Curr Anthropol 49:627–654
Baird RW, Webster DL, Swaim Z, Foley HJ, Anderson DB, Read AJ (2016) Spatial use by odontocetes satellite tagged off Cape Hatteras, North Carolina in 2015. Final report. Prepared for US Fleet Forces Command. Submitted to Naval Facilities Engineering Command Atlantic, Norfolk, Virginia, under Contract No. N62470-10-3011, Task Order 57 and N62470–15-8006, Task Order 07, issued to HDR Inc., Virginia Beach, VA
Baker I, O'Brien J, McHugh K, Ingram SN, Berrow S (2018) Bottlenose dolphin (*Tursiops truncatus*) social structure in the Shannon Estuary, Ireland, is distinguished by age- and area-related associations. Mar Mamm Sci 34:458–487
Ball L, Shreves K, Pilot M, Moura AE (2017) Temporal and geographic patterns of kinship structure in common dolphins (*Delphinus delphis*) suggest site fidelity and female-biased long-distance dispersal. Behav Ecol Sociobiol 71:123
Ballance LT (2018) Contributions of photographs to cetacean science. Aquat Mamm 44:668–682
Balmer B, Sinclair C, Speakman T, Quigley B, Barry K, Cush C, Hendon M, Mullin K, Ronje E, Rosel P, Schwacke L, Wells RS, Zolman E (2016) Extended movements of common bottlenose dolphins (*Tursiops truncatus*) along the Northern Gulf of Mexico's central coast. Gulf of Mexico Sci 33:8

Bejder L, Fletcher D, Bräger S (1998) A method for testing association patterns of social animals. Anim Behav 56:719–725

Bejder L, Samuels A, Whitehead H, Gales N, Mann J, Connor R, Heithaus M, Watson-Capps J, Flaherty C, Krützen M (2006) Decline in relative abundance of bottlenose dolphins exposed to long-term disturbance. Conserv Biol 20:1791–1798

Bekoff M, Daniels TJ (1984) Life history patterns and the comparative social ecology of carnivores. Annu Rev Ecol Syst 15:191–232

Bennett NC, Faulkes CG (2000) African mole-rates: ecology and eusociality. Cambridge University Press, Cambridge

Benoit-Bird KJ (2004) Prey coloric value and predator energy needs: foraging predictions of wild spinner dolphins. Mar Biol 145:435–444

Benoit-Bird KJ, Au WWL (2003) Prey dynamics affect foraging by a pelagic predator (*Stenella longirostris*) over a range of spatial and temporal scales. Behav Ecol Sociobiol 53:364–373

Benoit-Bird KJ, Würsig B, McFadden CJ (2004) Dusky dolphin (*Lagenorhynchus obscurus*) foraging in two different habitats: active acoustic detection of dolphins and their prey. Mar Mamm Sci 20:215–231

Bertram BCR (1978) Living in groups: predators and prey. In: Krebs CJ, Davies NB (eds) Behavioural ecology: an evolutionary approach. Blackwell Scientific, London

Bigg MA, Ellis GM, Ford JKB, Balcomb KC (1987) Killer whales: a study of their identification, genealogy and natural history in British Columbia and Washington State. Phantom Press, Nanaimo

Bräger S (1999) Association patterns in three populations of Hector's dolphin, *Cephalorhynchus hectori*. Can J Zool 77:13–18

Bro-Jørgensen J (2011) Intra- and intersexual conflicts and cooperation in the evolution of mating strategies: lessons learnt from Ungulates. Evol Biol 38:28–41

Buckstaff KC (2004) Effects of watercraft noise on the acoustic behavior of bottlenose dolphins, *Tursiops truncatus*, in Sarasota Bay, Florida. Mar Mamm Sci 20:709–725

Cantor M, Farine DR (2018) Simple foraging rules in competitive environments can generate socially structured populations. Ecol Evol 8:4978–4991

Charlton-Robb K, Taylor A, McKechnie S (2015) Population genetic structure of the Burrunan dolphin (*Tursiops australis*) in coastal waters of south-eastern Australia: conservation implications. Conserv Genet 16:195–207

Christal J, Whitehead H (2001) Social affiliations within sperm whale (*Physeter macrocephalus*) groups. Ethology 107:323–340

Christiansen F, McHugh KA, Bejder L, Siegal EM, Lusseau D, McCabe EB, Lovewell G, Wells RS (2016) Food provisioning increases the risk of injury in a long-lived marine top predator. R Soc Open Sci 3. https://doi.org/10.1098/rsos.160560

Claridge D (2013) Population ecology of Blainville's beaked whales (*Mesoplodon densirostris*). PhD dissertation, University of St. Andrews. St. Andrews, Scotland, 312 p

Connor RC, Krützen M (2015) Male dolphin alliances in Shark Bay: changing perspectives in a 30-year study. Anim Behav 103:223–235

Connor RC, Wells RS, Mann J, Read AJ (2000) The bottlenose dolphin: social relationships in a fission-fusion society. In: Mann J, Connor RC, Tyack P, Whitehead H (eds) Cetacean societies: field studies of dolphins and whales. University of Chicago Press, Chicago

Connor RC, Cioffi WR, Randic S, Allen SJ, Watson-Capps J, Krützen M (2017) Male alliance behaviour and mating access varies with habitat in a dolphin social network. Sci Rep 7:46354

Daura-Jorge FG, Cantor M, Ingram SN, Lusseau D, Simoes-Lopes PC (2012) The structure of a bottlenose dolphin society is coupled to a unique foraging cooperation with artisanal fishermen. Biol Lett 8:702–705

De Stephanis R, Verbourgh P, Pérez S, Minville-Sebastia L, Guinet C (2008) Long-term social structure of long-finned pilot whales (*Globicephala melas*) in the Strait of Gibralter. Acta Ethol 11:81–94

Defran RH, Weller DW (1999) Occurrence, distribution, site fidelity, and school size of bottlenose dolphins (*Tursiops truncatus*) off San Diego, California. Mar Mamm Sci 15:366–380

1 Grouping Behaviors of Dolphins and Other Toothed Whales

Elliser CR, Herzing DL (2011) Replacement dolphins? Social restructuring of a resident pod of Atlantic bottlenose dolphins, *Tursiops truncatus*, after two major hurricanes. Mar Mamm Sci 27:39–59

Elliser CR, Herzing DL (2014) Social structure of Atlantic spotted dolphins, *Stenella frontalis*, following environmental disturbance and demographic changes. Mar Mamm Sci 30:329–347

Falcone EA, Schorr GS, Douglas AB, Calambokidis J, Henderson E, McKenna MF, Hildebrand J, Moretti D (2009) Sighting characteristics and photo-identification of Cuvier's beaked whales (*Ziphius cavirostris*) near San Clemente Island, California: a key area for beaked whales and the military? Mar Biol 156:2631–2640

Fiori L, Doshi A, Martinez E, Orams MB, Bollard-Breen B (2017) The use of unmanned aerial systems in marine mammal research. Remote Sens 9:543

Foroughirad V, Mann J (2013) Long-term impacts of fish provisioning on the behavior and survival of wild bottlenose dolphins. Biol Conserv 160:242–249

Frankel AS, Zeddies D, Simard P, Mann D (2014) Whistle source levels of free-ranging bottlenose dolphins and Atlantic spotted dolphins in the Gulf of Mexico. J Acoust Soc Am 135:1624–1631

Frère CH, Krützen M, Mann J, Watson-Capps JJ, Tsai YJ, Patterson EM, Connor R, Bejder L, Sherwin WB (2010a) Home range overlap, matrilineal and biparental kinship drive female associations in bottlenose dolphins. Anim Behav 80:481–486

Frère CH, Krützen M, Mann J, Connor RC, Bejder L, Sherwin WB (2010b) Social and genetic interactions drive fitness variation in a free-living dolphin population. Proc Natl Acad Sci USA 107:19949–19954

Galezo AA, Krzyszczyk E, Mann J (2018) Sexual segregation in Indo-Pacific bottlenose dolphins is driven by female avoidance of males. Behav Ecol 29:377–386

Gawain B (1981) The dolphin's gift. Whatever Publications, Mill Valley, CA

Gazda S, Iyer S, Killingback T, Connor R, Brault S (2015) The importance of delineating networks by activity type in bottlenose dolphins (*Tursiops truncatus*) in Cedar Key, Florida. R Soc Open Sci 2. https://doi.org/10.1098/rsos.140263

Gero S, Bejder L, Whitehead H, Connor RC (2005) Behaviourally specific preferred associations in bottlenose dolphins, Tursiops spp. Can J Zool 83:1566–1573

Gero S, Engelhaupt D, Whitehead H (2008) Heterogeneous social associations within a sperm whale, *Physeter macrocephalus*, unit reflect pairwise relatedness. Behav Ecol Sociobiol 63:143–151

Gowans S, Whitehead H, Hooker SK (2001) Social organization in northern bottlenose whales (*Hyperoodon ampullatus*): not driven by deep water foraging? Anim Behav 62:369–377

Gowans S, Würsig B, Karczmarski L (2008) The social structure and strategies of delphinids: predictions based on an ecological framework. Adv Mar Biol 53:195–294

Hatchwell BJ (2009) The evolution of cooperative breeding in birds: kinship, dispersal and life history. Philos Trans R Soc B 364:3217–3227

Herzing DL, Johnson CM (1997) Interspecific interactions between Atlantic spotted dolphins (*Stenella frontalis*) and bottlenose dolphins (*Tursiops truncatus*) in the Bahamas, 1985-1995. Aquat Mamm 23:85–99

Hinde RA (1976) Interactions, relationships and social structure. Man 11:1–17

Hooker SK, Whitehead H, Gowans S, Baird RW (2002) Fluctuations in distribution and patterns of individual range use of northern bottlenose whales. Mar Ecol Prog Ser 225:287–297

Janik VM, Sayigh LS (2013) Communication in bottlenose dolphins: 50 years of signature whistle research. J Comp Physiol A Neuroethol Sens Neural Behav Physiol 199:479–489

Jensen FH, Beedholm K, Wahlberg M, Bejder L, Madsen PT (2012) Estimated communication range and energetic cost of bottlenose dolphin whistles in a tropical habitat. J Acoust Soc Am 131:582–592

Johnston DR, Rayment W, Slooten E, Dawson SM (2017) A time-based method for defining associations using photo-identification. Behaviour 154:1029–1050

Karczmarski L, Würsig B, Gailey G, Larson KW, Vanderlip C (2005) Spinner dolphins in a remote Hawaiian atoll: social grouping and population structure. Behav Ecol 16:675–685

Kovacs CJ, Perrtree RM, Cox TM (2017) Social differentiation in common bottlenose dolphins (*Tursiops truncatus*) that engage in human-related foraging behaviors. PLoS One 12:e0170151

Laplanche C, Marques TA, Thomas L (2015) Tracking marine mammals in 3D using electronic tag data. Methods Ecol Evol 6:987–996

Louis M, Fontaine MC, Spitz J, Schlund E, Dabin W, Deaville R, Caurant F, Cherel Y, Guinet C, Simon-Bouhet B (2014) Ecological opportunities and specializations shaped genetic divergence in a highly mobile marine top predator. Proc R Soc B Biol Sci 281. https://doi.org/10.1098/rspb.2014.1558

Louis M, Simon-Bouhet B, Viricel A, Lucas T, Gally F, Cherel Y, Guinet C (2018) Evaluating the influence of ecology, sex and kinship on the social structure of resident coastal bottlenose dolphins. Mar Biol 165:80

Lukas D, Clutton-Brock T (2018) Social complexity and kinship in animal societies. Ecol Lett 21:1129–1134

Lusseau D, Schneider K, Boisseau OJ, Haase P, Slooten E, Dawson D (2003) The bottlenose dolphin community of doubtful sound features a large proportion of long-lasting associations: can geographic isolation explain this unique trait? Behav Ecol Sociobiol 54:396–405

Madsen PT, Wahlberg M, Møhl B (2002) Male sperm whale (*Physeter macrocephalus*) acoustics in a high-latitude habitat: implications for echolocation and communication. Behav Ecol Sociobiol 53:31–41

Mann J (1999) Behavioural sampling methods for cetaceans: a review and critique. Mar Mamm Sci 15:102–122

Mann J, Sargeant B (2003) Like mother, like calf: the ontogeny of foraging traditions in wild Indian ocean bottlenose dolphins (*Tursiops* sp.). In: Fragazy DM, Perry S (eds) The biology of traditions: models and evidence. Cambridge University Press, Cambridge, UK

Mann J, Smuts BB (1998) Natal attraction: allomaternal care and mother-infant separations in wild bottlenose dolphins. Anim Behav 55:1–17

Mann J, Smuts B (1999) Behavioral development in wild bottlenose dolphin newborns (Tursiops sp.). Behaviour 136:529–566

Mann J, Stanton MA, Patterson EM, Bienenstock EJ, Singh LO (2012) Social networks reveal cultural behaviour in tool-using using dolphins. Nat Commun 3:980

Markowitz TM, Harlin AD, Würsig B (2003) Digital photography improves efficiency of individual dolphin identification: a reply to Mizroch. Mar Mamm Sci 19:608–612

Martin AR, Da Silva VMF (2004) Number, seasonal movements, and residency characteristics of river dolphins in an Amazonian floodplain lake system. Can J Zool 82:1307–1315

McHugh KA, Allen JB, Barleycorn AA, Wells RS (2011) Severe *Karenia brevis* red tides influence juvenile bottlenose dolphin (*Tursiops truncatus*) behavior in Sarasota Bay, Florida. Mar Mamm Sci 27:622–643

McSweeney DJ, Baird RW, Mahaffy SD (2007) Site fidelity, associations, and movements of Cuvier's (*Ziphius cavirostris*) and Blainville's (*Mesoplodon densirostris*) beaked whales off the island of Hawai'i. Mar Mamm Sci 23:666–687

Möller LM (2012) Sociogenetic structure, kin associations and bonding in delphinids. Mol Ecol 21:745–764

Norris KS, Dohl TP (1980) The structure and functions of cetacean schools. In: Herman LM (ed) Cetacean behavior: mechanisms and functions. Wiley-Interscience, New York

Norris KS, Schilt CR (1988) Cooperative societies in three dimensional space: on the origins of aggregations, flocks, and schools with special reference to dolphins and fish. Ethol Sociobiol 9:149–179

Nowacek DP, Christiansen F, Bejder L, Goldbogen JA, Friedlaender AS (2016) Studying cetacean behaviour: new technological approaches and conservation applications. Anim Behav 120:235–244

Owen ECG, Wells RS, Hofmann S (2002) Ranging and association patterns of paired and unpaired adult male Atlantic bottlenose dolphins, *Tursiops truncatus*, in Sarasota, Florida, provide no evidence for alternative male strategies. Can J Zool 80:2072–2089

Pearson HC, Markowitz TM, Weir JS, Würsig B (2017) Dusky dolphin (*Lagenorhynchus obscurus*) social structure characterized by social fluidity and preferred companions. Mar Mamm Sci 33:251–276

Pirotta E, New L, Harwood J, Lusseau D (2014) Activities, motivations and disturbance: an agent-based model of bottlenose dolphin behavioral dynamics and interactions with tourism in Doubtful Sound, New Zealand. Ecol Model 282:44–58

Pryor K, Kang Shallenberger I (1991) Social structure in spotted dolphins (*Stenella attenuata*) in the tuna purse seine fishery in the Eastern Tropical Pacific. In: Pryor K, Norris KS (eds) Dolphin societies: discoveries and puzzles. University of California Press, Berkley

Ramos EA, Baloney BM, Magnasco MO, Reiss D (2018) Bottlenose dolphins and Antillean manatees respond to small multi-rotor unmanned aerial systems. Front Mar Sci 5:316

Riesch R, Barrett-Lennard LG, Ellis GM, Ford JKB, Deecke VB (2012) Cultural traditions and the evolution of reproductive isolation: ecological speciation in killer whales? Biol J Linn Soc 106:1–17

Rossbach KA, Herzing DL (1997) Underwater observations of benthic feeding bottlenose dolphins (*Tursiops truncatus*) near Grand Bahamas Island, Bahamas. Mar Mamm Sci 13:498–503

Schulz TM, Whitehead H, Gero S, Rendell L (2008) Overlapping and matching of codas in vocal interactions between sperm whales: insights into communication function. Anim Behav 76:1977–1988

Scott MD (1991) The size and structure of pelagic dolphin herds. Dissertation, University of California, Los Angeles

Sellas AB, Wells RS, Rosel PE (2005) Mitochondrial and nuclear DNA analyses reveal fine scale geographic structure in bottlenose dolphins (*Tursiops truncatus*) in the Gulf of Mexico. Conserv Genet 6:715–728

Sørensen PM, Wisniewska DM, Jensen FH, Johnson M, Teilmann J, Madsen PT (2018) Click communication in wild harbour porpoises (*Phocoena phocoena*). Sci Rep 8:9702

Sterck EHM, Watts DP, Van Schaik CP (1997) The evolution of female social relationships in nonhuman primates. Behav Ecol Sociobiol 41:291–309

Tavares SB, Samarra FIP, Miller PJO (2017) A multilevel society of herring-eating killer whales indicates adaptation to prey characteristics. Behav Ecol 28:500–514

Titcomb EM, O'Corry-Crowe G, Hartel EF, Mazzoil MS (2015) Social communities and spatio-temporal dynamics of association patterns in estuarine bottlenose dolphins. Mar Mamm Sci 31:1314–1337

Wallen MM, Krzyszczyk E, Mann J (2017) Mating in a bisexually philopatric society: bottlenose dolphin females associate with adult males but not adult sons during estrous. Behav Ecol Sociobiol 71:153

Wang K, Wang D, Akamatsu T, Fujita K, Shiraki R (2006) Estimated detection distance of a baiji's (Chinese river dolphin, *Lipotes vexillifer*) whistles using a passive acoustic survey method. J Acoust Soc Am 120:1361–1365

Weir JS, Fiori L, Orbach DN, Piwetz S, Protheroe C, Würsig B (2018) Dusky dolphin (*Lagenorhynchus obscurus*) mother-calf pairs: an aerial perspective. Aquat Mamm 44:603–607

Wells RS (2003) Dolphin social complexity: lessons from long term study and life history. In: De Waal FBM, Tyack PL (eds) Animal social complexity: intelligence, culture and individualized societies. Harvard University Press, Cambridge MA

Wells RS (2014) Social structure and life history of bottlenose dolphins near Sarasota Bay, Florida: insights from four decades and five generations. In: Yamagiwa J, Karczmarski L (eds) Primates and cetaceans: field research and conservation of complex mammalian societies. Springer, Tokyo

Whitehead H (2008) Analysing animal societies: quantitative methods for vertebrate social analysis. University of Chicago Press, Chicago

Whitehead H (2009) SOCPROG programs: analysing animal social structures. Behav Ecol Sociobiol 63:765–778

Whitehead H, Dufault S (1999) Techniques for analyzing vertebrate social structure using identified individuals: review and recommendations. Adv Study Behav 28:33–74

Whitehead H, Ford JKB (2018) Consequences of culturally-driven ecological specialization: Killer whales and beyond. J Theor Biol 456:279–294

Whitehead H, Mann J (2000) Female reproductive strategies of cetaceans. In: Mann J, Connor R, Tyack PL, Whitehead H (eds) Cetacean societies: field studies of dolphins and whales. University of Chicago Press, Chicago

Whitehead H, Rendell R, Osbone RW, Würsig B (2004) Culture and conservation of non-humans with reference to whales and dolphins: review and new directions. Biol Conserv 120:431–441

Wilson EO (1971) The insect societies. Harvard University Press, Cambridge, MA

Wilson DRB (1995) The ecology of bottlenose dolphins in the Moray Firth, Scotland: a population at the northern extreme of the species' range. Dissertation, Aberdeen University, Aberdeen, Scotland

Wiszniewski J, Lusseau D, Moller LM (2010) Female bisexual kinship ties maintain social cohesion in a dolphin network. Anim Behav 80:895–904

Wittemyer G, Douglas-Hamilton I, Getz WM (2005) The socioecology of elephants: analysis of the processes creating multitiered social structures. Anim Behav 69:1357–1371

Wrangham RW (1980) An ecological model of female-bonded primate groups. Behaviour 75:262–300

Würsig B, Pearson HC (2014) Dusky dolphins: flexibility in foraging and social strategies. In: Yamagiwa J, Karczmarski L (eds) Primates and Cetaceans: field research and conservation of complex mammalian societies. Springer Press, Tokyo

Würsig B, Würsig M (1977) The photographic determination of group size, composition, and stability of coastal porpoises (*Tursiops truncatus*). Science 198:755–756

Würsig B, Würsig M (1980) Behavior and ecology of the dusky dolphin, *Lagenorhynchus obscurus*, in the South Atlantic. Fish Bull 77:871–890

Würsig B, Cipriano F, Würsig M (1991) Dolphin movement patterns: information from radio and theodolite tracking studies. In: Pryor K, Norris KS (eds) Dolphin societies: discoveries and puzzles. University of California Press, Berkeley

Würsig B, Wells RS, Würsig M, Norris KS (1994) A spinner dolphin's day. In: Norris KS, Würsig B, Wells RS, Würsig M (eds) The Hawaiian spinner dolphin. University of California Press, Berkeley

Chapter 2
Communication by Sound and by Visual, Tactile, and Chemical Sensing

Peter Tyack

Abstract Toothed whales use vision, chemical sensing, and touch for short-range communication, but they produce sounds to communicate over ranges of hundreds to thousands of meters. Sperm whales and porpoises communicate and echolocate using click sounds, and many toothed whales may eavesdrop on clicks. Many toothed whale species have two sound sources, one specialized for echolocation and the other for communication. Killer whales can independently modulate low-frequency clicks and a higher-frequency component to make complex communication signals. Dolphins click from the right sound source and produce tonal whistles from the left sound source. Bottlenose dolphins develop an individually distinctive whistle, called a signature whistle, through copying elements of sounds in their natal environment. Dolphins in the laboratory imitate synthetic whistle-like sounds, demonstrating their capacity to learn to produce vocalizations, a skill that is rare among nonhuman mammals. Adult dolphins can imitate the signature whistles of partners for use as a vocal label. These individual-specific labels are suited to the fission-fusion societies of most dolphins, in which group composition may change every few minutes. Killer whales, by contrast, live in stable matrilineal groups. They produce stereotyped calls that change slowly over generations, and a process of dialect formation leads members of each group to share a group-specific call repertoire. The process of call change suggests that killer whales learn to modify their calls based upon listening to other whales, but evidence for vocal learning is weaker than for dolphins. Sperm whales are usually sighted in temporary groupings formed of several more stable social units that may join for several days at a time. Sperm whales communicate with rhythmic patterns of clicks called codas. Codas are believed to have a short enough range to suggest a primary function for communication within a group, perhaps identifying group and individual identity. Social units that share the same coda repertoire are defined as members of the same vocal clan. Social units of the same vocal clan may join one another but joining is rare among

P. Tyack (✉)
Sea Mammal Research Unit, Scottish Oceans Institute, School of Biology, University of St. Andrews, St. Andrews, UK
e-mail: plt@st-andrews.ac.uk

© Springer Nature Switzerland AG 2019
B. Würsig (ed.), *Ethology and Behavioral Ecology of Odontocetes*, Ethology and Behavioral Ecology of Marine Mammals,
https://doi.org/10.1007/978-3-030-16663-2_2

sympatric units that have different repertoires, suggesting a role for codas mediating affiliation between units. The same repertoire may be recorded over tens of thousands of km, with the vocal clan comprising tens of thousands of whales. If codas are learned, then vocal clans would represent stable cultural traditions on a grand scale, but evidence for vocal learning is weak in sperm whales. Toothed whales have complex communication systems, but more work is needed to fully understand the role of learning, and we need more detailed longitudinal study of social relationships to fully understand functions of social communication.

Keywords Animal communication · Contact calls · Vocal learning · Vocal culture · Signature whistle · Sperm whale coda · *Tursiops* · *Orcinus* · *Physeter*

2.1 Introduction

Different sensory modalities enable different effective ranges for communication, defined as distances over which a signaler and recipient can effectively communicate. The sense of touch can be a powerful medium for communication, but requires sender and receiver to be in direct physical contact. Chemicals can spread from sender to receiver, but this occurs slowly underwater, limiting the speed and distance of communicating by chemical signals in the sea. This is particularly limiting for highly mobile species such as the toothed whales. By contrast, acoustic and visual signals can be transmitted rapidly, and acoustic signals can travel over large distances underwater. Marine organisms can create signals by generating chemicals (pheromones), light (bioluminescence), or sound (vocalizations), but toothed whales are only known to generate acoustic signals.

2.2 Communication by Touch

Toothed whales have been reported to touch one another in agonistic and affiliative interactions. It can be difficult to separate when touch might be used to signal vs harm an opponent in agonistic interactions, so this section will focus on affiliative interactions. Tavolga and Essapian (1957) noted affiliative rubbing as part of pre-copulatory behaviors of common bottlenose dolphins (*Tursiops truncatus*). Norris (1991) and Dudzinski (1998) noted that one toothed whale may rub a conspecific with its flipper, and they suggested that this may take place as part of affiliative behavior. Sakai et al. (2006) described flipper rubbing among wild Indo-Pacific bottlenose dolphins (*Tursiops aduncus*) in great detail, pointing out that two dolphins may alternate rubbing the leading edge of the other's flipper or one dolphin may use its flipper to rub the body of the other. The animal being rubbed may move so that a specific part of the body is being rubbed, and this may remove loose skin, suggesting a role similar to grooming in primates. However, Sakai et al. (2006) report seeing parasites attached to dolphins, but they did not observe one dolphin rub off a parasite that was attached to another dolphin. Rather, the way in which dolphins exchanged rubbing roles suggested to Sakai et al. (2006) that rubbing functions as an affiliative signal. This interpretation is strengthened by the observations of Tamaki et al. (2006) that after an aggressive interaction, captive common bottlenose dolphins took longer to engage in later aggression if they engaged in flipper rubbing with their former opponent. These observations suggest that not only is flipper rubbing an affiliative behavior but that it is also a signal for reconciliation after conflict (Weaver 2003).

2.3 Communication by Visual Sensing

The ability to recognize visual patterns of objects has limited range underwater. Light is attenuated and scattered as it passes through water, limiting vision to ranges of tens of meters underwater. The common bottlenose dolphin eye has a pupil with an unusual shape, which allows it to have relatively good visual acuity in air and in water. However, this limits the range for best underwater acuity to about 1 m (Herman et al. 1975). Dolphin mothers and calves may be able to recognize one another at ranges of 1m or less, but vision does not play the same role for long-range perception that it does in most terrestrial mammals. Some toothed whales such as the killer whale (*Orcinus orca*) and the Commerson's dolphin (*Cephalorhynchus commersonii*) have evolved high-contrast black and white body surface pigmentation patterns, with large patches that do not demand high acuity; these may have evolved for longer-range visual detection and discrimination. Pigmentation patterns are similar among matrilineally related clans of killer whales and differ across ecotypes (Baird and Stacey 1988), suggesting that pigmentation could provide recognition cues at relatively short ranges.

2.4 Integration of Information Across Senses

Animals can integrate information across sensory systems to solve communication problems. We know more about this process in terrestrial mammals than in toothed whales. For example, in settings where the young of several mothers can intermix, a female mammal faces the problem of assuring that the young she suckles is her own. The most important cue for a ewe (*Ovis aries*) is the visual appearance of her lamb, which allows her to reject distant lambs that do not have the right appearance. She can also recognize the voice of her lamb, but the most important adjunct to vision is the ability to smell her lamb once she approaches to <0.25 m (Alexander and Shillito 1977). We know very little about how cetaceans use chemical senses in this manner, but since they evolved from ungulates, they may have adapted chemical sensing to their aquatic environment.

2.5 Communication by Chemical Sensing

Most mammals have three ways to sense chemicals—the sense of taste discriminates compounds dissolved in water, olfaction senses a broader range of chemicals that are in air, and the vomeronasal sense primarily detects pheromones produced by conspecifics (Pihlström 2008). Toothed whales have a sense of taste that is similar to other mammals (Nachtigall and Hall 1984), but their sense of olfaction is different. Most mammals breathe relatively constantly, and this airflow passes odors across the olfactory mucosa in the nasal passages. Toothed whales hold their breath most of the time and breathe explosively. For example, bottlenose dolphins exhale more than $12\times$ faster than humans (Fahlman et al. 2015). Toothed whales also use nasal structures to produce sound. These high flow rates and adaptations for sound production are not consistent with terrestrial olfaction. Toothed whales have neither olfactory mucosa nor an olfactory bulb in the brain (Pihlström 2008). Odontocete cetaceans have lost about 2/3 of their olfactory receptor genes, consistent with loss of olfactory function (Kishida et al. 2007). The vomeronasal sense is the other major system used by mammals to sense chemicals—it detects pheromones produced by conspecifics along with some other odorants (Baxi et al. 2006). This accessory chemical sense plays an important role in the detection of reproductive state in many mammals, including ungulates that are phylogenetically related to cetaceans (Montgelard et al. 1997). Though there is strong evidence that odontocetes have greatly reduced olfactory capabilities compared to terrestrial mammals, they may have evolved specialized chemical sensory abilities better suited to their aquatic lifestyle, perhaps located away from nasal structures that have specialized for non-olfactory functions.

2.6 Communication by Acoustic Sensing

When the terrestrial ancestors of cetaceans entered the sea, their senses had to adapt to a new environment. Vision is the best way to sense things that are far away in air, but light is not as good as sound for sensing distant objects in water. Whales and dolphins have adapted to their underwater world by specializing in sound much as we humans specialize in vision. Humans have about 1,159,000 nerve fibers in the optic nerve, 38 for every one in the auditory nerve (30,500), while an Amazon river dolphin, *Inia geoffrensis*, which inhabits murky muddy waters has only 0.15 optic nerve fibers for every auditory fiber (Mass and Supin 1989). Most other odontocetes have about the same number of optical as auditory fibers, and they achieve this by

2 Communication by Sound and by Visual, Tactile, and Chemical Sensing 29

having 2.2 (finless porpoise, *Neophocoena phocoenoides*) to 5.3 (sperm whale, *Physeter macrocephalus*) times more auditory fibers than do humans (Ketten 1997).

2.6.1 Echolocation

Toothed whales not only have specialized hearing, but they also have evolved the ability to detect objects in the dark by listening for echoes from their own sounds. Echolocation in toothed whales and bats has required coevolution of sound production and reception capabilities. One way that human and animal sonars increase the level of sound is to direct sound energy in a narrow beam. They also can reduce interfering noise by aligning directional hearing in the same direction as the sound beam. Au et al. (2012) showed that echolocation clicks recorded in the center of the sound beam of bottlenose dolphins are much higher in level and in peak frequency (~120 kHz) than when recorded at off-axis angles. Not only are dolphins able to hear well at 120 kHz, but they also have highly directional hearing at this frequency, with sensitivity aligned along the axis of sound transmission (Au and Moore 1984). The evolution of a sophisticated biosonar created selection pressures for specialized high-frequency sound production and hearing in toothed whales (Au et al. 2009).

Echolocation allows toothed whales to orient and forage in the dark deep ocean and at night, an advantage for a top predator. As air-breathing mammals, they maintain a body temperature elevated above ambient water, carry oxygen to sustain high metabolism during underwater pursuit, and use air to power sound production. Their mammalian ears allow for hearing of high echolocation frequencies, giving them a predator's advantage. But by making echolocation sounds to forage, they also provide cues that eavesdroppers can detect. Four taxa of toothed whales have evolved a cryptic anti-predator strategy, producing echolocation signals that are so high in frequency that their primary predator, killer whales, apparently cannot hear them (Madsen et al. 2005; Morisaka and Connor 2007).

A form of communication occurs when members of the same species eavesdrop on each other's echolocation signals to keep track of one another (Jones and Siemers 2011). For example, Cuvier's beaked whales, *Ziphius cavirostris*, synchronize dives within a group. They disperse once they start to use echolocation to forage, perhaps to avoid distracting one another. When pairs of *Ziphius* are simultaneously tagged with acoustic recording tags, the clicks produced by one whale are often audible on the tag of the other whale, and they appear to use these clicks to reunite before they surface silently (Zimmer et al. 2005a).

2.6.2 Acoustic Communication

Some toothed whales, such as sperm whales and porpoises, make only click sounds and have adopted click signals derived from echolocation clicks for communication.

Porpoises have evolved narrow-beam high-frequency clicks, possibly to avoid detection by eavesdropping predators (Morisaka and Connor 2007). To maintain crypsis, they use high-frequency clicks in different rhythmic patterns for communication (Sørensen et al. 2018), with different patterns produced in different behavioral contexts (Clausen et al. 2010). The directionality of echolocation clicks may be a feature for communication by cryptic animals, allowing them to direct their message to an intended recipient while making it less likely that animals elsewhere will hear them. Sperm whales use rhythmic patterns of clicks for communication, relying more on social defense against predation than on crypsis. They have evolved a sound production apparatus that produces clicks that are highly directional at frequencies of 8–25 kHz (Madsen et al. 2002a) and much less directional at frequencies of about 1–3 kHz (Zimmer et al. 2005b).

The sperm whale and dwarf and pygmy sperm whales of the genus *Kogia* only have one pair of phonic lips to produce sound (Cranford et al. 1996). Other toothed whales produce sounds with two pairs of phonic lips, each one set above bony nares in the upper respiratory tract (Cranford et al. 1996). Delphinids make specialized communication sounds in addition to clicks, using the right set of phonic lips to produce echolocation clicks, and the left side to produce longer, tonal whistle sounds that function for communication (Madsen et al. 2013). The left phonic lips vibrate so rapidly under pneumatic pressure from the lungs that they can produce fundamental frequencies of <2 to >20 kHz (Madsen et al. 2011).

2.6.2.1 Contact Calls

A common function for communication involves maintaining contact between animals who share strong social bonds. For example, all mammals require a mechanism for lactating mothers to maintain contact with their young while they are dependent on suckling, a period that lasts for years in most toothed whales. Unlike terrestrial mammals, toothed whales live in an environment where mother and calf can seldom see one another if they separate by as little as 10 m, and their reduced olfaction hinders use of olfactory cues for recognition. This puts a higher priority on acoustic communication to maintain contact when mother and calf separate.

2.6.2.2 Signature Whistles in Bottlenose Dolphins

Some of the most detailed observations of communication signaling in toothed whales come from two species of bottlenose dolphin, the common and Indo-Pacific bottlenose dolphins, which have been observed closely in captivity and in the wild. Bottlenose dolphin calves produce tonal whistles and click-like sounds as early as the day they are born (Caldwell and Caldwell 1979). Young dolphins show a fascinating combination of precocial and altricial features. They are born highly mobile with well-developed senses and are able to surface independently to breathe within minutes of birth. Young calves often swim tens of meters from their mothers, well out of visual range. At the same time, dolphin calves have a remarkably long period of dependency. In the wild, they will suckle for many years and typically stay with their mother for 3–5 years, until the next calf is born (Wells 2003; for social structure details, see Wells for common bottlenose dolphin and Chap. 16).

The combination of high mobility, low visibility, and prolonged dependence has selected for the development of an acoustic communication system allowing mother and calf to maintain contact and to indicate a desire to reunite across long separations. Macfarlane (2016) used data from wild common bottlenose dolphin mother-calf pairs that were simultaneously tagged with acoustic recording tags (Johnson and Tyack 2003) to monitor sound production at different stages of separations and reunions. The tags were synchronized, enabling calculation of distance by timing how long it took for a sound emitted by one dolphin to be detected on the other tag. When a dolphin calf was swimming away from a mother who was producing echolocation clicks, the most likely detection range for the clicks was about 100 m, but when the mother was pointing toward the eavesdropping calf, the range was 300–450 m. These data suggest potential problems in maintaining contact as animals are separating >100 m. In addition, there is no evidence that a dolphin can tell which individual is making an echolocation click, so the listener could get confused if several dolphins are in acoustic detection range.

Each individual bottlenose dolphin produces a diversity of whistle calls, including a stereotyped individually distinctive whistle called the signature whistle, which comprises about 40–70% of the whistle repertoire of wild dolphins (Janik and Sayigh 2013). Signature whistles produced by dolphins have relatively omnidirectional fundamental frequencies and are individually distinctive, allowing closely bonded animals to keep track of their partners. Ever since signature whistles were first described in common bottlenose dolphins, they have been described as contact calls (Caldwell and Caldwell 1965), and these features make them better suited to this task than eavesdropping on echolocation clicks. Evidence from captivity (Janik and Slater 1998) and the wild (Smolker et al. 1993; Watwood et al. 2005) shows that bottlenose dolphins are more likely to produce individually distinctive signature whistles when separated from social partners than when together. The range at which dolphin whistles can be detected is better suited than clicks for reliable detection during separations. Quintana-Rizzo et al. (2006) followed hundreds of separations of common bottlenose dolphin mothers from their dependent calves and found a mean separation distance of about 100 m. The detection range of bottlenose dolphin whistles depends on habitat, but they are detectable in shallow habitats at ranges of many hundreds of meters and in deeper habitats to many kilometers (Quintana-Rizzo et al. 2006; Jensen et al. 2012), well beyond the maximum separation ranges observed. The shortest detection ranges were similar to those for echolocation clicks when a mother was pointing directly at the calf, but the omnidirectional whistles were detectable at any orientation of the whistler, an important feature for most communication.

Macfarlane (2016) identifies three different potential functions for contact calls: maintaining contact, regaining lost contact, and advertising identity. The pattern of call production predicted for maintaining contact is a spontaneous call rate that would allow animals to monitor each other's location while separated. The function of maintaining contact would not necessarily predict that a listener would modify its behavior upon being informed of the location of a partner. Calls used to regain lost contact would represent a motivation for reuniting; these calls would be predicted to cause a partner to call back and approach. The function of advertising identity is predicted to be important for the last stage of reunion, when animals come close enough to exchange resources or face risk of attack. A common setting where this becomes important is when a parent must ensure that it is feeding the correct offspring, and the offspring must be wary of approaching a non-parent that might attack it. Quick and Janik (2012) provided support for an identity advertisement function when they showed that groups of common bottlenose dolphins in the wild engage in exchanges of signature whistles just before meeting, which they interpret as an exchange of identity information just before the decision to join.

Macfarlane (2016) tested the predictions of these three functions using data from simultaneously tagged mother-calf pairs of common bottlenose dolphins (calf age ranged from 2 to 7 years) during separations of >30 m. Every signature whistle produced by each dolphin during separations was scored as "separating" when animals were moving apart and as "reuniting" when they were coming together again. Logistic regression modeling suggested that whistles were more likely to be produced during reunions than during the separation phase. Another important factor for predicting when a signature whistle would be produced was the delay since the tag on the whistler last detected an echolocation click from its partner. These results are more consistent with the hypothesis that signature whistles are used for reuniting rather than just maintaining contact, a function that can also be supported by eavesdropping on echolocation clicks. Cases of increased whistle rates as a pair joined support the identity advertisement function, but the majority of signature whistles in this dataset were produced when the pair was not separated, suggesting the potential for other whistle functions not examined by Macfarlane (2016).

Most mammals rely upon voice cues for individual identification. Differences in voice produced by differences in the configuration of the vocal tract of different individuals provide sufficient cues for individual discrimination in many species. However, when an animal dives, the air-filled cavities of the vocal tract change shape (Jensen et al. 2011), making voice cues unreliable (Madsen et al. 2011). Bottlenose dolphins do not rely on voice cues to classify a signature whistle, but rather they use the distinctive pattern of modulation of the fundamental frequency, called the contour (Janik et al. 2006), which is under voluntary control of the whistler. Infant common bottlenose dolphins tend to produce simple unstereotyped whistles, but by the end of the first 3 months of life, most develop a stereotyped individually distinctive signature whistle (Caldwell and Caldwell 1979). Newborn dolphins stay so close to their mother that this proximity may reduce the need for an acoustic signal for individual identification, but by 3 months of age, they separate for long enough and far enough to pose problems for reunification.

This need for individual identification is heightened by the fission-fusion society of bottlenose dolphins where each grouping tends to last only for minutes, and a mother and calf may associate with dozens of individuals. Dolphins learn to produce signature whistles that vary in contour more across individuals and less within an individual than is typical for mammals that rely upon voice cues for individual recognition (Tyack 2000). When a wild dolphin hears the signature whistle of an animal with which it shares a strong social bond, it responds more strongly than when it hears a familiar whistle from a less strongly bonded dolphin (Sayigh et al. 1999), and dolphins even recognize the contours of synthetic whistles devoid of voice cues (Janik et al. 2006), demonstrating the ability to recognize signatures by contour rather than voice.

Vocal learning is rare among mammals, which tend to inherit the motor programs that generate species-typical vocalizations (Janik and Slater 1997). The gold standard for demonstrating vocal learning is to measure the pre-exposure repertoire of a subject, create a novel stimulus, and then show that the subject can copy the stimulus after hearing it. The best evidence for vocal learning in dolphins stems from studies of adults trained to imitate novel synthetic contours in the laboratory (Richards et al. 1984). There is also evidence that common bottlenose dolphins learn to incorporate acoustic features of sounds they hear into their signature whistles. The strongest evidence comes from cases where a dolphin calf incorporates a sound that differs from normal dolphin whistles. One common artificial sound in a dolphinarium is the trainer's whistle, which indicates to a dolphin that it can get a reward. Calves that grow up in dolphinaria are more likely than wild dolphins to develop signature whistles with unmodulated frequency contours, similar to the trainer's whistle (Miksis et al. 2002).

Once a female common bottlenose dolphin develops a stereotyped signature whistle in the wild, the signature whistle tends to remain stable for the rest of the dolphin's life (Sayigh et al. 1990, 2007). However, development of a new social relationship can alter the signature whistle of male dolphins. As they mature, some male bottlenose dolphins form stable associations with one other male in the case of common bottlenose dolphins in coastal waters of Florida (Wells 1991) and one or two other males in the case of Indian Ocean bottlenose dolphins in the inshore waters of Western Australia (Connor et al. 1992). These male coalitions develop synchronized and coordinated behaviors that may function in improving foraging, protection from predators, territorial disputes between males from adjacent communities (Wells 2003), and competing with other males for access to females (Connor et al. 2001). Smolker and Pepper (1999) documented how the whistles of three Indo-Pacific bottlenose dolphins became more similar as their alliance bond developed, and Watwood et al. (2004) showed for common bottlenose dolphins that the whistles of established alliance partners are more similar to one another than to the whistles of other alliances. In these cases of convergence, each male retains some individually distinctive features of their whistle as the overall contour becomes more similar.

Dolphins not only use vocal learning in the development of their own signature whistles, but they also copy the signature whistles of dolphins with whom they share strong bonds. Tyack (1986) first discovered signature imitation in two captive bottlenose dolphins. Each dolphin primarily produced a stereotyped whistle contour that differed from the favored whistle of the other dolphin, but each dolphin also occasionally repeated a contour similar to the other dolphin's signature. Similar copying has been observed in whistle exchanges between pairs of captive dolphins and in wild dolphins temporarily restrained for health assessment (King et al. 2013). In this setting, each dolphin primarily produces its signature whistle, but when two dolphins that share a strong bond are held together, one may copy the signature whistle of the other. Figure 2.1a, b shows spectrograms of whistles from exchanges between two mother-calf pairs recorded during temporary restraint. Figure 2.1c shows spectrograms of whistles from exchanges between two adult males recorded in an aquarium pool. The top row of each sub-figure shows three examples of the signature whistle of the dolphin being copied; the middle row shows the copies; and the bottom row shows three examples of the signature whistle of the dolphin making the copies. In exchanges where copying was detected, the rate of copying was 0.18 copy/min/individual, much lower than the overall whistle rate of 5.3 whistles/min/individual. King et al. (2014) used an interactive playback design where they waited until a dolphin subject produced its own signature whistle, and then they would play back either a synthetic copy of its signature (copy) or a different signature whistle (control). The subjects were significantly more likely to respond with their signature whistle to the copy than to the control, with an optimal time interval of 1 s between the original signature whistle and the matching stimulus. King et al. (2014) conclude that whistle matching is an affiliative signal that allows one dolphin to direct a signal to a particular individual within a large communication network.

2.6.2.3 Group-Specific Dialects of Killer Whales

The evidence described above suggests that dolphins use and copy individually distinctive whistles to maintain associations between strongly bonded partners within a fission-fusion society. There is evidence for a different pattern of contact calls in killer whales (*Orcinus orca*), which have stable groups. Killer whales have some of the most stable groups known in any mammal. The fish-eating ecotype of killer whales in the northeast Pacific lives in matrilineal groups spanning 2–4 generations (Olesiuk et al. 2005). Both sexes remain with their mothers throughout their life, with neither sex dispersing from their natal group (Bigg et al. 1990). Different matrilineal groups often travel together in an assemblage called a pod, with more closely related matrilines tending to associate more frequently than more distant matrilines (see Chap. 11, for details of killer whale societies).

Killer whales produce stereotyped calls that may include high-frequency components similar to dolphin whistles and pulsed components with energy at lower frequencies corresponding to the repetition rate of the clicks (Ford 1987). When members of a group are separated, they often exchange shared calls with one another (Miller et al. 2004a). Each pod of killer whales studied in the Pacific Northwest, Alaska, and Norway has a group-specific repertoire of these stereotyped calls (Ford and Fisher 1983; Ford 1989, 1991; Yurk et al. 2002; Strager 1995). Each pod averaged about ten call types, with more closely related pods sharing more call types. When one individual separates from its group, recordings show that

2 Communication by Sound and by Visual, Tactile, and Chemical Sensing   35

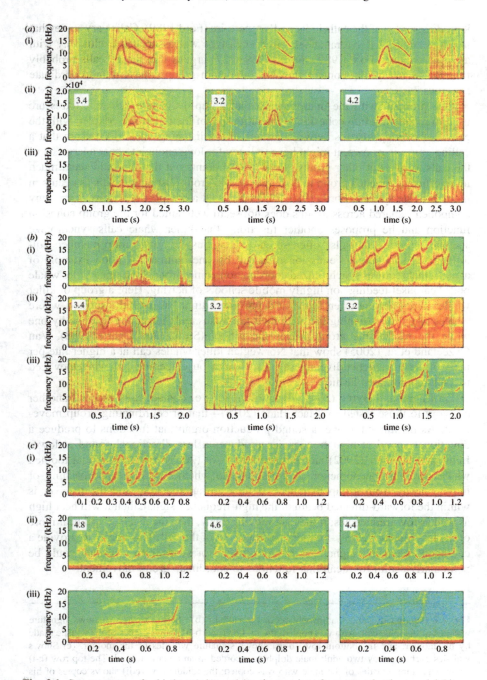

Fig. 2.1 Spectrograms of whistles exchanged by three pairs of common bottlenose dolphins, *Tursiops truncatus*. (**a**) and (**b**) Shows exchanges of whistles between mother-calf pairs recorded during temporary restraint in waters off Sarasota FL. In (**a**), the top row (**a-i**) shows three examples of the signature whistle of the mother, who was copied; the middle row (**a-ii**) shows three examples of copies of her whistle made by her male calf; and the bottom row (**a-iii**) shows three examples of

it produces most of the calls in its pod's repertoire (Ford 1989). Calls vary somewhat in usage among behavioral states, but no call type was associated exclusively with any one context (Ford 1989). Ford (1989: 727) argues that these "calls probably function as intragroup contact signals to maintain group cohesion and coordinate activities."

One problem with the interpretation that group-distinctive call repertoires are used to maintain group cohesion is the large size of the repertoire coupled with the extent of overlap of calls between groups. For a listening whale to be sure that a calling group is its own, it might have to listen for long enough to hear a large set of the calls to rule out other groups. Ford (1989) points out that it would seem much more efficient for each group to have one or two group-distinctive calls as occurs in many primate species. Ford (1989: 741) notes that a complex repertoire with many components shared across groups does not seem well suited for the group cohesion function and he proposes another function—that killer whale calls synchronize specific activities, as calls "often spread contagiously among group members following the spontaneous emission of a call by one animal." A good example of activity that requires synchronized coordination comes from Norwegian killer whale pods "carousel feeding" on highly mobile schools of herring. Here a group of killer whales herds herring often found at depth into a tight ball near the surface, where whales take turns stunning fish with their tail flukes and then eating fish one-by-one while the rest of the group circles the school (Similä and Ugarte 1993). Van Opzeeland et al. (2005) show that Norwegian killer whales call at a higher rate per individual with a higher diversity of calls when carousel feeding than when they feed in a less coordinated fashion.

Da Cunha and Byrne (2009) review functions of contact calls and add another category to those tested by Macfarlane (2016)—that of coordinating group movement. As mentioned above, a sound production organ that functions to produce a directional sound at high frequencies produces a less directional sound at lower frequencies. Miller (2002) and Lammers and Au (2003) hypothesize that toothed whales use this phenomenon to communicate whether they are approaching or moving away from group members. If they are approaching and the receiver is within the forward-directed beam of the high-frequency signal, then it will hear high as well as low frequencies; if they are moving away, then the receiver will hear more of the less directional low-frequency component of the call. This ability to provide a cue about whether a signaler is moving toward or away from a listener could be particularly useful for a call that functions to coordinate group movement.

Fig. 2.1 (continued) the signature whistle of the calf. In (**b**), the top row (**b-i**) shows signature whistles of the male calf, who was copied; the middle row (**b-ii**) shows copies of his whistle made by his mother; and the bottom row (**b-iii**) shows signature whistles of the mother. (**c**) Shows whistles exchanged by two adult male dolphins recorded in an aquarium pool. The top row (**c-i**) shows signature whistles of the male who was copied; the middle row (**c-ii**) shows copies of his whistle made by the other male; and the bottom row (**c-iii**) shows signature whistles of the copier. Each middle row has inset numbers that indicate how similar the copy is judged by human observers to be to the signature that is copied, with 1 being not similar and 5 being very similar. From Fig. 1 of King et al. (2013)

2 Communication by Sound and by Visual, Tactile, and Chemical Sensing

Call repertoires may function not only for cohesion of matrilineal groups but also to help killer whales identify their relationship to other pods. Ford (1991) compared call repertoires of 16 killer whale pods, and found that they formed four distinct vocal clans, with some calls shared within each clan and none shared between clans. Even though pods from different clans did not share calls, they often associated together close enough for prolonged acoustic contact between the pods. When pods within a clan shared a call, there often were acoustic differences in the call type that were distinctive for each pod. When Yurk et al. (2002) compared call repertoires of Alaskan pods to genetic relationships, they found two acoustically distinct groups, each of which had different maternal DNA. These results led to the hypothesis that when a pod grows too large, it may split into two and that vocal clans are formed of related pods descended from a common ancestral group. Figure 2.2 shows spectrograms of calls that are shared between Alaskan pods within a clan, showing pod-specific variation. The first two rows show variants of calls AKS01 and AKS05 which are produced by different pods in the AB vocal clan, and the third and fourth rows show variants of two calls produced by the AD clan (from Yurk et al. 2002). The overall similarity of the structure of each call type should be obvious, as well as the differences between variants. Deecke et al. (2000) measured fine-scale acoustic features of two calls from two matrilines within the same vocal clan and showed that one call changed consistently from year to year, with both groups tracking the same change, and the other call showed little change over the 14-year sample. Deecke et al. (2000) suggest that group-specific dialects could develop if as pods diverge, they add, drop, or modify calls and that the combination of differentiation coupled with maintenance of similar calls may result from cultural drift that is modulated by an active process in which matching of the acoustic structure of calls occurs across different pods within the same vocal clan but not across different acoustic clans. This pattern suggests that killer whales may be able to use call repertoires to assess genetic relatedness. Barrett-Lennard (2000) analyzed the DNA of fish-eating killer whales of the northeast Pacific and found that most mating was between rather than within sympatric vocal clans. Pods within a clan are closely enough related that selecting a mate from a different clan may represent a mechanism to avoid inbreeding.

The patterns of vocal change described above are most easily explained as resulting from a process of vocal learning in which one animal modifies its own vocalizations based on what it hears from others. This ability has been well established for common bottlenose dolphins (Janik and Slater 1997) but is less well established for killer whales. Ford (1991) describes evidence for killer whales learning calls from conspecifics. However, demonstrating that they are actually modifying their own calls to match the natural call of a conspecific is more difficult than for matching artificial calls or calls of other species. Janik and Slater (1997) distinguish vocal usage learning from vocal production learning. In vocal usage learning, an animal learns to produce a call already in its repertoire in a new context, while vocal production learning requires an animal to modify its pre-exposure vocal repertoire using imitation of a sound to produce a new vocalization that was not in its pre-exposure repertoire. Foote et al. (2006) provide evidence that a wild killer whale

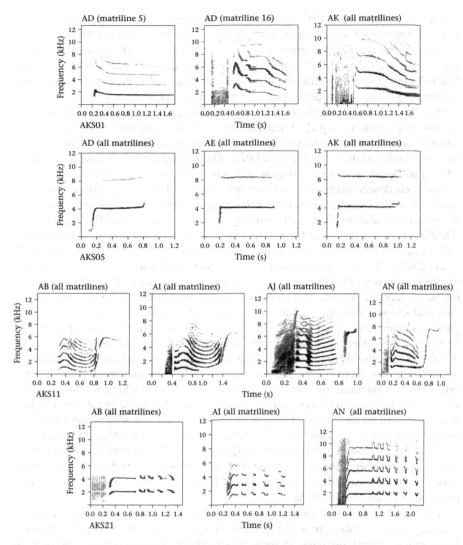

Fig. 2.2 Spectrograms of call types produced by different matrilines in different pods and vocal clans of Alaskan killer whales, *Orcinus orca*. Each of the top two rows shows a call type from the AB vocal clan, and each of the bottom two rows shows a call type from the AD clan, with different variants of each call as produced by different matrilines in the clan. From Fig. 5 of Yurk et al. (2002)

may have copied barks of sea lions; however, they do not show that these calls differ from normal killer whale calls nor can they rule out that they might have actually been produced by sea lions. Abramson et al. (2018) trained a captive killer whale to match sounds either from her own calf or from a human. Unfortunately, the Abramson study did not use the same methods to test for matches in the pre- and post-exposure repertoires, which hinders interpretation. All of these killer whale studies face problems in demonstrating that the matches represent vocal production rather than usage learning. None of the studies of vocal learning in killer whales fully meet the gold standard of quantifying a pre-exposure repertoire, designing signals that clearly differ from this repertoire, and demonstrating accurate matching in the exposure or post-exposure repertoires.

Most biologists studying the stereotyped calls of killer whales categorize them by the whole call—defined as an utterance separated by silence. However, Strager (1995) showed that some of the calls produced by Norwegian killer whales were formed of different combinations of call components. The relative positions of components in different calls had a fixed order. For example, call N22 was similar to a suffix, always being added at the end of a sequence of call components. This observation led Strager (1995) to suggest that some killer whale calls may be formed of a sequence of phoneme-like subunits. Yurk (2005) proposed that calls of killer whales from the Pacific Northwest are also made up of subunits. Shapiro et al. (2011) used methods developed for computer processing of human speech to test the pros and cons of analyzing killer whale calls as complete calls or as sequences of subunits. They defined subunit boundaries as a silent gap of >0.1 s or a 500 Hz spectral jump within 0.25 s. They then tested three different ways to classify calls: one based on the whole call, a second that assumed each call is made up of call-specific subunits, and a third that assumed that subunits could be shared across calls. All three methods to classify calls had error rates that were not statistically different. These results suggest that killer whales could construct their call repertoire either by memorizing all of the whole call types, or by memorizing the sequencing of a smaller set of subunits. Several strands of evidence supported the shared subunit model over the unshared subunits:

1. In the shared subunit analysis, 75% of all calls contained at least one subunit shared across calls.
2. The shared subunit analysis used only 1/3 the number of subunits generated by the unshared analysis, reducing memory requirements and enabling a more efficient representation.
3. Nearly half of variable calls analyzed matched a subunit generated by the shared subunit analysis of stereotyped calls, suggesting that some calls classified as variable may be made up of rare sequences of the same subunits as stereotyped calls.

In addition, analysis of calls from killer whales from the Northeast Pacific showed some subunits that matched with Norwegian subunits, suggesting that subunits may be shared across populations that do not share whole calls, just as phonemes may be shared across human languages that do not share words. This subunit view suggests a different type of vocal learning in killer whales. It suggests that auditory categorization may start by detecting subunits before categorizing the entire call as a unit and that if whales have pattern generators for subunits, then learning to produce a call may involve memorizing the correct sequence of subunits. Byrne (1999) describes a similar interpretation that nonvocal imitation may also be comprised of parsing a string of behaviors that are observed and learning the sequence of motor actions that produces this string. Even if sequence learning is a critical component of learning whole calls, Deecke et al.'s (2000) demonstration that pods may change the frequency of a section of the call over time suggests that subunits may be modified by classic vocal production learning, just as different human cultures may have slightly different versions of the same phoneme.

2.6.2.4 How Sperm Whales Use Clicks for Communication

Sperm whales have a vocal apparatus so specialized for echolocation that even though it takes up 1/3 of the body, sperm whales appear to be limited to using clicks for communication as well as for echolocation. Sperm whales produce clicks with one pair of phonic lips at the front of the head, with most of the energy directed backwards through the spermaceti organ (Møhl 2001). Some of this energy escapes into the water, but most reflects off an air-filled sac close to the skull and is directed in an intense narrow beam forward into the water (Møhl et al. 2003; Zimmer et al. 2005b). Some of the energy reflects back toward the skull again, and this reverberation causes the sperm whale click to be made up of a series of pulses, where the inter-pulse interval is proportional to the size of the spermaceti organ (Zimmer et al. 2005c). Bioacousticians can use the inter-pulse interval to estimate the size of the clicking whale (Bøttcher et al. 2018), and it is likely that sperm whales can extract the same information from these clicks.

Sperm whales typically echolocate during deep foraging dives, producing regular series of clicks every 0.5–2 s, and more rapid series of clicks, called a buzz, as they attempt to capture prey (Miller et al. 2004b; Watwood et al. 2006). The omnidirectional low-frequency component of these regular clicks is typically detectable at ranges of ~5 km (Barlow and Taylor 1997). Male sperm whales also produce clicks with a low-frequency emphasis (centroid frequencies from 2 to 4 kHz), less directionality, and longer duration (0.5–10 ms) than other sperm whale clicks (Madsen et al. 2002b). These clicks tend to have a longer inter-click interval than other sperm whale clicks, leading them to be called "slow clicks." Oliveira et al. (2013) show that these slow clicks tend not to be produced during echolocation-based foraging, but rather in contexts where communication is the more likely function. Oliveira et al. (2013) also recorded sperm whales engaged in exchanges of slow clicks, suggesting a communicative function. Madsen et al. (2002b) estimate that these slow clicks may be detected as far as 60 km away, an estimate that is supported by recordings of slow clicks from ranges of 37 km (Barlow and Taylor 1997).

Watkins and Schevill (1977) reported that sperm whales also make rhythmic series of clicks in stereotyped sequences either in exchanges between whales near one another or in sequences that seemed to be produced by the same whale. Whitehead and Weilgart (1991) showed that sperm whales tend to produce codas when socializing at the surface, in contrast to regular echolocation clicks that tend to be produced during deep foraging dives. Madsen et al. (2002a) showed that coda clicks differ from echolocation clicks in having less than 1/10th the source level and with less decay from the main pulse to the reverberation pulses, yielding a less powerful and more reverberant click. Madsen et al. (2002a) propose that when sperm whales use a click for echolocation, they direct most of the energy in a forward directed beam, but that to make coda clicks, they add air to a reflective sac in the front of the head, which causes more energy to reflect in the nasal complex. This results in clicks that have a longer pulse duration and lower directionality that appear better suited to a communicative function. Weilgart and Whitehead (1997) suggest that coda clicks have a 600 m detection range, and Schulz et al. (2011: 153) state that codas "are only clearly audible through near-surface hydrophones at ranges of a few hundred meters or less" detection ranges much lower than estimated for slow clicks or regular echolocation clicks. The context of coda production in combination with the short estimated range of detection led Weilgart and Whitehead (1993: 744) to suggest that codas function "to maintain social cohesion within stable groups of females following periods of dispersion during foraging." This interpretation is supported by acoustic localization of sperm whales exchanging codas, which estimated the distance between exchanging whales ranged from 1 to 324 m (Schulz et al. 2008).

Sperm whales have a prolonged period of maternal care, with the young remaining for years with their mother in groups of adult females with young that live in tropical or subtropical waters (Best 1979; see also Chap. 12). As males mature, they leave their natal group and form dispersed groups with other males, often moving seasonally to higher latitudes to feed. Whitehead et al. (1991) report that adult female sperm whales and their young in the Galapagos Archipelago typically are sighted in groups of about 24 whales. Photo identification of individual sperm whales shows that each of these groups tends to be formed of two stable social units that associate for about 1 week at a time. Even though whales within a unit may associate for years at a time, individuals have been observed to transfer between units (Christal et al. 1998). Sperm whale units are not as matrilineal as those of killer whales; they may contain a combination of clusters of related individuals and individuals with no close genetic relationship to any other member of the unit (Mesnick et al. 2003).

Rendell and Whitehead (2003) studied the coda repertoires of 64 groups of sperm whales recorded across the South Pacific and Caribbean. Focusing on 22 stable social units repeatedly sighted in the Galapagos Archipelago, they found that all of these units could be assigned to one of three coda-use patterns, which they called vocal clans. Figure 2.3a shows the three clusters of codas, whose distribution across the units is marked in Fig. 2.3b. One vocal clan produced codas with regularly spaced clicks ("R," marked green in Fig. 2.3), the second made codas where the last inter-click interval was longer than previous ones ("+1," marked purple in Fig. 2.3), and the final clan made short codas with just three clicks or rapid bursts of four clicks ("short," marked orange in Fig. 2.3). These three clans were sympatric, but units tended to associate only with other units from the same vocal clan. Of 26 encounters with groups containing 2 known units, only 1 involved a sighting of 2 units from different clans sighted within a few kilometers for at least 1 h (one was R and the other +1, but only +1 codas were recorded during this time). In the recordings of groups from the broader geographical range, only codas from one clan were typically

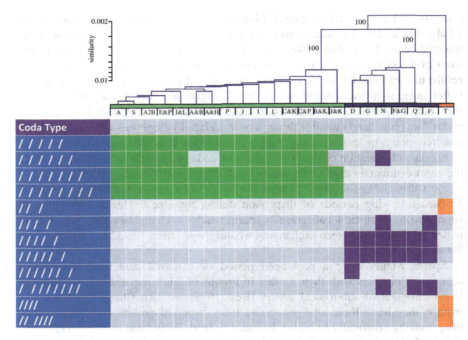

Fig. 2.3 Coda repertoires of 22 social units of sperm whales, *Physeter macrocephalus*, recorded near the Galapagos Islands. The top panel shows three clusters of codas that show stronger coda similarity within a cluster than between. The lower panel shows the distribution of the most common (>10%) codas for each social unit. Social units in the leftmost cluster, marked in green, produce codas with regular inter-click intervals. Social units from the middle cluster, marked in purple, produce a series of 4+ clicks either ending or starting with one slightly longer interval. The rightmost cluster, marked in orange, is comprised of just one social unit which produces either three click codas or codas with four clicks in rapid succession. From Fig. 1 of Rendell and Whitehead (2003)

recorded. The broader recordings yielded five vocal clans in the South Pacific and one vocal clan from the Caribbean. The vocal clans of sperm whales span a distance of about 10,000 km in the Pacific Ocean (including perhaps 10,000 whales on average); these clans are larger than those of killer whales that span about 1000 km and include about 100 whales.

Whitehead and Rendell (2004) took inspiration from the observation that different sympatric killer whale ecotypes had different foraging adaptations to study movement and foraging behaviors of two vocal clans of sperm whales in the Galapagos Islands. They show that the +1 clan moved in straighter lines offshore, while the R clan groups had more convoluted tracks and tended to be sighted inshore. Judging foraging success by defecation rates, they suggest that +1 clans were more successful during an El Niño year, while the R clan was more successful when surface waters were cooler. These observations led them to suggest that clans do not just share vocal behavior, but that they share other behavioral traits such as foraging strategies that have fitness consequences. Male sperm whales foraging in high latitudes vary their foraging behavior to take advantage of different kinds of prey at different depths (Teloni et al. 2008), but there is little evidence of different foraging strategies for female groups in tropical or temperate waters. Detailed analysis of foraging patterns of tagged sperm whales show differences in dive depth comparing the Gulf of Mexico, North Atlantic, and Mediterranean, but remarkably little difference in dive/surface durations, time spent foraging, and rate of attempts to capture prey (Watwood et al. 2006). Similar analyses will be required to test the hypothesis that clans differ in foraging strategies.

Whitehead (2003: 309) argues that "... for a sperm whale, membership in a clan has a connotation comparable to that of nationality in humans. ... Group identity has benefits for an animal: a well-proven way of behaving and a pool of companions who behave similarly who can be used as models and colleagues in cooperative endeavors." This interpretation hinges on the prediction that codas function to mediate inter-group interactions, such that only groups that share the same coda repertoire will join one another. This interpretation is at odds with Weilgart and Whitehead (1993) who use the short detection range of codas and the usage patterns of codas when whales within a group are socializing to argue for an intragroup rather than inter-group function. However, codas might serve an identity advertisement function described by Macfarlane (2016) in which groups might advertise their identity at close range just before a possible join. The hypothesis that social units use codas to decide which other units to join with, which I will call the "clan coda join" hypothesis, leads to a set of testable predictions:

1. Whales within a unit would be likely to produce codas when they hear regular echolocation clicks indicating the presence of another unit nearby.
2. Whales within a unit would be likely to produce codas when they hear codas from another unit.
3. A unit should be more likely to swim toward a group producing codas from their clan vs a different clan.
4. A unit should be more likely to swim toward playback of codas from their clan vs a different clan.

Visual observations and acoustic localization of clicking whales (Watkins and Schevill 1977) are well suited to testing these predictions, but most such observations focus on coda exchanges within, not between, groups (e.g., Schulz et al. 2008). The only tests of these predictions I am aware of involve tests of vocal reactions of sperm whales to coda playback. Initial results do not support the "clan coda join" hypothesis: that sperm whales exchange codas between groups to make decisions about which other groups to join. Rendell and Whitehead (2005) report no change in coda production for the majority of coda playbacks; when whales responded, they were more likely to stop making codas on playback than to start. One particularly problematic finding for the clan coda join hypothesis was the observation that the subjects were as likely to match the codas of a different clan as their own clan. However, there was only one instance of each kind of matching; more research on acoustic interactions between groups and responses to coda playbacks are necessary to fully test this hypothesis. Clearly it is equally important to test hypotheses about intragroup communication by codas as well.

Rendell and Whitehead (2003: 225) argue that the codas produced by vocal clans of sperm whales should be viewed as culture, "defined as group-level information or behavior transmitted by social learning." After some resistance from social scientists who wanted to reserve the term "culture" for exclusive application to humans, there has been growing interest in how social learning leads to the development of animal cultures such as the vocal traditions of songbirds or tool use in apes (Laland and Galef 2009). Vocal cultures, where animals learn to copy acoustic features of the vocalizations of others, "provide the largest body of evidence for cultural transmission of behavioral traits in the animal kingdom" (Laland and Janik 2006: 543). There is a well-established set of methods to demonstrate vocal production learning, which has helped to establish vocal cultures (Janik and Slater 1997). However, the evidence for vocal learning in sperm whales is quite weak.

Coda repertoires are typically categorized using a simple method of comparing codas as having regular inter-click intervals or as forming a long/short pattern similar to how Morse code intervals are distinguished as "dit" or "dah." Some regular codas are also distinguished on the basis of the absolute values for inter-click intervals (Antunes et al. 2011; Gero et al. 2016; Schulz et al. 2011). Control of inter-click interval is important for echolocation; all of the inter-click intervals used in codas are also used by sperm whales in slowly changing sequences of echolocation clicks, and the difference between usual and coda clicks is thought by Madsen et al. (2002a) to be caused by inflation of a sac in the sound production organ rather than by the same neural circuits responsible for generating timing of click intervals. These points suggest that if sperm whales learn codas, this can just involve remembering a sequence of relative or absolute inter-click intervals that are already part of its repertoire. This makes it hard to rule out vocal usage learning to explain the differences between the codas typical of the different vocal clans of sperm whales. Laland and Janik (2006) advocate that rather than thinking of traits to be defined categorically as genetic, learned asocially, or cultural (learned socially), it is more useful to ask how much of the variance in the trait can be attributed to social learning. They argue that this change in focus would help us to study interactions between genes, ecology, and learning in a way that helps answer questions about how cultural behavior affects evolutionary processes, an important reason for comparative studies of culture (e.g., Whitehead 1998). We still do not know how much of the variance in coda repertoires is a product of vocal production learning through matching social models.

2.7 Conclusion

When comparing the broad range of studies of communication in toothed whales to those of terrestrial mammals, marine mammalogists face difficulties in observing behavioral interactions in enough detail to sort out the pattern of signal and response that make up a communication system, and they face similar difficulties in tracking social relationships among individuals. Our understanding of chemical communication is hampered by our ignorance of chemical senses beyond the most basic taste sense; this suggests the need for more research studying how aquatic lifestyles may have selected for changes in the chemical senses of marine mammals. Toothed whales seldom produce visible cues that indicate when they are vocalizing, and they are often not visible, so it can even be difficult to tell which animal produces which signal, whether it is vocal, visual, or tactile. Advances in acoustic localization and development of acoustic, movement and image recording tags are helping to solve this problem for vocal signaling and for movement cues, even to observe tactile signaling (Aoki et al. 2013). Some of the most interesting patterns of communication involve learning to produce and use individual- and group- or clan-specific vocalizations, and this makes it essential to have long-term studies of identified individuals along with their association patterns to put communication in the context of changing social relationships. The more we know about the history of each individual, the better we can understand the context of communication and the problems communication must solve. Other important methods include field experiments—for example—use of human-controlled sound playbacks to test hypotheses about when and whether a particular signal will evoke a particular response in a particular individual (King 2015).

As we have learned to appreciate the diverse ways in which toothed whales rely upon sound to solve ecological and social problems, the more we have learned that noise from human activities may disrupt the behavior of toothed whales in diverse and surprising ways (Tyack 2009). Responses of cryptic species appear to be triggered by particularly low levels of sound. For example, harbor porpoises move away from human activities that generate noise at low enough levels that porpoises avoid pile driving at ranges of 20 km or more (Tougaard et al. 2009, 2014). Sperm whales do not show horizontal avoidance of pulses from air guns used in seismic surveys, but this exposure causes a reduction in foraging effort at ranges of <10 km (Miller et al. 2009). The most intense acute responses of toothed whales to anthropogenic noise involve atypical mass strandings in which two or more beaked whales strand over tens of kilometers during a few hours that coincide with naval sonar exercises (D'Amico et al. 2009). Even if lethal strandings were prevented, if disturbance causes whales to leave preferred habitats, this could affect the population if large numbers are affected (New et al. 2013). Our understanding of how toothed whales use and respond to sound thus has implications for their conservation (Wartzok et al. 2005).

References

Abramson JZ, Hernández-Lloreda MV, García L, Colmenares F, Aboitiz F, Call J (2018) Imitation of novel conspecific and human speech sounds in the killer whale (*Orcinus orca*). Proc R Soc B 285(1871):20172171

Alexander G, Shillito EE (1977) The importance of odour, appearance and voice in maternal recognition of the young in Merino sheep (*Ovis aries*). Appl Anim Behav Sci 3(2):127–135

Antunes R, Schulz T, Gero S, Whitehead H, Gordon J, Rendell L (2011) Individually distinctive acoustic features in sperm whale codas. Anim Behav 81(4):723–730

Aoki K, Sakai M, Miller PJ, Visser F, Sato K (2013) Body contact and synchronous diving in long-finned pilot whales. Behav Process 99:12–20

Au WW, Moore PW (1984) Receiving beam patterns and directivity indices of the Atlantic bottlenose dolphin *Tursiops truncatus*. J Acoust Soc Am 75(1):255–262

Au WW, Branstetter BK, Benoit-Bird KJ, Kastelein RA (2009) Acoustic basis for fish prey discrimination by echolocating dolphins and porpoises. J Acoust Soc Am 126(1):460–467

Au WW, Branstetter B, Moore PW, Finneran JJ (2012) The biosonar field around an Atlantic bottlenose dolphin (*Tursiops truncatus*). J Acoust Soc Am 131(1):569–576

Baird RW, Stacey PJ (1988) Variation in saddle patch pigmentation in populations of killer whales (*Orcinus orca*) from British Columbia, Alaska, and Washington State. Can J Zool 66 (11):2582–2585

Barlow J, Taylor B (1997) Acoustic census of sperm whales in the eastern temperate North Pacific. J Acoust Soc Am 102:3213

Barrett-Lennard LG (2000) Population structure and mating patterns of killer whales (*Orcinus orca*) as revealed by DNA analysis. PhD Thesis UBC, Vancouver

Baxi KN, Dorries KM, Eisthen HL (2006) Is the vomeronasal system really specialized for detecting pheromones? Trends Neurosci 29(1):1–7

Best PB (1979) Social organization in sperm whales, *Physeter macrocephalus*. In: Winn HE, Olla BL (eds) Behavior of marine animals, vol 3. Plenum, New York, pp 227–290

Bigg MA, Olesiuk PF, Ellis GM, Ford JK, Balcomb KC (1990) Social organization and genealogy of resident killer whales (*Orcinus orca*) in the coastal waters of British Columbia and Washington State. Report of the International Whaling Commission 12:383–405

Bøttcher A, Gero S, Beedholm K, Whitehead H, Madsen PT (2018) Variability of the inter-pulse interval in sperm whale clicks with implications for size estimation and individual identification. J Acoust Soc Am 144(1):365–374

Byrne RW (1999) Imitation without intentionality. Using string parsing to copy the organization of behaviour. Anim Cogn 2(2):63–72

Caldwell MC, Caldwell DK (1965) Individualized whistle contours in bottlenose dolphins (*Tursiops truncatus*). Nature 207:434–435

Caldwell MC, Caldwell DK (1979) The whistle of the Atlantic bottlenosed dolphin (*Tursiops truncatus*)—ontogeny. In: Winn HE, Olla BL (eds) Behavior of marine animals: current perspectives in research, vol 3. Plenum, New York, pp 369–401

Christal J, Whitehead H, Lettevall E (1998) Sperm whale social units: variation and change. Can J Zool 76(8):1431–1440

Clausen KT, Wahlberg M, Beedholm K, Deruiter S, Madsen PT (2010) Click communication in harbour porpoises (*Phocoena phocoena*). Bioacoustics 20:1–28

Connor RC, Smolker RA, Richards AF (1992) Dolphin alliances and coalitions. In: Harcourt AH, de Waal FBM (eds) Coalitions and alliances in humans and other animals. Oxford University Press, Oxford, pp 415–443

Connor RC, Heithaus MR, Barre LM (2001) Complex social structure, alliance stability and mating access in a bottlenose dolphin 'super-alliance'. Proc R Soc Lond B 268:263–267

Cranford TW, Amundin M, Norris KS (1996) Functional morphology and homology in the odontocete nasal complex: implications for sound generation. J Morphol 228(3):223–285

D'Amico AD, Gisiner R, Ketten DR, Hammock JA, Johnson C, Tyack P, Mead J (2009) Beaked whale strandings and naval exercises. Aquat Mamm 35:452–472

Da Cunha RGT, Byrne RW (2009) The use of vocal communication in keeping the spatial cohesion of groups: intentionality and specific functions. In: Garber PA, Estrada A, Bicca-Marques JC, Heymann EW, Strier KB (eds) South American primates. Springer, New York, pp 341–363

2 Communication by Sound and by Visual, Tactile, and Chemical Sensing 47

Deecke VB, Ford JK, Spong P (2000) Dialect change in resident killer whales: implications for vocal learning and cultural transmission. Anim Behav 60(5):629–638

Dudzinski KM (1998) Contact behavior and signal exchange in Atlantic spotted dolphins (*Stenella frontalis*). Aquat Mamm 24:129–142

Fahlman A, Loring SH, Levine G, Rocho-Levine J, Austin T, Brodsky M (2015) Lung mechanics and pulmonary function testing in cetaceans. J Exp Biol 215:2030–2038

Foote AD, Griffin RM, Howitt D, Larsson L, Miller PJ, Hoelzel AR (2006) Killer whales are capable of vocal learning. Biol Lett 2(4):509-12

Ford JKB (1987) A catalogue of underwater calls produced by killer whales (*Orcinus orca*) in British Columbia. Can Data Rep Fish Aquat Sci, No. 633

Ford JKB (1989) Acoustic behavior of resident killer whales (*Orcinus orca*) off Vancouver Island, British Columbia. Can J Zool 67:727–745

Ford JKB (1991) Vocal traditions among resident killer whales *Orcinus orca* in coastal waters of British Columbia. Can J Zool 69:1454–1483

Ford JKB, Fisher HD (1983) Group-specific dialects of killer whales (*Orcinus orca*) in British Columbia. In: Payne R (ed) Communication and behavior of whales. Westview, Boulder, CO, pp 129–161

Gero S, Whitehead H, Rendell L (2016) Individual, unit and vocal clan level identity cues in sperm whale codas. R Soc Open Sci 3(1):150372

Herman LM, Peacock MF, Yunker MP, Madsen CJ (1975) Bottle-nosed dolphin: double-slit pupil yields equivalent aerial and underwater diurnal acuity. Science 189(4203):650–652

Janik VM, Sayigh LS (2013) Communication in bottlenose dolphins: 50 years of signature whistle research. J Comp Physiol A 199(6):479–489

Janik VM, Slater PJB (1997) Vocal learning in mammals. Adv Study Behav 26:59–99

Janik VM, Slater PJB (1998) Context-specific use suggests that bottlenose dolphin signature whistles are cohesion calls. Anim Behav 56:829–838

Janik VM, Sayigh LS, Wells RS (2006) Signature whistle shape conveys identity information to bottlenose dolphins. Proc Natl Acad Sci 103(21):8293–8297

Jensen FH, Perez JM, Johnson M, Soto NA, Madsen PT (2011) Calling under pressure: short-finned pilot whales make social calls during deep foraging dives. Proc R Soc Lond B Biol Sci 23:20102604. https://doi.org/10.1098/rsbl.2011.0701

Jensen FH, Beedholm K, Wahlberg M, Bejder L, Madsen PT (2012) Estimated communication range and energetic cost of bottlenose dolphin whistles in a tropical habitat. J Acoust Soc Am 131:582–592

Johnson MP, Tyack PL (2003) A digital acoustic recording tag for measuring the response of wild marine mammals to sound. IEEE J Ocean Eng 28(1):3–12

Jones G, Siemers BM (2011) The communicative potential of bat echolocation pulses. J Comp Physiol A 197(5):447–457

Ketten DR (1997) Structure and function in whale ears. Bioacoustics 8(1–2):103–135

King SL (2015) You talkin' to me? Interactive playback is a powerful yet underused tool in animal communication research. Biol Lett 11(7):20150403

King SL, Sayigh LS, Wells RS, Fellner W, Janik VM (2013) Vocal copying of individually distinctive signature whistles in bottlenose dolphins. Proc R Soc Lond B 280:20130053

King SL, Harley HE, Janik VM (2014) The role of signature whistle matching in bottlenose dolphins, *Tursiops truncatus*. Anim Behav 96:79–86

Kishida T, Kubota S, Shirayama Y, Fukami H (2007) The olfactory receptor gene repertoires in secondary-adapted marine vertebrates: evidence for reduction of the functional proportions in cetaceans. Biol Lett 3(4):428–430

Laland KN, Galef BG (2009) The question of animal culture. Harvard University Press, Cambridge, MA

Laland KN, Janik VM (2006) The animal cultures debate. Trends Ecol Evol 21(10):542–547

Lammers MO, Au WW (2003) Directionality in the whistles of Hawaiian spinner dolphins (*Stenella longirostris*): a signal feature to cue direction of movement? Mar Mamm Sci 19(2):249–264

Macfarlane NBW (2016) The choreography of belonging: toothed whale spatial cohesion and acoustic communication. PhD Dissertation, MIT/WHOI Joint PhD Program, Cambridge, MA

Madsen PT, Payne R, Kristiansen NU, Wahlberg M, Kerr I, Møhl B (2002a) Sperm whale sound production studied with ultrasound time/depth-recording tags. J Exp Biol 205:1899–1906

Madsen PT, Wahlberg M, Møhl B (2002b) Male sperm whale (*Physeter macrocephalus*) acoustics in a high-latitude habitat: implications for echolocation and communication. Behav Ecol Sociobiol 53:31–41

Madsen PT, Carder DA, Bedholm K, Ridgway SH (2005) Porpoise clicks from a sperm whale nose—convergent evolution of 130 kHz pulses in toothed whale sonars? Bioacoustics 15:195–206

Madsen PT, Jensen FH, Carder D, Ridgway S (2011) Dolphin whistles: a functional misnomer revealed by heliox breathing. Biol Lett 8(2):211–213

Madsen PT, Lammers M, Wisniewska D, Beedholm K (2013) Nasal sound production in echolocating delphinids (*Tursiops truncatus* and *Pseudorca crassidens*) is dynamic, but unilateral: clicking on the right side and whistling on the left side. J Exp Biol 216:4091–4102

Mass AM, AYa S (1989) Distribution of ganglion cells in the retina of an Amazon river dolphin, *Inia geoffrensis*. Aquat Mamm 16:49–56

Mesnick SL, Evans K, Taylor BL, Hyde J, Escorza-Trevino S, Dizon AE (2003) Sperm whale social structure: why it takes a village to raise a child. In: De Waal FBM, Tyack PL (eds) Animal social complexity: intelligence, culture, and individualized societies. Harvard University Press, Cambridge, pp 444–464

Miksis JL, Tyack PL, Buck JR (2002) Captive dolphins, *Tursiops truncatus*, develop signature whistles that match acoustic features of human-made sounds. J Acoust Soc Am 112:728–739

Miller PJ (2002) Mixed-directionality of killer whale stereotyped calls: a direction of movement cue? Behav Ecol Sociobiol 52(3):262–270

Miller PJ, Shapiro AD, Tyack PL, Solow AR (2004a) Call-type matching in vocal exchanges of free-ranging resident killer whales, *Orcinus orca*. Anim Behav 67(6):1099–1107

Miller PJO, Johnson MP, Tyack PL (2004b) Sperm whale behaviour indicates the use of rapid echolocation click buzzes 'creaks' in prey capture. Proc R Soc B 271:2239–2247

Miller PJO, Johnson MP, Madsen PT, Biassoni N, Quero ME, Tyack PL (2009) Using at-sea experiments to study the effects of airguns on the foraging behavior of sperm whales in the Gulf of Mexico. Deep-Sea Res 56:1168–1181

Møhl B (2001) Sound transmission in the nose of the sperm whale, *Physeter catodon*. A post mortem study. J Comp Physiol A 187:335–340

Møhl B, Wahlberg M, Madsen PT, Heerfordt A, Lund A (2003) The monopulsed nature of sperm whale clicks. J Acoust Soc Am 114(2):1143–1154

Montgelard C, Catzeflis FM, Douzery E (1997) Phylogenetic relationships of artiodactyls and cetaceans as deduced from the comparison of cytochrome b and 12s rRNA mitochondrial sequences. Mol Biol Evol 14(5):550–559

Morisaka T, Connor RC (2007) Predation by killer whales (*Orcinus orca*) and the evolution of whistle loss and narrow-band high frequency clicks in odontocetes. J Evol Biol 20(4):1439–1458

Nachtigall PE, Hall RW (1984) Taste reception in the bottlenosed dolphin. Acta Zool Fenn 172:147–148

New LF, Moretti DJ, Hooker SK, Costa DP, Simmons SE (2013) Using energetic models to investigate the survival and reproduction of beaked whales (family Ziphiidae). PLoS One 8 (7):e68725

Norris KS (1991) Dolphin days: the life and times of the spinner dolphin. W. W. Norton & Company, New York

Olesiuk PF, Ellis GM, Ford JKB (2005) Life history and population dynamics of northern resident killer whales (*Orcinus orca*) in British Columbia. Canadian Science Advisory Secretariat Research Document 2005/045. ISSN 1499-3848

Oliveira C, Wahlberg M, Johnson M, Miller PJ, Madsen PT (2013) The function of male sperm whale slow clicks in a high latitude habitat: communication, echolocation, or prey debilitation? J Acoust Soc Am 133(5):3135–3144

2 Communication by Sound and by Visual, Tactile, and Chemical Sensing 49

Pihlström H (2008) Comparative anatomy and physiology of chemical senses in aquatic mammals. In: Thewissen JGM, Nummela S (eds) Sensory evolution on the threshold: adaptations in secondarily aquatic vertebrates. University of California Press, Berkeley, pp 95–109

Quick NJ, Janik VM (2012) Bottlenose dolphins exchange signature whistles when meeting at sea. Proc R Soc B 279(1738):2539–2545

Quintana-Rizzo E, Mann DA, Wells RS (2006) Estimated communication range of social sounds used by bottlenose dolphins (*Tursiops truncatus*). J Acoust Soc Am 120(3):1671–1683

Rendell LE, Whitehead H (2003) Vocal clans in sperm whales (*Physeter macrocephalus*). Proc R Soc Lond B Biol Sci 270(1512):225–231

Rendell L, Whitehead H (2005) Coda playbacks to sperm whales in Chilean waters. Mar Mamm Sci 21(2):307–316

Richards DG, Wolz JP, Herman LM (1984) Vocal mimicry of computer-generated sounds and vocal labeling of objects by a bottlenosed dolphin, *Tursiops truncatus*. J Comp Psychol 98:10–28

Sakai M, Hishii T, Takeda S, Kohshima S (2006) Flipper rubbing behaviors in wild bottlenose dolphins (*Tursiops aduncus*). Mar Mamm Sci 22(4):966–978

Sayigh LS, Tyack PL, Wells RS, Scott MD (1990) Signature whistles of free-ranging bottlenose dolphins, *Tursiops truncatus*: mother-offspring comparisons. Behav Ecol Sociobiol 26:247–260

Sayigh LS, Tyack PL, Wells RS, Solow AR, Scott MD, Irvine AB (1999) Individual recognition in wild bottlenose dolphins: a field test using playback experiments. Anim Behav 57:41–50

Sayigh LS, Esch HC, Wells RS, Janik VM (2007) Facts about signature whistles of bottlenose dolphins (*Tursiops truncatus*). Anim Behav 74:1631–1642

Schulz TM, Whitehead H, Gero S, Rendell L (2008) Overlapping and matching of codas in vocal interactions between sperm whales: insights into communication function. Anim Behav 76(6):1977–1988

Schulz TM, Whitehead H, Gero S, Rendell L (2011) Individual vocal production in a sperm whale (*Physeter macrocephalus*) social unit. Mar Mamm Sci 27(1):149–166

Shapiro AD, Tyack PL, Seneff S (2011) Comparing call-based versus subunit-based methods for categorizing Norwegian killer whale, *Orcinus orca*, vocalizations. Anim Behav 81(2):377–386

Similä T, Ugarte F (1993) Surface and underwater observations of cooperatively feeding killer whales in northern Norway. Can J Zool 71(8):1494–1499

Smolker R, Pepper JW (1999) Whistle convergence among allied male bottlenose dolphins (Delphinidae, *Tursiops* sp.). Ethology 105:595–617

Smolker R, Mann J, Smuts B (1993) Use of signature whistles during separations and reunions by wild bottlenose dolphin mothers and infants. Behav Ecol Sociobiol 33(6):393–402

Sørensen PM, Wisniewska DM, Jensen FH, Johnson M, Teilmann J, Madsen PT (2018) Click communication in wild harbour porpoises (*Phocoena phocoena*). Sci Rep 8(1):9702

Strager H (1995) Pod-specific call repertoires and compound calls of killer whales, *Orcinus orca* Linnaeus 1758, in the water of northern Norway. Can J Zool 73:1037–1047

Tamaki N, Morisaka T, Taki M (2006) Does body contact contribute towards repairing relationships? The association between flipper-rubbing and aggressive behavior in captive bottlenose dolphins. Behav Process 73(2):209–215

Tavolga MC, Essapian FS (1957) The behavior of bottle-nosed dolphin (*Tursiops truncatus*): mating, pregnancy, parturition and mother-infant behavior. Zoologica 42:11–31

Teloni V, Johnson MP, Miller PJ, Madsen PT (2008) Shallow food for deep divers: dynamic foraging behavior of male sperm whales in a high latitude habitat. J Exp Mar Biol Ecol 354:119–131

Tougaard J, Carstensen J, Teilmann J, Skov H, Rasmussen P (2009) Pile driving zone of responsiveness extends beyond 20 km for harbour porpoises (*Phocoena phocoena*, (L.)). J Acoust Soc Am 126:11–14

Tougaard J et al (2014) Cetacean noise criteria revisited in the light of proposed exposure limits for harbour porpoises. Mar Pollut Bull. https://doi.org/10.1016/j.marpolbul.2014.10.051

Tyack P (1986) Whistle repertoires of two bottlenosed dolphins, *Tursiops truncatus*: mimicry of signature whistles? Behav Ecol Sociobiol 18:251–257

Tyack PL (2000) Dolphins whistle a signature tune. Science 289(5483):1310–1311

Tyack PL (2009) Effects of human-generated sound on marine mammals. Phys Today 62:39–44

van Opzeeland I, Corkeron PJ, Leyssen T, Simila T, Van Parijs SM (2005) Vocal behaviour of Norwegian killer whales, *Orcinus orca*, during carousel and seiner foraging on spring-spawning herring. Aquat Mamm 31:110–119

Wartzok D, Altmann J, Au W, Ralls K, Starfield A, Tyack PL (2005) Marine mammal populations and ocean noise: determining when noise causes biologically significant effects. National Academy Press, Washington, DC

Watkins WA, Schevill WE (1977) Spatial distribution of *Physeter catodon* (sperm whales) underwater. Deep Sea Res 24:693–699

Watwood SL, Tyack PL, Wells RS (2004) Whistle sharing in paired male bottlenose dolphins, *Tursiops truncatus*. Behav Ecol Sociobiol 55:531–543

Watwood SL, Owen ECG, Tyack PL, Wells RS (2005) Signature whistle use by temporarily restrained and free-swimming bottlenose dolphins, *Tursiops truncatus*. Anim Behav 69:1373–1386

Watwood SL, Miller PJO, Johnson M, Madsen PT, Tyack PL (2006) Deep-diving foraging behavior of sperm whales (*Physeter macrocephalus*). J Anim Ecol 75:814–825

Weaver A (2003) Conflict and reconciliation in captive bottlenose dolphins, *Tursiops truncatus*. Mar Mamm Sci 19(4):836–846

Weilgart L, Whitehead H (1993) Coda communication by sperm whales (*Physeter macrocephalus*) off the Galapagos Islands. Can J Zool 71(4):744–752

Weilgart L, Whitehead H (1997) Group-specific dialects and geographical variation in coda repertoire in South Pacific sperm whales. Behav Ecol Sociobiol 40:277–285

Wells RS (1991) The role of long-term study in understanding the social structure of a bottlenose dolphin community. In: Pryor K, Norris KS (eds) Dolphin societies: discoveries and puzzles. University of California Press, Berkeley, pp 199–225

Wells RS (2003) Dolphin social complexity: lessons from long-term study and life-history. In: De Waal FBM, Tyack PL (eds) Animal social complexity: intelligence, culture, and individualized societies. Harvard University Press, Cambridge, pp 32–56

Whitehead H (1998) Cultural selection and genetic diversity in matrilineal whales. Science 282 (5394):1708–1711

Whitehead H (2003) Sperm whales: social evolution in the ocean. University of Chicago Press, Chicago

Whitehead H, Rendell L (2004) Movements, habitat use and feeding success of cultural clans of South Pacific sperm whales. J Anim Ecol 73(1):190–196

Whitehead H, Weilgart L (1991) Patterns of visually observable behaviour and vocalizations in groups of female sperm whales. Behaviour 118:275–296

Whitehead H, Waters S, Lyrholm T (1991) Social organization in female sperm whales and their offspring: constant companions and casual acquaintances. Behav Ecol Sociobiol 29:385–389

Yurk H (2005) Vocal culture and social stability in resident killer whales (*Orcinus orca*). Dissertation, University of British Columbia, Vancouver

Yurk H, Barrett-Lennard L, Ford JK, Matkin CO (2002) Cultural transmission within maternal lineages: vocal clans in resident killer whales in southern Alaska. Anim Behav 63(6):1103–1119

Zimmer WMX, Johnson M, Madsen PT, Tyack PL (2005a) Echolocation clicks of Cuvier's beaked whales (*Ziphius cavirostris*). J Acoust Soc Am 117:3919–3927

Zimmer WMX, Tyack PL, Johnson M, Madsen P (2005b) 3-Dimensional beam pattern of regular sperm whale clicks confirms bent-horn hypothesis. J Acoust Soc Am 117:1473–1485

Zimmer WMX, Madsen PT, Teloni V, Johnson MP, Tyack PL (2005c) Off-axis effects on the multi-pulse structure of sperm whale usual clicks with implications for the sound production. J Acoust Soc Am 118:3337–3345

Chapter 3
Social Ecology of Feeding in an Open Ocean

Robin Vaughn-Hirshorn

Abstract Odontocetes typically live in groups, and strengths of associations among individuals likely affect coordination of behaviors while avoiding predators, having sex, rearing young, and locating, containing, and capturing prey. In this chapter, I concentrate on food. In determining "good" locations to search for prey, odontocetes likely use prior knowledge about their environment, either newly acquired or shared by others. To locate prey, odontocetes use a combination of echolocation, hearing, and sight, and these sensory as well as other potential modalities are likely combined in subtle ways to facilitate searching for prey as a group. After locating prey, toothed dolphins and whales at times work together to contain prey. Social prey containment occurs if the costs of containing prey increase feeding efficiency sufficiently to make it worthwhile to expend the energy. Benefits of coordinating behaviors to capture prey include increasing capture efficiency by working together to briefly disorganize a fish school, trap prey between dolphins, or debilitate prey. While much of this chapter focuses on epipelagic and mesopelagic odontocetes, we are learning more about how deep-diving dolphins and whales forage, and coordination there seems likely as well.

Keywords Communication · Culture · Foraging · Learning · Prey capture · Prey containment · Prey searching · Roles

3.1 Introduction

We catch glimpses of the lives of dolphins and whales, glimpses in part made possible by technological advances such as data tags (DTAGs), passive acoustic arrays, sonar echo sounders, and crittercams (of video cameras and associated devices attached to animals) and in part due to serendipity. These glimpses, windows

R. Vaughn-Hirshorn (✉)
Texas A&M University, Galveston, TX, USA

© Springer Nature Switzerland AG 2019
B. Würsig (ed.), *Ethology and Behavioral Ecology of Odontocetes*, Ethology and Behavioral Ecology of Marine Mammals,
https://doi.org/10.1007/978-3-030-16663-2_3

into the lives of toothed cetaceans, allow us to momentarily step into their worlds, to share their lives with them. In doing so, we discover more about how they forage, the social fabric of their society, the role of communication in social foraging, and more. This knowledge is a patchwork process, but the glimpses of researchers across diverse niches within the ocean of cetacean science allow us to piece together a more comprehensive picture of how odontocetes live their lives. But, we can only follow, observe, and be a part of the lives of dolphins to a small extent. They spend most of their being underwater, and we are not good at this "underwater" thing. Here, I explore insights relative to foraging of social odontocetes in an open ocean, including those that forage in coastal waters and those that forage in deeper, pelagic waters.

Odontocetes typically live in groups; these groups may in turn be part of a larger dolphin or whale community. Two common types of odontocete communities are the fission–fusion societies of small delphinids such as bottlenose dolphins (*Tursiops* spp; e.g., Lusseau et al. 2003), dusky dolphins (*Lagenorhynchus obscurus*; Würsig et al. 2007), and spinner dolphins (*Stenella longirostris*; Gowans et al. 2008) and the matrilineal societies of killer whales (*Orcinus orca*; Baird 2000), sperm whales (*Physeter macrocephalus*; Whitehead 2003), and pilot whales (*Globicephala* spp; Amos et al. 1993). One factor that affects group living decisions is costs versus benefits of foraging decisions (Gowans et al. 2008), and foraging in groups potentially facilitates locating, containing, or capturing prey. For example, dusky dolphins join into larger groups during foraging in Admiralty Bay, New Zealand (Pearson 2009), and off Argentina (Würsig and Würsig 1980), suggesting potential benefits relative to coordination of foraging behaviors.

Foraging benefits for dolphins are difficult to measure, although one example is bottlenose dolphins in Sarasota Bay, Florida, using echolocation less often when they feed in groups than when they feed individually (Nowacek 1999), which suggests a reduced echolocation cost per individual when feeding in groups. Similarly, it is difficult to determine if and how coordination of behaviors occurs during underwater foraging, although such coordination is likely to occur. In captivity, pairs of bottlenose dolphins at times cooperate to solve unfamiliar problem-solving tasks (Kuczaj et al. 2015), and this type of cooperation likely also occurs in nature.

Strength of associations between individuals in groups differs by species and location, and the nature of associations provides clues about their social or ecological benefits. Long-term close associations between individuals occur in small-medium odontocetes such as dusky dolphins (Pearson et al. 2017; Würsig and Bastida 1986), bottlenose dolphins (Connor et al. 2000), killer whales (e.g., Baird and Whitehead 2000), and false killer whales (*Pseudorca crassidens*; Baird et al. 2008) and in medium-large, deep-diving odontocetes such as pilot whales (Heimlich-Boran 1993) and sperm whales (e.g., Gero et al. 2015; Whitehead et al. 2012). These long-term associations may facilitate individuals working together more effectively in a coordinated foraging fashion (Würsig et al. 1989). For example, dusky dolphins in Admiralty Bay, NZ, more frequently associate with preferred associates during foraging than during other behaviors (Pearson 2008), and bottlenose dolphins in Shark Bay, Australia, most often associate with preferred companions during

3 Social Ecology of Feeding in an Open Ocean

foraging and socializing behaviors (Gero et al. 2005). Long-term associations between individuals—familial or "friendship" based—may occur more often when foraging is more difficult, but likely when there is still a benefit to foraging in a group. For dusky dolphins off Argentina, long-term associations occur more often during winter (Degrati et al. 2019), when dolphins appear to feed on prey that is less abundant, located deeper in the water column, and lower in nutritional value (Degrati et al. 2012).

The present examination of foraging of social odontocetes will delve into the above topics as I explore how toothed dolphins and whales locate, contain, and capture their prey and the role of learning in those tasks.

3.2 Where Should We Look for Prey?

Distributions of odontocetes are in large part determined by presence of predators (see Chap. 7), as well as where odontocetes can find suitable prey (e.g., Hastie et al. 2003), which is affected by temperature, depth (Joyce et al. 2017), gradient (Garaffo et al. 2007; Heimlich-Boran 1988), and other oceanographic variables. Presence of a sufficiently dense patch of prey is also key (e.g., Benoit-Bird 2004), as well as presence of prey that is sufficiently rich in energy (e.g., Spitz et al. 2012; Giorli and Au 2017), as these factors affect energetic costs and benefits of finding and capturing prey and therefore determine if a predator can meet its metabolic needs when feeding on particular prey. For example, killer whales off the coast of British Columbia select larger, fat-rich Chinook salmon (*Oncorhynchus tshawytscha*) over other abundant salmon species (Ford et al. 1998).

Odontocetes eat a variety of fish, cephalopods, and crustaceans (e.g., Spitz et al. 2011). The foraging niche of a population or species can be categorized (Table 3.1), although some populations seasonally switch between niches, and different populations of a species may occupy different niches. Foraging occurs during day or night, depending on species and habitat. However, epipelagic feeding generally occurs during day and by small-medium odontocetes. Larger odontocetes often feed at depth in or below the deep scattering layer (DSL), largely because they are able to dive a bit deeper than the smaller ones.

But, how do odontocetes determine where to look for prey? Knowledge about their environment, including how environmental features relate to presence of prey, appears to be a key factor. Blainville's beaked whales (*Mesoplodon densirostris*) seem to use knowledge of their deep-sea environment to decide where to dive for prey, as they locate prey soon after they start echolocating at depth (Arranz et al. 2011). And foraging individuals such as male sperm whales off Norway use prior information that they obtain from previous foraging dives to guide the depth and location of their subsequent foraging dives (Fais et al. 2015). Odontocetes that forage in groups likely share information to facilitate more efficient locating of prey. This knowledge may be especially important for species that dive over 1000 m deep for food but must return to the surface for air. Additionally,

Table 3.1 Odontocete species that feed in groups occupy diverse foraging niches that can be generally categorized as shown below

Oceanic zone	Prey type	Odontocete species	References
Epipelagic	Small, schooling fish (*e.g., anchovies, herring*)	Dusky dolphins	Vaughn et al. (2007), Würsig and Würsig (1980)
		Common dolphins	Young and Cockcroft (1994)
		Killer whales	Similä and Ugarte (1993)
	Larger pelagic fish (*e.g., salmon, catfish, mahi-mahi*)	Bottlenose dolphins	Janik (2000), Ronje et al. (2017)
		Killer whales	Ford et al. (1998)
		False killer whales	Baird et al. (2008)
	Benthic fish	Bottlenose dolphins	Santos et al. (2001)
		Atlantic spotted dolphins	Herzing (1996)
		Commerson's dolphins	Loziaga de Castro et al. (2013)
		Atlantic humpback dolphins	Weir (2009)
	Schooling cephalopods (*infrequent*)	Bottlenose dolphins	Finn et al. (2009)
		Dusky dolphins	Degrati et al. (2019)
Mesopelagic	DSL[a] cephalopods and fish	Dusky dolphins	Benoit-Bird et al. (2004)
		Spinner dolphins	Benoit-Bird and Au (2003)
Bathypelagic (*and deeper*)	Cephalopods and fish (*often including DSL*)	Melon-headed whales	Joyce et al. (2017)
		Pilot whales	Abecassis et al. (2015), Aguilar de Soto et al. (2008)
		Beaked whales	Johnson et al. (2007), Arranz et al. (2011), Tyack et al. (2006)
		Sperm whales	Watwood et al. (2006), Teloni et al. (2008), Miller et al. (2013), Guerra et al. (2017)
		Bottlenose whales	Hooker and Baird (1999)
		Narwhals	Laidre et al. (2003)

Dusky dolphin, *Lagenorhynchus obscurus*; common dolphin, *Delphinus delphis*; killer whale, *Orcinus orca*; bottlenose dolphin, *Tursiops* spp.; false killer whale, *Pseudorca crassidens*; Atlantic spotted dolphin, *Stenella frontalis*; Commerson's dolphin, *Cephalorhynchus commersonii*; Atlantic humpback dolphin, *Sousa teuszii*; spinner dolphin, *Stenella longirostris*; melon-headed whales, *Peponocephala electra*; pilot whales, *Globicephala* spp.; beaked whales, Ziphiidae; sperm whales, *Physeter macrocephalus*; bottlenose whales, *Hyperoodon* spp.; narwhals, *Monodon monoceros*; anchovies, Engraulidae; herring, *Clupea* spp.; salmon, Salmoninae; catfish, Siluriformes; mahi-mahi, *Coryphaena hippurus*
[a]*DSL* deep scattering layer

3 Social Ecology of Feeding in an Open Ocean

environmental knowledge may be particularly important in more complex, heterogeneous environments (Lewis et al. 2013a) and during times when prey is scarce (Brent et al. 2015). Knowledge of how environmental features relate to presence of prey is probably important to decide where to look for prey. For example, beaked whales identify topographic features such as underwater canyons that are likely locations for their prey before they begin searching for prey (Hazen et al. 2011). And bottlenose dolphins off Australia that carry sponges on their rostra to feed continue to improve their ability to forage effectively until at least middle age (Patterson et al. 2016). This improvement in sponge foraging may in part be due to an increased ability to learn to locate sponges (Patterson et al. 2016).

Individuals likely accrue knowledge about where to search for prey both via learning via trial and error about their environment, as well as by learning from others. And, for some species, cultural knowledge that individuals accrue over a lifetime of searching for prey with others appears to have an important role in helping individuals of their groups find prey. For example, bottlenose dolphins in the Lower Florida Keys often follow a leader dolphin to a prey site (Lewis et al. 2013a). However, some dolphins are able to lead the group to prey faster and in a more direct fashion than others (Lewis et al. 2013a). This dynamic is reminiscent of elephant matriarchs acquiring knowledge about locations of water sources that help their family group to survive during severe droughts (Payne 2003). Similar roles appear to be occupied by postmenopausal resident killer whales, who appear to use their learned ecological knowledge to lead matrilineal groups to patchy, seasonal salmon prey resources off the Pacific Northwest (Brent et al. 2015) (Fig. 3.1a). Pilot whales, whose matrilineal groups also contain postmenopausal females (Amos et al. 1993), likely also depend upon long-term knowledge to locate patchily distributed prey (Fig. 3.1b). And given the short, high-energy, fast chases that pilot whales use to capture their prey (Aguilar de Soto et al. 2008), pilot whales likely need to be able to find high-energy prey at depth relatively efficiently. Close-knit matrilineal groups of sperm whales (e.g., Whitehead et al. 2012) likely also incorporate a knowledge component into their foraging tactics, as may other species that exhibit long-term, tightly bonded social relationships (e.g., false killer whales, Baird et al. 2008).

Hypothesized benefits of sharing knowledge between group members relative to where to find prey include kin selection (Lewis et al. 2013b; Brent et al. 2015) and reciprocal altruism (Trivers 1971). The group benefits of working together to contain or capture prey probably compel individuals to also work together to locate prey.

3.3 How Will We Find Prey?

After deciding on productive locations in which to search for prey, or perhaps following a knowledgeable leader to a good location, how do odontocetes then find their prey? It makes sense that dolphins searching for prey in groups would find prey faster than dolphins that search individually, and data support this hypothesis (Connor 2000). And groups at times search for prey in a coordinated fashion—for

Fig. 3.1 Long-term knowledge, including locations of seasonally dependent patchily distributed prey, is likely one evolutionary benefit of closely bonded matrilineal groups that contain postmenopausal females which occur in some populations of killer whales, *Orcinus orca*, and pilot whales, *Globicephala* spp. Top photo part of AG pod off Juneau, Alaska, that includes three generations of killer whales—grandmother (second from left), son (far left), daughter (second from right), and granddaughter (far right) (courtesy of Heidi Pearson). The AG pod is part of the AB clan, which is part of the southern Alaska resident population (Matkin et al. 2014). Bottom photo part of a group of short-finned pilot whales, *Globicephala macrorhynchus*, off Tenerife, Spain (courtesy of Sergio Hanquet)

Fig. 3.2 Foraging can be divided into the stages of locating, containing, and capturing prey. Here, dusky dolphins, *Lagenorhynchus obscurus*, use echolocation to locate fish prey. Illustration by Katia Hapgood

example, resident killer whales off the Pacific Northwest search for salmon prey in line formations (Heimlich-Boran 1988). However, besides the added efficiency of having more individuals searching for patchily distributed underwater prey, some species may exhibit role differentiation during prey searching, similar to how bottlenose dolphins differentiate behaviors when capturing fish off Florida (Gazda et al. 2005). Individual dolphins potentially use at least four different tactics to locate prey from a distance, and these as well as other tactics are likely combined in subtle ways to facilitate searching for prey as a group (Fig. 3.2).

Some species locate prey by using echolocation to find areas of greater prey density or preferred prey species; examples include sperm whales (Watwood et al. 2006; Miller et al. 2004), beaked whales (Hazen et al. 2011; Madsen et al. 2005; Johnson et al. 2008), and pilot whales (Aguilar de Soto et al. 2008). The echolocation clicks of some species can detect prey across astonishingly large distances. For example, modeling studies have suggested that sperm whales can detect prey as far as 500 m away (Møhl et al. 2003), and bottlenose dolphin clicks are detectable at distances of several hundred meters away (Dos Santos and Almada 2004). For smaller odontocetes, body and head movements may accompany echolocation emissions to increase search efficacy. Finless porpoise (*Neophocaena phocaenoides*) echolocation clicks occur during rolling dives that include head movements, which appear to increase the search area for that individual (Akamatsu et al. 2013).

Dolphins and porpoises also passively listen to the sound of their prey. Bottlenose dolphins in Sarasota Bay, Florida, preferentially capture soniferous prey (McCabe et al. 2010; Gannon et al. 2005). Mammal-feeding killer whales also are hypothesized to locate their prey via listening, as the whales tend to be remarkably silent during hunts (Deecke et al. 2005).

At times toothed whales likely locate fish schools by using hearing or vision to cue off other feeding odontocetes (e.g., Würsig and Würsig 1980), similar to how seabirds use dolphins to locate fish schools (e.g., Vaughn et al. 2008). In some locations and perhaps especially for species who live in fission–fusion societies, dolphins of different species aggregate around prey to feed (Quérouil et al. 2008; Clua and Grosvalet 2001), suggesting that dolphins may be using each other as a proxy to locate prey. A wide range of species intentionally attract others to a feeding bout via calls (Clay et al. 2012), and this calling potentially has diverse direct and indirect benefits to the caller (e.g., Slocombe et al. 2010; Brown et al. 1991). Feeding dolphins or whales potentially signal others about the location of a particularly good prey school via vocalizations or leaps, although this signaling could be either intentional or unintentional. High rates of communication vocalizations such as calls often occur during dolphin foraging (e.g., killer whales in the northeast Atlantic, Simon et al. 2007; dusky dolphins off New Zealand, Vaughn-Hirshorn et al. 2012), and these as well as other vocalizations such as whistles or clicks can travel hundreds of meters underwater. Whistles produced by bottlenose dolphins may be detectable up to 5.7 km away (Jensen et al. 2012), although transmission distance is dependent on acoustic power as well as environmental factors such as ambient noise levels. And clicks produced by deep-diving Blainville's beaked whales can travel up to 6.5 km at depth (Ward et al. 2008). Loud percussive sounds, such as those produced by breaches or noisy leaps, may also inform others of locations of feeding bouts, as these leaps appear to function in signaling between dolphins (Würsig and Würsig 1980; Pearson 2017). On some occasions, this signaling could relate to foraging, and the sound of a dolphin noisy leap or breach could communicate to another dolphin that is over 400 m away (Finneran et al. 2000).

Some odontocetes use vision or hearing to cue off other taxa of feeding predators such as birds, as do Cape gannets (*Morus capensis*; Tremblay et al. 2014); cueing off fishing boats is an analogous tactic (e.g., Leatherwood 1975; Powell and Wells 2011). Feeding aggregations that include odontocetes and seabirds occur for a diversity of species including white-beaked dolphins (*Lagenorhynchus albirostris*; Mehlum et al. 1998), common dolphins (*Delphinus* spp.; Filby et al. 2013), bottlenose dolphins (Acevedo-Gutiérrez and Parker 2000), killer whales (Ridoux 1987), dusky dolphins (Vaughn et al. 2007), and various tropical Delphinidae (Thiebot and Weimerskirch 2013). It has been hypothesized that gannets (*Morus* spp.) can see other plunge-diving feeding gannets from distances up to 15–17 km away (Tremblay et al. 2014), in part due to the high contrast between these white birds and the dark ocean. Leaping dolphins would be able to spot feeding birds from less far away than would flying gannets, but they do have good vision not only in water but also in air (Wartzok and Ketten 1999), and visual cueing off light-colored seabirds could occur when dolphins do high leaps (e.g., clean leaps, Würsig and Würsig 1980; Pearson 2017). In a similar fashion, aggregations of light-colored and vocal seabirds also make it easier for researchers and fishermen to locate fish schools (e.g., in the eastern tropical Pacific, Au and Pitman 1986).

Less is known about coordination of foraging behaviors among deep-diving odontocetes, including how they locate prey. Deep divers frequently live in close-knit societies, and they often synchronize their foraging bouts and dives (e.g., pilot whales, Visser et al. 2014; Aoki et al. 2013; beaked whales, Johnson et al. 2007; northern bottlenose whales, *Hyperoodon ampullatus*, Hooker and Baird 1999; sperm whales, Whitehead 2003). Most deep divers appear to separate as they dive to depth, but some species use frequent communication vocalizations throughout foraging dives (e.g., pilot whales, Visser et al. 2017; Jensen et al. 2011; Pérez et al. 2017; sperm whales, Oliveira et al. 2013; Blainville's beaked whales, Aguilar de Soto et al. 2012). Production of vocalizations by odontocetes incurs a clear metabolic cost, both during and after the vocalization is produced (e.g., Noren et al. 2013), suggesting a distinct benefit to communicating while at depth. Perhaps communication vocalizations at depth function in a manner analogous to long-distance communication between elephants (McComb et al. 2003), by maintaining social cohesion during the dive (Jensen et al. 2011; Pérez et al. 2017). Some vocalizations could also function in communicating information about the location of prey.

Deep divers are selective when deciding what prey types to search for. However, they incur a greater cost when diving deeply for prey (e.g., sprint dives of pilot whales, Aoki et al. 2017; Aguilar de Soto et al. 2008) and thus may benefit to a greater extent from information sharing relative to locating prey. Some deep divers seek denser prey patches and will dive deeper for such (e.g., sperm whales, Teloni et al. 2008; mysticete humpback whales, *Megaptera novaeangliae*, Friedlaender et al. 2006). Other deep divers select for energy-rich prey (e.g., pilot whales, Aoki et al. 2017; Aguilar de Soto et al. 2008). Communication about locations of optimal prey sites may have an important role in efficient foraging for some species, and such communication may occur at depth (Fig. 3.3).

3.4 Should We Try to Contain Prey? If So, How?

Optimal foraging theory predicts that diving mammals seek to maximize benefits associated with foraging such as amount of prey captured per dive and minimize costs such as energy expended to capture a prey item or diving depth (e.g., Thompson and Fedak 2001). Trapping prey against an obstacle such as the surface of the water or the shore reduces escape options for prey and can therefore increase food intake per individual. Further, for air-breathing mammals such as dolphins or whales, keeping prey at the surface means that an individual can spend more time capturing prey and less time traveling from a source of oxygen at the surface down to deeper prey (Kramer 1988). Conversely, feeding at depth would allow prey to escape more easily and leave less time for feeding. This difference can be significant, even at relatively shallow depths. For example, it was estimated that bottlenose dolphins feeding deeper than 50 m would have only 53 s to capture prey before needing to return to the surface for oxygen, whereas an individual feeding at a depth of 10 m would have 110 s (Hastie et al. 2006). Thus, whether or not odontocetes expend energy to contain prey

Fig. 3.3 Containing prey often occurs before the foraging stage of capturing prey, or it may occur simultaneous with capturing prey, as is seen for dusky dolphins, *Lagenorhynchus obscurus*, feeding on fish prey in Admiralty Bay, New Zealand. Above, the dolphin swimming under the prey ball with belly-facing prey is likely conducting a prey-herding behavior, at the same time as other dolphins conduct prey capture behaviors. Illustration by Katia Hapgood

prior to capturing it likely depends on energetic tradeoffs. Is the cost of working to contain prey worth the benefit of having prey in a more accessible feeding location at the surface?

Herding schooling fishes closer to the surface is a common containment tactic in the ocean. For example, tuna appear to herd prey to the surface (Au and Pitman 1986), as do some alcid seabirds (Anderwald et al. 2011), and humpback whales use bubble nets to herd schooling herring and crustaceans into a denser cluster at the surface (D'Vincent et al. 1985). Among odontocetes, killer whales off Norway (Similä and Ugarte 1993) and Iceland (Simon et al. 2005) work together to corral herring prey into a tight school at the surface using bubble emissions and swimming under prey balls with their white bellies facing the fish (Similä and Ugarte 1993). Dusky dolphins herd anchovies off Argentina (Würsig and Würsig 1980) and pilchard off New Zealand (Vaughn et al. 2008) toward the surface of the water and possibly into tighter schools, using coordinated swimming behaviors that include swimming under and by the bottom of prey balls with their white bellies facing toward fish (Vaughn-Hirshorn et al. 2013). Similar herding and swimming behaviors have been observed for common dolphins (Gallo Reynoso 1991) and Atlantic spotted dolphins (*Stenella frontalis*, Fertl and Würsig 1995).

Odontocetes also herd prey against other barriers, including the shore and underwater slopes. Bottlenose dolphins herd fish onto mud banks off South Carolina by working together as a group to make a bow wave to wash fish onto the shore (Duffy-

3 Social Ecology of Feeding in an Open Ocean

Echevarria et al. 2008), and humpback dolphins (*Sousa* spp.) in the Indian Ocean use a similar coordinated tactic to herd fish onto shore (Peddemors and Thompson 1994). Killer whales in the northeastern Pacific herd groups of 50 or more Pacific white-sided dolphins (*Lagenorhynchus obliquidens*) by driving them into a confined area of shallow water (Ford et al. 1998). This allows them to isolate an individual dolphin (Ford et al. 1998), which likely makes it easier to chase and capture that individual. In the three above cases, the shore appears to represent a barrier that can serve to contain prey and therefore make it easier to isolate and capture their food. Land may also be used as a barrier at a greater depth. For example, killer whales herd Chinook salmon into more dense aggregations by using steep underwater slopes as a barrier (Heimlich-Boran 1988).

Containment or herding behaviors can also occur at greater depth. In mesopelagic waters off Hawaii, spinner dolphins work together to concentrate DSL-related organisms into a denser aggregation before capture, and line and then circle formations appear to facilitate this tightening of prey aggregations as dolphins push and then encircle the prey (Benoit-Bird and Au 2009a). Thus, spinner dolphins may use each other as a barrier to increase density of DSL prey. Dusky dolphins off Kaikoura, New Zealand, also feed on the mesopelagic DSL layer, and they do so in groups of up to five dolphins (Benoit-Bird et al. 2004). However, the nature of their coordinated behaviors is largely unknown. These dusky dolphins more often feed in small groups at shallower depths (Benoit-Bird et al. 2004), suggesting that as they feed at greater depths, their more limited time and energy needs to be focused on capturing prey rather than herding it. Less is known about foraging behaviors of deep-diving odontocetes such as pilot whales and beaked whales and if any prey containment behaviors occur. However, Blainville's beaked whales may conduct circling J-shaped prey capture tracks, in which they swim by a fish school before turning back toward the school to capture individual fish (Johnson et al. 2008); this swim-by may serve as a way to facilitate fish schooling and thus increase the density of prey (Johnson et al. 2008).

3.5 Capturing Prey: Work Together, Or Go It Alone?

The end goal of foraging is, of course, to capture prey, and this typically occurs for odontocetes via echolocation. When capturing prey, odontocete echolocation clicks typically change to clicks that are very close together such as buzzes in porpoises (DeRuiter et al. 2009), bottlenose dolphins (Wisniewska et al. 2014), pilot whales (Aguilar de Soto et al. 2008), and beaked whales (Madsen et al. 2005; Johnson et al. 2008), or creaks in sperm whales (Miller et al. 2004; Miller et al. 2013), just before prey capture. Vision likely also plays a role in capturing prey (e.g., DeRuiter et al. 2009), probably important even when foraging at depth, for odontocetes feeding on bioluminescent prey.

While prey most often is captured directly, debilitation of prey at times occurs before prey is captured, which increases accessibility of that prey. For example, killer whales off Norway and Iceland use tail slaps to stun Atlantic herring (*Clupea harengus*) before capturing them (Similä and Ugarte 1993; Simon et al. 2005), with tail slaps made more effective by preceding low frequency calls that cause herring to cluster together tightly (Simon et al. 2006). Similarly, bottlenose dolphins use tail slaps to stun fish in Sarasota Bay, Florida (Nowacek 2002), before capture. Although stunning efforts are at times an individual endeavor, at other times they are group efforts, such as when killer whales stun fish that are then also captured by other whales (Similä and Ugarte 1993). Killer whales also combine their efforts to subdue larger prey, such as sea lions in the Pacific Northwest (Ford et al. 1998) and the young of migrating gray whales (*Eschrichtius robustus*) off California (Goley and Straley 1994); these prey are then shared between group members. And killer whales in Antarctica work together to create waves that wash their seal prey off ice floes (Visser et al. 2008).

Prey capture occurs both individually and in a coordinated fashion, and yet it is often difficult to tell if coordination is occurring. For example, gannets feeding off South Africa seem to individually capture fish prey and simply to aggregate at a fish school due to prey being concentrated there. However individual gannets may benefit from feeding together, as those that make prey capture attempts directly after another gannet's attempt are more successful at capturing a fish (Thiebault et al. 2016). Each gannet prey capture attempt appears to disorganize the fish school, such that successive attempts likely allow gannets to take advantage of the relative ease of capturing fish briefly swimming in a disorganized fashion—making it easier to focus on individuals, rather than a tightly schooling mass of fish.

For foraging odontocetes, simply feeding at the same time on a school of fish may therefore increase capture success rates by briefly disorganizing a fish school, making it easier to isolate individual fish. And, when feeding at the same time, conducting prey capture attempts in close proximity to another dolphin or whale may increase success rates. Dolphins who feed near the surface frequently capture prey in close proximity to other feeding dolphins, and they at times do so in quite orderly formations. These formations likely serve to disorganize fish and therefore facilitate prey capture, while also reducing escape options for the prey. Dusky dolphins feeding on schools of fish in Admiralty Bay, New Zealand, at times form a pinwheel formation around a prey ball as they simultaneously capture prey (Vaughn-Hirshorn et al. 2013); at other times, they swim directly toward each other as they capture prey that are sandwiched between them (Vaughn-Hirshorn et al. 2013). This latter prey capture tactic appears to also occur for spinner dolphins when they feed on concentrated DSL prey, as pairs of dolphins seem to conduct prey captures from opposite sides of a prey aggregation (Benoit-Bird and Au 2009a). A similar feeding tactic occurs off Florida for bottlenose dolphins, in which a "driver" dolphin herds fish toward a line of dolphins; the dolphin group then feeds on the fish as they are sandwiched between dolphins (Gazda et al. 2005). Another feeding tactic that makes it easier to isolate individual fish is exhibited by bottlenose dolphins in Florida Bay and involves the creation of mud plumes, which isolate individual mullet by causing them to jump out of the water and over the plume, allowing the dolphins to then catch the mullet in-air (Torres and Read 2009).

Individual prey capture behaviors are at times reminiscent of group prey capture tactics and provide further clues as to ways that feeding in a group may increase capture efficiency. For example, some bottlenose dolphins in Sarasota Bay spin around the midpoint of their body in a pinwheel movement to make a tight turn just before capturing prey (Nowacek 2002), which likely allows them to rapidly get in front of the fish they are chasing and in doing so perhaps to have the same disorganization effect on fish as in the above group examples. And killer whales off Argentina and the Crozet Islands capture young elephant seals (*Mirounga* spp.) and South American sea lions (*Otaria byronia*) at the interface of the surf and the shore, as the predators beach themselves in a rapid prey capture attempt (Lopez and Lopez 1985; Guinet and Bouvier 1995). Doing so allows them to briefly trap seals at the interface, which appears to make it easier to capture them, akin to how dusky dolphins and bottlenose dolphins sandwich fish between themselves during the above group examples.

Odontocetes probably use a diversity of behaviors including acoustic and visual signals to coordinate prey captures, as well as to coordinate other foraging behaviors. Burst pulse vocalizations seem to have a prominent role in communication during coordinated behaviors for some odontocetes (e.g., bottlenose dolphins, Eskelinen et al. 2016; dusky dolphins, Vaughn-Hirshorn et al. 2012), as do whistles (e.g., bottlenose dolphins, King and Janik 2015); clicks also at times function in communication (e.g., spinner dolphins feeding in coordination on DSL prey, Benoit-Bird and Au 2009b). Some of these vocalizations likely function in prey capture coordination, akin to how mysticete humpback whales feeding off Alaska use a unique vocalization type at the start of vertical lunge feeding (D'Vincent et al. 1985). Visually, bubble emissions also appear to have a role in coordination of feeding behaviors; for example, dusky dolphins in Admiralty Bay, New Zealand, emit bubbles just before the start of a coordinated prey capture attempt—these emissions thus seem to function in synchronizing prey capture behaviors (Trudelle 2010). Bubble emissions may also have a communication role during foraging by killer whales off Antarctica (e.g., Visser et al. 2008).

Some very deep divers typically separate when they are at depth (e.g., Aguilar de Soto et al. 2008; Aguilar de Soto et al. 2012) and are thus unlikely to coordinate prey capture behaviors, but members of other species feed in association with others. Northern bottlenose whales have been observed foraging at depth in association with another whale (Hooker and Baird 1999). And a pair of long-finned pilot whales (*Globicephala melas*) was observed synchronizing underwater dives (Aoki et al. 2013). The difficulty of using frequent communication calls due to increasing pressure with depth may be one reason why deep-diving odontocetes coordinate foraging behaviors to a lesser extent than epipelagic and mesopelagic species (Jensen et al. 2011). However, deep-diving odontocetes who feed individually provide additional clues about tactics that facilitate successful prey capture and ways that coordination between individuals in less deep waters could increase efficiency of these prey captures. For example, pilot whales (Aguilar de Soto et al. 2008) and sperm whales (Miller et al. 2013; Fais et al. 2016) chase down squid prey individually, but sperm whales may attempt to line up several prey items before beginning a prey capture bout (Miller et al. 2013). This sequential prey capture tactic would have the same effect as prey herding, in that both tactics allow an individual to feed on a concentration of prey and so to spend more time capturing prey and less time searching. Other odontocetes, such as the Blainville's beaked whale, appear to reduce energetic expenditure to capture prey by targeting slower moving prey in oxygen minimum layers below the DSL (Arranz et al. 2011). In a similar manner, a group of dolphins that exhaust a school of fish by conducting a large number of prey captures make it easier to capture their prey.

3.6 What Is My Role?

Unanswered questions relative to prey capturing include those that pertain to individual roles during coordinated behaviors. Do some individuals specialize in particular roles such as leading prey capture charges, driving fish toward other dolphins, herding the bottom of a prey ball, communicating with other dolphins in a bay relative to particularly good prey resources, or determining when to let go of a prey ball that is particularly difficult to contain? Odontocetes often exhibit population-level specialization relative to foraging tactics (e.g., Heimlich-Boran 1988), as well as individual-level specialization within a population (e.g., Sargeant et al. 2005; Smolker et al. 1997). As such, specialization relative to roles during coordinated foraging behaviors would not be surprising, and some evidence of role differentiation in cetaceans has been discovered. Individuals within groups of mysticete humpback whales at times occupy particular roles during feeding lunges (D'Vincent et al. 1985), as indicated by their place within lunge feeding formations. Bottlenose dolphins off Florida exhibit specialization relative to which dolphin occupies the role of "driver dolphin" versus which dolphins occupy the role of "barrier dolphins" (Gazda et al. 2005).

3.7 How Do Odontocetes Learn to Forage?

For odontocetes, learning is a key component of diverse behaviors (Whitehead and Rendell 2014), including those that relate to foraging (e.g., killer whales, Patterson et al. 2016; bottlenose dolphins, Sargeant and Mann 2009; dusky dolphins, Deutsch et al. 2014). Relatives including moms often have a key role in this learning early in an odontocete's life (e.g., Colbeck et al. 2013; Rossman et al. 2015), although individuals engaged in at least some foraging behaviors continue to increase their foraging effectiveness throughout much of their life (Patterson et al. 2016). Learning can lead to the transmission of complex behaviors such as coordinated foraging, and this cultural transmission can include learning new solutions to old problems, but the extent of learning varies with respect to structure of social networks including long-range movements of individuals within cetacean societies (Cantor and Whitehead 2013).

Learning to forage occurs via vertical and horizontal transmission. Vertical transmission often occurs via instruction from moms. In Shark Bay, Australia, bottlenose dolphin moms teach their young how to forage via sponging (Krützen et al. 2005), rooster-tail foraging and mill foraging (Sargeant and Mann 2009), and beach hunting (Sargeant et al. 2005). However, offspring also learn how to forage via horizontal transmission or from peers (Sargeant and Mann 2009). One example that illustrates how teaching can occur is exhibited by Atlantic spotted dolphins, who appear to teach their calves to capture prey via observational learning (Bender et al. 2009). Spotted dolphin moms take longer to catch prey and use referential body movements to a greater extent when foraging with their calves, than they do when calves are not present (Bender et al. 2009). A second example that shows observational learning is exhibited by killer whales, who appear to teach their young how to capture seal prey, in part by accompanying and assisting their offspring as they practice this foraging tactic (Guinet and Bouvier 1995).

These learning processes can result in cultural transmission of behaviors that spread between close associates within a population or even throughout an entire population. For example, only a portion of Icelandic killer whales use low frequency calls before a prey-debilitating tail slap (e.g., Simon et al. 2006), which suggests a learning-related cultural transmission component to this effective prey-herding behavior. Such spread of foraging behaviors can lead to the use of a particular foraging specialization by an entire population, or it can lead to a diversity of specializations used by different individuals within a population (e.g., Rossman et al. 2015), which can correlate with the use of different microhabitats. An example of the former type of specialization is shown by resident and transient killer whale populations in the Pacific Northwest, who overlap in their habitat use, but residents specialize in foraging on salmon while transients specialize in foraging on harbor seals (Heimlich-Boran 1988). An example of the latter type of specialization is shown by individual bottlenose dolphins in Florida Bay, Florida, who specialize in one of two general foraging tactics, which consequently affects the use of this bay on a microhabitat scale (Torres and Read 2009). Within-population specialization likely reduces competition between individuals or populations, but it could also reduce the rate of cultural transmission of useful foraging behaviors.

3.8 Conclusions

The diversity is wide of behaviors and types of coordination that odontocetes exhibit when locating, containing, and capturing. The role of knowledge in knowing where to look for prey is impressive, especially considering seasonal, annual, and long-term changes in prey patterns. As we learn more about foraging tactics of social odontocetes via new technology such as video attachment devices (e.g., Tremblay et al. 2014), tagging options that allow us to learn about underwater behaviors by analyzing movements in relation to acoustics (e.g., Miller et al. 2013; Aguilar de Soto et al. 2008), and analysis-related innovations such as those that relate to movement ecology (e.g., Hays et al. 2016) or association patterns (e.g., Whitehead 2009), we will continue to learn more about these underwater worlds. And yet, it is helpful to periodically take a step back. To what uses are we putting our newfound knowledge? In what ways is it facilitating conservation of species and their ecosystems? Research on the foraging behaviors of deep-diving odontocetes such as pilot whales, beaked whales, and sperm whales is particularly compelling, as we know so little about what happens at depths. Do individuals coordinate behaviors at depth, and if so, in what ways?

Further, can a deeper appreciation of the underwater worlds of dolphins and whales lead us to at least partial solutions of environmental degradations? May we dive as deeply and with such astute awareness as foraging pilot whales, as we search for solutions to ocean conservation problems. And may we have the collective will to implement solutions, in a human cultural fashion, perhaps analogous to that found in matrilineal marine mammal ocean communities.

Acknowledgments I thank Steven Hirshorn for his endless support, Douglas Van Houten for his deep inspiration, and Kathy Joseph for her insightful comments and suggestions on an earlier draft. And I thank the dolphins and whales, for being there, in the depths.

References

Abecassis M, Polovina J, Baird RW, Copeland A, Drazen JC, Domokos R, Oleson E, Jia Y, Schorr GS, Webster DL, Andrews RD (2015) Characterizing a foraging hotspot for short-finned pilot whales and Blainville's beaked whales located off the west side of Hawai'i Island by using tagging and oceanographic data. PLoS One 10(11):e0142628. https://doi.org/10.1371/journal.pone.014628

Acevedo-Gutiérrez A, Parker N (2000) Surface behavior of bottlenose dolphins is related to spatial arrangement of prey. Mar Mamm Sci 16:287–298

Aguilar de Soto NA, Johnson MP, Madsen PT, Díaz F, Domínguez I, Brito A, Tyack P (2008) Cheetahs of the deep sea: deep foraging sprints in short-finned pilot whales off Tenerife (Canary Islands). J Anim Ecol 77:936–947

Aguilar de Soto NA, Madsen PT, Tyack P, Arranz P, Marrero J, Fais A, Revelli E, Johnson M (2012) No shallow talk: cryptic strategy in the vocal communication of Blainville's beaked whales. Mar Mamm Sci 28:E75–E92

Akamatsu T, Wang D, Wang K, Li S, Dong S (2013) Scanning sonar of rolling porpoises during prey capture dives. J Exp Biol 213:146–152

Amos B, Schlotterer C, Tautz D (1993) Social structure of pilot whales revealed by analytical DNA profiling. Science 260:670–672

Anderwald P, Evans PGH, Gygax L, Hoelzel AR (2011) Role of feeding strategies in seabird-minke whale associations. Mar Ecol Prog Ser 424:219–224

Aoki K, Sakai M, Miller PJO, Visser F, Sato K (2013) Body contact and synchronous diving in long-finned pilot whales. Behav Process 99:12–20

Aoki K, Sato K, Isojunno S, Narazaki T, Miller PJO (2017) High diving metabolic rate indicated by high-speed transit to depth in negatively buoyant long-finned pilot whales. J Exp Biol 220:3802–3811

Arranz P, Aguilar de Soto N, Madsen PT, Brito A, Bordes F, Johnson MP (2011) Following a foraging fish-finder: diel habitat use of Blainville's beaked whales revealed by echolocation. PLoS One 6(12):e28353. https://doi.org/10.1371/journal.pone.0028353

Au DWK, Pitman RL (1986) Seabird interactions with dolphins and tuna in the eastern tropical Pacific. Condor 88:304–317

Baird RW (2000) The killer whale: foraging specializations and group hunting. In: Mann J, Connor RC, Tyack PL, Whitehead H (eds) Cetacean societies: field studies of dolphins and whales. University of Chicago Press, Chicago, pp 127–153

Baird RW, Whitehead H (2000) Social organization of mammal-eating killer whales: group stability and dispersal patterns. Can J Zool 78:2096–2015

Baird RW, Gorgone AM, McSweeney DJ, Webster DL, Salden DR, Deakos MH, Ligon AD, Schorr GS, Barlow J, Mahaffy SD (2008) False killer whales (*Pseudorca crassidens*) around the main Hawaiian Island: long-term site fidelity, inter-island movements, and association patterns. Mar Mamm Sci 24:591–612

Bender CE, Herzing DL, Bjorklund DF (2009) Evidence of teaching in Atlantic spotted dolphins (*Stenella frontalis*) by mother dolphins foraging in the presence of their calves. Anim Cogn 12:43–53

Benoit-Bird KJ (2004) Prey caloric value and predator energy needs: foraging predictions for wild spinner dolphins. Mar Biol 145:435–444

Benoit-Bird KJ, Au WWL (2003) Prey dynamics affect foraging by a pelagic predator (*Stenella longirostris*) over a range of spatial and temporal scales. Behav Ecol Sociobiol 53:364–373

Benoit-Bird KJ, Au WWL (2009a) Cooperative prey herding by the pelagic dolphin, *Stenella longirostris*. J Acoust Soc Am 125:539–546

Benoit-Bird KJ, Au WWL (2009b) Phonation behavior of cooperatively foraging spinner dolphins. J Acoust Soc Am 125:125–137

Benoit-Bird KJ, Würsig B, McFadden CJ (2004) Dusky dolphin (*Lagenorhynchus obscurus*) foraging in two different habitats: active acoustic detection of dolphins and their prey. Mar Mamm Sci 20:215–231

Brent LJN, Franks DW, Foster EA, Balcomb KC, Cant MA, Croft DP (2015) Ecological knowledge, leadership, and the evolution of menopause in killer whales. Curr Biol 25:746–750

Brown CR, Brown MB, Shaffer ML (1991) Food-sharing signals among socially foraging cliff swallows. Anim Behav 42:551–564

Cantor M, Whitehead H (2013) The interplay between social networks and culture: theoretically and among whales and dolphins. Philos Trans R Soc B 368:20120340. https://doi.org/10.1098/rstb.2012.0340

Clay Z, Smith CL, Blumstein DT (2012) Food-associate vocalizations in mammals and birds: what do these calls really mean? Anim Behav 83:323

Clua E, Grosvalet F (2001) Mixed-species feeding aggregation of dolphins, large tunas, and seabirds in the Azores. Aquat Living Resour 14:1–8

Colbeck GJ, Duchesne P, Postma LD, Lesage V, Hammill MO, Turgeon J (2013) Groups of related belugas (*Delphinapterus leucas*) travel together during their seasonal migrations in and around Hudson Bay. Proc R Soc B 280:20122552. https://doi.org/10.1098/rspb.2012.2552

Connor RC (2000) Group living in whales and dolphins. In: Mann J, Connor RC, Tyack PL, Whitehead H (eds) Cetacean societies: field studies of dolphins and whales. University of Chicago Press, Chicago, pp 199–218

Connor RC, Wells RS, Mann J, Read AJ (2000) The bottlenose dolphin: social relationships in a fission-fusion society. In: Mann J, Connor RC, Tyack PL, Whitehead H (eds) Cetacean societies: field studies of dolphins and whales. University of Chicago Press, Chicago, pp 91–126

D'Vincent CG, Nilson RM, Hanna RE (1985) Vocalization and coordinated feeding behavior of the humpback whale in southeastern Alaska. Sci Rep Whal Res Inst 36:41–47

Deecke VB, Ford JKB, Slater PJB (2005) The vocal behavior of mammal-eating killer whales: communicating with costly calls. Anim Behav 69:395–405

Degrati M, Dans SL, Garaffo GV, Crespo EA (2012) Diving for food: a switch of foraging strategy of dusky dolphins in Argentina. J Ethol 30:361–367

Degrati M, Coscarella MA, Crespo EA, Dans SL (2019) Dusky dolphin group dynamics and association patterns in Peninsula Valdes, Argentina. Mar Mamm Sci 35:416–433

DeRuiter SL, Bahr A, Blanchet MA, Hansen SF, Kristensen JH, Madsen PT, Tyack PL, Wahlberg M (2009) Acoustic behavior of echolocating porpoises during prey capture. J Exp Biol 212:3100–3107

Deutsch S, Pearson H, Würsig B (2014) Development of leaps in dusky dolphin (*Lagenorhynchus obscurus*) calves. Behaviour 151:1555–1577

Dos Santos ME, Almada VC (2004) A case for passive sonar: analysis of click train production patterns by bottlenose dolphins in a turbid estuary. In: Thomas JA, Moss CF, Vater M (eds) Echolocation in bats and dolphins. University of Chicago Press, Chicago, pp 400–403

Duffy-Echevarria EE, Connor RC, Aubin DJS (2008) Observations of strand-feeding behavior by bottlenose dolphins (*Tursiops truncatus*) in Bull Creek, South Carolina. Mar Mamm Sci 24:202–206

Eskelinen HC, Winship KA, Jones BL, Ames AEM, Kuczaj SA (2016) Acoustic behavior associated with cooperative task success in bottlenose dolphins (*Tursiops truncatus*). Anim Cogn 19:789–797

Fais A, Aguilar Soto N, Johnson M, Miller PJO, Pérez-González C, Madsen PT (2015) Sperm whale echolocation behavior reveals a directed, prior-based search strategy informed by prey distribution. Behav Ecol Sociobiol 69:663–674

Fais A, Johnson M, Wilson M, Aguilar Soto N, Madsen PT (2016) Sperm whale predator-prey interactions involve chasing and buzzing, but no acoustic stunning. Sci Rep 6:28562. https://doi.org/10.1038/srep28562

Fertl D, Würsig B (1995) Coordinated feeding by Atlantic spotted dolphins (*Stenella frontalis*) in the Gulf of Mexico. Aquat Mamm 21:3–5

Filby NE, Bossley M, Stockin KA (2013) Behaviour of free-ranging short-beaked common dolphins (*Delphinus delphis*) in Gulf St Vincent, South Australia. Aust J Zool 61:291–300

Finn J, Tregenza T, Norman M (2009) Preparing the perfect cuttlefish meal: complex prey handling by dolphins. PLoS One 4(1):e4217. https://doi.org/10.1371/journal.pone.0004217

Finneran JJ, Oliver CW, Schaefer KM, Ridgway SH (2000) Source levels and estimated yellowfin tuna (*Thunnus albacares*) detection ranges for dolphin jaw pops, breaches, and tail slaps. J Acoust Soc Am 107:649–656

Ford JKB, Ellis GM, Barrett-Lennard LG, Morton AB, Palm RS, Balcomb KC III (1998) Dietary specialization in two sympatric populations of killer whales (*Orcinus orca*) in coastal British Columbia and adjacent waters. Can J Zool 76:1456–1471

Friedlaender AS, Halpin PN, Qian SS, Lawson GL, Wiebe PH, Thiele D, Read AJ (2006) Whale distribution in relation to prey abundance and oceangraphic processes in shelf waters of the Western Antarctic Peninsula. Mar Ecol Prog Ser 317:297–310

Gallo Reynoso JP (1991) Group behavior of common dolphins (*Delphinus delphis*) during prey capture. Ser Zool 62:253–262

Gannon DP, Barros NB, Nowacek DP, Read AJ, Waples DM, Wells RS (2005) Prey detection by bottlenose dolphins, *Tursiops truncatus*: an experimental test of the passive listening hypothesis. Anim Behav 69:709–720

Garaffo GV, Dans SL, Pedraza SN, Crespo EA, Degrati M (2007) Habitat use by dusky dolphin in Patagonia: how predictable is their location? Mar Biol 152:165–177

Gazda SK, Connor RC, Edgar RK, Cox F (2005) A division of labour with role specialization in group-hunting bottlenose dolphins (*Tursiops truncatus*) off Cedar Key, Florida. Proc R Soc B 272:135–140

Gero S, Bejder L, Whitehead H, Mann J, Connor RC (2005) Behaviourally specific preferred associations in bottlenose dolphins, *Tursiops* spp. Can J Zool 83:1566–1573

Gero S, Gordon J, Whitehead H (2015) Individualized social preferences and long-term social fidelity between social units of sperm whales. Anim Behav 102:15–23

Giorli G, Au WWL (2017) Combining passive acoustics and imaging sonar techniques to study sperm whales' foraging strategies (L). J Acoust Soc Am 142:1428–1431

3 Social Ecology of Feeding in an Open Ocean 69

Goley PD, Straley JM (1994) Attack on gray whales (*Eschrichtius robustus*) in Monterey Bay, California, by killer whales (*Orcinus orca*) previously identified in Glacier Bay, Alaska. Can J Zool 72:1528–1530

Gowans S, Würsig B, Karczmarski L (2008) The social structure and strategies of delphinids: predictions based on an ecological framework. Adv Mar Biol 53:195–294

Guerra M, Hickmott L, van der Hoop J, Rayment W, Leunissen E, Slooten E, Moore M (2017) Diverse foraging strategies by a marine top predator: sperm whales exploit pelagic and demersal habitats in the Kaikoura submarine canyon. Deep-Sea Res Part I 128:98–108. https://doi.org/10.1016/j.dsr.2017.08.012

Guinet C, Bouvier J (1995) Development of intentional stranding hunting techniques in killer whale (*Orcinus orca*) calves at Crozet Archipelago. Can J Zool 73:27–33

Hastie GD, Wilson B, Wilson LJ, Parsons KM, Thompson PM (2003) Functional mechanisms underlying cetacean distribution patterns: hotspots for bottlenose dolphins are linked to foraging. Mar Biol 144:397–403

Hastie GD, Wilson B, Thompson PM (2006) Diving deep in a foraging hotspot: acoustic insights into bottlenose dolphin dive depths and feeding behavior. Mar Biol 148:1181–1188

Hays GC, Ferreira LC, Sequeira AMM, Meekan MG, Duarte CM, Bailey H, Bailleul F, Bowen WD, Caley MJ, Costa DP, Eguíluz VM, Fossette S, Friedlaender AS, Gales N, Gleiss AC, Gunn J, Thums M (2016) Key questions in marine megafauna movement ecology. Trends Ecol Evol 31:463–475

Hazen EL, Nowacek DP, Laurent LS, Halpin PN, Moretti DJ (2011) The relationship among oceanography, prey fields, and beaked whale foraging habitat in the tongue of the ocean. PLoS One 6(4):e19269. https://doi.org/10.1371/journal.pone.0019269

Heimlich-Boran JR (1988) Behavioral ecology of killer whales (*Orcinus orca*) in the Pacific Northwest. Can J Zool 66:565–578

Heimlich-Boran JR (1993) Social organization of the short-finned pilot whales, *Globicephala macrorhynchus*, with special reference to the comparative social ecology of delphinids. Doctoral dissertation, University of Cambridge

Herzing DL (1996) Vocalizations and associated underwater behavior of free-ranging Atlantic spotted dolphins, *Stenella frontalis* and bottlenose dolphins, *Tursiops truncatus*. Aquat Mamm 22:61–79

Hooker SK, Baird RW (1999) Deep-diving behavior of the northern bottlenose whale, *Hyperoodon ampullatus* (Cetacea: Ziphiidae). Proc R Soc Lond B 266:671–676

Janik VM (2000) Food-related bray calls in wild bottlenose dolphins (*Tursiops truncatus*). Proc R Soc Lond B 267:923–927

Jensen FH, Perez JM, Johnson M, Soto NA, Madsen PT (2011) Calling under pressure: short-finned pilot whales make social calls during deep foraging dives. Proc R Soc B 278:3017–3025

Jensen FH, Beedholm K, Wahlberg M, Bejder L, Madsen PT (2012) Estimated communication range and energetic cost of bottlenose dolphin whistles in a tropical habitat. J Acoust Soc Am 131:582–592

Johnson M, Madsen PT, Zimmer WMX, Aguilar de Soto N, Tyack PL (2007) Beaked whales echolocate on prey. Proc R Soc Lond B 271:S383–S386

Johnson M, Hickmott LS, Soto NA, Madsen PT (2008) Echolocation behavior adapted to prey in foraging Blainville's beaked whale (*Mesoplodon densirostris*). Proc R Soc B 275:133–139

Joyce TW, Durban JW, Claridge DE, Dunn CA, Fearnbach H, Parsons KM, Andrews RD, Ballance LT (2017) Physiological, morphological, and ecological tradeoffs influence vertical habitat use of deep-diving toothed-whales in the Bahamas. PLoS One 12(10):e0185113. https://doi.org/10.1371/journal.pone.0185

King SL, Janik VM (2015) Come dine with me: food-associated social signaling in wild bottlenose dolphins (*Tursiops truncatus*). Anim Cogn 18:969–974

Kramer DL (1988) The behavioral ecology of air breathing by aquatic animals. Can J Zool 66:89–94

Krützen M, Mann J, Heithaus MR, Connor RC, Bejder L, Sherwin WB (2005) Cultural transmission of tool use in bottlenose dolphins. Proc Natl Acad Sci 102:8939–8943

Kuczaj SA, Winship KA, Eskelinen HI (2015) Can bottlenose dolphins (*Tursiops truncatus*) cooperate when solving a novel task? Anim Cogn 18:543–550

Laidre KL, Heidi-JØrgensen MP, Dietz R, Hobbs RC, JØrgensen OA (2003) Deep-diving by narwhals *Monodon monoceros*: differences in foraging behavior between wintering areas? Mar Ecol Prog Ser 261:269–281

Leatherwood S (1975) Some observations of feeding behavior of bottle-nosed dolphins (*Tursiops truncatus*) in the northern Gulf of Medico and (*Tursiops* cf *T. gilli*) off southern California, Baja California, and Nayarit, Mexico. Marine Fisheries Review Paper 1157, vol 37

Lewis JS, Wartzok D, Heithaus MR (2013a) Individuals as information sources: could followers benefit from leaders' knowledge? Behaviour 150:635–657

Lewis JS, Wartzok D, Heithaus M, Krützen M (2013b) Could relatedness help explain why individuals lead in bottlenose dolphin groups. PLoS One 8(3):e58162. https://doi.org/10.1371/journal.pone.0058162

Lopez JC, Lopez D (1985) Killer whales (*Orcinus orca*) of Patagonia, and their behavior of intentional stranding while hunting nearshore. J Mammal 66:181–183

Loziaga de Castro R, Dans SL, Coscarell MA, Crespo EA (2013) Living in an estuary: Commerson's dolphin (*Cephalorhynchus commersonii* (Lacépedè, 1804)), habitat use and behavioural pattern at the Santa Cruz River, Patagonia, Argentina. Lat Am J Aquat Res 41:985–991

Lusseau D, Schneider K, Boisseau OJ, Haase P, Slooten E, Dawson SM (2003) The bottlenose dolphin community of doubtful sound features a large proportion of long-lasting associations. Behav Ecol Sociobiol 54:396–405

Madsen PT, Johnson M, Aguilar de Soto N, Zimmer WMX, Tyack P (2005) Biosonar performance of foraging beaked whales (*Mesoplodon densirostris*). J Exp Biol 208:181–194

Matkin CO, Ward Testa J, Ellis GM, Saulitis EL (2014) Life history and population dynamics of southern Alaska resident killer whales (*Orcinus orca*). Mar Mamm Sci 30:460–479

McCabe EJB, Gannon DP, Barros NB, Wells RS (2010) Prey selection by resident common bottlenose dolphins (*Tursiops truncatus*) in Sarasota Bay, Florida. Mar Biol 157:931–942

McComb K, Reby D, Baker L, Moss C, Sayialel S (2003) Long-distance communication of acoustic cues to social identity in African elephants. Anim Behav 65:317–329

Mehlum F, Hunt GL Jr, Decker MB, Nordlund N (1998) Hydrographic features, cetaceans and the foraging of thick-billed murres and other marine birds in the northwestern Barents Sea. Arctic 51:243–252

Miller PJO, Johnson MP, Tyack PL (2004) Sperm whale behavior indicates the use of echolocation click buzzes 'creaks' in prey capture. Proc R Soc Lond B 271:2239–2247

Miller B, Dawson S, Vennell R (2013) Underwater behavior of sperm whales off Kaikoura, New Zealand, as revealed by a three-dimensional hydrophone array. J Acoust Soc Am 134:2690–2700

Møhl B, Wahlberg M, Madsen PT, Heerfordt A, Lund A (2003) The monopulsed nature of sperm whale clicks. J Acoust Soc Am 114:1143–1154

Noren DP, Holt MM, Dunkin RC, Williams TM (2013) The metabolic cost of communicative sound production in bottlenose dolphins (*Tursiops truncatus*). J Exp Biol 216:1624–1629

Nowacek DP (1999) Sound use, sequential behavior and ecology of foraging bottlenose dolphins, *Tursiops truncatus*. Doctoral dissertation, Massachusetts Institute of Technology & Wood Holes Oceanographic Institute

Nowacek DP (2002) Sequential foraging behavior of bottlenose dolphins, *Tursiops truncatus*, in Sarasota Bay, FL. Behaviour 139:1125–1145

Oliveira C, Wahlberg M, Johnson M, Miller PJO, Madsen PT (2013) The function of male sperm whale slow clicks in a high latitude habitat: communication, echolocation, or prey debilitation? J Acoust Soc Am 133:3135–3144

3 Social Ecology of Feeding in an Open Ocean 71

Patterson EM, Krzyszczyk E, Mann J (2016) Age-specific foraging performance and reproduction in tool-using wild bottlenose dolphins. Behav Ecol 27:401–410

Payne K (2003) Sources of social complexity in the three elephant species. In: de Waal FBM, Tyack PL (eds) Animal social complexity: intelligence, culture, and individualized societies. Harvard University Press, Cambridge, pp 57–85

Pearson HC (2008) Fission-fusion sociality in dusky dolphins (*Lagenorhynchus obscurus*), with comparisons to other dolphins and great apes. Doctoral dissertation, Texas A&M University

Pearson HC (2009) Influences on dusky dolphin (*Lagenorhynchus obscurus*) fission-fusion dynamics in Admiralty Bay, New Zealand. Behav Ecol Sociobiol 63:1437–1446

Pearson HC (2017) Unravelling the function of dolphin leaps using the dusky dolphin (*Lagenorhynchus obscurus*) as a model species. Behaviour 154:563–581

Pearson HC, Markowitz TM, Weir JS, Würsig B (2017) Dusky dolphin (*Lagenorhynchus obscurus*) social structure characterized by social fluidity and preferred companions. Mar Mamm Sci 33:251–276

Peddemors VM, Thompson G (1994) Beaching behavior during shallow water feeding by humpback dolphins *Sousa plumbea*. Aquat Mamm 20:65–67

Pérez JM, Jensen FH, Rogano-Doñate L, Aguilar de Soto N (2017) Different modes of acoustic communication in deep-diving short-finned pilot whales (*Globicephala macrorhynchus*). Mar Mamm Sci 33:59–79

Powell JR, Wells RS (2011) Recreational fishing depredation and associated behaviors involving common bottlenose dolphins (*Tursiops truncatus*). Mar Mamm Sci 27:111–129

Quérouil S, Silva MA, Cascão I, Magalhães S, Seabra MI, Machete MA, Santos RS (2008) Why do dolphins form mixed-species associations in the Azores. Ethology 114:1183–1194

Ridoux V (1987) Feeding association between seabirds and killer whales, *Orcinus orca*, around subantarctic Crozet Islands. Can J Zool 65:2113–2115

Ronje EI, Sinclair C, Grace MA, Barros N, Allen J, Balmer B, Panike A, Toms C, Mullin KD, Wells RS (2017) A common bottlenose dolphin (*Tursiops truncatus*) prey handling technique for marine catfish (Ariidae) in the northern Gulf of Mexico. PLoS One. https://doi.org/10.1371/journal.pone.0181179

Rossman S, Ostrom PH, Stolen M, Barros NB, Gandhi H, Stricker CA, Wells RS (2015) Individual specialization in the foraging habits of female bottlenose dolphins living in a trophically diverse and habitat rich estuary. Oecologia 178:415–425

Santos MB, Pierce GJ, Reid RJ, Patterson IAP (2001) Stomach contents of bottlenose dolphins (*Tursiops truncatus*) in Scottish waters. J Mar Biol Assoc UK 81:873–878

Sargeant BL, Mann J (2009) Developmental evidence for foraging traditions in wild bottlenose dolphins. Anim Behav 78:715–721

Sargeant BL, Mann J, Berggren P, Krützen M (2005) Specialization and development of beach hunting, a rare foraging behavior, by wild bottlenose dolphins (*Tursiops* sp.). Can J Zool 83:1400–1410

Similä T, Ugarte F (1993) Surface and underwater observations of cooperatively feeding killer whales in Northern Norway. Can J Zool 71:1494–1499

Simon M, Wahlberg M, Ugarte F, Miller LA (2005) Acoustic characteristics of underwater tail slaps used by Norwegian and Icelandic killer whales (*Orcinus orca*) to debilitate herring (*Clupea harengus*). J Exp Biol 208:2459–2466

Simon M, Ugarte F, Wahlberg M, Miller LA (2006) Icelandic killer whales *Orcinus orca* use a pulsed call suitable for manipulating the schooling behavior of herring *Clupea harengus*. Bioacoustics 16:57–74

Simon M, McGregor PK, Ugarte F (2007) The relationship between the acoustic behavior and surface activity of killer whales (*Orcinus orca*) that feed on herring (*Clupea harengus*). Acta Ethol. https://doi.org/10.1007/s10211-007-0029-7

Slocombe KE, Kaller T, Turman L, Townsend SW, Papworth S et al (2010) Production of food-associated calls in wild male chimpanzees is dependent on the composition of the audience. Behav Ecol Sociobiol 64:1959–1966

Smolker R, Richards A, Connor R, Mann J, Berggren P (1997) Sponge carrying by dolphins (Delphinidae, *Tursiops* sp.): a foraging specialization involving tool use? Ethology 103:454–465

Spitz J, Cherel Y, Bertin S, Kiszka J, Dewez A, Ridoux V (2011) Prey preferences among the community of deep-diving odontocetes from the Bay of Biscay, Northeast Atlantic. Deep-Sea Res Part I 58:273–282

Spitz J, Trites AW, Becquet B, Brind'Amour A, Cherel Y, Galois R, Ridoux V (2012) Cost of living dictates what whales, dolphins and porpoises eat: The importance of prey quality on predator foraging strategies. PLoS One 7(11):e50096. https://doi.org/10.1371/journal/pone.0050096

Teloni V, Johnson MP, Miller POJ, Madsen PT (2008) Shallow food for deep divers: Dynamic foraging behavior of male sperm whales in a high latitude habitat. J Exp Mar Biol Ecol 354:119–131

Thiebault A, Semeria M, Lett C, Tremblay Y (2016) How to capture fish in a school? Effect of successive predator attacks on seabird feeding success. J Anim Ecol 85:157–167

Thiebot JB, Weimerskirch H (2013) Contrasted associations between seabirds and marine mammals across four biomes of the southern Indian Ocean. J Ornithol 154:441–453

Thompson D, Fedak MA (2001) How long should a dive last? A simple model of foraging decisions by breath-hold divers in a patchy environment. Anim Behav 61:287–296

Torres LG, Read AJ (2009) Where to catch a fish? The influence of foraging tactics on the ecology of bottlenose dolphins (*Tursiops truncatus*) in Florida Bay, Florida. Mar Mamm Sci 25:797–815

Tremblay Y, Thiebault A, Mullers R, Pistorius P (2014) Bird-borne video-cameras show that seabird movement patterns relate to previously unrevealed proximate environment, not prey. PLoS One 9(2):ee88424. https://doi.org/10.1371/journal.pone.0088424

Trivers RL (1971) The evolution of reciprocal altruism. Q Rev Biol 46:35–57

Trudelle L (2010) Dusky dolphin bubble emissions during foraging: potential functions. Master's II Internship, Oceanography. Centre d'Océanologie de Marseille, Université Aix-Marseille II. June

Tyack PL, Johnson M, Aguilar Soto N, Sturlese A, Madsen PT (2006) Extreme diving of beaked whales. J Exp Biol 209:4238–4253

Vaughn RL, Shelton DE, Timm LL, Watson LA, Würsig B (2007) Dusky dolphin (*Lagenorhynchus obscurus*) feeding tactics and multi-species associations. NZ J Mar Freshw Res 41:391–400

Vaughn RL, Würsig B, Shelton DS, Timm LL, Watson LA (2008) Dusky dolphins influence prey accessibility for seabirds in Admiralty Bay, New Zealand. J Mammal 89:1051–1058

Vaughn-Hirshorn RL, Hodge K, Würsig B, Sappenfield R, Lammers MO, Dudzinski K (2012) Characterizing dusky dolphin sounds from Argentina and New Zealand. J Acoust Soc Am 132:498–506

Vaughn-Hirshorn RL, Muzi E, Richardson JL, Fox GJ, Hansen LN, Salley AM, Dudzinski KM, Würsig B (2013) Dolphin underwater bait-balling behaviors in relation to group and prey ball sizes. Behav Process 98:1–8

Visser IN, Smith TG, Bullock ID, Green GD, Carlsson OGL, Imberti S (2008) Antarctic peninsula killer whales (*Orcinus orca*) hunt seals and a penguin on floating ice. Mar Mamm Sci 24:225–234

Visser F, Miller PJO, Antunes RN, Oudejans MG, Mackenzie ML, Aoki K, Lam FPA, Kvadsheim PH, Huisman J, Tyack PL (2014) The social context of individual foraging behavior in long-finned pilot whales (*Globicephala melas*). Behaviour 151:1453–1477

Visser F, Kok ACM, Oudejans MG, Scott-Hayward LAS, DeRuiter SL, Alves AC, Antunes RN, Isojunno S, Pierce GJ, Slabbekoorn H, Huisan J, Miller PJO (2017) Vocal foragers and silent crowds: context-dependent vocal variation in Northeast Atlantic long-finned pilot whales. Behav Ecol Sociobiol 71:170. https://doi.org/10.1007/s00265-017-2397-y

Ward J, Morrissey R, Moretti D, DiMarzio N, Jarvis S, Johnson M, Tyack P, White C (2008) Passive acoustic detection and localization of *Mesoplodon densirostris* (Blainville's beaked

3 Social Ecology of Feeding in an Open Ocean

whale) vocalizations using distributed bottom-mounted hydrophones in conjunction with a digital tag (DTag) recording. Can Acoust 36:60–66

Wartzok D, Ketten DR (1999) Marine mammal sensory systems. In: Reynolds JE III, Rommel SA (eds) Biology of marine mammals. Smithsonian Institution, Washington, pp 117–175

Watwood SL, Miller PJO, Johnson M, Madsen PT, Tyack PL (2006) Deep-diving foraging behavior of sperm whales (*Physeter macrocephalus*). J Anim Ecol 75:814–825

Weir CR (2009) Distribution, behavior and photo-identification of Atlantic humpback dolphins *Sousa teuszii* off Flamingos, Angola. Afr J Mar Sci 31:319–331

Whitehead H (2003) Sperm whales: social evolution in the oceans. University of Chicago Press, Chicago

Whitehead H (2009) SOCPROG programs: analyzing animal social structures. Behav Ecol Sociobiol 63:765–778

Whitehead H, Rendell L (2014) The cultural lives of whales and dolphins. University of Chicago Press, Chicago

Whitehead H, Antunes R, Gero S, Wong SNP, Engelhaupt D, Rendell L (2012) Multilevel societies of female sperm whales (*Physeter macrocephalus*) in the Atlantic and Pacific: why are they so different? Int J Primatol 33:1142–1164

Wisniewska DM, Johnson M, Nachtigall PE, Madsen PT (2014) Buzzing during biosonar-based interception of prey in the delphinids *Tursiops truncatus* and *Pseudorca crassidens*. J Exp Biol 217:4279–4282

Würsig B, Bastida R (1986) Long-range movement and individual associations of two dusky dolphins (*Lagenorhynchus obscurus*) off Argentina. J Mammal 67:773–774

Würsig B, Würsig M (1980) Behavior and ecology of the dusky dolphin, *Lagenorhynchus obscurus*, in the South Atlantic. Fish Bull 77:871–890

Würsig B, Würsig M, Cipriano F (1989) Dolphins in different worlds. Oceanus 32:71–75

Würsig B, Duprey N, Weir J (2007) Dusky dolphins (*Lagenorhynchus obscurus*) in New Zealand waters: present knowledge and research goals. DOC Res Dev Ser 270:1–28

Young DD, Cockcroft VG (1994) Diet of common dolphins (*Delphinus delphis*) off the south-east coast of southern Africa: opportunism or specialization? J Zool 234:41–53

Chapter 4
Sexual Strategies: Male and Female Mating Tactics

Dara N. Orbach

Abstract Conflicting interests between the sexes to enhance their fitness potentials have resulted in several sexual strategies used by odontocetes under various social and ecological contexts. Mating tactics are diverse and non-mutually exclusive and can entail both precopulatory and postcopulatory mechanisms. Males typically rove between females, and their mating tactics include display, contest, endurance, scramble, and sperm competition. Female mating tactics to maintain mate choice and control paternity are less well documented but may include signal discrimination, mate choice copying, evasive behaviors, polyestry, multiple mating, and modified genitalia. Species-specific examples of mating tactics are reviewed, as are potential costs and benefits, to better understand the fitness trade-offs associated with odontocete sociosexual relationships.

Keywords Competition · Mate choice · Mating tactic · Odontocetes · Sexual selection

The mating behaviors of most species of cetaceans have not been described due to the logistical challenges of directly observing opportunistic copulation events in a clade that is submerged beneath the surface of the water most of the time (Schaeff 2007; Lanyon and Burgess 2014; Orbach et al. 2015a). Reproductive patterns are instead generally derived from studies of captive animals, anatomy, and endocrinology or inferred based on similarities to terrestrial models (Lanyon and Burgess 2014). The sociosexual behaviors of odontocetes frequently occur year-round and outside the breeding season. In addition to conception, the sociosexual behaviors of dolphins, toothed whales, and porpoises may facilitate social learning, play, and the establishment of social bonds and dominance relationships (Mann 2006). Accordingly, some caution has been warranted when considering anecdotal reports of copulations, and such reports may have hindered efforts to explore mating strategies

---

D. N. Orbach (✉)
Department of Biology, Dalhousie University, Halifax, NS, Canada

Department of Life Sciences, Texas A&M Corpus Christi, Corpus Christi, TX, USA

© Springer Nature Switzerland AG 2019   75
B. Würsig (ed.), *Ethology and Behavioral Ecology of Odontocetes*, Ethology and Behavioral Ecology of Marine Mammals,
https://doi.org/10.1007/978-3-030-16663-2_4

of cetaceans. However, inferred mating tactics are continually being supported by new field evidence, and systematic studies of the mating behaviors of free-ranging cetaceans have been possible in some populations with conducive environments and overt behavioral traits. Through evaluation of the costs and benefits associated with varied mating tactics, it is possible to better understand fitness trade-offs and evolutionary constraints and thereby predict mating patterns.

A reproductive asymmetry exists between the sexes, likely related to anisogamy (females produce larger and more energetically costly gametes than males) and variation in parental investment (Bateman 1948; Trivers 1972). Females are generally limited in their reproductive success by the availability of resources for—and large temporal and energetic investments in—parental care and are consequently discriminatory of mate quality (Trivers 1972). Females distribute themselves relative to resources necessary for offspring survival, such as food and breeding site availability, while also balancing ecological factors with costs and benefits of group living, such as predation pressure and resource competition (Trivers 1972). Accordingly, the operational sex ratio (the male/female ratio among individuals searching for mates) is often male-biased in mammals, which leads to intrasexual variation in reproductive success among males and strong sexual selection (Daly and Wilson 1983). Males have the potential to increase their fitness by mating with multiple fertile females. Thus, males disperse themselves relative to the temporal and spatial distributions of receptive females and invest more in mating effort than parental effort (Emlen and Oring 1977), especially in internal fertilizing species where paternity is uncertain because of sperm competition. Paternal care of offspring has not been reported in any cetacean species (Connor et al. 2000a), although in some matriarchal societies where males do not disperse from their natal groups, males may help non-descendant young to survive (e.g., long-finned pilot whales, *Globicephala melas*, Augusto et al. 2017).

Across sexually reproducing animals, mating systems can be defined broadly as monogamous (one male and one female mate exclusively), polygynous (one male mates with multiple females), polyandrous (one female mates with multiple males), or polygynandrous (males and females mate with multiple partners; multi-mate or promiscuous). Cetaceans have a polygynous or polygynandrous mating system (Wells et al. 1999) with no confirmed example of obligate monogamy (Connor et al. 2000a). Odontocetes are polyestrous and give birth to one offspring per calving event (Chittleborough 1958). These factors, in addition to gregarious lifestyles and extensive behavioral plasticity, have resulted in a broad and diverse array of cetacean sexual strategies in both sexes. Male and female precopulatory and postcopulatory mating tactics (the behavioral, morphological, or physiological phenotype of a genetically based strategy) are reviewed for odontocetes to demonstrate mechanisms of paternity control in addition to potential costs and benefits experienced by both sexes.

4.1 Male Mating Tactics

In general, males compete for access to receptive females by defending territories of value to females (resource defense), following and defending females, or by roving (roaming between females) and mating briefly before departing to find more mates

4 Sexual Strategies: Male and Female Mating Tactics

(Clutton-Brock 1989). Territorial defense is not known to occur among cetaceans, likely because of the highly mobile nature of their prey in the marine environment (Connor et al. 2000a). Evidence is lacking of males defending mates from predators. In some populations of bottlenose dolphins (*Tursiops* sp.), males guard receptive females from mating with rival males (e.g., Connor et al. 1992, 1996). Male cetaceans most commonly rove for receptive females. Because females are highly mobile, a male's ability to prevent extra-pair copulations and assure paternity is limited (Boness et al. 2002). The benefits of roving and potentially mating with more females are predicted to exceed the costs of traveling and losing fertilizations to other males when the duration of estrus is longer than the time required to travel between females (Whitehead 1990). The duration of estrus is unknown for most wild odontocetes and only known for a few species in captivity (common bottlenose dolphins, *T. truncatus*, Robeck et al. 2005; Pacific white-sided dolphins, *Lagenorhynchus obliquidens*, Robeck et al. 2009; beluga whales, *Delphinapterus leucas*, Steinman et al. 2012; killer whales, *Orcinus orca*, Robeck et al. 1993). Other variables, such as the predicted number of females encountered by a roving male, may factor into the decision of when to leave a female (Magnusson and Kasuya 1997).

Whether defending resources, defending females, or roving between females, the male intrasexual behavioral-sperm competition spectrum can be broad. Several mating tactics have evolved across and within odontocete families that optimize reproductive success while decreasing associated costs. Connor et al. (2000a) assessed the relationship between testes size and sexual size dimorphism across the family Delphinidae to predict the intensity of fighting and/or sperm competition. For example, genera with high sexual size dimorphism but small testes-to-body mass ratios, such as pilot whales (*Globicephala* sp.), were predicted to fight to monopolize females more than compete by sperm competition and to be less promiscuous than species with large testes sizes (Connor et al. 2000a). However, mating tactics are not mutually exclusive and likely depend on environmental and social contexts. Five prominent male mating tactics and their associated costs and benefits discussed herein are display, contest, scramble, endurance, and sperm competition (Table 4.1).

Display Competition Males engage in courtship displays and compete for the attention of females using morphological or behavior signals that are assumed to reflect genetic quality, dominance, readiness to breed, or access to resources. Darwin's (1871) theory of sexual selection suggests that the presence of seemingly maladaptive male secondary sexual characteristics is a mechanism for epigamic selection (female mate choice) and differential male reproductive success. Sexually dimorphic morphological traits may not be used exclusively for displays and may also directly assist males to win battles by contest competition.

Sexual dimorphism of morphological characteristics is largely limited to variations in body shape and size, as toothed whales, dolphins, and porpoises have evolved fusiform bodies that increase laminar flow in the marine environment. Morphological variations that increase drag forces may significantly increase the energetic costs of swimming and reduce fitness. Social and ecological constraints may also affect the development of sexually dimorphic traits. For example, male

Table 4.1 General male mating tactics of odontocetes for intrasexual competition and potential fitness costs

Male mating tactic	Potential costs for males	Potential costs for females	Example	Species	References
Display competition	- Increased conspicuousness to predators - Time/energy not spent foraging or detecting predators	- Time/energy evaluating males - "Dishonest" signals - Lower offspring survival if inexperienced in evaluating males	Object carrying and posturing	Humpback dolphins (*Sousa sahulensis*)	Allen et al. (2017)
Contest competition	- High risk of injury - Increased conspicuousness to predators - Time/energy not spent foraging or detecting predators	- Not directly choosing mate - Risk of injury if male aggressive to female	Weaponry- "battle teeth"	Some beaked whale (family Ziphiidae)	McCann (1974), Heyning (1984), Pitman (2018)
Endurance competition	- Extensive temporal investment - Lost mating opportunities with other females - Reduced paternity in alliance	- Prevented from mating with preferred mate - Reduced socializing opportunity - Risk of injury if coerced	Mate guarding/consortships	Indo-Pacific bottlenose dolphins (*Tursiops aduncus*)	Connor et al. (1996)
Scramble competition	- Extensive energetic expenditure on challenging chases - Conditional on competitive sperm	- Time/energy actively evaluating male and avoiding rejected males	Maneuverability during mating chases	Dusky dolphins (*Lagenorhynchus obscurus*)	Markowitz et al. (2010), Orbach et al. (2014)
Sperm competition	- Tissue costly to maintain - Conditional on male copulation success	- Limited choice of copulation partners	Strong seasonal testes mass and cellular activity patterns	Short-beaked common dolphins (*Delphinus delphis*)	Murphy et al. (2005)

Table modified from Orbach (2016)

sperm whales (*Physeter macrocephalus*) are considerably larger and forage in more productive waters than females (Whitehead 2018). In contrast, dusky dolphins (*Lagenorhynchus obscurus*) are relatively sexually monomorphic, and as both sexes live and forage together (Würsig et al. 2007), they have likely evolved similar morphological adaptations that optimize foraging success in their shared habitat.

Display competitions often occur in odontocete species with strong sexual dimorphism and in which males have small relative testes sizes. Consistent with game theory predictions that males have fixed energy budgets to allocate toward traits that aid in reproduction (Parker et al. 2013), male cetaceans experience trade-offs between investments in precopulatory and postcopulatory traits (Dines et al. 2015). For example, Dall's porpoises (*Phocoenoides dalli*) have relatively small testes sizes, and compared to females, males have enlarged postanal humps, forward-canted dorsal fins, and deepened caudal peduncles, which could signal mate quality to females (Jefferson 1990). Other commonplace sexually dimorphic visual characteristics that may signal sexual maturity or quality to females include variations in the size or shape of fins, flukes, postanal humps, rostrums, and teeth, in addition to differences in thoracic girth, colorations, and ossification of the skull (Ralls and Mesnick 2018).

In many terrestrial animals, secondary sexual characteristics associated with male display competitions often increase and recede relative to the breeding period, suggesting that maintenance of morphological signals are metabolically expensive to males or increase predation risks and reduce survival probabilities (Daly and Wilson 1983). Similar trends are uncommon among odontocetes, suggesting that the displayed morphological traits are not costly for males to maintain, that the mechanisms to conditionally diminish secondary sexual characteristics have not evolved, or that it is more energetically expensive for males if the traits increase and recede seasonally than remain unchanged. Behavioral sociosexual displays, however, have been reported to peak and wane seasonally among odontocetes, potentially signifying associated costs for survival, such as predator detection. For example, some male Amazon River dolphins (*Inia geoffrensis*) carry sticks in their mouths, and the seasonal peaks in these apparent sociosexual displays to females may be correlated with increased conceptions (Martin et al. 2008). Adult male Australian humpback dolphins (*Sousa sahulensis*) display large marine sponges on their rostra/melons to adult females and pair these presentations with physical posturing (e.g., "banana pose"; Allen et al. 2017). Sexual displays enriched by object carrying are rare among nonhuman mammals, yet they may be more common than recognized among dolphins, as several species have complex cognitive abilities (Marino et al. 2007).

Sexual displays by sounds can also be an important mating tactic for cetaceans, as the marine environment provides an excellent medium for sound transmission. The click vocalizations of sperm whales (*P. macrocephalus*) may provide females with indicators of mate quality if whales have the capacity to measure each other's body lengths by the interpulse intervals of successive clicks (Norris and Harvey 1972; Growcott et al. 2011). Sperm whale "slow clicks" are typically produced only by large sexually mature males on breeding grounds and have been hypothesized to attract mates and establish dominance hierarchies among males (Weilgart and

Whitehead 1988; Whitehead 1993). If information about male quality is transmitted by acoustic communication, it can reduce female energy and time investments to travel to and assess roving males.

Display competition confers benefits to females at minimal costs if cues are "honest" and females are not coerced to mate with rejected males. When cues are not transmitted acoustically, some energetic and temporal costs of evaluating prospective mates may be incurred, in addition to decreased predator detection efforts. Females may discriminately select mates based on some social signal that reflects genetic quality, such as availability of resources, defense capabilities, or other indicators of survival potential. Elaborate male displays may increase conspicuousness to predators or require time and energy investments in signaling that could otherwise be directed toward foraging. The ability of males to survive despite their "handicap" may provide females with "honest" indicators of heritable genetic quality (Zahavi 1975). Selective forces often eliminate "dishonest" signals of quality. Thus, display competition is a relatively beneficial mating tactic for experienced females that can discriminate male quality from signals. Young female odontocetes, however, may make poor mate choice decisions based on inexperience assessing male displays. There is a general trend of low survival rates of firstborn offspring among mammals (Clutton-Brock 1984). This trend may reflect a young mother's inexperience in choosing "high quality" mates or at rearing offspring. Alternatively or concurrently, there may be physiological causes such as the inability of small primiparous female odontocetes to sufficiently meet the metabolic demands of lactation or the high transfer rates of lipophilic organochlorine to firstborn offspring (Cockcroft et al. 1989; Wells 2014).

Contest Competition Contest competitions involve one or more males limiting the access of other males to reproductive females through fights or aggressive behaviors. Females are considered high value resources that warrant "risky" behavior to establish dominance hierarchies and "win" access to mates. Males often develop armaments that aid in combat. High stakes intrasexual physical combats are common among male odontocetes, as evident through the evolution of "weaponry" in several species and scarring patterns in older males (MacLeod 1998). For example, male narwhals (*Monodon monoceros*) develop tusks (elongated upper left canines) used for jousting and intra- or intersexual displays. Fragments of tusks have been found embedded in conspecific males (Gerson and Hickie 1985). Beaked whales (family Ziphiidae) have cephalopod-rich diets and do not require teeth for feeding. Females are toothless, while males develop "battle teeth" (1–2 pairs of mandibular teeth) that they may use to attack other males and occasionally females (McCann 1974; Heyning 1984; Pitman 2018). Male northern bottlenose whales (*Hyperoodon ampullatus*) have large, squared, and flattened melons compared to females and headbutt each other with this enlarged surface area (Gowans and Rendell 1999). In several species of oceanic dolphins (family Delphinidae), males aggressively bite or injure each other, as evident from scarring patterns, rake marks, and opportunistic sightings of violent intrasexual interactions (e.g., Visser 1998; Parsons et al. 2003). The prevalence of physical combats may be underestimated among cetaceans, as

4 Sexual Strategies: Male and Female Mating Tactics 81

agonistic encounters involving strikes with flukes, peduncles, melons, or other body parts leave internal wounds not visible to observers. For example, Ross and Wilson (1996) reported that 36% of deceased harbor porpoises (*Phocoena phocoena*) with fractured bones or organ damage from antagonistic interactions with common bottlenose dolphins (*T. truncatus*) showed no external signs of skin damage.

Among cetacean species in which males evolved "weapons" or enhanced combat skills, the net payoff of increased reproductive success appears to exceed the costs of injury and potential mortality. Like most mammals, female cetaceans generally have longer lifespans than males (Ralls et al. 1980), and the costs of combats may contribute to reductions in male longevity. When the stakes are high, male cetaceans may discern and fight more aggressively for females in estrus (Tyack and Whitehead 1982). The number of violent intrasexual battles are predicted to be inversely related to the associated costs (Clutton-Brock et al. 1979), and males in species with "dangerous weapons" may signal their quality to each other rather than engage in physical combat. Body scars may also signal fighting abilities and serve as badges of status. Males may evaluate scars and determine their opponents' dominance ranks to avoid costly or lethal battles (MacLeod 1998). For example, the slow rate of scar tissue re-pigmentation in Risso's dolphins (*Grampus griseus*) has been hypothesized to accentuate scars for intrasexual competitor evaluation (MacLeod 1998). Females may also use scars to evaluate mate quality, as males with many scars tend to be older and have strong immune systems (MacLeod 1998).

Females gain more reproductive benefits than costs when males compete among each other to establish dominance relationships, as females may choose mates directly based on fighting abilities. Alternatively, if females choose mates indirectly by allowing males to establish social hierarchies and mate with the "winners" of contests (i.e., a lek, Emlen 1976), females potentially benefit through genetic transfer of "high quality" traits to their offspring. However, when contest competitions involve aggressive behaviors directed toward females, or when epigamic selection of a preferred mate is restricted, fitness costs can become substantial.

Endurance Competition In endurance competition, males attempt to outlast their rivals for the duration of a "war of attrition" despite accruing costs. For example, males may defend a female for at least the duration of her estrus and ovulation cycle (e.g., Indo-Pacific bottlenose dolphins, *T. aduncus*, Connor et al. 1996). Mate guarding occurs when males monopolize a female and competitively exclude rivals from mating with her and her from mating with a preferred male. In mate guarding contests, the female cannot be abandoned after copulation without risking the loss of a fertilization already won to another male. Thus, endurance competition costs females lost mating opportunities with preferred mates in addition to altered energy budgets, such as reduced group socializing opportunities. Endurance competition is the most temporally costly of the five reviewed male mating tactics. The time spent guarding one female to ensure paternity is time lost courting additional females or engaging in other activities such as foraging. For example, male Dall's porpoises (*P. dalli*) observed guarding females dove for shorter durations than non-mate guarding males, suggesting potentially reduced food intake (Willis and Dill 2007).

Although endurance competition is not known to occur among most species of odontocetes, it has been well-documented in one of the best studied populations of cetacean mating behavior. Male Indo-Pacific bottlenose dolphins (*T. aduncus*) off Shark Bay, Western Australia, engage in endurance competition paired with mate coercion, in which males chase a female, isolate her for periods that can span several months, and aggressively force her to copulate (Connor et al. 1992, 1996). Males form alliances of stable long-term associations that can last over seasons or years and coordinate activities to "herd" and sexually coerce reproductive females (Connor et al. 1992; Connor and Krützen 2015). Males can aggressively sequester and control females by charging, biting, or colliding into them (Connor et al. 1992) and use threatening "pop" vocalizations to constrain movement (Connor and Smolker 1996). As there is limited sexual size dimorphism in this population of dolphins, males in alliances benefit by their collective ability to overpower individual females, which is more challenging for a lone male (Connor et al. 2000b). However, paternity is skewed toward certain males within alliances, and fertilizations are not divisible (Krützen et al. 2004). Subordinate males appear to gain inclusive fitness benefits by kin selection, as genetic evidence indicates that many males in stable first order alliances are strongly related (Krützen et al. 2003).

Scramble Competition During scramble competition, males compete to quickly find and mate with as many fertile females as possible within typically brief time constraints. Males jostle for a proximate position near a sexually receptive female, followed by a series of actions directed toward her. Females subject to scramble competition are usually in estrus for short durations and are spatially aggregated. Often the most maneuverable or fastest male succeeds. Scramble competition is a prevalent mating tactic in odontocete species lacking strong sexual dimorphism, with the exceptions of enlarged caudal peduncles or pectoral fin sizes in males, which could be adaptive for increased agility. Females may benefit by overtly discriminating between different males. The mating chases are generally energetically expensive, although they are brief. However, scramble competition alone may not be an effective male mating tactic to secure paternity, as ejaculation does not guarantee siring offspring if other males have more competitive sperm. Similarly, a large sperm count does not ensure paternity if there are limited opportunities to copulate (Frasier et al. 2007).

Dusky dolphins (*L. obscurus*) engage in high speed and energetic mating chases, in which four sexually mature males typically chase one sexually mature female for around 10 min while some catch up and copulate with her (Orbach et al. 2015a). Both sexes incur substantial short-term energetic costs, as females lead males on challenging three-dimensional chases that include several leaps through the air, deep dives, and sudden changes in swimming speeds and directions (Orbach et al. 2014, 2015a). Males are not aggressive toward each other and appear to take turns mating with females but do not cooperate with each other (Orbach et al. 2015b). Dusky dolphins have a highly fission–fusion social structure (Würsig et al. 2007; Orbach et al. 2018), and new mating groups form throughout the day. The same female has been observed mating with different groups of males within and across days (pers. observ.).

Sperm Competition Sperm competition is a male postcopulatory tactic that occurs inside the female reproductive tract. Males with higher quality or quantities of sperm that can displace or dilute their rivals' sperm succeed in fertilizing the most ova and have the highest reproductive success (Parker 1970). Sperm competition is particularly prevalent among cetacean species in which females mate with multiple males, male intrasexual aggressive interactions are limited, and males have large testes-to-body mass ratios and corresponding high sperm counts (Brownell and Ralls 1986). Allometric relationships indicate that testes sizes vary greatly among mammals and that odontocetes generally have higher testes-to-body mass ratios than similarly sized terrestrial mammals (Kenagy and Trombulak 1986; Aguilar and Monzon 1992). However, there is a large range of relative testes sizes among odontocetes (Aguilar and Monzon 1992; Connor et al. 2000a; Dines et al. 2015), some of which invest >5% of their body weight into testes mass (e.g., dusky dolphins, *L. obscurus*, Van Waerebeek and Read 1994).

In cetacean species hypothesized to use sperm competition as a prominent mating tactic, such as short-beaked common dolphins (*Delphinus delphis*), testes mass and cellular activity regress considerably outside the seasonal reproductive period (Murphy et al. 2005). Sperm storage appears to be very metabolically costly to males, as large testes masses are not maintained year-round. Gonadal tissue maintenance and sperm development can account for 5–10% of basal metabolic rates (Kenagy and Trombulak 1986). The seasonal cycle of testicular growth and recession, combined with the males' substantial energetic investments in sperm volume, is indicative of a rut, which may be induced by females having brief estrus periods (Murphy et al. 2005). Females can benefit from sperm competition through increased genetic compatibility (Olsson and Madsen 2001) and increased fertilization success (Marconato and Shapiro 1996) or if good sperm competitors sire "higher quality" offspring or produce sons that are good at sperm competition.

4.2 Female Mating Tactics

Female mating strategies and tactics to control paternity remain unknown for most species of cetaceans (Connor et al. 2000a; Boness et al. 2002; Schaeff 2007; Mesnick and Ralls 2018). Females are often perceived to have passive roles in paternity control as sexual coercion and intrasexual male competition can mask female preferences (Clutton-Brock and McAuliffe 2009). Research on female reproduction in odontocetes has instead focused on the temporal and energetic investments used by mothers to raise reproductively viable offspring (Whitehead and Mann 2000). However, as the costs of mating with a "poor quality" mate are high (gestation and lactation can be lengthy in odontocetes), females likely demonstrate more mating tactics than recognized to confer advantages to their offspring. Females can respond to prospective mates by selectively copulating with preferred males, copulating with any male to reduce harassment, or using evasive tactics. Females may also engage in postcopulatory sperm selection or control the seasonality of conception and

Table 4.2 Hypothesized general female mating tactics of odontocetes for intersexual selection

Female mating tactic	Example/evidence	Species	References
Signal discrimination	Extended mating chases led by females that may be used to evaluate male maneuverability	Dusky dolphin (*Lagenorhynchus obscurus*)	Markowitz et al. (2010)
Mate choice copying	Suggested by patterns of paternal relatedness within matrilineal groups	Sperm whale (*Physeter macrocephalus*)	Richard et al. (1996)
Evasive behaviors	Females fled from pursuant males, moved to shallow waters where males could not fit beneath them, rolled ventrum-up, and raised flukes in the air so their genital groove was inaccessible	Dusky dolphin (*Lagenorhynchus obscurus*)	Orbach et al. (2015a)
Polyestry/ multiple matings	Hypothesized mechanism to improve fertility, reduce sexual harassment costs, and obscure paternity	Indo-Pacific bottlenose dolphin (*Tursiops aduncus*)	Connor et al. (1996)
Modified genitalia	Complex vaginal folds that occlude penetration of the penis	Harbor porpoise (*Phocoena phocoena*)	Orbach et al. (2017)

Table modified from Orbach (2016)

parturition to ensure that sufficient resources are available to support the physiological demands of reproduction (Whitehead and Mann 2000). Five female mating tactics that increase female control over paternity are reviewed for odontocetes—signal discrimination, mate choice copying, evasive behaviors, polyestry/multiple matings, and modified genitalia (Table 4.2).

Signal Discrimination Females evaluate prospective mates based on signaled cues that reflect heritable attributes including morphological characteristics, behavioral displays, competitive abilities, and access to resources. This form of epigamic selection is hypothesized to be the predominant female mating tactic across animals, a driving force behind the evolution of secondary sexual characteristics, and likely plays an important role in mate selection for odontocetes. Specialized adaptions to overcome challenges associated with living in marine environments may enable females to choose from among males and control paternity. For example, female dusky dolphins (*L. obscurus*) have the advantage of ventrum-down positioning during mating chases and ventrum-to-ventrum copulations (Markowitz et al. 2010), which allow them to breathe without major repositioning. Male dusky dolphins have the disadvantage of ventrum-up positioning and must curtail their chase because of the need to breathe. Accordingly, females may evaluate the vigor and agility of disadvantaged (ventrum-up) potential mates during extended chases. Although dusky dolphins engage in predominantly scramble and sperm competition (Markowitz et al. 2010), the mating chase also appears to function as a display competition subject to female mate choice. Despite the energetic investments associated with mating chases, the chase itself may benefit the female by providing her with "honest" indicators of potential mate fitness, as similar maneuverability attributes are necessary to capture prey and evade predators (Markowitz et al. 2010). Among sexually monomorphic delphinid species, agility may be the preferred quality in a mate rather than large body size, unlike in many terrestrial breeding marine mammals (e.g., northern elephant seals, *Mirounga angustirostris*, Haley et al. 1994).

Mate Choice Copying Females may increase or decrease their likelihood of mating with a particular male based on observing the mating behaviors of other females. While mature female odontocetes may have experience selecting higher quality mates and discerning honest signals of their capabilities, young and inexperienced females risk making poorer mate selections that could result in nonviable or lower-quality offspring. Inexperienced females may develop the skills to select higher-quality mates through mate choice copying. Female sperm whales synchronize their estrus cycles within their social groups (Best and Butterworth 1980), and patterns of paternal relatedness occur (Richard et al. 1996), suggesting that multiple females within a group mate with a single male. While there is no direct evidence that older and more experienced females mate first followed by younger females, it is conceivably possible as cultural transmission has been well documented within matrilineal sperm whale clans (Rendell and Whitehead 2001), and communication codas within female groups increase in the presence of mature males on breeding grounds (Whitehead 1993).

Evasive Behaviors A female's ability to assess the quality of prospective mates increases her reproductive success if she can ensure that she copulates with her preferred suitor and evades copulations with rejected suitors. However, costs imposed by rejected males to resistant females can be substantial and include harassment, aggression, injury, and occasional infanticide (Watson 2005). Female cetaceans appear to actively avoid some attempted copulations through body positioning or use of their habitat. For example, during mating chases, female dusky dolphins (*L. obscurus*) evaded pursuant males with deep dives and inverted (ventrum-up) swims at the surface so that their genital opening was inaccessible (Orbach et al. 2015a; Fig. 4.1). Females also moved to shallow waters and raised their flukes in the air, as has been observed among female southern right whales (*Eubalaena australis*, Payne 1995). Females may cooperate to evade males. For example, the formation of dusky dolphin nursery groups (mothers with calves) is hypothesized to reduce male harassment (Weir et al. 2008). Female sperm whales (*P. macrocephalus*) adopt a group "marguerite" or "wagon wheel" formation to defend themselves and their young against heterospecific attacks (Whitehead and Weilgart 2000). Female sperm whales have the behavioral capacity to use this group formation to evade sexual advances from ardent males, although it has not been documented.

Polyestry and Multiple Matings When female odontocetes are coerced to mate with non-preferred males, they may have subtle mechanisms to control who sires their offspring, including polyestrous cycling and multiple matings. After being monopolized by an undesirable male, repeated estrus cycles with short durations of ovulation can facilitate mating and conceiving with a preferred mate. This hypothesis has been proposed as a mechanism for female Indo-Pacific bottlenose dolphins

Fig. 4.1 Female dusky dolphin, *Lagenorhynchus obscurus*, evades male by inverting her body at the surface so her genital opening is inaccessible to the male beneath her. (Photo by Dara Orbach)

(*T. aduncus*) trapped in aggressive consortships to control paternity (Connor et al. 1996). Repeated estrus cycles and mating with multiple males can induce sperm competition, prevent inbreeding, reduce the risk of mating exclusively with infertile mates, reduce sexual harassment, and confuse paternity (Furuichi et al. 2014). Non-parental infanticide (intentional killing of non-descendant young) has been reported in multiple species of dolphins (Towers et al. 2018) and can benefit males by prompting non-receptive females to resume estrus cycling sooner (Hrdy 1979). At minimum, female costs include non-recuperative time and energy invested in the killed offspring. Among some mammals with high infanticide rates, non-ovulating females have been observed actively soliciting copulations from multiple males (Hrdy 1977). Such proceptive behavior by females may obscure paternity and thereby deter infanticide (Connor et al. 1996; Wolff and Macdonald 2004).

Modified Genitalia Cryptic female choice is any behavioral, morphological, or physiological mechanism by which females bias paternity to favor particular mates after copulation (Eberhard 1996; Firman et al. 2017). It is a postcopulatory female

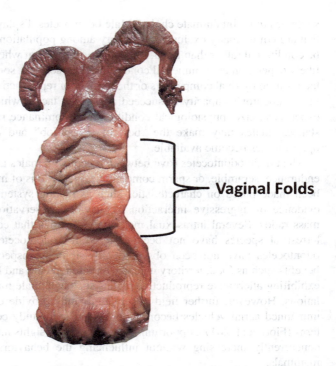

Fig. 4.2 Ventral view of dissected adult female harbor porpoise, *Phocoena phocoena*, highlighting the elaborate and extensive vaginal folds. (Photo by Dara Orbach)

mating tactic that occurs inside the female reproductive tract and is a counter-adaptation to male sperm competition. Females may favor sperm with the best reproductive potential by having elongated and convoluted reproductive tracts. For example, a positive correlation was found between testes weight and oviduct length in 33 mammalian genera (Anderson et al. 2006), and studies of cetacean reproductive anatomy may yield similar patterns. Female cetaceans possess vaginal folds, which are muscular protrusions of the vaginal wall into the vaginal lumen that are unique to cetartiodactyls (Orbach et al. 2016; Fig. 4.2). It does not appear that common bottlenose dolphins (*T. truncatus*) can discriminately expel sperm from undesirable males by contracting reproductive tract muscles (Orbach et al. 2016), as has been demonstrated for some birds (Wagner et al. 2004) and moths (Curril and LaMunyon 2006). However, vaginal folds can occlude the penis and curtail the depth of penetration, thereby increasing the distance semen must travel to fertilize the ova (e.g., harbor porpoise, *P. phocoena*, Orbach et al. 2017).

4.3 Summary

Cetaceans have multi-mate mating systems, in which the sexes behave differently to optimize their respective reproductive success. Sexual selection and conflict entail males employing varied strategies to fertilize females, while females use counterstrategies to maintain mate choice. Male odontocetes display diverse mating tactics that are not mutually exclusive, may vary among populations and species, and may be conditional rather than fixed. The "decision rules" of when to use a certain tactic likely depend on environmental conditions (e.g., season), social conditions (e.g., the tactics used by rival competitors or the number of reproductive females nearby), and the competitor's capacity to succeed at a given tactic, which may vary with age, experience, size, physiological condition, and dominance rank. In certain circumstances, males may make the "best of a bad job" and change tactics as new opportunities become available.

Most male odontocetes rove between receptive females and use display, contest, endurance, scramble, or sperm competition. Predictions of male mating tactics have been made based on characteristics such as mating systems, sexual dimorphism, evidence of aggressive interactions, behavioral observations, and testes-to-body mass ratios. Several intrasexual male mating tactics that commonly occur among terrestrial species have not been reported for odontocetes. For example, male odontocetes have not been observed provisioning prospective mates with direct benefits such as food, territory defense, or parental care and have not been observed exhibiting alternative reproductive tactics such as female mimicry or sneaky copulations. However, further field observations may provide evidence otherwise. As unmanned aerial vehicles become more popular to study cetacean behavioral patterns (Fiori et al. 2017), opportunities to gain new insights into mating behaviors are concurrently increasing without influencing the behaviors of the target marine mammals.

Female mating strategies are not typically as overt as male strategies and have generally been overlooked in cetacean research. Instead, counter-adaptations to male tactics are occasionally described, including evaluating cues that signify heritable fitness benefits, actively avoiding rejected males through behavior and body positioning, inducing pre- or postcopulatory male–male competition, and repeated estrus cycling. Known female mating tactics likely underrepresent the repertoire of mechanisms odontocetes use to control paternity. As our understanding of the diversity and complexity of cetacean mating strategies continues to expand in exciting novel directions, it is critical to consider how the behaviors of one sex alter the other, as male and female strategies can be strongly interdependent (Bro-Jørgensen 2011), and offer a window to explore underlying processes of evolution (Orbach 2016).

Acknowledgments I thank Andreas Fahlman, Bernd Würsig, Melissa Brewer, and Jesse Farruggella for their insightful edits and suggestions on an earlier draft of this manuscript.

References

Aguilar A, Monzon F (1992) Interspecific variation of testis size in cetaceans: a clue to reproductive behaviour. Eur Res Cetacean 6:162–164

Allen SJ, King SL, Krützen M, Brown AM (2017) Multi-modal sexual displays in Australian humpback dolphins. Sci Rep 7(1):13644

Anderson MJ, Dixson AS, Dixson AF (2006) Mammalian sperm and oviducts are sexually selected: evidence for co-evolution. J Zool 270:682–686. https://doi.org/10.1111/j.1469-7998.2006.00173.x

Augusto JF, Frasier TR, Whitehead H (2017) Characterizing alloparental care in the pilot whale (*Globicephala melas*) population that summers off Cape Breton, Nova Scotia, Canada. Mar Mamm Sci 33(2):440–456. https://doi.org/10.1111/mms.12377

Bateman AJ (1948) Intra-sexual selection in *Drosophila*. Heredity 2:349–368. https://doi.org/10.1038/hdy.1948.21

Best PB, Butterworth DS (1980) Timing of oestrus within sperm whale schools. Rep Int Whal Comm 2:137–140

Boness DJ, Clapham PJ, Mesnick SL (2002) Life history and reproductive strategies. In: Hoelzel AR (ed) Marine mammal biology: an evolutionary approach. Blackwell Science, Oxford, pp 278–324

Bro-Jørgensen J (2011) Intra- and intersexual conflicts and cooperation in the evolution of mating strategies: lessons learnt from ungulates. Evol Biol 38:28–41. https://doi.org/10.1007/s11692-010-9105-4

Brownell RL Jr, Ralls K (1986) Potential for sperm competition in baleen whales. In: Donovan GP (ed) Behaviour of whales in relation to management. International Whaling Commission, Cambridge, pp 97–112 (Rep Int Whal Comm, Special Issue 8)

Chittleborough RG (1958) The breeding cycle of the female humpback whale, *Megaptera nodosa* (Bonnaterre). Aust J Mar Freshwat Res 9(1):18

Clutton-Brock TH (1984) Reproductive effort and terminal investment in iteroparous animals. Am Nat 132:212–229. https://doi.org/10.1086/284198

Clutton-Brock TH (1989) Mammalian mating systems. Proc R Soc Lond B 236:339–372

Clutton-Brock TH, McAuliffe K (2009) Female mate choice in mammals. Q Rev Biol 84:3–27. https://doi.org/10.1086/596461

Clutton-Brock TH, Albon SD, Gibson RM, Guinness FE (1979) The logical stag: adaptive aspects of fighting in red deer (*Cervus elaphus L.*). Anim Behav 27:211–255. https://doi.org/10.1016/0003-3472(79)90141-6

Cockcroft VG, De Kock AC, Lord DA, Ross GJB (1989) Organochlorines in bottlenose dolphins *Tursiops truncatus* from the east coast of South Africa. S Afr J Mar Sci 8:207–217. https://doi.org/10.2989/02577618909504562

Connor RC, Krützen M (2015) Male dolphin alliances in Shark Bay: changing perspectives in a 30-year study. Anim Behav 103:223–235. https://doi.org/10.1016/j.anbehav.2015.02.019

Connor RC, Smolker RA (1996) "Pop" goes the dolphin: a vocalization male bottlenose dolphins produce during consortships. Behaviour 133:643–662. https://doi.org/10.1163/156853996X00404

Connor RC, Smolker RA, Richards AF (1992) Two levels of alliance formation among bottlenose dolphins (*Tursiops* sp.). Proc Natl Acad Sci USA 89:987–990. https://doi.org/10.1073/pnas.89.3.987

Connor RC, Richards AF, Smolker RA, Mann J (1996) Patterns of female attractiveness in Indian Ocean bottlenose dolphins. Behaviour 133:37–69. https://doi.org/10.1163/156853996X00026

Connor RC, Read AJ, Wrangham R (2000a) Male reproductive strategies and social bonds. In: Mann J, Connor RC, Tyack PL, Whitehead H (eds) Cetacean societies: field studies of dolphins and whales. University of Chicago Press, Chicago, pp 247–269

Connor RC, Wells RS, Mann J, Read AJ (2000b) The bottlenose dolphin: social relationships in a fission-fusion society. In: Mann J, Connor RC, Tyack PL, Whitehead H (eds) Cetacean societies: field studies of dolphins and whales. University of Chicago Press, Chicago, pp 91–126

Curril IM, LaMunyon CW (2006) Sperm storage and arrangement within females of the arctiid moth *Utetheisa ornatrix*. J Insect Physiol 52:1182–1188. https://doi.org/10.1016/j.jinsphys.2006.08.006

Daly M, Wilson M (1983) Sex, evolution, and behavior, 2nd edn. Wadsworth Publishing, Belmont

Darwin C (1871) The descent of man, and selection in relation to sex. Murray, London

Dines JP, Mesnick SL, Ralls K et al (2015) A trade-off between precopulatory and postcopulatory trait investment in male cetaceans. Evolution 69(6):1560–1572. https://doi.org/10.1111/evo.12676

Eberhard WG (1996) Female control: sexual selection by cryptic female choice. Princeton University Press, Princeton

Emlen ST (1976) Lek organization and mating strategies in the bullfrog. Behav Ecol Sociobiol 1 (3):283–313

Emlen ST, Oring LW (1977) Ecology, sexual selection, and the evolution of mating systems. Science 197:215–223

Fiori L, Doshi A, Martinez E et al (2017) The use of unmanned aerial systems in marine mammal research. Remote Sens (Basel) 9(6):543. https://doi.org/10.3390/rs9060543

Firman RC, Gasparini C, Manier MK, Pizzari T (2017) Postmating female control: 20 years of cryptic female choice. Trends Ecol Evol 32(5):368–382. https://doi.org/10.1016/j.tree.2017.02.010

Frasier TR, Hamilton PK, Brown MW et al (2007) Patterns of male reproductive success in a highly promiscuous whale species: the endangered North Atlantic right whale. Mol Ecol 16:5277–5293. https://doi.org/10.1111/j.1365-294X.2007.03570.x

Furuichi T, Connor R, Hashimoto C (2014) Non-conceptive sexual interactions in monkeys, apes, and dolphins. In: Yamagiwa J, Karczmarski L (eds) Primates and cetaceans: field research and conservation of complex mammalian societies. Springer, Tokyo, pp 385–408

Gerson HB, Hickie JP (1985) Head scarring on male narwhals (*Monodon monoceros*): evidence for aggressive tusk use. Can J Zool 63:2083–2087. https://doi.org/10.1139/z85-306

Gowans S, Rendell L (1999) Head-butting in northern bottlenose whales (*Hyperoodon ampullatus*): a possible function for big heads? Mar Mamm Sci 15(4):1342–1350

Growcott A, Miller B, Sirguey P et al (2011) Measuring body length of male sperm whales from their clicks: the relationship between inter-pulse intervals and photogrammetrically measured lengths. J Acoust Soc Am 130(1):568–573. https://doi.org/10.1121/1.3578455

Haley MP, Deutsch CJ, Le Boeuf BJ (1994) Size, dominance and copulatory success in male northern elephant seals, *Mirounga angustirostris*. Anim Behav 48(6):1249–1126. https://doi.org/10.1006/anbe.1994.1361

Heyning JE (1984) Functional morphology involved in intraspecific fighting of the beaked whale, *Mesoplodon carlhubbsi*. Can J Zool 62:1645–1654. https://doi.org/10.1139/z84-239

Hrdy SB (1977) The langurs of Abu: female and male strategies of reproduction. Harvard University Press, Cambridge

Hrdy SB (1979) Infanticide among animals: a review, classification, and examination of the implications for the reproductive strategies of females. Ethol Sociobiol 1(1):13–40. https://doi.org/10.1016/0162-3095(79)90004-9

Jefferson TA (1990) Sexual dimorphism and development of external features in Dall's porpoise *Phocoenoides dalli*. Fish Bull 88:119–132

Kenagy GJ, Trombulak SC (1986) Size and function of mammalian testes in relation to body size. J Mammal 67(1):1–22. https://doi.org/10.2307/1380997

Krützen M, Sherwin WB, Connor RC et al (2003) Contrasting relatedness patterns in bottlenose dolphins (*Tursiops* sp.) with different alliance strategies. Proc R Soc Lond B 270:497–502. https://doi.org/10.1098/rspb.2002.2229

Krützen M, Barré LM, Connor RC et al (2004) "O father: where art thou?" – paternity assessment in an open fission-fusion society of wild bottlenose dolphins (*Tursiops* sp.) in Shark Bay, Western Australia. Mol Ecol 13:1975–1990. https://doi.org/10.1111/j.1365-294X.2004.02192.x

Lanyon JM, Burgess EA (2014) Methods to examine reproductive biology in free-ranging, fully-marine mammals. In: Holt WV, Brown JL, Comizzoli P (eds) Reproductive sciences in animal conservation: progress and prospects. Springer, New York, pp 241–274. https://doi.org/10.1007/978-1-4939-0820-2_11

MacLeod CD (1998) Intraspecific scarring in odontocete cetaceans: an indicator of male 'quality' in aggressive social interactions? J Zool 244(1):71–77. https://doi.org/10.1111/j.1469-7998.1998.tb00008.x

Magnusson KG, Kasuya T (1997) Mating strategies in whale populations: searching strategy vs. harem strategy. Ecol Model 102:225–242. https://doi.org/10.1016/S0304-3800(97)00058-6

Mann J (2006) Establishing trust: socio-sexual behaviour and the development of male-male bonds among Indian Ocean bottlenose dolphins. In: Sommer V, Vasey PL (eds) Homosexual behaviour in animals: an evolutionary perspective. Cambridge University Press, Cambridge, pp 107–130

Marconato A, Shapiro DY (1996) Sperm allocation, sperm production and fertilization rates in the bucktooth parrotfish. Anim Behav 52(5):971–980. https://doi.org/10.1006/anbe.1996.0245

Marino L, Connor RC, Fordyce RE, Herman LM, Hof PR, Lefebvre L, Lusseau D, McCowan B, Nimchinsky EA, Pack AA, Rendell L, Reidenberg JS, Reiss D, Uhen MD, Van der Gucht E, Whitehead H (2007) Cetaceans have complex brains for complex cognition. PLoS Biol 5(5): e139. https://doi.org/10.1371/journal.pbio.0050139

Markowitz TM, Markowitz WJ, Morton LM (2010) Mating habits of New Zealand dusky dolphins. In: Würsig B, Würsig M (eds) The dusky dolphin: master acrobat off different shores. Elsevier Academic, Amsterdam, pp 151–176

Martin AR, da Silva VMF, Rothery P (2008) Object carrying as socio-sexual display in an aquatic mammal. Biol Lett 4:243–245. https://doi.org/10.1098/rsbl.2008.0067

McCann TS (1974) Body scarring on cetacea-odontocetes. Sci Rep Whales Res Inst Tokyo 26:145–155

Mesnick SL, Ralls K (2018) Mating systems. In: Würsig B, Thewissen JGM, Kovacs KM (eds) Encyclopedia of marine mammals, 3rd edn. Elsevier, London, pp 586–592

Murphy S, Collet A, Rogan E (2005) Mating strategy in the male common dolphin (*delphinus delphis*): what gonadal analysis tells us. J Mammal 86(6):1247–1258. https://doi.org/10.1644/1545-1542(2005)86[1247:MSITMC]2.0.CO;2

Norris KS, Harvey GW (1972) A theory for the function of the spermaceti organ of the sperm whale (*Physeter catodon* L.). In: Galler SR, Schmidt-Koenig K, Jacobs GJ, Belleville RE (eds) Animal orientation and navigation, vol 262. NASA, Washington, pp 397–417

Olsson M, Madsen T (2001) Promiscuity in sand lizards (*Lacerta agilis*) and adder snakes (*Vipera berus*): causes and consequences. J Hered 92(2):190–197. https://doi.org/10.1093/jhered/92.2.190

Orbach DN (2016) Mating strategies of female cetaceans. Dissertation, Texas A&M University at Galveston

Orbach DN, Packard JM, Würsig B (2014) Mating group size in dusky dolphins (*Lagenorhynchus obscurus*): costs and benefits of scramble competition. Ethology 120(8):804–815. https://doi.org/10.1111/eth.12253

Orbach DN, Packard JM, Kirchner T, Würsig B (2015a) Evasive behaviours of female dusky dolphins (*Lagenorhynchus obscurus*) during exploitative scramble competition. Behaviour 152:1953–1977. https://doi.org/10.1163/1568539X-00003310

Orbach DN, Rosenthal GG, Würsig B (2015b) Copulation rate declines with mating group size in dusky dolphins (*Lagenorhynchus obscurus*). Can J Zool 93(6):503–507. https://doi.org/10.1139/cjz-2015-0081

Orbach DN, Marshall CD, Würsig B, Mesnick SL (2016) Variation in female reproductive tract morphology of the common bottlenose dolphin (*Tursiops truncatus*). Anat Rec 299 (4):520–537. https://doi.org/10.1002/ar.23318

Orbach DN, Kelly DA, Solano M, Brennan PLR (2017) Genital interactions during simulated copulation amongst marine mammals. Proc R Soc Lond B 284(1864):20171265. https://doi.org/10.1098/rspb.2017.1265

Orbach DN, Pearson HC, Beier-Engelhaupt A, Deutsch S, Srinivasan M, Weir JS, Yin S, Würsig B (2018) Long-term assessment of spatio-temporal association patterns of dusky dolphins (*Lagenorhynchus obscurus*) off Kaikoura, New Zealand. Aquat Mamm 44(6):608–619. https://doi.org/10.1578/AM.44.6.2018.608

Parker GA (1970) Sperm competition and its evolutionary consequences in the insects. Biol Rev 45:525–567. https://doi.org/10.1111/j.1469-185X.1970.tb01176.x

Parker GA, Lessells CM, Simmons LW (2013) Sperm competition games: a general model for pre-copulatory male-male competition. Evolution 67(1):95–109. https://doi.org/10.1111/j.1558-5646.2012.01741.x

Parsons KM, Durban JW, Claridge DE (2003) Male-male aggression renders bottlenose dolphin (*Tursiops truncatus*) unconscious. Aquat Mamm 29(3):360–362

Payne R (1995) Among whales. Scribner, New York

Pitman R (2018) Mesoplodont beaked whales (*Mesoplodon* spp.). In: Würsig B, Thewissen JGM, Kovacs KM (eds) Encyclopedia of marine mammals, 3rd edn. Elsevier, London, pp 595–602

Ralls K, Mesnick SL (2018) Sexual dimorphism. In: Würsig B, Thewissen JGM, Kovacs KM (eds) Encyclopedia of marine mammals, 3rd edn. Elsevier, London, pp 848–853

Ralls K, Brownell RL Jr, Ballou J (1980) Differential mortality by sex and age in mammals, with specific reference to the sperm whale. Rep Int Whal Comm 2:233–243

Rendell L, Whitehead H (2001) Culture in whales and dolphins. Behav Brain Sci 24(2):309–324. https://doi.org/10.1017/S0140525X0100396X

Richard KR, Dillon MC, Whitehead H, Wright JM (1996) Patterns of kinship in groups of free-living sperm whales (*Physeter macrocephalus*) revealed by multiple molecular genetic analyses. Proc Natl Acad Sci 93(16):8792–8795. https://doi.org/10.1073/pnas.93.16.8792

Robeck TR, Schneyer AL, McBain JF, Dalton LM, Walsh MT, Czekala NM, Kraemer DC (1993) Analysis of urinary immunoreactive steroid metabolites and gonadotropins for characterization of the estrous cycle, breeding period, and seasonal estrous activity of captive killer whales (*Orcinus orca*). Zoo Biol 12(2):173–187. https://doi.org/10.1002/zoo.1430120204

Robeck TR, Steinman KJ, Yoshioka M, Jensen E, O'Brien JK, Katsumata E, Gili C, McBain JF, Sweeney J, Monfort SL (2005) Estrous cycle characterisation and artificial insemination using frozen–thawed spermatozoa in the bottlenose dolphin (*Tursiops truncatus*). Reproduction 129 (5):659–674. https://doi.org/10.1530/rep.1.00516

Robeck TR, Steinman KJ, Greenwell M, Ramirez K, Van Bonn W, Yoshioka M, Katsumata E, Dalton L, Osborn S, O'Brien JK (2009) Seasonality, estrous cycle characterization, estrus synchronization, semen cryopreservation, and artificial insemination in the Pacific white-sided dolphin (*Lagenorhynchus obliquidens*). Reproduction 138(2):391–405. https://doi.org/10.1530/rep-08-0528

Ross HM, Wilson B (1996) Violent interactions between bottlenose dolphins and harbor porpoises. Proc R Soc Lond B 263:283–286. https://doi.org/10.1098/rspb.1996.0043

Schaeff CM (2007) Courtship and mating behavior. In: Miller DL (ed) Reproductive biology and phylogeny of cetacean. CRC, Boca Raton, pp 349–370

Steinman KJ, O'Brien JK, Monfort SL, Robeck TR (2012) Characterization of the estrous cycle in female beluga (*Delphinapterus leucas*) using urinary endocrine monitoring and transabdominal ultrasound: evidence of facultative induced ovulation. Gen Comp Endocrinol 175(3):389–397. https://doi.org/10.1016/j.ygcen.2011.11.008

Towers JR, Hallé MJ, Symonds HK et al (2018) Infanticide in a mammal-eating killer whale population. Sci Rep 8(1):4366. https://doi.org/10.1038/s41598-018-22714-x

Trivers RL (1972) Parental investment and sexual selection. In: Campbell BG (ed) Sexual selection and the descent of man 1871–1971. Aldine, Chicago, pp 136–179

Tyack P, Whitehead H (1982) Male competition in large groups of wintering humpback whales. Behaviour 83(1):132–154. https://doi.org/10.1163/156853982X00067

Van Waerebeek K, Read AJ (1994) Reproduction of dusky dolphins, *Lagenorhynchus obscurus*, from coastal Peru. J Mammal 75(4):1054–1062. https://doi.org/10.2307/1382489

Visser IN (1998) Prolific body scars and collapsing dorsal fin on killer whales (*Orcinus orca*) in New Zealand waters. Aquat Mamm 24(2):71–81

Wagner RH, Helfenstein F, Danchin E (2004) Female choice of young sperm in a genetically monogamous bird. Proc R Soc Lond B 271:S134–S137. https://doi.org/10.1098/rsbl.2003.0142

Watson JJ (2005) Female mating behavior in the context of sexual coercion and female ranging behavior of bottlenose dolphins (*Tursiops* sp.) in Shark Bay, Western Australia. Dissertation. Georgetown University

Weilgart LS, Whitehead H (1988) Distinctive vocalizations from mature male sperm whales (*Physeter macrocephalus*). Can J Zool 66(9):1931–1937. https://doi.org/10.1139/z88-282

Weir JS, Duprey NMT, Würsig B (2008) Dusky dolphin (*Lagenorhynchus obscurus*) subgroup distribution: are shallow waters a refuge for nursery groups? Can J Zool 86(11):1225–1234. https://doi.org/10.1139/Z08-101

Wells RS (2014) Social structure and life history of bottlenose dolphins near Sarasota Bay, Florida: insights from four decades and five generations. In: Yamagiwa J, Karczmarski L (eds) Primates and cetaceans: field research and conservation of complex mammalian societies. Springer, Tokyo, pp 149–172

Wells RS, Boness DJ, Rathbun GB (1999) Behavior. In: Reynolds IIIJE, Rommel SA (eds) Biology of marine mammals. Smithsonian Institution Press, Washington, pp 324–422

Whitehead H (1990) Rules for roving males. J Theor Biol 145:355–368

Whitehead H (1993) The behaviour of mature male sperm whales on the Galápagos Islands breeding grounds. Can J Zool 71(4):689–699. https://doi.org/10.1139/z93-093

Whitehead H (2018) Sperm whale. In: Würsig B, Thewissen JGM, Kovacs KM (eds) Encyclopedia of marine mammals, 3rd edn. Elsevier, London, pp 919–925

Whitehead H, Mann J (2000) Female reproductive strategies of cetaceans: life histories and calf care. In: Mann J, Connor RC, Tyack PL, Whitehead H (eds) Cetacean societies: field studies of dolphins and whales. University of Chicago Press, Chicago, pp 219–246

Whitehead H, Weilgart L (2000) The sperm whale. In: Mann J, Connor RC, Tyack PL, Whitehead H (eds) Cetacean societies: field studies of dolphins and whales. University of Chicago Press, Chicago, pp 154–172

Willis PM, Dill LM (2007) Mate guarding in male Dall's porpoises (*Phocoenoides dalli*). Ethology 113(6):587–597. https://doi.org/10.1111/j.1439-0310.2007.01347.x

Wolff JO, Macdonald DW (2004) Promiscuous females protect their offspring. Trends Ecol Evol 19 (3):127–134. https://doi.org/10.1016/j.tree.2003.12.009

Würsig B, Duprey N, Weir J (2007) Dusky dolphins (*Lagenorhynchus obscurus*) in New Zealand waters: present knowledge and research goals. DOC research and development series, vol 270. Science & Technical Publishing, Wellington, pp 1–28

Zahavi A (1975) Mate selection: a selection of a handicap. J Theor Biol 53:205–214. https://doi.org/10.1016/0022-5193(75)90111-3

Chapter 5
Maternal Care and Offspring Development in Odontocetes

Janet Mann

Abstract Odontocetes are characterized by slow life histories and extensive maternal care, where offspring nurse for years in some species. Among some of the largest toothed whales, the mother and offspring of one or both sexes stay together for a lifetime, forming the basis of strong matrilineal social units and transmission of culture along maternal lines. Mother and calf face a series of challenges from the moment of birth. The newborn must quickly learn to follow and breathe alongside the mother—and wait for her while she dives for food. Within months the calf transitions to infant position for much of the time, although their swimming ability allows them to associate with others in the mother's network. Because calves can easily become separated from their mothers, an effective communication system is necessary, and signature whistles and pod-specific dialects appear to serve this function. The mother plays a central role in the development of calf social and foraging tactics. Where this has been studied, calves adopt maternal behaviors, including foraging specializations, and share the mother's network post-weaning. Although difficult to demonstrate "teaching" per se, dolphins are particularly good candidates given their exquisite learning ability and social tolerance. The role of non-mothers is clearly important in calf development, but whether calf interactions with non-mothers constitute "allomothering" remains unclear for most species. What is clear is that group living by cetaceans affords the calf protection from predators and possibly from infanticidal males. The causes of calf mortality are generally not known, as carcasses are rarely retrieved, but disease, predation, poor maternal condition, and anthropogenic causes (pollutants, provisioning, bycatch, boat strikes), and—rarely—infanticide, are all implicated. Weaning occurs when the calf no longer nurses, evident by cessation of infant position swimming. Interbirth intervals are also used as a proxy for weaning, though the calf frequently nurses during the mother's subsequent pregnancy. Post-weaning, mothers and daughters continue to have preferential bonds, but in killer whales and pilot whales, sons also continue to have a strong relationship with the mother.

J. Mann (✉)
Department of Biology, Georgetown University, Washington, DC, USA
e-mail: mannj2@georgetown.edu

© Springer Nature Switzerland AG 2019
B. Würsig (ed.), *Ethology and Behavioral Ecology of Odontocetes*, Ethology and Behavioral Ecology of Marine Mammals,
https://doi.org/10.1007/978-3-030-16663-2_5

96 J. Mann

Keywords Maternal care · Development · Weaning · Lactation · Culture · Social learning · Predation · Babysitting

5.1 Introduction

Outside of humans, no other mammal has such prolonged and intensive maternal investment as members of the dolphin family. Paternal care has not been reported for any wild cetacean species, although data are insufficient for some taxa, especially beaked whales. Among several species of odontocetes, such as sperm whales, killer whales, pilot whales, and false killer whales, at least some offspring remain with their mothers for life, enabling investment that extends well beyond nursing, likely entailing protection, support, food sharing, social learning, and transmission of culture. Strong matrilineal kin bonds are the foundation of most odontocete societies, with such bonds lasting for decades. At its core, there is no greater influence on an individual cetacean's life than his or her mother. It begins with the birth of a relatively large, precocial calf—always one at a time—that follows the mother wherever she goes.

5.2 Birth and the Newborn Period

For most odontocete cetaceans, births are seasonal with peaks during spring or summer. Unlike baleen whales, which have marked periods of breeding and feeding, all odontocetes feed year-round and throughout lactation. Prey availability does not seem to drive birth seasonality. Predation could be a factor, but some of the most common predators (e.g., tiger sharks) also prefer warm waters. Some populations calve during months when predation is less likely (Fearnbach et al. 2012). Reduced energetic costs for mother and calf are plausible factors, since neither mother nor calf would need to maintain thick blubber stores during the early stages of lactation.

All cetacean calves are born tail first and have wobbly dorsal fins, tail flukes, and pectoral fins (flippers) that quickly become more rigid within hours after birth. The distinct fetal lines wrapping their midsection are from being curled up in the womb (Fig. 5.1), and these lines are an excellent visual cue defining the newborn period; in Indo-Pacific bottlenose dolphins (*Tursiops aduncus*), they are visible for up to 3 months (Mann and Smuts 1999) but can remain faint for longer if the calf is in poor condition. Parturition, known mostly from bottlenose dolphins and killer whales in aquaria, is relatively quick (20–30 min). The difficult part comes right after birth, when mother and calf face a series of challenges. First, following the mother means swimming and breathing, the latter of which is under conscious

Fig. 5.1 Indo-Pacific bottlenose dolphin, *Tursiops aduncus*, newborn swimming in echelon with her mother. Fetal lines are visible. Calf is doing a "chin up" surfacing characteristic of very young calves as they lift the blowhole out of the water. Photo by Ewa Krzyszczyk, Shark Bay Dolphin Research Project monkeymiadolphins.org

control. Unlike terrestrial mammals, cetaceans sleep with half of their brain at a time, and newborn calves hardly sleep at all (Lyamin et al. 2005, 2007). Similarly, their mothers forgo rest (Lyamin et al. 2005, 2007) and hunt very little in the early days (Mann and Smuts 1999), presumably so that the neonate can tag alongside her in echelon position and not be left alone at the surface (Fig. 5.1). Echelon aids the calf energetically as it can get a hydrodynamic boost in her slipstream (Noren 2008; Noren et al. 2008). Neonate dolphins, although precocious in locomotion relative to most terrestrial mammals, lack aerobic stamina and, as compensation, have relatively more blubber for buoyancy (Dearolf et al. 2000). Presumably, the characteristic of fast swimming at the surface by neonates and their mothers facilitates calf physiological development and following the mother. Behavior during the newborn period is gleaned from a few studies of wild odontocetes. Here, the first few months of life can be characterized as a time mother–calf synchrony and social contact (petting and rubbing) is high (Mann and Smuts 1999; Sakai et al. 2013), but calves also learn to socialize with others, separate from and rejoin with the mother, and experiment in catching fish (Mann and Smuts 1998, 1999; Krasnova et al. 2014). By the end of the newborn period, the calf rarely swims in echelon with his/her mother and can swim well on its own, although diving proficiency takes years (Noren 2008; Noren et al. 2008).

5.3 Infant Position and Nursing

In most species studied, newborn calves transition from predominantly echelon position to infant position by the end of 3 months. Infant position, where the calf swims under the mother's tailstock, by her abdomen, is widespread in toothed dolphins and whales, from narwhals (Charry et al. 2018) to dolphins and porpoises (Mann 2017; Xian 2012). Like echelon, this position provides a hydrodynamic boost to the calf (Noren 2008; Noren et al. 2008; Noren and Edwards 2011), in addition to protection and nursing access. Infant position is akin to "carrying" because of the close contact and energetic cost to the mother. Infant position is also thought to help stimulate milk production, as the calf intermittently bumps the mother's abdomen and mammary glands located on both sides of the genital slit. Calves nurse from infant position by tilting onside and inserting the tongue into the mammary slit. When waters are murky, observers have a difficult time seeing the calf in this position, but it is evident upon surfacing, where the calf is slightly staggered behind the mother and angles in under her after a breath. For example, in *Sotalia guianensis*, this was referred to as the "longitudinal position" (Tardin et al. 2013). Among killer whales in the northwest Pacific, infant position is also called "lateral position" (Karenina et al. 2013). In sperm whales, this has been called "peduncle diving" (Gero et al. 2009), or at the cow's tail or below her tail in beluga whales (Krasnova et al. 2006). In Shark Bay, calves spend about 39% of their time on average in infant position, but this can vary widely from calf to calf (10–80%; Foroughirad and Mann 2013). Under stress, calves stay in infant position more often, such as when the mother is being consorted by males (Watson 2005) or when in poor condition (Mann and Watson-Capps 2005). On rare occasion, calves swim in infant position with a non-mother (typically juvenile females, Mann and Smuts 1998), but this is brief and uncommon. Infant position is a valuable tool for determining lactation length, as calves sometimes continue to associate closely with the mother well after weaning, but not in infant position (e.g., Grellier et al. 2003; Tsai and Mann 2013).

Milk Composition

A critical component of maternal care is nursing, which defines the calf or infancy period in mammals. Compared to baleen whales, odontocetes have lower fat but still very high-energy milk. Lactation strategies have received little attention, in large part because of the difficulty in sampling cetacean milk and knowing the rate, quantity, and composition of milk delivered to offspring. However, based on terrestrial and non-cetacean marine mammals, a pattern emerges that is consistent with theories of maternal investment. Two primary lactational strategies are evident in cetaceans: the baleen whale pattern of very high fat milk (30–60% fat), fast growth, and early weaning versus the odontocete pattern of lower fat (<40%, albeit far higher than terrestrial mammals), slower growth, and late weaning (Oftedal 1997). Consistent with this is the pattern of maternal fasting during most of lactation for baleen whales and minimal or no fasting during lactation for odontocetes. This is also known as capital versus income breeding. At the one extreme, blue whales are estimated to produce and transfer about 500 lbs (220 kg) of milk per day to their

5 Maternal Care and Offspring Development in Odontocetes 99

calves (Oftedal 1997). Maternal reserves become substantially depleted for all cetaceans during lactation, but baleen whales dedicate their blubber (capital) to lactation, while odontocete mothers replenish (income) by food intake needs 40–50% higher than normal (Cheal and Gales 1991; Williams et al. 2011). Further, the smaller odontocetes expend more energy per unit of mass than the larger whales, such that they would be incapable of storing enough fat to completely support lactation (Oftedal 2000). A better understanding of odontocete lactation is needed given their prolonged nursing that can extend 8 years or more, with substantial variation within even the same population (Karniski et al. 2018).

5.4 Communication and Coordination

In the marine environment where a calf can quickly swim out of visual contact, communication and spatial coordination prove vital between mother and calf. Calves generally do not stray far from the mother when in groups and can easily or readily associate with others. However, when the mother is foraging and repeatedly diving, the calf cannot always stay with her given its limited diving ability. Options vary depending on species and context. For deep-diving species, mothers might shorten their dives to accommodate calves, but data for most species are lacking. Beluga whale mothers spend more time near the surface than females without calves (Heide-Jørgensen et al. 2001). Observations of a single female narwhal with its newborn indicated that she did not adjust her dive depth or duration compared with other females (Heide-Jørgensen and Dietz 1995). The narwhal pattern is unexpected given that among species that do not dive deeply (<15 m), such as bottlenose dolphins, mothers shorten their dive times when calves are young (Miketa et al. 2018). Similarly, pregnant pantropical spotted dolphins (*Stenella attenuata*) fed mostly on squid, which require deep dives, whereas lactating females ate flying fish, presumably so they could stay near the surface with their calves (Bernard and Hohn 1989).

Acoustic signaling is also critical. As detailed elsewhere (see Chap. 2), signature whistles in some delphinids have emerged as a key mechanism for mother and calf to maintain acoustic contact and navigate separations and reunions (King et al. 2016a; Smolker et al. 1993). We (Mann and Smuts 1998, 1999) proposed the "imprinting hypothesis"—that calves must learn the mothers' signature whistle in the first week (or two) of life so that there is no confusion on who to follow. This hypothesis was based on the observation that mothers were whistling almost constantly postpartum and that newborns did not venture away from their mothers in the first week of life. Furthermore, mothers were intolerant of others associating closely with calves in the first week and would chase females that attempted to associate with their calves. After the first week, mothers are much more tolerant of mother–calf separations and allow the calf to swim freely with others. The fact that pregnant females and newly parturient females emit signature whistles at very high rates is consistent with the imprinting hypothesis (Fripp and Tyack 2008; King et al. 2016b). Outside of delphinids, a variety of contact calls likely exist, but additional study is needed.

5.5 The Babysitting Debate

The belief that babysitting is a fundamental feature of odontocete societies is so widespread that it persists in popular and scientific literature despite little evidence in support. As the dictum "absence of evidence is not evidence of absence" may well apply, rigorous research on the topic is needed. A few studies have investigated babysitting empirically, most notably in sperm whales (Whitehead 1996; Gero et al. 2009) and bottlenose dolphins (Mann and Smuts 1998) with mixed results. Some studies assume babysitting based on simple association with non-mothers (e.g., Augusto et al. 2017). Clear and rigorous definitions are needed. The theoretical literature on this point is useful and dates back more than 40 years. Hrdy (1976) distinguished between allomaternal behavior, care, and abuse. Allomaternal behavior is when non-mothers interact with offspring, where no costs or benefits to the allomother, offspring, or mother are assumed. Both abuse and care are types of allomaternal behavior. Allomaternal care is when non-mothers nurture, guard, or protect offspring, providing some benefit to the offspring and mother. Babysitting is a special type of allomaternal care that necessitates absence of the mother. The allomother in this case might benefit (e.g., learn to care for offspring, foment a bond with the mother or infant) or incur a cost (e.g., reduced foraging time). Some studies find a fitness cost to breeders, raising the question of whether instances of allomaternal care are adaptive for the mother and offspring (see Gilchrist 2007). Allomaternal abuse is always at a cost to the mother and infant, typically by physically harming the offspring or keeping it away from the mother. Infanticide (see below) is clearly an example of the latter.

In considering cetacean calves, and their precocial ability, calves might readily associate with a wide range of individuals, including other calves, when not with their mothers. Since maternal care is behavior that increases the offspring's fitness, allomaternal care should do the same. But, maternal care is also defined by its directionality. The mother feeds the offspring, not vice versa. The mother carries the offspring, not vice versa. The mother protects and guards the offspring, not vice versa. Social behaviors and associations of the calf do not in of themselves qualify as they are mutual or reciprocal. Allonursing would obviously qualify but is difficult to demonstrate; spontaneous lactation has been reported in captivity when the "allomother" or foster mother has extensive exposure to the calf (Ridgway et al. 1995). Allomaternal care would also allow the mother to benefit in the proximate sense through resource acquisition (e.g., foraging more or taking longer dives) and, in the ultimate sense, by being able to wean her calf earlier and produce another. Social play, for example, might benefit the calf and the partner, but it is bi-directional and not a form of allomaternal care in of itself. This helps differentiate between instances where two calves are playing at some distance from their mothers. Who is babysitting whom?

Association can qualify as allomaternal care if it entails some form of guarding or protection—i.e., the mother is not nearby. Deep-diving species such as sperm whales and beaked whales appear to fit such criteria. In his original study, Whitehead (1996) found that sperm whale units were less synchronous in their foraging dives when calves were present, enhancing the chance that at least one whale would accompany the calf at the surface. This is suggestive of "guarding" behavior and where the allomother(s) or "guard(s)" would time dives to minimize the period that the calf is alone at the surface. Subsequent research (Gero et al. 2009) found population differences where either one female or multiple individuals of both sexes were responsible for guarding the calf. Although the trade-off between calf care and deep diving is thought to be a driving force in the matrilineal structure of sperm whale groups, this is not the case for other deep-diving species. Bottlenose whales, *Hyperoodon ampullatus*, have a different fission–fusion social structure and do not show evidence of babysitting (Gowans et al. 2001). Similarly, observations to date on beaked whales do not suggest babysitting, as calves are mostly left alone at the surface during foraging dives (MacLeod and D'Amico 2006). More observations of these elusive subjects are needed.

5.6 Protection from Predators

Habitat use is one way mothers might adjust their behavior to protect their calves. For example, among Risso's dolphins *Grampus griseus*, females stay closer to shore with young calves, foraging in shallower waters and possibly focusing on different species of squid that do not require such deep dives (Hartman et al. 2014). A number of studies have suggested a preference for shallow, nearshore habitat in several dolphin species (e.g., Pine et al. 2017; Mann et al. 2000; Gibson et al. 2013; Weir et al. 2008; Mann and Watson-Capps 2005), possibly to protect calves from predators, but there are other benefits of shallow habitats, such as more fish for mother and calf, protection from rough seas, or less interaction with conspecific males. Dolphins might generally change their habitat use in response to shark predation pressure (Wells et al. 1987; Heithaus 2001; Heithaus and Dill 2002). Other studies find little evidence for such nearshore preferences among mother–calf groups (e.g., Elwen et al. 2010), but this might be a consequence of large group sizes that mitigate the impact of sharks. Large sharks (tiger shark *Galeocerdo cuvier*), white pointer (*Carcharodon carcharias*), bull sharks (*Carcharhinus leucas*), and killer whales (*Orcinus orca*) are the main predators of odontocetes even if odontocetes are not their major prey source (see Heithaus 2001). Cookiecutter shark scars have been observed on beaked and sperm whales (e.g., McSweeney et al. 2007; Best and Photopoulou 2016), but they are not lethal predators. Killer whales have no known predators except humans.

Grouping is an important way for mothers to reduce calf predation risk, either because of dilution, detection, deterrence, or defense. Dilution just means that by grouping, everyone has a reduced chance of being taken (e.g., 10% chance of being taken in a group of ten as opposed to a 50% chance in a group of two). Detection means that more eyes are available to detect predators. Deterrence refers to the fact that a shark is less likely to attack a group of dolphins than a lone individual. Defense is characterized by active, even cooperative defense by a group against sharks. There is good evidence that grouping is a common strategy to reduce predation risk. Group sizes tend to be larger when calves, especially newborns, are present (*Tursiops* spp. Mann et al. 2000; Wells et al. 1987; tucuxi, *Sotalia fluviatilis* Azevedo et al. 2005; Indo-Pacific humpback dolphins *Sousa chinensis* Karczmarski 1999). This is also suggested for beaked whales (Chap. 14). Defense by the mother or group against tiger sharks has been observed multiple times in Shark Bay, Australia. Responses include forming a tight ball and facing the shark, mobbing the shark (getting on top of it until it leaves the area), chasing the shark, attacking the shark, and fleeing—depending on how startled they are (Mann and Barnett 1999; Mann and Watson-Capps 2005; personal observation). Even calves (but not neonates) become involved in mobbing and chasing—alongside their mothers.

5.7 Protection from Conspecifics

Conspecific aggression toward calves is rare in cetaceans. One exception is infanticide, which has been reported in at least four odontocete species including bottlenose dolphins (*Tursiops truncatus*; Patterson et al. 1998; Dunn et al. 2002; Kaplan et al. 2009; Robinson 2014), Indo-Pacific humpback dolphins (*Sousa chinensis*; Zheng et al. 2016), guiana dolphins (*Sotalia guianensis*; Nery and Simão 2009), and, most recently, killer whales (*Orcinus orca*; Towers et al. 2018). The classic and mostly supported sexual selection hypothesis (Hrdy 1979) predicts that infanticide by males can enhance their reproductive success (at a cost to the female's) when there is a very low chance the victim is their offspring, the lactation period is long, male tenure or access to the female is short, and there is a high chance of achieving mating success following the infanticide. Additionally, female counterstrategies are expected, such as mating multiply to confuse paternity and "faking" estrous—also to confuse paternity. In some species, females fight back.

Recently, a killer whale estimated to be at least 46 years old assisted her adult son (32 years old) in killing a young calf of another pod (Towers et al. 2018). This was the first report of its kind and is notable because a prominent hypothesis of post-reproductive lifespans in killer whales is that the mother enhances the fitness of their adult offspring, particularly sons. Although one dramatic case, this was the only observation in a marine mammal of a mother assisting her son in killing a conspecific (and likely unrelated) infant. This instance is consistent with Hrdy's (1979) sexual selection hypothesis, but the mother's role as a co-perpetrator in the infanticide is what stands out. It is also notable how vigorously the mother of the calf fought back, with assistance from kin. Although the defense was ultimately unsuccessful, the case highlights why the behavior has rarely been seen in killer whales. Killer whales are typically with their maternal group, and that might act as a deterrent.

Among bottlenose dolphins, evidence for infanticide varies by site. In areas with a considerable influx of males during the breeding season (births and conceptions given a 12-month pregnancy), infanticide has been reported, although the

5 Maternal Care and Offspring Development in Odontocetes

perpetrators, victims, and relatedness are rarely known (*T. truncatus*: Patterson et al. 1998; Dunn et al. 2002; Kaplan et al. 2009; Robinson 2014; Perrtree et al. 2016). In wild Indo-Pacific bottlenose dolphins (*T. aduncus*), there is good evidence that pregnancy is detected early, evidenced by a marked reduction in male–female association post-conception (Wallen et al. 2017). This suggests that female dolphins could not benefit from counterstrategies widely seen in the mammalian literature (see Hrdy 1979). In addition, some of the most intensively studied research sites, such as Shark Bay, Australia, and Sarasota, Florida, USA, have *not* reported aggression by adult males toward young infants (Wallen et al. 2017; Wells 2014). Paternity data suggest that females in these residential populations mate with local males, those that they have associated with throughout their lives (Krützen et al. 2004; unpublished). Consequently, infanticide might be absent or rare. Albeit difficult to observe, male–female bonds might be maintained, indirectly, which favor paternities within a subcommunity and protection by local males. Although sex segregation is a common feature of bottlenose dolphin societies (Galezo et al. 2018), males and females do associate, and this familiarity might both benefit males in terms of mating success and benefit females in reducing infanticide risk. That said, in locations with periodic influx of unfamiliar males, extra-community *and* local males might gain an advantage with infanticide if it occurs soon after calf birth, the perpetrator is not the father, and the male has a chance of fathering the next offspring. Once the calf has grown (>3 months) and maternal investment is considerable, the chances of a male fathering subsequent offspring declines as the female is unlikely to resume cycling until the next breeding season (Mann et al. 2000).

5.8 Calf Mortality

Both the rate and causes of calf mortality are difficult to identify in cetaceans as perinatal mortality is likely to be missed and mortality rates are surely higher than those reported. Causes are rarely known, other than extreme events such as morbillivirus, when carcasses are retrieved (e.g., Fauquier et al. 2017), or reasonably inferred based on fishing practices (e.g., Noren and Edwards 2007). It is notable that poor maternal care has *rarely* been identified as cause of calf mortality, although human provisioning is linked to maternal neglect (Foroughirad and Mann 2013; Mann et al. 2018).

Bottlenose Dolphins and Other Delphinids
As bottlenose dolphins (common, *T. truncatus* and Indo-Pacific *T. aduncus*) are the best-studied species, the variation in calf mortality and its causes between sites is informative. In Shark Bay, Australia, with 35 years of longitudinal data on Indo-Pacific bottlenose dolphins, the calf mortality rate, defined as survival to age 3, is consistently ~34% ($N = 676$ calves), lower than originally reported in 2000 (Mann et al. 2000). We use age 3 because over 91% of calves are weaned after age 3 ($N = 410$ calves), no calf has been weaned before age 2, and only one (blind)

nursing calf died after age 3 but possibly before weaning. Causes range from human impacts such as provisioning, bycatch, and boat strikes (Foroughirad and Mann 2013; Moore and Read 2008; Stone and Yoshinaga 2000) to shark attack (Mann and Barnett 1999), to maternal condition, which is linked to habitat such as prey-rich seagrass beds (Mann et al. 2000; Mann and Watson-Capps 2005; Miketa et al. in review).

A comparison between Shark Bay and Bunbury, Western Australia, showed that Shark Bay has much higher reproductive and vital rates (Manlik et al. 2016), which is not surprising given that the Bunbury dolphins are exposed to much higher human impact. In the Bay of Islands, New Zealand, bottlenose dolphin (*T. truncatus*) calf mortality is at least 50% by age 2 (>34% before age 1 and an additional 15–59% died by age 2; Tezanos-Pinto et al. 2015). As this population is declining, the high mortality rate and low calving rate are of concern. Although tourism is implicated, the precise causes are not known. In the Moray Firth, Scotland, common bottlenose dolphins appear to have lower calf mortality than all other sites, losing only 17% of their calves before age 2 (Robinson et al. 2017). Predation is virtually absent in the Moray Firth, but there are other environmental stressors, including contaminants (Wilson et al. 1999; Wells et al. 2005). Unobserved mortality (perinatal mortality especially) is also possible and might explain some of the differences between sites, depending on the intensity of monitoring. Outside of *Tursiops*, there are few reports of calf mortality rates. Herzing (1997) reports 24% mortality for Atlantic spotted dolphins (*Stenella frontalis*) in the first year of life. Killer whale calf mortality was originally estimated as between 30 and 50% in the first 6 months of life, but later estimates were that 39.2% of juveniles did not survive to adulthood (Olesiuk et al. 2005).

Most mammals with extensive maternal care, including humans, have a parabolic pattern of high mortality with firstborn offspring, lower mortality with middle born, and then high mortality again with last or later born (Bérubé et al. 1999; Sharp and Clutton-Brock 2010; Nussey et al. 2009; Mar et al. 2012). Firstborn mortality is generally attributed to smaller body size and/or poorer condition and lactational ability of young mothers combined with a lack of experience (Lang et al. 2011; but see Nuñez et al. 2015). Later-born mortality is generally attributed to reproductive senescence, which includes both fertility senescence and maternal effect senescence, declining physiological condition, and ability to nurture offspring (Karniski et al. 2018).

Patterns of firstborn mortality in wild populations are not well-documented for most species, but among three of the longest running bottlenose dolphin study sites, the pattern differs. In Sarasota, Florida, high mortality of firstborn common bottlenose dolphin offspring appears to be linked to maternal depuration of perfluoroalkyl compounds (PFCs) through milk. Later-born offspring receive fewer PFCs from their mothers (Houde et al. 2006; Wells et al. 2005). Moray Firth common bottlenose dolphins also show higher firstborn mortality at 45% (Robinson et al. 2017), but the causes are not identified. In Shark Bay, Indo-Pacific bottlenose dolphins do not show a firstborn effect. In fact there is a strong linear decline in calf survivorship with maternal age where calves born early in a mother's reproductive career have the best chance of survival (Karniski et al. 2018). Shark Bay is a relatively pristine environment, and the lack of contaminants could be a factor. Also, females have their first calf later in Shark

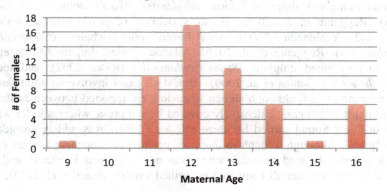

Fig. 5.2 $N = 52$ Indo-Pacific bottlenose dolphin, *Tursiops truncatus*, females in Shark Bay, Australia, with known age of first birth (see also Karniski et al. 2018). These data are conservative and biased toward births at early maternal ages because we excluded any female where we could have missed a late pregnancy (>6 months) or had perinatal mortality. Average is 12.8 years (SD = 1.6); 79% of females have their first calf at 12 years of age or later

Bay than the other sites, with 85% of females having their first calf between the ages of 11 and 14 (Fig. 5.2). With a prolonged juvenile period, typically lasting 10 years, immature females have ample opportunity to associate with calves in their network, and appear to be quite interested in doing so (Mann and Smuts 1998; Gibson and Mann 2008). At other sites, females produce their first calves between 6 and 13 years of age, with a mean age of 8 (Moray Firth, Robinson et al. 2017). At Sarasota, Florida, females produced their first calves between 6 and 8 years of age (Wells and Scott 1999; Wells 2003). Patterns of calf mortality might be attributed to maternal experience given that both sites with higher firstborn mortality also have shorter juvenile periods, but the anthropogenic stressors are also greater at Moray Firth and Sarasota than Shark Bay. We are also comparing different species. Herzing (1997) reports first parturition between 9 and 11 years of age for Atlantic spotted dolphins. Killer whale age at first reproduction is also around 11–14 for females (Matkin et al. 2014; Olesiuk et al. 2005). Better demographic data across populations and species will help us better understand how reproductive maturity and calf survival are linked.

Maternal Reactions to Calf Death

Another line of evidence in support of the intensity of the mother–offspring bond is based on responses to calf death, where the mother (or presumed mother) continues to provide care postmortem. Supportive behavior and postmortem attending have been observed with stillborn or dead calves among many toothed whale species—including bottlenose dolphins (*Tursiops* spp.) (Tayler and Saayman 1972; Cockcroft and Sauer 1990; Wells 1991; Mann and Barnett 1999; Fertl and Schiro 1994; Connor and Smolker 1989), sperm whales (*Physeter macrocephalus*) (Reggente et al. 2016), spinner dolphins (*Stenella longirostris*), beluga whale (*Delphinapterus leucas*)

(Smith and Sleno 1986; Krasnova et al. 2014), killer whales (*Orcinus orca*), Australian humpback dolphins (*Sousa sahulensis*), Risso's dolphins (*Grampus griseus*) (Reggente et al. 2016), Hector's dolphin (*Cephalorhynchus hectori*) (Stone and Yoshinaga 2000), short-finned pilot whales (*Globicephala macrorhynchus*) (Reggente et al. 2016), Atlantic spotted dolphin (Alves et al. 2015), rough-toothed dolphins (*Steno bredanensis*) (Ritter 2007), and tucuxi (*Sotalia fluviatilis*) (Santos et al. 2000). Almost all cases involve an adult female with a neonate or calf, although similar behaviors are reported between adults. A recent (August 2018) case, extensively covered by the press, where a killer whale mother in Puget Sound carried her dead calf for 17 days was widely considered intense "grief." Although difficult to interpret, postmortem attending is most characteristic of larger-brained mammals where maternal investment is intense, and it is comparably rare in balaenids compared to toothed whales (Bearzi et al. 2018).

5.9 Weaning

Even though weaning marks the end of infancy and is a critical transition for the calf, determination of weaning in odontocetes remains a methodological challenge. Actual nursing is difficult to observe, and although nursing from the same individual likely involves milk transfer, the amount of milk does not correlate well with nursing frequency, even for terrestrial mammals (Cameron 1998). In Shark Bay bottlenose dolphins, infant position is perfectly correlated with nursing, as all nursing occurs from this position. That, with observations of swollen mammaries, has enabled us to determine fairly precise weaning ages (to the week or month) for a large number of calves (Karniski et al. 2018). Cessation of infant position swimming is also well-timed with precipitous decline in mother–calf association, dropping well below 50%. Of over 450 weaned calves we have tracked, only one resumed nursing after being weaned, when her mother lost her next newborn. Interestingly, this calf was weaned before age 3 and was quite small but visibly caught up in size by nursing for another 1.5 years after her sibling died. Her second weaning occurred a few months after her fourth birthday. Sarasota reports weaning ages occasionally under 2 years and up to 7 years, with interbirth intervals typically 3–6 years (Wells and Scott 1999; Wells 2003). Among killer whales, the average calving interval was 4.88 years between viable calves, although there are some 2-year intervals (Olesiuk et al. 2005).

Herzing (1997) was able to observe both nursing and swollen mammaries of Atlantic spotted dolphins and found that most calves nursed for 3 or more years and up to 5 years. In contrast, average weaning age for pantropical spotted dolphins was estimated at just under 2 years (Myrick et al. 1986). This might be due to pressures from the tuna purse seine fishery (see Noren and Edwards 2007) or methodological differences. Based on longitudinal study, Amazon River dolphins (*Inia geoffrensis*) nurse for 1.5–5.8 years (Martin and Da Silva 2018). Beaked whales likely nurse their calves for 2 or more years (MacLeod and D'Amico 2006; New et al. 2013). Weaning

5 Maternal Care and Offspring Development in Odontocetes

ages are grossly underestimated for most toothed whales given the paucity of longitudinal observations on mother–calf pairs.

Most sites do not collect infant position data and use a marked reduction mother–calf association or the birth of the next calf as a proxy (e.g., Wells 2014). Another method is using isotope profiles of dentin growth layer groups (GLGs) in teeth (Matthews and Ferguson 2015), but this involves collection of teeth, and one downside is that low rates of nursing might not be captured by this method. Finding milk in the stomachs of calves either stranded or captured during whaling has also been used, but this method is not particularly reliable as the calf would have to have nursed within the last couple of hours for milk to be detected (Oftedal 1997).

Weaning ages in odontocetes can range from under a year for porpoises and river dolphins (e.g., Franciscana dolphin, *Pontoporia blainvillei*, Denuncio et al. 2013) to >8 years in bottlenose dolphins *Tursiops aduncus* (Mann et al. 2000; Karniski et al. 2018) and possibly 15 years in short-finned pilot whales, *Globicephala macrorhynchus*, based on harvested pods (Kasuya and Marsh 1984). Stable isotope profiles in tooth layers suggest that Canadian Arctic belugas wean their offspring typically by the end of the second year (Matthews and Ferguson 2015). As is likely for other species, calves were consuming fish well before weaning. Weaning is sometimes inferred from calving intervals or modeling. The tucuxi (*Sotalia fluviatilis*) has calving intervals of 2–3 years, so presumably they nurse for close to 2 years (if they wean just before the next calf; de Oliveira Santos et al. 2001). This species is particularly interesting because of its large brain relative to body size. Consistent with life history theory that amply applies to terrestrial systems, one would expect larger-brained species to have later weaning than smaller brained, due to the energetic investment in offspring brain growth—in addition to body growth, by the mother—but also because of the extensive learning period (Barton and Capellini 2011; Street et al. 2017).

What is clear is that odontocete calves nurse for very long periods of time, especially when contrasted with the much larger baleen whales, where weaning takes place within the year under most circumstances. Only primates and elephants have such late weaning ages, testimony to the substantial investment that mothers have in each offspring.

5.10 Post-Weaning

There is little research on the juvenile period (Krzyszczyk et al. 2017; McHugh et al. 2011) and mother–offspring associations post-weaning, in part because the decline in mother–offspring associations is used to define weaning. In bottlenose dolphins, mother–daughter associations remain stronger post-weaning than mother–son (e.g., Tsai and Mann 2013; Krzyszczyk et al. 2017), and home ranges continue to overlap extensively with the mother for both sexes (McHugh et al. 2011; Tsai and Mann 2013). With the exception of killer whales and possibly long-finned pilot whales— where both sexes remain with the mother's group (*Globicephala melas*, Amos et al. 1993; Ottensmeyer and Whitehead 2003; *Orcinus orca*, Baird 2000)—the relationship between mother and daughter persists post-weaning more than between mother and son, either in a stable group (sperm whales, Engelhaupt et al. 2009) or in dynamic fission–fusion systems (McHugh et al. 2011; Tsai and Mann 2013).

5.11 Maternal and Calf Hunting, Food-Sharing and Cultural Transmission

Unlike baleen whales that fast during much of the calf's development, odontocete mothers have the challenge of hunting while looking after their calves, a problem compounded by the energetic demands of lactation. Diving presents particular challenges (detailed above).

Prey sharing has been observed in some odontocetes (notably killer whales, Hoelzel 1991; Ford and Ellis 2006), which might mitigate or reduce lactation demands, potentially explaining the shorter than expected weaning times seen in tucuxi dolphins (*Sotalia fluviatilis*) and killer whales (Spinelli et al. 2008; Olesiuk et al. 2005). Notably, despite over 30 years of intensive observation of wild bottlenose dolphins in Shark Bay, prey sharing has only been observed once between a mother and calf—and it may have been incidental, as the dolphins often toss their fish, this one just happened to be tossed into the mouth of her offspring (personal obs.). Through prey sharing and extensive observation of their mothers, calves undoubtedly learn what to eat and what not to eat. Some fish are poisonous, difficult to swallow or digest, spiny or tough, or downright dangerous. Fish have a variety of defensive strategies that make them difficult to catch.

The calf period is undoubtedly a critical learning stage where the offspring can begin to master finding, capturing, and processing prey and navigate a complex environment (e.g., follow migration pathways, avoid stranding, maintain contact with conspecifics, know about seasonal changes in habitat and prey). Calves tend to adopt maternal foraging strategies (Sargeant and Mann 2009; Mann and Sargeant 2003; Mann et al. 2008). Calves learn about their network, and maintaining position in that network can be critical for survival (Stanton and Mann 2012). They might learn displays and a range of social behaviors (dusky dolphins, *Lagenorhynchus obscurus*, Deutsch et al. 2014). In addition to the changing relationship with the mother, the calf develops his or her own social bonds that are critical for survival after independence. Although the calf learns from others besides the mother (e.g., what fish to eat, Mann et al. 2007), it is the mother who provides the foundation for these social contacts. As such, the distinction between vertical transmission (from the parent) and lateral transmission (from non-parents) becomes muddled.

Social transmission and the importance of culture have received a great deal of attention in the field of animal behavior and for cetaceans specifically (Rendell and Whitehead 2001; Whitehead 2017). The most prominent examples of cultural transmission are primarily vertical from mother to calf or from the maternal lineage. Killer whales adopt the maternal pod-specific dialect (e.g., Yurk et al. 2002); sperm whales also communicate with codas of matrilineal origin (Rendell and Whitehead 2003); bottlenose dolphins adopt maternal foraging tactics and home range of their mother (Tsai and Mann 2013; Mann et al. 2008).

Teaching

In a seminal paper, Tim Caro and Marc Hauser (1992) argued for a rigorous definition of "teaching" as distinct from other forms of social learning, specifically where the "teacher" modifies his/her behavior in ways that costs the "teacher" and expressly benefit naïve individual(s) such that they can learn the behavior. Demonstration that the behavior benefits the learner is also needed. Only a handful of nonhuman studies have met these criteria (e.g., Thornton and McAuliffe 2006), in large part because the last part can rarely be shown in wild animals. That is, the improved skill of the learner must be tied to the teaching experience, not just a product of practice or maturation. Bender et al. (2009) were able to show that Atlantic spotted dolphin mothers modified their hunting behavior in the presence of calves, suggestive of teaching, but it is unclear whether calves improved their hunting skills as a result. Calves might also hinder maternal foraging as they follow the mother in pursuit of prey, but the fact that mothers occasionally "toyed" with prey in the presence of calves is more convincing. Killer whales at times assist calves in beaching to catch pinniped prey in the Crozet Islands (Guinet and Bouvier 1995). Zefferman (2016) argues that sponge dolphins are a good candidate for teaching. In fact, our data suggest that mothers modify their sponging behavior with sons, who are far less likely to become spongers. That is, mothers sponge less when with sons than daughters, and the result is that those sons with less exposure are less likely to sponge. With daughters, mothers do not modify their sponging behavior (unpublished). Similarly, we found (Miketa et al. 2018) that mothers adjusted their diving behavior to accommodate daughters more than sons, as they altered their diving only when daughters were close by. We suggest that mothers were affording their daughters learning opportunities with respect to foraging as daughters are more likely to adopt maternal foraging tactics than sons (e.g., Mann et al. 2008; Sargeant et al. 2005). Although experimental challenges in the field would make it exceedingly difficult to meet Caro and Hauser's criteria for cetaceans, evidence demonstrating social transmission, learning, and innovative ability (Patterson and Mann 2015) is abundant.

Regardless, the mother plays the central role in calf learning the intricacies of the social and physical environment. By following their mothers for many months, or even decades, the calf acquires the necessary physiological, social, and ecological skills to survive and reproduce. Among most odontocetes, this entails detailed knowledge of their network, kin and non-kin, allies, acquaintances, and even whom to avoid. Like other long-lived mammals, the mother provides essential guidance in the dynamic, intricate, and captivating world of cetaceans.

References

Alves F, Nicolau C, Dinis A, Ribeiro C, Freitas L (2015) Supportive behavior of free-ranging Atlantic spotted dolphins (*Stenella frontalis*) toward dead neonates, with data on perinatal mortality. Acta Ethol 18(3):301–304

Amos B, Schlotterer C, Tautz D (1993) Social structure of pilot whales revealed by analytical DNA profiling. Science 260(5108):670–672

Augusto JF, Frasier TR, Whitehead H (2017) Characterizing alloparental care in the pilot whale (*Globicephala melas*) population that summers off Cape Breton, Nova Scotia, Canada. Mar Mamm Sci 33(2):440–456

Azevedo AF, Viana SC, Oliveira AM, Van Sluys M (2005) Group characteristics of marine tucuxis (*Sotalia fluviatilis*) (Cetacea: Delphinidae) in Guanabara Bay, south-eastern Brazil. J Mar Biol Assoc UK 85(1):209–212

Baird RW (2000) The killer whale. In: Mann J, Connor R, Tyack P, Whitehead H (eds) Cetacean societies: field studies of dolphins and whales. University of Chicago Press, Chicago, pp 127–153

Barton RA, Capellini I (2011) Maternal investment, life histories, and the costs of brain growth in mammals. Proc Natl Acad Sci USA 108:6169–6174

Bearzi G, Kerem D, Furey NB, Pitman RL, Rendell L, Reeves RR (2018) Whale and dolphin behavioural responses to dead conspecifics. Zoology 128:1–15

Bender CE, Herzing DL, Bjorklund DF (2009) Evidence of teaching in Atlantic spotted dolphins (*Stenella frontalis*) by mother dolphins foraging in the presence of their calves. Anim Cogn 12 (1):43–53

Bernard HJ, Hohn AA (1989) Differences in feeding habits between pregnant and lactating spotted dolphins (*Stenella attenuata*). J Mammal 70(1):211–215

Bérubé CH, Festa-Bianchet M, Jorgenson JT (1999) Individual differences, longevity, and reproductive senescence in bighorn ewes. Ecology 80:2555–2565

Best PB, Photopoulou T (2016) Identifying the "demon whale-biter": patterns of scarring on large whales attributed to a cookie-cutter shark *Isistius* sp. PLoS One 11(4):e0152643

Cameron EZ (1998) Is suckling behaviour a useful predictor of milk intake? A review. Anim Behav 56(3):521–532

Caro TM, Hauser MD (1992) Is there teaching in nonhuman animals? Q Rev Biol 67(2):151–174

Charry B, Marcoux M, Humphries MM (2018) Aerial photographic identification of narwhal (*Monodon monoceros*) newborns and their spatial proximity to the nearest adult female. Arctic Sci 4:1–12

Cheal AJ, Gales NJ (1991) Body mass and food intake in captive, breeding bottlenose dolphins, *Tursiops truncatus*. Zoo Biol 10(6):451–456

Cockcroft VG, Sauer W (1990) Observed and inferred epimeletic (nurturant) behaviour in bottlenose dolphins. Aquat Mamm 16(1):31–32

Connor RC, Smolker RA (1989) Quantitative description of a rare behavioral event: a bottlenose dolphin's behavior toward her deceased offspring. In: Leatherwood S, Reeves R (eds) The bottlenose dolphin. Academic, New York, pp 355–360

de Oliveira Santos MC, Acuña LB, Rosso S (2001) Insights on site fidelity and calving intervals of the marine tucuxi dolphin (*Sotalia fluviatilis*) in South-eastern Brazil. J Mar Biol Assoc UK 81 (6):1049–1052

Dearolf JL, McLellan WA, Dillaman RM, Frierson D Jr, Pabst DA (2000) Precocial development of axial locomotor muscle in bottlenose dolphins (*Tursiops truncatus*). J Morphol 244(3):203–215

Denuncio PE, Bastida RO, Danilewicz D, Morón S, Rodríguez-Heredia S, Rodríguez DH (2013) Calf chronology of the Franciscana Dolphin (*Pontoporia blainvillei*): birth, onset of feeding, and duration of lactation in coastal waters of Argentina. Aquat Mamm 39(1):73–80. https://doi.org/10.1578/AM.39.1.2013.73

Deutsch S, Pearson H, Würsig B (2014) Development of leaps in dusky dolphin (*Lagenorhynchus obscurus*) calves. Behaviour 151(11):1555–1577

5 Maternal Care and Offspring Development in Odontocetes

Dunn DG, Barco SG, Pabst DA, McLellan WA (2002) Evidence for infanticide in bottlenose dolphins of the Western North Atlantic. J Wildl Dis 38(3):505–510

Elwen SH, Thornton M, Reeb D, Best PB (2010) Near-shore distribution of Heaviside's (*Cephalorhynchus heavisidii*) and dusky dolphins (*Lagenorhynchus obscurus*) at the southern limit of their range in South Africa. Afr Zool 45(1):78–91

Engelhaupt D, Rus Hoelzel A, Nicholson C, Frantzis A, Mesnick S, Gero S, Whitehead H, Rendell L, Miller P, De Stefanis R, Canadas A (2009) Female philopatry in coastal basins and male dispersion across the North Atlantic in a highly mobile marine species, the sperm whale (*Physeter macrocephalus*). Mol Ecol 18(20):4193–4205

Fauquier DA, Litz J, Sanchez S, Colegrove K, Schwacke LH, Hart L, Saliki J, Smith C, Goldstein T, Bowen-Stevens S, McFee W (2017) Evaluation of morbillivirus exposure in cetaceans from the northern Gulf of Mexico 2010-2014. Endanger Species Res 33:211–220

Fearnbach H, Durban J, Parsons K, Claridge D (2012) Seasonality of calving and predation risk in bottlenose dolphins on Little Bahama Bank. Mar Mamm Sci 28(2):402–411

Fertl D, Schiro A (1994) Carrying of dead calves by free-ranging Texas bottlenose dolphins (*Tursiops truncatus*). Aquat Mamm 20:53–56

Ford JK, Ellis GM (2006) Selective foraging by fish-eating killer whales *Orcinus orca* in British Columbia. Mar Ecol Prog Ser 316:185–199

Foroughirad V, Mann J (2013) Human fish provisioning has long-term impacts on the behaviour and survival of bottlenose dolphins. Biol Conserv 160:242–249

Fripp D, Tyack P (2008) Postpartum whistle production in bottlenose dolphins. Mar Mamm Sci 24 (3):479–502

Galezo A, Krzyszczyk E, Mann J (2018) Sexual segregation in Indo-Pacific bottlenose dolphins is driven by female avoidance of males. Behav Ecol 29(2):377–386

Gero S, Engelhaupt D, Rendell L, Whitehead H (2009) Who cares? Between-group variation in alloparental caregiving in sperm whales. Behav Ecol 20(4):838–843

Gibson QA, Mann J (2008) The size and composition of wild bottlenose dolphin (*Tursiops* sp.) mother-calf groups in Shark Bay, Australia. Anim Behav 76:389–405

Gibson QA, Howells EM, Lambert JD, Mazzoil MM, Richmond JP (2013) The ranging patterns of female bottlenose dolphins with respect to reproductive status: testing the concept of nursery areas. J Exp Mar Biol Ecol 445:53–60

Gilchrist JS (2007) Cooperative behaviour in cooperative breeders: costs, benefits, and communal breeding. Behav Process 76(2):100–105

Gowans S, Whitehead H, Hooker SK (2001) Social organization in northern bottlenose whales, *Hyperoodon ampullatus*: not driven by deep-water foraging? Anim Behav 62(2):369–377

Grellier K, Hammond PS, Wilson B, Sanders-Reed CA, Thompson PM (2003) Use of photo-identification data to quantify mother calf association patterns in bottlenose dolphins. Can J Zool 81(8):1421–1427

Guinet C, Bouvier J (1995) Development of intentional stranding hunting techniques in killer whale (*Orcinus orca*) calves at Crozet Archipelago. Can J Zool 73(1):27–33

Hartman KL, Fernandez M, Azevedo JM (2014) Spatial segregation of calving and nursing Risso's dolphins (*Grampus griseus*) in the Azores, and its conservation implications. Mar Biol 161 (6):1419–1428

Heide-Jørgensen MP, Dietz R (1995) Some characteristics of narwhal, Monodon monoceros, diving behaviour in Baffin Bay. Can J Zool 73(11):2120–2132

Heide-Jørgensen MP, Hammeken N, Dietz R, Orr J, Richard PR (2001) Surfacing times and dive rates for narwhals (*Monodon monoceros*) and belugas (*Delphinapterus leucas*). Arctic:284–298

Heithaus MR (2001) Shark attacks on bottlenose dolphins (*Tursiops aduncus*) in Shark Bay, Western Australia: attack rate, bite scar frequencies, and attack seasonality. Mar Mamm Sci 17(3):526–539

Heithaus MR, Dill LM (2002) Food availability and tiger shark predation risk influence bottlenose dolphin habitat use. Ecology 83(2):480–491

Herzing DL (1997) The life history of free-ranging Atlantic spotted dolphins (*Stenella frontalis*): age classes, color phases, and female reproduction. Mar Mamm Sci 13(4):576–595

Hoelzel AR (1991) Killer whale predation on marine mammals at Punta Norte, Argentina; food sharing, provisioning and foraging strategy. Behav Ecol Sociobiol 29(3):197–204

Houde M, Pacepavicius G, Wells RS, Fair PA, Letcher RJ, Alaee M, Bossart GD, Hohn AA, Sweeney J, Solomon KR, Muir DC (2006) Polychlorinated biphenyls and hydroxylated polychlorinated biphenyls in plasma of bottlenose dolphins (*Tursiops truncatus*) from the Western Atlantic and the Gulf of Mexico. Environ Sci Technol 40(19):5860–5866

Hrdy SB (1976) Care and exploitation of nonhuman primate infants by conspecifics other than the mother. In: Rosenblatt JS, Hinde RA, Shaw E, Beer C (eds) Advances in the study of behavior, vol 6. Academic, New York, pp 101–158

Hrdy SB (1979) Infanticide among animals: a review, classification, and examination of the implications for the reproductive strategies of females. Ethol Sociobiol 1(1):13–40

Kaplan JD, Lentell BJ, Lange W (2009) Possible evidence for infanticide among bottlenose dolphins (*Tursiops truncatus*) off St. Augustine, Florida. Mar Mamm Sci 25(4):970–975

Karczmarski L (1999) Group dynamics of humpback dolphins *Sousa chinensis* in the Algoa Bay region, South Africa. J Zool 249:283–293

Karenina K, Giljov A, Ivkovich T, Burdin A, Malashichev Y (2013) Lateralization of spatial relationships between wild mother and infant orcas, *Orcinus orca*. Anim Behav 86 (6):1225–1231

Karniski C, Krzyszczyk E, Mann J (2018) Senescence impacts reproduction and maternal investment in bottlenose dolphins. Proc R Soc B 285:20181123

Kasuya T, Marsh H (1984) Life history and reproductive biology of the short-finned pilot whale, *Globicephala macrorhynchus*, off the Pacific coast of Japan. Rep Int Whal Comm 6:259–310

King SL, Guarino E, Keaton L, Erb L, Jaakkola K (2016a) Maternal signature whistle use aids mother-calf reunions in a bottlenose dolphin, *Tursiops truncatus*. Behav Process 126:64–70

King SL, Guarino E, Donegan K, Hecksher J, Jaakkola K (2016b) Further insights into postpartum signature whistle use in bottlenose dolphins (*Tursiops truncatus*). Mar Mamm Sci 32:1458–1469

Krasnova VV, Bel'Kovich VM, Chernetsky AD (2006) Mother-infant spatial relations in wild beluga (*Delphinapterus leucas*) during postnatal development under natural conditions. Biol Bull 33(1):53–58

Krasnova VV, Chernetsky AD, Zheludkova AI, Bel'kovich VM (2014) Parental behavior of the beluga whale (*Delphinapterus leucas*) in natural environment. Biol Bull 41(4):349–356

Krützen M, Barre LM, Connor RC, Mann J, Sherwin WB (2004) O father: where art thou? – paternity assessment in an open fission-fusion society of wild bottlenose dolphins (*Tursiops* sp.) in Shark Bay, Western Australia. Mol Ecol 13:1975–1990

Krzyszczyk E, Stanton MA, Patterson E, Mann J (2017) The transition to independence: sex differences in social and behavioral development of wild bottlenose dolphins. Anim Behav 129:43–59

Lang SL, Iverson SJ, Bowen WD (2011) The influence of reproductive experience on milk energy output and lactation performance in the grey seal (*Halichoerus grypus*). PLoS One 6(5):e19487

Lyamin O, Pryaslova J, Lance V, Siegel J (2005) Animal behaviour: continuous activity in cetaceans after birth. Nature 435(7046):1177

Lyamin O, Pryaslova J, Kosenko P, Siegel J (2007) Behavioral aspects of sleep in bottlenose dolphin mothers and their calves. Physiol Behav 92(4):725–733

MacLeod CD, D'Amico A (2006) A review of beaked whale behaviour and ecology in relation to assessing and mitigating impacts of anthropogenic noise. J Cetacean Res Manag 7(3):211–221

Manlik O, McDonald JA, Mann J, Raudino HC, Bejder L, Krützen M, Connor RC, Heithaus MR, Lacy RC, Sherwin WB (2016) The relative importance of reproduction and survival for the conservation of two dolphin populations. Ecol Evol 6(11):3496–3512

Mann J (2017) Parental behavior. In: Würsig B, Thewissen HGM, Kovacs K (eds) Encyclopedia of marine mammals, 3rd edn. Elsevier/Academic, San Diego, 1488p

5 Maternal Care and Offspring Development in Odontocetes

Mann J, Barnett H (1999) Lethal tiger shark (*Galeocerdo cuvieri*) attack on bottlenose dolphin (*Tursiops* sp.) calf: defense and reactions by the mother. Mar Mamm Sci 15(2):568–575

Mann J, Sargeant B (2003) Like mother, like calf: the ontogeny of foraging traditions in wild Indian Ocean bottlenose dolphins (*Tursiops* sp.). In: Fragaszy D, Perry S (eds) The biology of traditions: models and evidence. Cambridge University Press, Cambridge, pp 236–266

Mann J, Smuts BB (1998) Natal attraction: allomaternal care and mother-infant separations in wild bottlenose dolphins. Anim Behav 55:1097–1113

Mann J, Smuts BB (1999) Behavioral development in wild bottlenose dolphin newborns (*Tursiops* sp.). Behaviour 136:529–566

Mann J, Watson-Capps J (2005) Surviving at Sea: ecological and behavioural predictors of calf mortality in Indian Ocean bottlenose dolphins (*Tursiops* sp.). Anim Behav 69:899–909

Mann J, Connor RC, Barre LM, Heithaus MR (2000) Female reproductive success in bottlenose dolphins (*Tursiops* sp.): life history, habitat, provisioning, and group size effects. Behav Ecol 11:210–219

Mann J, Sargeant BL, Minor M (2007) Calf inspection of fish catches: opportunities for oblique social learning? Mar Mamm Sci 23(1):197–202

Mann J, Sargeant BL, Watson-Capps J, Gibson Q, Heithaus MR, Connor RC, Patterson E (2008) Why do dolphins carry sponges? PLoS One 3(12):e3868

Mann J, Senigaglia V, Jacoby A, Bejder L (2018) A comparison of tourism and feeding wild dolphins at monkey Mia and Bunbury, Australia. In: Carr N, Broom D (eds) Animal welfare and tourism. CABI, Oxfordshire

Mar KU, Lahdenperä M, Lummaa V (2012) Causes and correlates of calf mortality in captive Asian elephants (*Elephas maximus*). PLoS One 7(3):e32335

Martin AR, Da Silva VMF (2018) Reproductive parameters of the Amazon river dolphin or boto, *Inia geoffrensis* (Cetacea: Iniidae); an evolutionary outlier bucks no trends. Biol J Linn Soc 123 (3):666–676

Matkin CO, Ward Testa J, Ellis GM, Saulitis EL (2014) Life history and population dynamics of southern Alaska resident killer whales (*Orcinus orca*). Mar Mamm Sci 30(2):460–479

Matthews CJD, Ferguson SH (2015) Weaning age variation in beluga whales (*Delphinapterus leucas*). J Mammal 96(2):425–437

McHugh KA, Allen JB, Barleycorn AA, Wells RS (2011) Natal philopatry, ranging behavior, and habitat selection of juvenile bottlenose dolphins in Sarasota Bay, Florida. J Mammal 92 (6):1298–1313

McSweeney DJ, Baird RW, Mahaffy SD (2007) Site fidelity, associations, and movements of Cuvier's (*Ziphius cavirostris*) and Blainville's (*Mesoplodon densirostris*) beaked whales off the island of Hawai'i. Mar Mamm Sci 23(3):666–687

Miketa M, Patterson EM, Krzyszczyk E, Foroughirad V, Mann J (2018) Calf age and sex affects maternal diving behavior in Shark Bay bottlenose dolphins. Anim Behav 137:107–117

Moore JE, Read AJ (2008) A Bayesian uncertainty analysis of cetacean demography and bycatch mortality using age-at-death data. Ecol Appl 18(8):1914–1931

Myrick AC, Hohn AA, Barlow J, Sloan PA (1986) Reproductive-biology of female spotted dolphins, *Stenella attenuata*, from the Eastern Tropical Pacific. Fish Bull 84(2):247–259

Nery MF, Simão SM (2009) Sexual coercion and aggression towards a newborn calf of marine tucuxi dolphins (*Sotalia guianensis*). Mar Mamm Sci 25(2):450–454

New LF, Moretti DJ, Hooker SK, Costa DP, Simmons SE (2013) Using energetic models to investigate the survival and reproduction of beaked whales (family Ziphiidae). PLoS One 8 (7):e68725

Noren SR (2008) Infant carrying behaviour in dolphins: costly parental care in an aquatic environment. Funct Ecol 22(2):284–288

Noren SR, Edwards EF (2007) Physiological and behavioral development in delphinid calves: implications for calf separation and mortality due to tuna purse-seine sets. Mar Mamm Sci 23 (1):15–29

Noren SR, Edwards EF (2011) Infant position in mother-calf dolphin pairs: formation locomotion with hydrodynamic benefits. Mar Ecol Prog Ser 424:229–236

Noren SR, Biedenbach G, Redfern JV, Edwards EF (2008) Hitching a ride: the formation locomotion strategy of dolphin calves. Funct Ecol 22(2):278–283

Nuñez CL, Grote MN, Wechsler M, Allen-Blevins CR, Hinde K (2015) Offspring of primiparous mothers do not experience greater mortality or poorer growth: revisiting the conventional wisdom with archival records of Rhesus. Am J Primatol 77(9):963–973

Nussey DH, Kruuk LEB, Morris A, Clements MN, Pemberton JM, Clutton-brock TH (2009) Inter- and intrasexual variation in aging patterns across reproductive traits in a wild red deer population. Am Nat 174:342–357

Oftedal OT (1997) Lactation in whales and dolphins: evidence of divergence between baleen- and toothed-species. J Mammary Gland Biol Neoplasia 2:205–230

Oftedal OT (2000) Use of maternal reserves as a lactation strategy in large mammals. Proc Nutr Soc 59:99–106

Olesiuk PF, Ellis GM, Ford JKB (2005) Life history and population dynamics of resident killer whales *Orcinus orca* in the coastal waters of British Columbia Research Document 2005/045. Fisheries and Oceans Canada, Nanaimo

Ottensmeyer CA, Whitehead H (2003) Behavioural evidence for social units in long-finned pilot whales. Can J Zool 81(8):1327–1338

Patterson EM, Mann J (2015) Cetacean innovation. In: Kaufman A, Kaufman J (eds) Animal creativity and innovation. Elsevier, San Diego, pp 73–120

Patterson IAP, Reid RJ, Wilson B, Grellier K, Ross HM, Thompson PM (1998) Evidence for infanticide in bottlenose dolphins: an explanation for violent interactions with harbour porpoises? Proc R Soc Lond B Biol Sci 265(1402):1167–1170

Perrtree RM, Sayigh LS, Williford A, Bocconcelli A, Curran MC, Cox TM (2016) First observed wild birth and acoustic record of a possible infanticide attempt on a common bottlenose dolphin (*Tursiops truncatus*). Mar Mamm Sci 32(1):376–385

Pine MK, Wang K, Wang D (2017) Fine-scale habitat use in Indo-Pacific humpback dolphins, *Sousa chinensis*, may be more influenced by fish rather than vessels in the Pearl River Estuary, China. Mar Mamm Sci 33(1):291–312

Reggente MAL, Alves F, Nicolau C, Freitas L, Cagnazzi D, Baird RW, Galli P (2016) Nurturant behavior toward dead conspecifics in free-ranging mammals: new records for odontocetes and a general review. J Mammal 97(5):1428–1434

Rendell L, Whitehead H (2001) Culture in whales and dolphins. Behav Brain Sci 24(2):309–324

Rendell LE, Whitehead H (2003) Vocal clans in sperm whales (*Physeter macrocephalus*). Proc R Soc Lond B Biol Sci 270(1512):225–231

Ridgway S, Kamolnick T, Reddy M, Curry C, Tarpley RJ (1995) Orphan-induced lactation in *Tursiops* and analysis of collected milk. Mar Mamm Sci 11(2):172–182

Ritter F (2007) Behavioral responses of rough-toothed dolphins to a dead newborn calf. Mar Mamm Sci 23(2):429–433

Robinson KP (2014) Agonistic intraspecific behavior in free-ranging bottlenose dolphins: calf-directed aggression and infanticidal tendencies by adult males. Mar Mamm Sci 30(1):381–388

Robinson KP, Sim TMC, Culloch RM, Bean TS, Cordoba AI, Eisfeld SM, Filan M, Haskins GN, Williams G, Pierce GJ (2017) Female reproductive success and calf survival in a North Sea coastal bottlenose dolphin (*Tursiops truncatus*) population. PLoS One 12(9):e0185000

Sakai M, Morisaka T, Iwasaki M, Yoshida Y, Wakabayashi I, Seko A, Kasamatsu M, Kohshima S (2013) Mother–calf interactions and social behavior development in Commerson's dolphins (*Cephalorhynchus commersonii*). J Ethol 31(3):305–313

Santos MCO, Rosso S, Siciliano S, Zerbini AN, Zampirolli E, Vicente A, Alvarenga F (2000) Behavioral observations of the marine tucuxi dolphin (*Sotalia fluviatilis*) in São Paulo estuarine waters, Southeastern Brazil. Aquat Mamm 26:260–267

Sargeant BL, Mann J (2009) Developmental evidence for foraging traditions in wild bottlenose dolphins. Anim Behav 78:715–721

Sargeant BL, Mann J, Berggren P, Krützen M (2005) Specialization and development of beach hunting, a rare foraging behavior, by wild Indian Ocean bottlenose dolphins (*Tursiops* sp.). Can J Zool 83(11):1400–1410

Sharp SP, Clutton-Brock TH (2010) Reproductive senescence in a cooperatively breeding mammal. J Anim Ecol 79:176–183

Smith TG, Sleno GA (1986) Do white whales, *Delphinapterus leucas*, carry surrogates in response to early loss of their young? Can J Zool 64(7):1581–1582

Smolker RA, Mann J, Smuts BB (1993) The use of signature whistles during separations and reunions among wild bottlenose dolphin mothers and calves. Behav Ecol Sociobiol 33:393–402

Spinelli LHP, Jesus AH, Nascimento LF, Yamamoto ME (2008) Prey-transfer in the marine tucuxi dolphin, *Sotalia fluviatilis*, on the Brazilian coast. JMBA2, Marine Biodiversity Records 1 (published online)

Stanton MA, Mann J (2012) Early social networks predict survival in wild bottlenose dolphins. PLoS One 7(10):e47508

Stone GS, Yoshinaga A (2000) Hector's Dolphin *Cephalorhynchus hectori* calf mortalities may indicate new risks from boat traffic and habituation. Pac Conserv Biol 6(2):162–170

Street SE, Navarrete AF, Reader SM, Laland KN (2017) Coevolution of cultural intelligence, extended life history, sociality, and brain size in primates. Proc Natl Acad Sci 114 (30):7908–7914

Tardin RH, Espécie MA, Lodi L, Simão SM (2013) Parental care behavior in the Guiana dolphin, *Sotalia guianensis* (Cetacea: Delphinidae), in Ilha Grande Bay, southeastern Brazil. Zoologia (Curitiba) 30(1):15–23

Tayler CK, Saayman GS (1972) The social organization and behaviour of dolphins (*Tursiops truncatus*) and baboons (*Papio ursinus*): some comparisons and assessments. Ann Cape Prov Mus (Nat Hist) 9(2):11–49

Tezanos-Pinto G, Constantine R, Berghan J, Baker CS (2015) High calf mortality in bottlenose dolphins in the Bay of Islands, New Zealand—a local unit in decline. Mar Mamm Sci 31:540–559

Thornton A, McAuliffe K (2006) Teaching in wild meerkats. Science 313:227–229

Towers JR, Hallé MJ, Symonds HK, Sutton GJ, Morton AB, Spong P, Borrowman JP, Ford JK (2018) Infanticide in a mammal-eating killer whale population. Sci Rep 8(1):4366

Tsai YJ, Mann J (2013) Dispersal, philopatry and the role of fission-fusion dynamics in bottlenose dolphins. Mar Mamm Sci 29(2):261–279

Wallen MM, Krzyszczyk E, Mann J (2017) Mating in a bisexually philopatric society: bottlenose dolphin females associate with adult males but not adult sons during estrous. Behav Ecol Sociobiol 71:153–165

Watson JJ (2005) Female mating behavior in the context of sexual coercion and female ranging behavior of bottlenose dolphins (*Tursiops* Sp.) in Shark Bay, Western Australia. Doctoral dissertation, Georgetown University

Weir JS, Duprey NMT, Würsig B (2008) Dusky dolphin (*Lagenorhynchus obscurus*) subgroup distribution: are shallow waters a refuge for nursery groups? Can J Zool 86(11):1225–1234

Wells RS (1991) The role of long-term study in understanding the social structure of a bottlenose dolphin community. In: Dolphin societies: discoveries and puzzles. University of California Press, Berkeley, pp 199–225

Wells RS (2003) Dolphin social complexity: lessons from long-term study and life-history. In: de Waal FBM, Tyack PL (eds) Animal social complexity: intelligence, culture, and individualized societies. Harvard University Press, Cambridge, pp 32–56

Wells RS (2014) Social structure and life history of bottlenose dolphins near Sarasota Bay, Florida: insights from four decades and five generations. In: Yamagiwa J, Karczmarski L (eds) Primates and cetaceans. Springer, Tokyo, pp 149–172

Wells RS, Scott MD (1999) Bottlenose dolphin *Tursiops truncatus* (Montagu, 1821). In: Ridgway SH, Harrison R (eds) Handbook of marine mammals. The second book of dolphins and porpoises, vol 6. Academic, San Diego, pp 137–182

Wells RS, Scott MD, Irvine AB (1987) The social structure of free-ranging bottlenose dolphins. In: Genoways HH (ed) Current mammalogy. Springer, Boston, pp 247–305

Wells RS, Tornero V, Borrell A, Aguilar A, Rowles TK, Rhinehart HL, Hofmann S, Jarman WM, Hohn AA, Sweeney JC (2005) Integrating life-history and reproductive success data to examine potential relationships with organochlorine compounds for bottlenose dolphins (*Tursiops truncatus*) in Sarasota Bay, Florida. Sci Total Environ 349(1–3):106–119

Whitehead H (1996) Babysitting, dive synchrony, and indications of alloparental care in sperm whales. Behav Ecol Sociobiol 38(4):237–244

Whitehead H (2017) Gene–culture coevolution in whales and dolphins. Proc Natl Acad Sci 114 (30):7814–7821

Williams R, Krkošek M, Ashe E, Branch TA, Clark S, Hammond PS, Hoyt E, Noren DP, Rosen D, Winship A (2011) Competing conservation objectives for predators and prey: estimating killer whale prey requirements for Chinook salmon. PLoS One 6:e26738

Wilson B, Arnold H, Bearzi G, Fortuna CM, Gaspar R, Ingram S, Liret C, Pribanic S, Read AJ, Ridoux V, Schneider K (1999) Epidermal diseases in bottlenose dolphins: impacts of natural and anthropogenic factors. Proc R Soc Lond B Biol Sci 266(1423):1077–1083

Xian Y (2012) The development of spatial positions between mother and calf of Yangtze finless porpoises (*Neophocaena asiaeorientalis asiaeorientalis*) maintained in captive and seminatural environments. Aquat Mamm 38:127–135

Yurk H, Barrett-Lennard L, Ford JKB, Matkin CO (2002) Cultural transmission within maternal lineages: vocal clans in resident killer whales in southern Alaska. Anim Behav 63(6):1103–1119

Zefferman MR (2016) Mothers teach daughters because daughters teach granddaughters: the evolution of sex-biased transmission. Behav Ecol 27(4):1172–1181

Zheng R, Karczmarski L, Lin W, Chan SC, Chang WL, Wu Y (2016) Infanticide in the Indo-Pacific humpback dolphin (*Sousa chinensis*). J Ethol 34(3):299–307

Chapter 6
Movement Patterns of Odontocetes Through Space and Time

Stefan Bräger and Zsuzsanna Bräger

Abstract Odontocetes are constantly on the move, but resulting movement patterns have rarely been analyzed comparatively. Within the continuum of space and time, several distinct patterns appear: diel, short-term, seasonal, and exceptionally long-distance movements, together delineating home ranges that may be occupied for extended periods or repeatedly. A variety of methods have been used to study movement patterns of odontocetes, such as absence/presence, photo-identification of individuals, static acoustic monitoring, and marking and tagging of individuals (e.g., with radio tags or satellite transmitters). These methods can provide a wealth of information at a variety of scales and thus not always comparable. We highlight examples of recent research on selected taxa: *Physeter, Ziphiidae and Globicephalinae, Orcinus, Monodontidae, Stenella, Grampus, Delphinus, Tursiops, Sousa, Sotalia, Cephalorhynchus, Phocoena, Neophocaena, Phocoenoides, Inia*, and *Platanista*. Prey availability and predation pressure are main drivers of movement patterns, although social factors also shape movement patterns, to date much overlooked.

Keywords Diel · Dispersal · Displacement · Distance · Home range · Migration · Residency · Seasonal · Site fidelity · Transit speed

6.1 Introduction

The ability to move is quintessential to the behavior of vertebrates, and dolphins are the epitome of motion. Their speed and perceived elegance have fascinated human observers for millennia. Movements provide the fabric for vital processes such as habitat selection, foraging and food acquisition, predator avoidance, mate choice, reproduction, and propagation (or avoidance) of diseases and parasites. Movements

S. Bräger (✉) · Z. Bräger
German Oceanographic Museum (DMM), Stralsund, Germany

© Springer Nature Switzerland AG 2019
B. Würsig (ed.), *Ethology and Behavioral Ecology of Odontocetes*, Ethology and Behavioral Ecology of Marine Mammals,
https://doi.org/10.1007/978-3-030-16663-2_6

constitute a key process that connects and enables most other processes that shape life histories. The ability to move is so important that injured immobilized individuals perish.

The extent of movements in time and space represents a continuum from tens of meters to thousands of kilometers, from minutes to decades. Nonetheless, it appears possible to discern a general pattern from small-scale diel movements, displacements within short- to medium-term home ranges, to large-scale seasonal migrations. There are also rare irruptive movements—then perhaps joining populations or expanding the range of a population. More confined home ranges and site fidelities are also expressions of movement patterns—rather the absence of movements—and are considered as well.

6.2 Ethology of Movements

Swimming and diving require highly specialized adaptations over evolutionary time, to allow odontocetes to move efficiently through a medium 800 times denser than air. It is not only a set of physiological and ethological adaptations but also has far-reaching consequences for behavioral ecology. The ability to change location provides the fundamental connective fabric for essential social and ecological processes such as habitat selection, mate choice, social connectivity and segregation, prey selection, predator avoidance, and more. We review the existence and extent of movement patterns among toothed whales, whereas the functions of these patterns will be discussed in several other chapters.

Movement extends from the short term and short range (e.g., diel movements) via the seasonal and midrange (e.g., migrations) to the long term and long range (e.g., population dispersal or range extension). The sums of short-term and short-range movements define a "seasonal home range" of an individual, whereas the sums of seasonal and midrange movements define an individual's lifetime home range. A population extending its range into previously unoccupied areas may be a rare event but perhaps more and more likely as necessary adaptation to a changing environment and a changing climate (e.g., Lambert et al. 2014).

For odontocete movements, comparative literature is limited: Shane et al. (1986), Connor et al. (2000), and Reynolds et al. (2000) focused exclusively on (coastal) bottlenose dolphins, whereas Hooker and Baird (2001) reviewed mostly three-dimensional diving behavior. More general overviews were provided by Stevick et al. (2002) and Stern and Friedlaender (2018). Stevick et al. (2002) describe the available methods and use humpback whales and pinnipeds as model organisms, whereas Stern and Friedlaender (2018) provide theoretical approaches to migrations and movement patterns of all cetaceans. In their review of the social structure and strategies of delphinids, Gowans et al. (2008) analyzed ranging patterns of *Tursiops*, *Stenella*, *Delphinus*, *Lagenorhynchus*, *Sousa*, *Orcinus*, and *Physeter* species. We attempt to capture recent advances in this swiftly developing research field.

6.3 Methods of Spatiotemporal Scales

While diurnal movements are easiest to observe directly from a boat, plane, or a cliff nearby, these observations usually end with waning daylight (Würsig et al. 1991). Diel (and nocturnal) observations generally require attachment of an instrument to the animal to measure and store or relay positions. Over the years, a multitude of tags have been used successfully in the marine realm, with ever longer and more sophisticated data acquisition. The study of migration over months and years is now possible and is being linked with sighting and re-sighting either of naturally (e.g., photo-ID) or artificially marked individuals (e.g., freeze-branding, colored spaghetti tags, etc.). Offshore (pelagic) populations are more difficult to observe and thus almost always require instrumentation with satellite tags (Read 2018).

Latest tags reduce underwater resistance and extend working life, thus increasing number of days with location data. Sample size and possible lack of representation of few samples to the group habits of all continue to pose concerns for the study of movement patterns (cf. Gendron et al. 2015).

Static acoustic monitoring is an ideal method to record presence—especially of rather secretive species—as it operates independently of weather condition and daylight hours. It does not provide information on individual movement patterns, however.

Besides the use of direct and indirect methods to study movement pattern, a few proxies for markings of individuals are indicators of movements: distribution of parasites, diseases, absorbed trace elements such as heavy metals, and accumulation of pollutants.

6.4 Movement Patterns, from Site Fidelity to Range Expansion

We summarize published knowledge on odontocete movements of functional groups, neither taxonomically nor ecologically defined: sperm whales, deep divers (*Ziphiidae* and *Globicephalinae*), killer whales (*Orcinus orca*), high-arctic *Monodontidae*, oceanic dolphins, bottlenose dolphins (*Tursiops* spp.), small coastal dolphins, and river dolphins. More details can be found in Table 6.1.

6.4.1 Sperm Whale (Physeter macrocephalus)

Sperm whales inhabit deep waters, and extensive study took place in the southern hemisphere, where 3558 individuals were marked between 1945 and 1979 with embedded "Discovery" tags (Brown 1981; Ivashin 1981). Several males moved over 5500 km (measured in a straight line) over several years, and females usually moved

120   S. Bräger and Z. Bräger

Table 6.1 Summarized details on the movement patterns of odontocetes using displacement distance, site fidelity, seasonality, transit speed and diel movements as key parameters

Species	Region	Study duration	Study method	Number of re-sighted individuals	Mean displacement distance	Maximum distance
Physeter macrocephalus	Worldwide	1945–1979	Discovery tags	93	A few hundred km	Males: >5500 km, females: 2700 km (after 15 years)
Physeter macrocephalus	Mediterranean Sea	1991–2005	Photo-identification	5 of 44	500 km	
Physeter macrocephalus	Mediterranean Sea	1999–2011	Photo-identification			1600–2100 km
Physeter macrocephalus	Norway– Azores	1993–2008	Photo-identification	3	4400 km	4400 km
Physeter macrocephalus	Eastern tropical Pacific	1985–2004	Photo-identification	322 females + immatures and 5 mature males	4 km after 1 h, 50 km after 1 day, 200 km after 3 day, 1000 km after 1 year	500–5000 km
Globicephala melas and *Mesoplodon densirostris*	Hawai'i	2006–2011	Satellite tagging	46 short-finned pilot whales and 12 Blainville's beaked whales	Mostly within 100–200 km	Approx. 600 km (*Globicephala*)
Feresa attenuata	Hawai'i	2008–2009	Satellite tagging	2		79–106 km
Feresa attenuata	Hawai'i	1985–2007	Photo-identification	112 of 250 (40 ind. re-sighted)	No movements among islands documented	
Ziphius cavirostris	California	2010–2012	Satellite tagging	8	56 km	450 km
Orcinus orca	Antarctica and New Zealand	2014–2015	Photo-identification	8		4660 km
Orcinus orca	Alaska	1984–2010	Photo-identification and satellite tagging (on four occasions)	88		
Orcinus orca	Baffin Island	2009	Satellite tagging	2		5400 km

6 Movement Patterns of Odontocetes Through Space and Time

Cumulative distance and home range (HR) size	Site fidelity or residency	Seasonality	Mean/max speed (displacement velocity)	Diel movements	Source
	Young males leave maternal groups for polar waters when 10.7–11.0 m long				Brown (1981), Ivashin (1981)
	Re-sighted over several summer seasons 1–7 years apart		500 km in 5 days		Drouot-Dulau and Gannier (2007)
	Social groups site-faithful in Hellenic Trench for 9–15 years				Frantzis et al. (2011, 2014), Carpinelli et al. (2014)
					Steiner et al. (2012)
Northern Chile to Sea of Cortez within up to 15 years		No indication found			Whitehead et al. (2008)
	High fidelity to depths of 1000–2500 m (*Globicephala*) and to 250–2000 m (*Mesoplodon*)			Deep scattering layer more inshore during nighttime (up to 2.5 km from shore)	Abecassis et al. (2015)
	Re-sightings over up to 21 years		2.7–3.1 km/h	Median distance from shore: 4–5 km	Baird et al. (2011)
	High degree of site fidelity: up to 12 separate years over up to 21.5 years	Resident; year-round use of the study area			McSweeney et al. (2009)
Four whales left the Southern California Bight, two of which returned	After an extra-regional excursion over 450 km away, the ind. returned to within 5 km of tagging location				Schorr et al. (2014)
	8 ind. re-sighted between years in McMurdo Sound with 11–42 km between re-sighting locations covering a convex polygon of 843 km^2				Eisert et al. (2015)
7107 km for 3 ind.	"AT1" pop. resident to Prince William Sound–Kenai Fjords only; "GOA" pop. ranges over up to 270,503 km^2 in 30 days		Mean: 97 km/day		Matkin et al. (2012)
		Migrating south in October	96–120 km/day; max.: 252 km/day avoiding ice formation		Matthews et al. (2011)

(continued)

Table 6.1 (continued)

Species	Region	Study duration	Study method	Number of re-sighted individuals	Mean displacement distance	Maximum distance
Orcinus orca	Alaska	2006–2014	Satellite tagging	37		
Orcinus orca	Prince Edward Islands		Satellite tagging	9		1333 km
Delphinapterus leucas	Alaska	2002–2011	Satellite tagging	31		
Delphinapterus leucas	Beaufort Sea and Chukchi Sea	1993–2007	Satellite tagging	64		1103 km
Delphinapterus leucas	Beaufort Sea	1993–1997	Satellite tagging	30		Approx. 2000 km
Monodon monoceros	Somerset Island, Canada	2000–2001	Satellite tagging	16 females		Approx. 1500 km
Delphinus delphis	Adriatic and Ionian Sea	2008–2011	Photo-identification	1		>1000 km
Delphinus delphis	Australia	2007–2014	Photo-identification	13		
Stenella frontalis	Texas	1995	Satellite tagging	1 rehabilitated		300 km
Grampus griseus	Wales, UK	1997–2007	Photo-identification	183	2–17 km	319 km
Grampus griseus	Azores	2004–2007	Photo-identification	42 of 1250		
Grampus griseus	Mediterranean Sea	1989–2012	Photo-identification	82	33 km	332 km
Grampus griseus	Florida	2006	Satellite tagging and freeze-branding	1 rehabilitated		
Tursiops truncatus	Greece	1993–2010	Photo-identification	9	187 km	265 km
Tursiops spp.	Australia	1996–1997	Satellite tagging	2	43 km	146 km
Tursiops truncatus ponticus	Russia	2004–2014	Photo-identification	91		325 km (2 ind.)
Tursiops truncatus	Bermuda (Atlantic)	2003	Satellite-telemetry	3		

6 Movement Patterns of Odontocetes Through Space and Time

Cumulative distance and home range (HR) size	Site fidelity or residency	Seasonality	Mean/max speed (displacement velocity)	Diel movements	Source
		Seasonal shift in core areas following salmon	106 km/day		Olsen et al. (2018)
416–4470 km			Mean: 83 km/ day		Reisinger et al. (2015)
	Frequent switching of bays between years	Mean distance to river mouths in winter: 46 km, in salmon season: 9 km			Citta et al. (2016)
Summer HR: 7000–24,000 km^2 (Chukchi) and 17,000–36,000 km^2 (Beaufort)		Seasonal migration with shifting home ranges	Mean: 52–57 km/day; max.: 71.2 km/ day		Hauser et al. (2014)
In July, many males moved clock-wise >400 km north into permanent pack ice		Migrate westward in September (through Bering Strait in November)	4.2–6.4 km/h		Richard et al. (2001)
95% kernel summering ground = 9500 km^2, wintering = 25,800 km^2	High degree of site fidelity to summering and wintering grounds	Distinct focal areas in summer, winter and spring			Heide-Jørgensen et al. (2003)
					Genov et al. (2012)
	7 ind. sighted over all 8 years, 5 over 7 years, and 1 over 3; mean dist. from coast: 2 km (max.: 9.3 km)	12 ind. sighted in all seasons			Mason et al. (2016)
1711 km in 24 days			Mean: 72 km/ day; max.: 5.69 m/s (=20 km/h)		Davis et al. (1996)
257 km	Long-term and seasonal site fidelity				De Boer et al. (2013)
HR: 172–212 km^2 for males, 151–195 km^2 for females (95% UD and MCP, respectively)	Males showed the highest degree of site fidelity				Hartman et al. (2015)
	Over up to 18 years				Labach et al. (2015)
3300 km in 23 days			Mean: 7.2 km/h		Wells et al. (2009)
(Repeated) return trips (2 ind.)	Up to 15 years (1 ind.)				Bearzi et al. (2011)
4-mo HR: 95% Kernel = 778 km^2; 50% Kernel = 86 km^2			102 km in 2 days		Corkeron and Martin (2004)
			135 km in 3 days		Gladilina et al. (2016)
			Mean: 28.3 km/ day ($\sim$8 km/h); max.: 98.3 km/ day	40% of night >100 m depth; <2% of day >100 m depth	Klatsky et al. (2007)

(continued)

124 S. Bräger and Z. Bräger

Table 6.1 (continued)

Species	Region	Study duration	Study method	Number of re-sighted individuals	Mean displacement distance	Maximum distance
Tursiops truncatus	St. Paul (Atlantic)	2011–2013	Line-transect surveys and photo-identification	19		
Tursiops cf. *australis*	Australia	2013–2015	Photo-identification	125	0.7–4.7 km/day	
Tursiops truncatus	Croatia	2005–2014	Photo-identification	44		
Tursiops truncatus	Ireland and UK	2001–2010	Photo-identification	8	870 km	1277 km (1 ind.)
Tursiops aduncus	Australia	2007–2013	Photo-identification	56		
Tursiops truncatus	Argentina	2007–2013	Photo-identification	47	200 km (20 ind.)	290 km (2 ind.)
Tursiops truncatus	California	1979–1988	Photo-identification	11		
Tursiops truncatus	Florida	1997	Satellite tagging and freeze-branding	2 rehabilitated		
Tursiops truncatus	Scotland	1990–1993	Photo-identification			220–400 km
Sousa sahulensis	Australia	2012–2014	Photo-identification			Low levels of gene flow over >200 km
Sousa sahulensis	Australia	2004–2007	Photo-identification	98	Exchange between northern and southern are possible	
Sousa plumbea	South Africa	2000–2016	Photo-identification	61 over multiple sites	120 km	500 km (1 ind.)
Sousa chinensis	Taiwan and China	2007–2014	Photo-identification	71 + 61, respectively	No re-sightings across Taiwan Strait (130–400 km)	
Sotalia guianensis	Paraguacu River, Brazil	2005–2007	Photo-identification	14		
Sotalia guianensis	Ilha Grande Bay, Brazil	2007–2008	Photo-identification	295		
Sotalia guianensis	Cananeia Estuary, Brazil	2000–2010	Photo-identification	31 (identified 20–44 times)		

6 Movement Patterns of Odontocetes Through Space and Time

Cumulative distance and home range (HR) size	Site fidelity or residency	Seasonality	Mean/max speed (displacement velocity)	Diel movements	Source
Population HR size: MCP = 0.5 km^2; 50% Kernel = 0.13 km^2; 95% Kernel = 0.99 km^2	Strong site fidelity; 5 ind. re-sighted after 9.25 years				Milmann et al. (2017)
Mean HR size (112 ind.): 95% Kernel = 15.2 km^2	46% (82 ind.) were seen in all seasons and 71% in all years; 67% (119 ind.) were also recorded in 2010				Passadore et al. (2017)
Mean HR size: MCP = 406 km^2; 50% KDE = 241 km^2; 95% KDE = 1294 km^2		Winter HR generally smaller than summer HR			Rako-Gospić et al. (2017)
					Robinson et al. (2012)
Mean HR size (22 males + 34 females): 95% UD = 94.8 km^2 for males and 65.6 km^2 for females					Sprogis et al. (2016)
Repeated return trips	52 ind. re-sighted annually	Highest re-sighting rate in winter	200 km in 8 days		Vermeulen et al. (2017)
1340 km round trip			670 km in 74 days and 66 days		Wells et al. (1990)
2050 km in 43 days; 4200 km in 47 days			89 km/day		Wells et al. (1999)
		Numbers increase in summer in inner Moray Firth	218 km in 2 days; 7.6 km/h		Wilson et al. (2004)
	Strong site fidelity				Brown et al. (2016)
Population HR size: 95% Kernel = 148–325-km^2 for southern and northern population, respectively	Strong site fidelity	41% (44 ind.) sighted every season; average sighting rate = 3 mo			Cagnazzi et al. (2011)
	Over 18 years in same area (3 ind.); 25% (61 ind.) re-sighted in different areas		45 km in 1 day (1 ind.); 135 km in 24 days (mother + calf)		Vermeulen et al. (2017)
					Wang et al. (2016)
Mean HR size: MCP = 3.88 km^2 (range = 1.0–8.8 km^2)	At least 5–6 years (2 ind. each)				Batista et al. (2014)
	21% (99 ind.) with high degree of residence				Espécie et al. (2010)
Mean HR size: MCP = 17.5 km^2; 50% Kernel = 4–14 km^2; 95% Kernel = 14–72 km^2					Oshima and Santos (2016)

(continued)

Table 6.1 (continued)

Species	Region	Study duration	Study method	Number of re-sighted individuals	Mean displacement distance	Maximum distance
Cephalorhynchus hectori	New Zealand	1993–1997	Photo-identification	143	4–18 km between sightings; 51.7% ≤5 km; restricted by Kaikoura Canyon	61.4 km
Cephalorhynchus hectori	New Zealand	2010–2012	Genetic analysis	6		≥400 km inter-island
Cephalorhynchus commersonii	Argentina	1999–2001	Photo-identification			250 km
Cephalorhynchus heavisidii	South Africa	1997	Satellite tagging and freeze-branding	3		
Cephalorhynchus heavisidii	South Africa	2004	Satellite tagging	6	36.6 km	
Cephalorhynchus eutropia	Chile	2001–2004	Photo-identification	42	32.1 km alongshore	45 km (70 km, 1 ind.); no movement among study areas
Phocoena phocoena	Baltic Sea	2011–2013	Static Acoustic Monitoring	297 hydrophone stations		
Phocoena phocoena	Salish Sea, Washington	2014–2017	Photo-identification	19		
Phocoena phocoena	North Atlantic, Greenland	2012–2017	Satellite tagging	30		>2000 km
Phocoena phocoena	Baltic Sea, Germany	2006–2007	Static Acoustic Monitoring	5 hydrophone stations		
Phocoena phocoena	San Francisco Bay, California	2011–2014	Surveys from bridge			
Phocoena phocoena	Baltic and North Seas, Denmark	1997–2007	Satellite tagging	64		Approx. 1000 km
Neophocaena phocaenoides	Japan	2007–2009	Passive acoustic monitoring	226 detections		
Inia geoffrensis	Brazil	2000–2003	Freeze-branding	161 of 235		

Cumulative distance and home range (HR) size	Site fidelity or residency	Seasonality	Mean/max speed (displacement velocity)	Diel movements	Source
Rare exchange of individuals across deep Kaikoura Canyon recorded	Revisiting places after 4–5 days		Mean: 0.1–4.0 km/day; max: 46.8 km in 30 h (and briefly 9.6 km/h)		Bräger and Bräger (2018)
					Hamner et al. (2014)
	On average, 15 days in study area and 73 days outside before returning		250 km in 5 days		Coscarella et al. (2011)
Mean HR size: MCP = 1027–2347 km²; 50% Kernel = 123–230 km²	3 years later within 13 km of tagging site				Davis et al. (2014)
HR: MCP = 1000–2500 km²; 90% LCH = 300–1030 km²			25 km in 8 h	Close to shore morning till noon, moving offshore afternoon till night	Elwen et al. (2006)
Mean HR size: 50% UD = 8.4 km²; 95% UD = 35.6 km²	Tentative year-round residents				Heinrich (2006)
		Seasonal redistributions: May–Oct and Nov–April			Carlén et al. (2018), Benke et al. (2014)
	15% (8 ind.) in >1 year	Presence lower in summer			Elliser et al. (2018)
Total combined MCP area = 4,145,000 km²	6 of 15 ind. returned to tag site after 490 days	50% (15 ind.) moved offshore toward S or SE thus avoiding ice cover	36 km/day (March–June) (June = 53 km/day); 22 km/day (July–Feb)		Nielsen et al. (2018)
		Seasonal changes in activity and feeding peaks		Four stations showed nocturnal or crepuscular peaks in feeding	Schaffeld et al. (2016)
Re-population after 65 years of absence		High sighting rates in Sep–Dec + Feb–Mar, lowest SPUE May–June			Stern et al. (2017)
Population-wide HR: 50% Kernel = approx. 10–15,000 km²		Significant seasonal movements in both populations	0.85 km/h		Sveegaard et al. (2011)
		73% of detections in March + April		76% of detections at night	Akamatsu et al. (2010)
	49% (79 ind.) "permanent residents"	Seasonal population-level movements in and out of the lake system depending on water level			Martin and da Silva (2004)

shorter distances. Females and young stayed in warm and temperate waters where they formed stable groups and only occasionally performed local migrations up to 3700 km. Young males leave maternal groups to migrate toward polar waters (e.g., 61°–65° S) when less than 12 m long. There are frequently seasonal migrations toward polar waters during summer. In the subequatorial Pacific, several stable populations occur, with mark-recapture distances of only 500–1300 km (Ivashin 1981) (Fig. 6.1).

Commercial whaling provided evidence for long-distance movements between Nova Scotia and Spain, through the recovery of traditional harpoons and other tags (Aguilar and Sanpera 1982). Photo-identification studies in Norway and the Azores confirmed matches 4400 km apart (Steiner et al. 2012). Acoustic monitoring off the North American East Coast showed distinct seasonal patterns in prevalence of sperm whale clicks (Wong and Whitehead 2014; Stanistreet et al. 2018). A migration toward higher latitudes for summer mirrored the one in the southern hemisphere.

In isolated sperm whale populations of smaller basins, movement patterns differ from those found in open-ocean populations. Extensive photo-identification studies took place in the Mediterranean Sea, Gulf of Mexico, and eastern Caribbean Sea (Drouot-Dulau and Gannier 2007; Gero et al. 2007; Frantzis et al. 2011; Carpinelli et al. 2014). Comparisons of mitochondrial DNA and other genetic markers across global and North Atlantic populations (including in the Mediterranean Sea and the Gulf of Mexico) indicated a site fidelity of females to coastal basins, whereas males appear to move among populations for breeding (Engelhaupt et al. 2009).

Social groups of female and immature sperm whales remain in tropical waters where they also perform movements over large spatial and temporal scales. In the

Fig. 6.1 A sperm whale, *Physeter macrocephalus*, in the Azores lifts its fluke high before an almost vertical descent, indicating a deep dive to forage. (Photo by Zsuzsanna Bräger)

6 Movement Patterns of Odontocetes Through Space and Time

Eastern Pacific (including waters between Chile, Galapagos, and Mexico), groups tracked by photo-identification moved frequently up to 2000 km and occasionally up to 4000 km, within 15 years (Whitehead et al. 2008). Average displacements were about 4 km after 1 h of movement, 50 km after 1 day, 200 km after 3 days, and 1000 km after 1 year or more. A home range of 2000 km may be what an individual or a social unit can keep track of, and such movements may lead to genetic homogenization within ocean basins (Whitehead et al. 2008).

6.4.2 Deep Divers (Ziphiidae and Globicephalinae)

Most information on movement patterns of deep-diving *Ziphiidae* and *Globicephalinae* has been collected from island-associated populations. Due to upwelling regimes of nutrient-rich waters, offshore islands can provide unusually abundant and predictable prey resources that may not be representative of the foraging habitats of more pelagic populations. Off the western side of Hawai'i Island, Blainville's beaked whales (*Mesoplodon densirostris*) displayed high site fidelity to the narrow strip between the 250 m and the 2000 m isobaths (Abecassis et al. 2015). Similar to other cetacean species in the region, these beaked whales appeared to forage on an island-associated deep mesopelagic boundary community of the deep scattering layer (DSL). Blainville's beaked whales off El Hierro, Canary Islands, showed the same foraging behavior (Arranz et al. 2011). Adult females of Blainville's and Cuvier's beaked whales (*Ziphius cavirostris*) also showed long-term site fidelity with repeated re-sightings of individuals along the west coast of Hawai'i. However, Cuvier's beaked whales off San Clemente Island, California, undertook excursions of 250–450 km within 10–90 days (Schorr et al. 2014; see also Chap. 14).

Similarly, short-finned pilot whales (*Globicephala macrorhynchus*) showed a strong site fidelity and appeared to forage on predators associated with the DSL (Abecassis et al. 2015), and pygmy killer whales (*Feresa attenuata*) of the Hawaiian islands were re-sighted over 22 years (McSweeney et al. 2009). Two satellite-tagged pygmy killer whales confirmed this strong site fidelity (Baird et al. 2011; Chap. 14).

6.4.3 Killer Whales (Orcinus orca)

Killer whales are a sexually dimorphic almost cosmopolitan species with sympatric ecotypes, and the species' taxonomy is currently under revision, likely to split into multiple species (Morin et al. 2010; Chap. 11). The piscivorous ecotype feeds on salmon (*Salmonidae*), Pacific halibut (*Hippoglossus stenolepis*), Atlantic herring (*Clupea harengus*), bluefin tuna (*Thunnus thynnus*), or Patagonian toothfish (*Dissostichus eleginoides*), whereas bird and mammal hunters prefer seals, cetaceans, and penguins. A third ecotype lives offshore and feeds predominantly on *Chondrichthyes* (sharks and rays). To satisfy high energy requirements, mammal-eating killer whales generally roam considerably farther (often termed "transients") than members of fish-eating populations (often termed "residents").

Fish-eating residents of the NE Pacific have stable summer home ranges and are exposed to anthropogenic impacts such as overfishing or pollution (Olsen et al. 2018). Piscivorous populations tend to be in synchrony with their migratory prey in Alaska, Canada, Iceland, Norway, and Spain, covering linear distances of 1000–4500 km (with transit speeds of 50–100 km/day) in the off-season. Mammal-eating populations cover similar distances (800–5400 km), albeit more continuously and thus in much larger home ranges of up to 270,000 km^2 (Matkin et al. 2012). Members of some polar populations undertake rapid excursions over thousands of kilometers (with up to 252 km/day) into subtropical waters. These excursions may be physiological maintenance migrations, as described by Matthews et al. (2011), Eisert et al. (2015), and Reisinger et al. (2015).

6.4.4 High-Arctic Odontocetes (Monodontidae)

Knowledge of movement patterns of beluga and narwhal (the Monodontidae) is largely based on tagging studies with satellite transmitters, as they live in the inhospitable climate of the high Arctic and lack dorsal fins with individually identifying marks.

During the seasonal salmon run, satellite-tagged belugas (*Delphinapterus leucas*) in Bristol Bay, Alaska, stayed near river entrances and dispersed once the run ended. However, even during winter, when they were farthest away, they were on average only 46.0 km from the nearest river mouth (Citta et al. 2016). Belugas in the Sea of Okhotsk also concentrated at river mouths feeding on anadromous eulachon (*Thaleichthys pacificus*; Solovyev et al. 2015).

In contrast, belugas from the eastern Beaufort Sea ranged much farther, e.g., over hundreds of kilometers either northward into dense pack ice or southwestward to Wrangel Island, Russia, or to the Bering Sea by late November (Richard et al. 2001). Individuals from the eastern Beaufort Sea and eastern Chukchi Sea performed extensive seasonal migrations and displayed distinct population differences in the east-west shifts of their monthly home ranges. Largest daily displacements occurred in September and October during westward fall migration. Individual summer core areas of both populations were 31,000–53,000 km^2 (Hauser et al. 2014).

Narwhals (*Monodon monoceros*) in the Canadian High Arctic had high site fidelity to summering grounds and followed specific migratory routes through sea ice formation and recession (Heide-Jørgensen et al. 2003).

6.4.5 Oceanic Dolphins (Stenella, Grampus, Delphinus)

Oceanic dolphins are frequently difficult to access, and some species are poorly marked and extremely gregarious (e.g., Hupman et al. 2018). Therefore, most information is derived from seasonal changes in species distribution [e.g., of striped dolphins (*Stenella coeruleoalba*) in the Western Mediterranean Sea or of common dolphins (*Delphinus delphis*) off South Africa (Arcangeli et al. 2017; Caputo et al. 2017, respectively)] or from single individuals under special circumstances such as a rehabilitated Atlantic spotted dolphin (*Stenella frontalis*) released and tracked with a satellite transmitter over 1711 km in 24 days (Davis et al. 1996) and a lone common dolphin photo-identified over 1000 km across the Adriatic Sea (Genov et al. 2012) (Figs. 6.2 and 6.3). Similarly, a small group of common dolphins exhibited atypical residency in a small bay in SE Australia over many years (Mason et al. 2016). It remains uncertain, however, whether these unusual results can be generalized.

Off Wales, Risso's dolphins (*Grampus griseus*) showed long-term and seasonal site fidelity and had long-distance re-sightings (de Boer et al. 2013), whereas they showed site fidelity in relatively restricted home ranges in the Azores (Hartman et al. 2015). During long-term photo-identification programs in the NW Mediterranean Sea (1989–2012), individuals covered distances up to 132 km in 39 days, but only an average of 59 km between years, and only two individuals covered 332 km (Labach et al. 2015). Wells et al. (2009) satellite-tagged a rehabilitated male and tracked it over 3300 km in 23 days, which shows the ability of this species to cover long distances in short periods.

Seasonal common dolphin movements in New Zealand correlate with a minimum sea surface temperature (SST) of 14 °C and thus commonly occur there in summer (Gaskin 1968; Neumann 2001; Stockin et al. 2008). Along the South Island,

Fig. 6.2 A striped dolphin, *Stenella coeruleoalba*, in the Greek Mediterranean Sea leaping as part of exuberant high-speed swimming. (Photo by Stefan Bräger)

Fig. 6.3 A common dolphin, *Delphinus delphis*, in the Azores, leaping beside a research vessel, in a common display of social activity for this species. (Photo by Zsuzsanna Bräger)

common dolphins occurred at SST >16 °C when another predator, the albacore tuna (*Thunnus alalunga*), also migrated southward (Bräger and Schneider 1998).

6.4.6 Bottlenose Dolphins (Tursiops *spp.*)

Bottlenose dolphins are at least two species (*Tursiops truncatus* and *T. aduncus*) and have coastal and offshore ecotypes (e.g., Tezanos-Pinto et al. 2009; Lowther-Thieleking et al. 2015; Oudejans et al. 2015) with diverse movement patterns. We treat them here as one general entity.

One of the variables to influence bottlenose dolphin movements appears to be the habitat they occupy in accordance with availability of prey. Long stretches of open (exposed) coast with little shallow water (<10 m) such as along the Mexican and US West Coast tend to facilitate longer and faster movements of up to >1000 km and transit speeds >50 km/day. There is little evidence for site fidelity or (seasonal) home range. This contrasts with the emarginated US coastline along the Gulf of Mexico that is rich in large, shallow-water bays. Each of these bay systems hosts its own dolphin population (Würsig 2017) and displays a high degree of site fidelity—frequently with diel movements and seasonal range shifts. Only rarely have individuals been re-sighted in another bay system some 140–800 km away, with most home range sizes for individuals approximately 50–150 km^2 and with cores of 10–30 km^2. Such small home range estimates are supported by similar results from bay systems along the US Atlantic coast. Both ocean basins (NW Atlantic and northern GoM), however, are also home to larger, more mobile populations with near-shore distribution. These populations appear to be genetically distinct (Rosel et al. 2009; Toth et al. 2012; Richards et al. 2013), although they mingle at times with the more sedentary ones. Many of the Gulf of Mexico studies thus contain a majority of transients in their ID-catalogues that are never re-sighted.

Other bottlenose dolphin populations, e.g., along coasts of the SW Atlantic Ocean, Mediterranean Sea, and Australia and New Zealand, are between these extremes, with regular maximum displacement distances of 130–400 (exceptionally up to 710) km (e.g., Bearzi et al. 2011; Gladilina et al. 2016; Vermeulen et al. 2017). Site fidelity usually occurs, but—depending on habitat—home ranges vary in size from about 10 to 1000 km^2 (with cores from 5 to 240 km^2), with the more sedentary populations in estuaries occupying small ranges and high site fidelity (e.g., Sprogis et al. 2016; Passadore et al. 2017; Rako-Gospić et al. 2017). Mean transit speeds usually vary between 10 and 50 km/day. The populations around the United Kingdom and Ireland are unusually mobile, frequently moving over 400–1280 km (sometimes at relatively high transit speeds; Robinson et al. 2012).

Movements of bottlenose dolphins around small (offshore) islands or archipelagos vary with extent of available coastline. The population around St. Paul, Brazil, for example, uses a home range of less than 1 km^2 (Milmann et al. 2017), whereas populations of larger islands may range over larger areas (comparable to mainland *Tursiops* mentioned above) and over longer distances (130–390 km), with low site fidelity.

Two rehabilitated satellite-tracked dolphins of the offshore ecotype separately travelled 2050 km in 43 days and 4200 km in 47 days after their release off Florida (Wells et al. 1999). Two individuals satellite-tracked off Queensland, Australia remained in shallow waters (mostly <50 m water depth) for the duration of the tags' transmissions (up to 143 days) and may thus not have been truly pelagic (Corkeron and Martin 2004). Some island-based offshore populations (Klatsky et al. 2007; Milmann et al. 2017) may be using deeper waters for foraging but perhaps rely on shallow waters nearshore for daytime rest.

6.4.7 Small Coastal Dolphins (Sousa, Sotalia, Cephalorhynchus)

Movements of humpback dolphins (*Sousa* sp.) were investigated in Asia, Australia, and Africa. *Sousa chinensis* appears to have the smallest home ranges of about 100 km^2, with year-round presence but some seasonal distribution shifts. The width of the Taiwan Strait (130–400 km) does not appear to be crossed by the populations on either side (Wang et al. 2016). Home ranges of *Sousa sahulensis* in Australia appear to be larger (148–325 km^2; Cagnazzi et al. 2011), but individuals nonetheless display strong site fidelity and a low level of gene flow over >200 km (Brown et al. 2016). South African *Sousa plumbea* are more mobile, moving up to 500 km (or 45 km/day), even though some individuals have been re-sighted in the same area for up to 18 years (Vermeulen et al. 2018).

Guiana dolphins (*Sotalia guianensis*) in Brazilian estuaries show a high degree of residence and have been re-sighted for 5–11 years. Their home range size depends on habitat, ranging from 4 to 18 km^2 with high site fidelity in enclosed estuaries (Batista et al. 2014; Oshima and Santos 2016), to over 60% transients (sighted ≤2 times) in an open bay (Espécie et al. 2010).

The four *Cephalorhynchus* species show a gradient of increasing mobility with increasing "openness" of coastal habitat (Bräger and Bräger 2018). According to photo-id and radiotelemetry studies, Chilean and Hector's dolphins (*C. eutropia* and *C. hectori*) live in rather small home ranges of <50 km^2 (mean 95% utilization distribution) that may extend 32–67 km along shore (Heinrich 2006). Along the northeastern coast of the South Island of New Zealand, a deepwater canyon comes close to shore, creating an environmental barrier that severely limits Hector's dolphin movements so that that northern and southern subpopulation are genetically differentiated (Bräger and Bräger 2018 and references therein). Avoiding deep water, however, has not prevented several individuals from the West Coast of the South Island to show up at North Island shores after a journey of about 400 km (Hamner et al. 2014) (Fig. 6.4).

Fig. 6.4 Hector's dolphin, *Cephalorhynchus hectori*, female and calf off Kaikoura, New Zealand. These small animals are usually shy delphinids but will surface next to vessels that approach slowly and cautiously. (Photo by Bernd Würsig)

The other two *Cephalorhynchus* species, Commerson's and Heaviside's dolphins (*C. commersonii* in South America and *C. heavisidii* in Southern Africa) appear to lead more mobile and pelagic lifestyles. *C. commersonii* off Argentina repeatedly occurred in areas 250 km apart (covering the distance in as little as 5 days) and, on average, spent only 15 days in the main study area and 73 days outside before returning (Coscarella et al. 2011). Three and six individuals of *C. heavisidii* satellite-tagged off western South Africa (Davis et al. 2014; Elwen et al. 2006, respectively) had home range sizes of 1000–2500 km^2. These home ranges included diel onshore-offshore movements along a stretch of coastline of about 37 km, presumably to feed on prey rising closer to the surface at night.

6.4.8 Porpoises (Phocoena, Neophocaena, Phocoenoides)

Porpoises are difficult to photo-identify, albeit not impossible, as shown for harbor porpoises (*Phocoena phocoena*) along the Pacific coast of the USA (Stern et al. 2017; Elliser et al. 2018). A Danish long-term satellite-telemetry study (Sveegaard et al. 2011) and a multinational static acoustic monitoring of the Baltic Sea harbor porpoise population (Benke et al. 2014; Carlén et al. 2018) provided insights into year-round movements and home range use. A year-round diel pattern of increased acoustic activity at night (or dawn and dusk) is very likely to correlate with short movements to preferred feeding areas (Schaffeld et al. 2016).

Although different harbor porpoise populations differ in extent of movements, almost all studies show a strong seasonality with likely seasonal migrations. In acoustic monitoring and line-transect surveys, such movements may only show up as changes in seasonal densities, whereas satellite-tracked porpoises in Denmark and Greenland moved up to 1000 km or >2000 km, respectively, correlating with prey availability and/or ice cover (Sveegaard et al. 2011; Nielsen et al. 2018). In extreme cases, Greenland porpoises crossed halfway across the North Atlantic—with a total combined range of >4 million km^2—returning with maximum transit speeds of 53 km/day (Nielsen et al. 2018).

Two studies also showed strong seasonality and diel patterns for finless porpoises (*Neophocaena phocaenoides*) in Japan and Dall's porpoises (*Phocoenoides dalli*) in Alaska. Japanese finless porpoises in Hario Strait, Nagasaki, cumulated in March and April (73%) and at night (76%), due to prey migration (Akamatsu et al. 2010). Alaskan surveys showed a clear seasonal distribution of Dall's porpoises in Prince William Sound apparently linked to the occurrence of overwintering and spawning herring (*Clupea pallasii*) as well as to predation risk by killer whales (Moran et al. 2018).

6.4.9 River Dolphins (Inia and Platanista)

Brazilian Amazon river dolphins (*Inia geoffrensis*) are site-faithful but show seasonal movements in and out of a study area in accordance with water level changes, i.e., an annual exodus out of a lake system with falling water level (Martin and da Silva 2004; Mintzer et al. 2016). South Asian river dolphins (*Platanista gangetica*) from India mostly moved at night and were more resident during the day (Sasaki-Yamamoto et al. 2013).

The literature cited for this section contains far more information than summarized here, and again we refer the reader to Table 6.1 for some (but not all) increased detail.

6.5 General Movements of Odontocetes

Odontocete movements frequently come in several spatial and temporal scales: Diel movements over kilometers within a day, roaming over tens of kilometers over weeks (to months) within a (seasonal) home range, and migration-like movements between (seasonal) ranges over hundreds to thousands of kilometers over weeks to months. These patterns, however, may be heavily influenced by our better knowledge of life histories of nearshore species. Offshore populations, especially with tropical distributions, may have no seasonally different ranges.

Prey availability is a central driver for odontocete movement patterns, but there is no simple relationship between body mass and home range size as in terrestrial mammals, as demonstrated (for example) in the wide dispersal of North Atlantic harbor porpoises (Nielsen et al. 2018). Among some coastal species including *Tursiops* and *Cephalorhynchus*, general mobility appears to correlate with "openness" of the coastline (cf. Shane et al. 1986). Populations along fractal coastlines with bays and frequently extensive shallow-water areas occupy smaller ranges than those along straight stretches of exposed coast.

Occasional to rare dispersal-like long-distance movements by individuals over hundreds of kilometers, e.g., to other populations, may sometimes not be reversed, possibly leading to permanent exchange. There are also several photo-ID studies that documented individuals switching—more or less frequently—their home ranges after a few years (e.g., *Tursiops* in NW Greece, around Scotland and Ireland, along the Texan coastline and along the Argentine coastline, and probably others). These irregular movements may be most difficult to understand as their drivers are unknown and possibly of a social nature.

Range expansions of odontocete populations have been documented in a limited number of cases: *Tursiops* moving northward into Northern California and possibly into British Columbia (Wells et al. 1990; Halpin et al. 2018), southward into eastern England (Wilson et al. 2004), and into southern Chile (Olavarría et al. 2010) and recently *Phocoena* re-colonizing San Francisco Bay (Stern et al. 2017). However, not all extralimital observations announce an imminent population expansion. Most extralimital observations are only of individual vagrants that—for one reason or another—have come off course and ended up "in the wrong ocean basin." A gray whale (*Eschrichtius robustus*) in the Mediterranean Sea in 2010 provided a spectacular example (Scheinin et al. 2011).

6.6 Factors Impacting Movement Patterns Directly or Indirectly

Movement patterns of odontocetes adapt to environmental conditions. The occurences of prey and predators are assumed to be the strongest natural drivers for habitat selection and movements (e.g., Corkeron and Connor 1999; Heithaus and Dill 2006; Ward et al. 2016; Sprogis et al. 2018). Their occurrence, however, is also influenced by numerous natural and anthropogenic factors. Fishing and aquaculture; shipping; construction; chemical, oil, noise and other pollution; climate change; tourism; and conservation measures change cetacean behaviors and odontocete movement patterns (Pirotta et al. 2015; see also Chap. 10).

As far as predation, it is not only the actual attack and consumption mostly by sharks, seals and killer whales but the likelihood of potential predation events that can instill fear and may lead to changes in movement behavior (Wirsing et al. 2008; Chap. 7). Tides, currents, and impeding ice cover are likely to impact movements, as are stochastic events such as hurricanes or flooding. Health aspects such as the need for physiological maintenance (Durban and Pitman 2012; Reisinger et al. 2015) can alter movement patterns.

There is also a suite of social factors that likely influences movement patterns, including group size (Gygax 2002; Owen et al. 2002), sexual segregation (Martin and da Silva 2004; Webster et al. 2009; Fury et al. 2013; Morteo et al. 2014; Hartman et al. 2015; Sprogis et al. 2016; Louis et al. 2018), relatedness (Lewis et al. 2013; Louis et al. 2018), immatures becoming independent (Krzyszczyk et al. 2017), and decision-making or consensus-building (Whitehead 2016; see Chap. 5). Furthermore, reproductive needs may play a more important role in shaping movement patterns than identified so far. These factors together with navigational abilities and geographic memory may constitute some of the social and cultural roots of brain development in odontocetes (see also Connor 2007; Fox et al. 2017; Chap. 16).

The current knowledge of the ethology of migration and movement patterns among odontocetes raises numerous questions that may be fundamental to other social and ecological studies, for example:

(a) Beyond general explanations such as prey availability, predator avoidance, or reproductive needs, do overarching patterns exist that help to explain the ethology of migration and movement patterns, perhaps akin to bird migration?
(b) Are migratory behaviors (such as timing, orientation, etc.) inherited or passed on through traditions? How important is the local knowledge of older group members for orientation?
(c) How can we imagine home ranges in truly pelagic odontocetes? Can we learn from other pelagic megafauna such as sharks, turtles, and seabirds?
(d) What is going to happen to movement patterns in times of progressing climate change? Are migratory species more resilient to habitat loss than those more stationary?

Answers to these and similar questions will help to expand our knowledge of odontocete movements.

References

Abecassis M, Polovina J, Baird RW, Copeland A, Drazen JC, Domokos R, Oleson E, Jia Y, Schorr GS, Webster DL, Andrews RD (2015) Characterizing a foraging hotspot for short-finned pilot whales and Blainville's beaked whales located off the west side of Hawai'i Island by using tagging and oceanographic data. PLoS One 10(11):e0142628

Aguilar A, Sanpera C (1982) Reanalysis of Spanish sperm, fin and sei whale catch data (1957–1980). Rep IWC 32:465–470

Akamatsu T, Nakamura K, Kawabe R, Furukawa S, Murata H, Kawakubo A, Komaba M (2010) Seasonal and diurnal presence of finless porpoises at a corridor to the ocean from their habitat. Mar Biol 157:1879–1887

Arcangeli A, Campana I, Bologna MA (2017) Influence of seasonality on cetacean diversity, abundance, distribution and habitat use in the western Mediterranean Sea: implications for conservation. Aquat Conserv Mar Freshwat Ecosyst 27:995–1010

Arranz P, Aguilar de Soto N, Madsen PT, Brito A, Bordes F, Johnson MP (2011) Following a foraging fish-finder: diel habitat use of Blainville's beaked whales revealed by echolocation. PLoS One 6(12):e28353

Baird RW, Schorr GS, Webster DL, McSweeney DJ, Hanson MB, Andrews RD (2011) Movements of two satellite-tagged pygmy killer whales (*Feresa attenuata*) off the island of Hawai'i. Mar Mamm Sci 27(4):E332–E337

Batista RLG, Alvarez MR, dos Reis M d SS, Cremer MJ, Schiavetti A (2014) Site fidelity and habitat use of the Guiana dolphin, *Sotalia guianensis* (Cetacea: Delphinidae), in the estuary of the Paraguaçú River, northeastern Brazil. NW J Zool 10:93–100

Bearzi G, Bonizzoni S, Gonzalvo J (2011) Mid-distance movements of common bottlenose dolphins in the coastal waters of Greece. J Ethol 29:369–374

Benke H, Bräger S, Dähne M, Gallus A, Hansen S, Honnef CG, Jabbusch M, Koblitz JC, Krügel K, Liebschner A, Narberhaus I, Verfuß UK (2014) Baltic Sea harbour porpoise populations: status and conservation needs derived from recent survey results. Mar Ecol Prog Ser 495:275–290

Bräger S, Bräger Z (2018) Range utilization and movement patterns of the coastal Hector's dolphins (*Cephalorhynchus hectori*). Aquat Mamm 44:633–642

Bräger S, Schneider K (1998) Near-shore distribution and abundance of dolphins along the West Coast of the South Island, New Zealand. NZ J Mar Freshw Res 32:105–112

Brown SG (1981) Movements of marked sperm whales in the southern hemisphere. Rep Int Whal Comm 31:835–837

Brown AM, Bejder L, Pollock KH, Allen SJ (2016) Site-specific assessments of the abundance of three inshore dolphin species to inform conservation and management. Front Mar Sci 3:4

Cagnazzi DB, Harrison PL, Ross GJB, Lynch P (2011) Abundance and site fidelity of Indo-Pacific Humpback dolphins in the Great Sandy Strait, Queensland, Australia. Mar Mamm Sci 27 (2):255–281

Caputo M, Froneman PW, du Preez D, Thompson G, Plön S (2017) Long-term trends in cetacean occurrence during the annual sardine run off the Wild Coast, South Africa. Afr J Mar Sci 39:83–94

Carlén I, Thomas L, Carlström J, Amundin M, Teilmann J, Tregenza N, Tougaard J, Koblitz JC, Sveegaard S, Wennerberg D, Loisa O, Dähne M, Brundiers K, Kosecka M, Anker Kyhn L, Tiberi Ljungqvist C, Pawliczka I, Koza R, Arciszewski B, Galatius A, Jabbusch M, Laaksonlaita J, Niemi J, Lyytinen S, Gallus A, Benke H, Blankett P, Skóra KE, Acevedo-Gutiérrez A (2018) Basin-scale distribution of harbour porpoises in the Baltic Sea provides basis for effective conservation actions. Biol Conserv 226:42–53

Carpinelli E, Gauffier P, Verborgh P, Airoldi S, David L, Di-Meglio N, Cañadas A, Frantzis A, Rendell L, Lewis T, Mussi B, Pace DS, de Stephanis R (2014) Assessing sperm whale (*Physeter macrocephalus*) movements within the western Mediterranean Sea through photo-identification. Aquat Conserv Mar Freshwat Ecosyst 24(Suppl 1):23–30

Citta JJ, Quakenbush LT, Frost KJ, Lowry L, Hobbs RC, Aderman H (2016) Movements of beluga whales (*Delphinapterus leucas*) in Bristol Bay, Alaska. Mar Mamm Sci 32:1272–1298

Connor RC (2007) Dolphin social intelligence: complex alliance relationships in bottlenose dolphins and a consideration of selective environments for extreme brain size evolution in mammals. Philos Trans R Soc B 362:587–602

Connor RC, Wells RW, Mann J, Read AJ (2000) The bottlenose dolphin – social relationships in a fission-fusion society. In: Mann J, Connor RC, Tyack PL, Whitehead H (eds) Cetacean societies – field studies of dolphins and whales. University of Chicago Press, Chicago, pp 91–126

Corkeron PJ, Connor RC (1999) Why do baleen whales migrate? Mar Mamm Sci 15:1228–1245

Corkeron PJ, Martin AR (2004) Ranging and diving behaviour of two 'offshore' bottlenose dolphins, *Tursiops* sp., off eastern Australia. J Mar Biol Assoc UK 84:465–468

Coscarella MA, Gowans S, Pedraza SN, Crespo EA (2011) Influence of body size and ranging patterns on delphinid sociality: associations among Commerson's dolphins. J Mammal 92:544–551

Davis RW, Worthy GAJ, Würsig B, Lynn SK, Townsend FI (1996) Diving behavior and at-sea movements of an Atlantic spotted dolphin in the Gulf of Mexico. Mar Mamm Sci 12:569–581

Davis RW, David JHM, Meÿer MA, Sekiguchi K, Best PB, Dassis M, Rodriguez DH (2014) Home range and diving behaviour of Heaviside's dolphins monitored by satellite off the west coast of South Africa. Afr J Mar Sci 36:455–466

De Boer MN, Clark J, Leopold MF, Simmonds MP, Reijnders PJH (2013) Photo-identification methods reveal seasonal and long-term site-fidelity of Risso's dolphins (*Grampus griseus*) in shallow waters (Cardigan Bay, Wales). Open J Mar Sci 3:65–74

Drouot-Dulau V, Gannier A (2007) Movements of sperm whale in the western Mediterranean Sea: preliminary photo-identification results. J Mar Biol Assoc UK 87:195–200

Durban JW, Pitman RL (2012) Antarctic killer whales make rapid, round-trip movements to subtropical waters: evidence for physiological maintenance migrations? Biol Lett 8:274–277

Eisert R, Ovsyanikova E, Visser I, Ensor P, Currey R, Sharp B (2015) Seasonal site fidelity and movement of type-C killer whales between Antarctica and New Zealand. In: International Whaling Commission SC/66a/SM/9, pp 1–13

Elliser CR, MacIver KH, Green M (2018) Group characteristics, site fidelity, and photo-identification of harbor porpoises, *Phocoena phocoena*, in Burrows Pass, Fidalgo Island, Washington. Mar Mamm Sci 34:365–384

Elwen S, Meyer MA, Best PB, Kotze PGH, Thornton M, Swanson S (2006) Range and movements of female Heaviside's dolphins (*Cephalorhynchus heavisidii*), as determined by satellite-linked telemetry. J Mammal 87:866–877

Engelhaupt D, Hoelzel AR, Nicholson C, Frantzis A, Mesnick S, Gero S, Whitehead H, Rendell L, Miller P, de Stefanis R, Cañadas A, Airoldi S, Mignucci-Giannoni AA (2009) Female philopatry in coastal basins and male dispersion across the North Atlantic in a highly mobile marine species, the sperm whale (*Physeter macrocephalus*). Mol Ecol 18:4193–4205

Espécie MA, Tardin RHO, Simão SM (2010) Degrees of residence of Guiana dolphins (Sotalia guianensis) in Ilha Grande Bay, south-eastern Brazil: a preliminary assessment. J Mar Biol Assoc U K 90(08):1633–1639

Fox KCR, Muthukrishna M, Shultz S (2017) The social and cultural roots of whale and dolphin brains. Nat Ecol Evol 1:1699–1705

Frantzis A, Airoldi S, Notarbartolo-di-Sciara G, Johnson C, Mazzariol S (2011) Inter-basin movements of Mediterranean sperm whales provide insight into their population structure and conservation. Deep-Sea Res 58:454–459

Frantzis A, Alexiadou P, Gkikopoulou KC (2014) Sperm whale occurrence, site fidelity and population structure along the Hellenic Trench (Greece, Mediterranean Sea). Aquat Conserv Mar Freshwat Ecosyst 24(S1):83–102

Fury CA, Ruckstuhl KE, Harrison PL (2013) Spatial and social sexual segregation patterns in Indo-Pacific bottlenose dolphins (*Tursiops aduncus*). PLos One 8(1):e52987

Gaskin DE (1968) Distribution of Delphinidae (Cetacea) in relation to sea surface temperatures off eastern and southern New Zealand. NZ J Mar Freshwat Res 2:527–534

Gendron D, Martinez Serrano I, Ugalde de la Cruz A, Calambokidis J, Mate B (2015) Long-term individual sighting history database: an effective tool to monitor satellite tag effects on cetaceans. Endanger Species Res 26:235–241

Genov T, Bearzi G, Bonizzoni S, Tempesta M (2012) Long-distance movement of a lone short-beaked common dolphin *Delphinus delphis* in the central Mediterranean Sea. Mar Biodivers Rec 5:1–3

Gero S, Gordon J, Carlson C, Evans P, Whitehead H (2007) Population estimate and inter-island movement of sperm whales, *Physeter macrocephalus*, in the eastern Caribbean Sea. J Cetacean Res Manag 9:143–150

Gladilina E, Shpak O, Serbin V, Kryukova A, Glazov D, Gol'din P (2016) Individual movements between local coastal populations of bottlenose dolphins (*Tursiops truncatus*) in the northern and eastern Black Sea. J Mar Biol Assoc UK 98:223–229

Gowans S, Würsig B, Karczmarski L (2008) The social structure and strategies of delphinids: predictions based on an ecological framework. Adv Mar Biol 53:195–294

Gygax L (2002) Evolution of group size in the dolphins and porpoises: interspecific consistency of intraspecific patterns. Behav Ecol 13:583–590

Halpin LR, Towers JR, Ford JKB (2018) First record of common bottlenose dolphin (*Tursiops truncatus*) in Canadian Pacific waters. Mar Biodivers Rec 11:3

Hamner RM, Constantine R, Oremus M, Stanley M, Brown P, Baker CS (2014) Long-range movement by Hector's dolphins provides potential genetic enhancement for critically endangered Maui's dolphin. Mar Mamm Sci 30:139–153

Hartman KL, Fernandez M, Wittich A, Azevedo JMN (2015) Sex differences in residency patterns of Risso's dolphins (*Grampus griseus*) in the Azores: causes and management implications. Mar Mamm Sci 31:1153–1167

Hauser DDW, Laidre KL, Suydam RS, Richard PR (2014) Population-specific home ranges and migration timing of Pacific Arctic beluga whales (*Delphinapterus leucas*). Polar Biol 37:1171–1183

Heide-Jørgensen MP, Dietz R, Laidre KL, Richard P, Orr J, Schmidt HC (2003) The migratory behaviour of narwhals (*Monodon monoceros*). Can J Zool 81:1298–1305

Heinrich S (2006) Ecology of Chilean dolphins and Peale's dolphins at Isla Chiloe, southern Chile. PhD thesis, University of St. Andrews, UK, 258 p

Heithaus MR, Dill LM (2006) Does tiger shark predation risk influence foraging habitat use by bottlenose dolphins at multiple spatial scales? Oikos 114:257–264

Hooker SK, Baird RW (2001) Diving and ranging behaviour of odontocetes: a methodological review and critique. Mamm Rev 31:81–105

Hupman K, Stockin KA, Pollock K, Pawley MDM, Dwyer SL, Lea C, Tezanos-Pinto G (2018) Challenges of implementing mark-recapture studies on poorly marked gregarious delphinids. PLoS One 13(7):e0198167

Ivashin MV (1981) Some results of the marking of sperm whales (*Physeter macrocephalus*) in the southern hemisphere under the Soviet marking programme. Rep Int Whal Comm 31:707–718

Klatsky LJ, Wells RS, Sweeney JC (2007) Offshore bottlenose dolphins (*Tursiops truncatus*): movement and dive behavior near the Bermuda Pedestal. J Mammal 88:59–66

Krzyszczyk E, Patterson EM, Stanton MA, Mann J (2017) The transition to independence: sex differences in social and behavioural development of wild bottlenose dolphins. Anim Behav 129:43–59

Labach H, Dhermain F, Bompar J-M, Dupraz F, Couvat J, David L, Di-Méglio N (2015) Analysis of 23 years of Risso's dolphins photo-identification in the north-western Mediterranean Sea, first results on movements and site fidelity. Sci Rep Port-Cros National Park 29:263–266

Lambert E, Pierce GJ, Hall K, Brereton T, Dunn TE, Wall D, Jepson PD, Deaville R, Macleod CD (2014) Cetacean range and climate in the eastern North Atlantic: future predictions and implications for conservation. Glob Chang Biol 20:1782–1793

Lewis JS, Wartzok D, Heithaus M, Krützen M (2013) Could relatedness help explain why individuals lead in bottlenose dolphin groups? PLoS One 8(3):e58162

Louis M, Simon-Bouhet B, Viricel A, Lucas T, Gally F, Cherel Y, Guinet C (2018) Evaluating the influence of ecology, sex and kinship on the social structure of resident coastal bottlenose dolphins. Mar Biol 165:80

Lowther-Thieleking JL, Archer FI, Lang AR, Weller DW (2015) Genetic differentiation among coastal and offshore common bottlenose dolphins, *Tursiops truncatus*, in the eastern North Pacific Ocean. Mar Mamm Sci 31:1–20

Martin AR, da Silva VMF (2004) Number, seasonal movements, and residency characteristics of river dolphins in an Amazonian floodplain lake system. Can J Zool 82:1307–1315

Mason S, Salgado Kent C, Donnelly D, Weir J, Bilgmann K (2016) Atypical residency of short-beaked common dolphins (*Delphinus delphis*) to a shallow, urbanized embayment in south-eastern Australia. R Soc Open Sci 3:160478

Matkin CO, Straley JM, Durban JW, Matkin DR, Saulitis EL, Ellis GM, Andrews RD (2012) Contrasting abundance and residency patterns of two sympatric populations of transient killer whales (*Orcinus orca*) in the northern Gulf of Alaska. Fish Bull 110:143–155

Matthews CJD, Luque SP, Petersen SD, Andrews RD, Ferguson SH (2011) Satellite tracking of a killer whale (*Orcinus orca*) in the eastern Canadian Arctic documents ice avoidance and rapid, long-distance movement into the North Atlantic. Polar Biol 34:1091–1096

McSweeney DJ, Baird RW, Mahaffy SD, Webster DL, Schorr GS (2009) Site fidelity and association patterns of a rare species: pygmy killer whales (*Feresa attenuata*) in the main Hawai'ian Islands. Mar Mamm Sci 25:557–572

Milmann LC, Danilewicz D, Baumgarten J, Ott PH (2017) Temporal–spatial distribution of an island-based offshore population of common bottlenose dolphins (*Tursiops truncatus*) in the equatorial Atlantic. Mar Mamm Sci 33:496–519

Mintzer VJ, Lorenzen K, Frazer TK, da Silva VMF, Martin AR (2016) Seasonal movements of river dolphins (*Inia geoffrensis*) in a protected Amazonian floodplain. Mar Mamm Sci 32:664–681

Moran JR, O'Dell MB, Arimitsu ML, Straley JM, Dickson DMS (2018) Seasonal distribution of Dall's porpoise in Prince William Sound, Alaska. Deep-Sea Res Part II 147:164–172

Morin PA, Archer FI, Foote AD, Vilstrup J, Allen EE, Wade P, Durban J, Parsons K, Pitman R, Li L, Bouffard P, Abel Nielsen SC, Rasmussen M, Willerslev E, Gilbert MTP, Harkins T (2010) Complete mitochondrial genome phylogeographic analysis of killer whales (*Orcinus orca*) indicates multiple species. Genome Res 20:908–916

Morteo E, Rocha-Olivares A, Abarca-Arenas LG (2014) Sexual segregation of coastal bottlenose dolphins (*Tursiops truncatus*) in the Southwestern Gulf of Mexico. Aquat Mamm 40:375–385

Neumann DR (2001) Seasonal movements of short-beaked common dolphins (*Delphinus delphis*) in the north-western Bay of Plenty, New Zealand: influence of sea surface temperatures and El Niño/La Niña. NZ J Mar Freshwat Res 35:371–374

Nielsen NH, Teilmann J, Sveegaard S, Hansen RG, Sinding M-HS, Dietz R, Heide-Jørgensen MP (2018) Oceanic movements, site fidelity and deep diving in harbour porpoises from Greenland show limited similarities to animals from the North Sea. Mar Ecol Prog Ser 597:259–272

Olavarría C, Acevedo J, Vester HI, Zamorano-Abramson J, Viddi FA, Gibbons J, Newcombe E, Capella J, Hoelzel AR, Flores M, Hucke-Gaete R, Torres-Flórez JP (2010) Southernmost distribution of common bottlenose dolphins (*Tursiops truncatus*) in the Eastern South Pacific. Aquat Mamm 36:288–293

Olsen DW, Matkin CO, Andrews RD, Atkinson S (2018) Seasonal and pod-specific differences in core use areas by resident killer whales in the northern Gulf of Alaska. Deep-Sea Res Part II 147:196–202

Oshima JEF, Santos MCO (2016) Guiana dolphin home range analysis based on 11 years of photo-identification research in a tropical estuary. J Mammal 97(2):599–610

Oudejans MG, Visser F, Englund A, Rogan E, Ingram SN (2015) Evidence for distinct coastal and offshore communities of bottlenose dolphins in the North East Atlantic. PLoS One 10(4): e0122668

Owen ECG, Wells RS, Hofmann S (2002) Ranging and association patterns of paired and unpaired adult male Atlantic bottlenose dolphins, *Tursiops truncatus*, in Sarasota, Florida, provide no evidence for alternative male strategies. Can J Zool 80:2072–2089

Passadore C, Möller L, Diaz-Aguirre F, Parra GJ (2017) High site fidelity and restricted ranging patterns in southern Australian bottlenose dolphins. Ecol Evol 8:242–256

Pirotta E, Harwood J, Thompson PM, New L, Cheney B, Arso M, Hammond PS, Donovan C, Lusseau D (2015) Predicting the effects of human developments on individual dolphins to understand potential long-term population consequences. Proc R Soc B 282:2109–2115

Rako-Gospić N, Radulović M, Vučur T, Pleslić G, Holcer D, Mackelworth P (2017) Factor associated variations in the home range of a resident Adriatic common bottlenose dolphin population. Mar Pollut Bull 124:234–244

Read AJ (2018) Biotelemetry. In: Würsig B, Thewissen JGM, Kovacs KM (eds) Encyclopedia of marine mammals, 3rd edn. Academic, San Diego, pp 103–106

Reisinger RR, Keith M, Andrews RD, de Bruyn PJN (2015) Movement and diving of killer whales (*Orcinus orca*) at a Southern Ocean archipelago. J Exp Mar Biol Ecol 473:90–102

Reynolds JE, Wells RS, Eide SD (2000) The bottlenose dolphin – biology and conservation. University Press of Florida, Gainesville

Richard PR, Martin AR, Orr JR (2001) Summer and autumn movements of belugas of the eastern Beaufort Sea stock. Arctic 54:223–236

Richards VP, Greig TW, Fair PA, McCulloch SD, Politz C, Natoli A, Driscoll CA, Hoelzel AR, David V, Bossart GD, Lopez JV (2013) Patterns of population structure for inshore bottlenose dolphins along the Eastern United States. J Hered 104:765–778

Robinson KP, O'Brien JM, Berrow SD, Cheney B, Costa M, Eisfeld SM, Haberlin D, Mandleberg L, O'Donovan M, Oudejans MG, Ryan C, Stevick PT, Thompson PM, Whooley P (2012) Discrete or not so discrete: long distance movements by coastal bottlenose dolphins in UK and Irish waters. J Cetacean Res Manag 12:365–371

Rosel PE, Hansen L, Hohn AA (2009) Restricted dispersal in a continuously distributed marine species: common bottlenose dolphins *Tursiops truncatus* in coastal waters of the western North Atlantic. Mol Ecol 18:5030–5045

Sasaki-Yamamoto Y, Akamatsu T, Ura T, Sugimatsu H, Kojima J, Bahl R, Behera S, Kohshima S (2013) Diel changes in the movement patterns of Ganges River dolphins monitored using stationed stereo acoustic data loggers. Mar Mamm Sci 29:589–605

Schaffeld T, Bräger S, Gallus A, Dähne M, Krügel K, Herrmann A, Jabbusch M, Ruf T, Verfuß UK, Benke H, Koblitz JC (2016) Diel and seasonal patterns in acoustic presence and foraging behaviour of free-ranging harbour porpoises. Mar Ecol Prog Ser 547:257–272

Scheinin AP, Kerem D, MacLeod CD, Gazo M, Chicote CA, Castellote M (2011) Gray whale (Eschrichtius robustus) in the Mediterranean Sea: anomalous event or early sign of climate-driven distribution change? Mar Biodivers Rec 4:1–5

Schorr GS, Falcone EA, Moretti DJ, Andrews RD (2014) First long-term behavioral records from Cuvier's beaked whales (*Ziphius cavirostris*) reveal record-breaking dives. PLoS One 9(3): e92633

Shane SH, Wells RS, Würsig B (1986) Ecology, behavior and social organization of the bottlenose dolphin: a review. Mar Mamm Sci 2:34–63

Solovyev BA, Shpak OV, Glazov DM, Rozhnov VV, Kuznetsova DM (2015) Summer distribution of beluga whales (*Delphinapterus leucas*) in the Sea of Okhotsk. Russ J Theriol 14:201–215

Sprogis KR, Raudino HC, Rankin R, Macleod CD, Bejder L (2016) Home range size of adult Indo-Pacific bottlenose dolphins (*Tursiops aduncus*) in a coastal and estuarine system is habitat and sex-specific. Mar Mamm Sci 32:287–308

Sprogis KR, King C, Bejder L, Loneragan NR (2018) Frequency and temporal trends of shark predation attempts on bottlenose dolphins (*Tursiops aduncus*) in temperate Australian waters. J Exp Mar Biol Ecol 508:35–43

Stanistreet JE, Nowacek DP, Bell JT, Cholewiak DM, Hildebrand JA, Hodge LEW, Van Parijs SM, Read AJ (2018) Spatial and seasonal patterns in acoustic detections of sperm whales *Physeter macrocephalus* along the continental slope in the western North Atlantic Ocean. Endanger Species Res 35:1–13

Steiner L, Lamoni L, Acosta Plata M, Jensen S-K, Lettevall E, Gordon J (2012) A link between male sperm whales, *Physeter macrocephalus*, of the Azores and Norway. J Mar Biol Assoc UK 92:1751–1756

Stern SJ, Friedlaender AS (2018) Migration and movement. In: Würsig B, Thewissen JGM, Kovacs KM (eds) Encyclopedia of marine mammals, 3rd edn. Academic, San Diego, pp 602–608

Stern SJ, Keener W, Szczepaniak ID, Webber MA (2017) Return of harbor porpoises (*Phocoena phocoena*) to San Francisco Bay. Aquat Mamm 43:691–702

Stevick PT, McConnell BJ, Hammond PS (2002) Patterns of movement. In: Hoelzel RA (ed) Marine mammal biology: an evolutionary approach. Blackwell, Oxford, pp 185–216

Stockin KA, Pierce GJ, Binedell V, Wiseman N, Orams MB (2008) Factors affecting the occurrence and demographics of common dolphins (*Delphinus* sp.) in the Hauraki Gulf, New Zealand. Aquat Mamm 34:200–211

Sveegaard S, Teilmann J, Tougaard J, Dietz R, Mouritsen KN, Desportes G, Siebert U (2011) High-density areas for harbor porpoises (*Phocoena phocoena*) identified by satellite tracking. Mar Mamm Sci 27:230–246

Tezanos-Pinto G, Baker CS, Russell K et al (2009) A worldwide perspective on the population structure and genetic diversity of bottlenose dolphins (*Tursiops truncatus*) in New Zealand. J Hered 100:11–24

Toth JL, Hohn AA, Able KW, Gorgone AM (2012) Defining bottlenose dolphin (*Tursiops truncatus*) stocks based on environmental, physical, and behavioral characteristics. Mar Mamm Sci 28:461–478

Vermeulen E, Balbiano A, Belenguer F, Colombil D, Failla M, Intrieri E, Bräger S (2017) Site-fidelity and movement patterns of bottlenose dolphins (*Tursiops truncatus*) in central Argentina: essential information for effective conservation. Aquat Conserv Mar Freshwat Ecosyst 27:282–292

Vermeulen E, Bouveroux T, Plön S, Atkins S, Chivell W, Cockcroft V, Conry D, Gennari E, Hörbst S, James BS, Kirkman S, Penry G, Pistorius P, Thornton M, Vargas-Fonseca OA, Elwen SH (2018) Indian Ocean humpback dolphin (*Sousa plumbea*) movement patterns along the South African coast. Aquat Conserv Mar Freshwat Ecosyst 28:231–240

Wang X, Wu F, Chang W-L, Hou W, Chou L-S, Zhu Q (2016) Two separated populations of the Indo-Pacific humpback dolphin (*Sousa chinensis*) on opposite sides of the Taiwan Strait: evidence from a larger-scale photo-identification comparison. Mar Mamm Sci 32:390–399

Ward EJ, Dahlheim ME, Waite JM, Emmons CK, Marshall KN, Chasco BE, Balcomb KC (2016) Long-distance migration of prey synchronizes demographic rates of top predators across broad spatial scales. Ecosphere 7(2):e01276

Webster TA, Dawson SM, Slooten E (2009) Evidence of sex segregation in hector's dolphin (*Cephalorhynchus hectori*). Aquat Mamm 35:212–219

Wells RS, Hansen LJ, Baldridge A, Dohl TP, Kelly DL, Defran RH (1990) Northward extension of the range of bottlenose dolphins along the California Coast. In: Leatherwood S, Reeves RR (eds) The bottlenose dolphin. Academic, San Diego, pp 421–431

Wells RS, Rhinehart HL, Cunningham P, Whaley J, Baran M, Koberna C, Costa DP (1999) Long distance offshore movements of bottlenose dolphins. Mar Mamm Sci 15:1098–1114

Wells RS, Manire CA, Byrd L, Smith DR, Gannon JG, Fauquier D, Mullin KD (2009) Movements and dive patterns of a rehabilitated Risso's dolphin, *Grampus griseus*, in the Gulf of Mexico and Atlantic Ocean. Mar Mamm Sci 25:420–429

Whitehead H (2016) Consensus movements by groups of sperm whales. Mar Mamm Sci 32:1402–1415

Whitehead H, Coakes A, Jaquet N, Lusseau S (2008) Movements of sperm whales in the tropical Pacific. Mar Ecol Prog Ser 361:291–300

Wilson B, Reid RJ, Grellier K, Thompson PM, Hammond PS (2004) Considering the temporal when managing the spatial: a population range expansion impacts protected areas-based management for bottlenose dolphins. Anim Conserv 7:331–338

Wirsing AA, Heithaus MR, Frid A, Dill LM (2008) Seascapes of fear: evaluating sublethal predator effects experienced and generated by marine mammals. Mar Mamm Sci 24:1–15

Wong, Whitehead (2014) Seasonal occurrence of sperm whales (*Physeter macrocephalus*) around Kelvin Seamount in the Sargasso Sea in relation to oceanographic processes. Deep-Sea Res Part I 91:10–16

Würsig B (2017) Marine mammals of the Gulf of Mexico. In: Ward CH (ed) Habitats and biota of the Gulf of Mexico: before the deepwater horizon oil spill. Springer, New York, pp 1489–1588

Würsig B, Cipriano F, Würsig M (1991) Dolphin movement patterns: information from radio and theodolite tracking studies. In: Pryor K, Norris KS (eds) Dolphin societies: discoveries and puzzles. University of California Press, Berkeley, pp 79–111

Chapter 7
Predator/Prey Decisions and the Ecology of Fear

Mridula Srinivasan

Abstract Animal behaviors are governed by the intrinsic need to survive and reproduce. Even when sophisticated predators and prey are involved, these tenets of behavioral ecology hold. Similar to humans, fear can be a strong motivator for change in animals. The terrestrial ecology literature is replete with examples of fear-mediated behavioral effects on species and community networks. In contrast, the marine mammal literature is sparse in its recognition and consideration of nonconsumptive effects or risk effects arising from powerful and lethal predators, such as killer whales and large sharks. This chapter encapsulates the ecology of fear concept by providing representative examples from the marine mammal literature with consideration of prey and predator perspectives. Additionally, research data gaps and new avenues for scientific examination are highlighted within documented examples. Lastly, conservation practitioners and marine mammal scientists are encouraged to adapt theoretical concepts and methods from predation risk studies to better understand the effects of nonbiological stressors on marine mammal species.

Keywords Ecology of fear · Killer whales · Sharks · Predation risk · Risk effects · Trait-mediated indirect interactions · Trophic cascades

7.1 Introduction

Fear—noun \ 'fir \ is "an unpleasant often strong emotion caused by anticipation or awareness of danger" (©Merriam-Webster Dictionary 2018).

In animal societies, fear is a subtle but enduring force, which can alter animal behavior with consequences that rival or exceed consumptive predation effects because of the significant number of individuals simultaneously affected by apparent or actual

---

M. Srinivasan (✉)
Office of Science and Technology, National Marine Fisheries Service, National Oceanic and Atmospheric Administration, Silver Spring, MD, USA
e-mail: mridula.srinivasan@noaa.gov

© Springer Nature Switzerland AG 2019
B. Würsig (ed.), *Ethology and Behavioral Ecology of Odontocetes*, Ethology and Behavioral Ecology of Marine Mammals,
https://doi.org/10.1007/978-3-030-16663-2_7

predation risk (Lima and Dill 1990; Schmitz et al. 1997; Preisser et al. 2005). In this well-established ecology of fear (Brown et al. 1999), predators affect prey behavioral ecology through intimidation that compels prey to optimize decision-making related to food, avoiding predators, habitat, and mates, even if predation rates are minimal (Creel and Christianson 2008). Prey decision-making can involve a series of calculated trade-offs that vary spatiotemporally depending on the environment and their evolutionary history (Sih 1980; Dill 2017). These trade-offs, resulting from antipredator behaviors, are associated with costs and benefits that can extend from the species to the community level (Werner and Peacor 2003; Schmitz et al. 2008).

Fear responses are a measurable quantity and can include fixed and flexible reactions. Thus, the immediate or perceived threat of a predator can elicit flight, fight, or freeze responses in prey (Heithaus 2004; Eilam 2005; Reeves et al. 2006; Ford and Reeves 2008)—the latter a physiological and neurological response. Alternatively, animals conditioned to varying levels of predation risk and predator types have evolved counteractive maneuvers to outsmart, minimize, or avoid risky encounters. These can include strategies such as reduced activity, choosing land-scapes with complex features, group living and defense, mobbing, intra- and inter-specific signaling, predator inspection, and vigilance (Bertram 1978; Inman and Krebs 1987; Lima 1998; Heithaus 2004; Sorato et al. 2012). The fear response initiated and subsequent ecological and evolutionary impact on prey are functions of the cost of predation (Brown 2010) and animal state (e.g., hunger level or body condition).

At the other end of this complex interaction, predators are not silent spectators. Indeed, predators change tactics in response to their motivational state, environmental fluctuations, and prey behavior (Lima 2002). This dynamic decision-making and predator and prey responses can be likened to behavioral games that are continually evolving (Brown et al. 2001; Bouskila 2010) and form the basis of game theory models (Hugie 2001).

Marine mammals are not immune from predation risk despite occupying high trophic levels. Large sharks (generally, >2 m) and killer whales (*Orcinus orca*) are the principal predators of marine mammals (Ford and Reeves 2008; Weller 2009). Marine mammals, like their terrestrial counterparts, exhibit antipredator behaviors and consequently can experience risk effects in response to spatiotemporal variability in predation risk (Heithaus and Dill 2002; Wirsing et al. 2007; Srinivasan et al. 2010) (Fig. 7.1).

Sharks and killer whales exert top-down effects in the marine food chain through Density Mediated Indirect Interaction (DMII), wherein changes transmit across trophic levels via numerical reduction of primary prey. As well, through behavior or Trait-Mediated Indirect Interactions (TMII) (Werner and Peacor 2003; Schmitz et al. 2004), killer whales and sharks by their mere presence can provoke habitual or panicked behavioral reactions in their (multiple) prey, which in turn can activate cascading effects through the marine community (Heithaus and Dill 2005; Wirsing et al. 2007; Frid et al. 2008; Heithaus et al. 2008). Despite compelling evidence and widespread acceptance of risk effects in the ecological literature, as well as studies on marine mammal risk effects and TMII (Frid et al. 2007; Macleod et al. 2007;

7 Predator/Prey Decisions and the Ecology of Fear

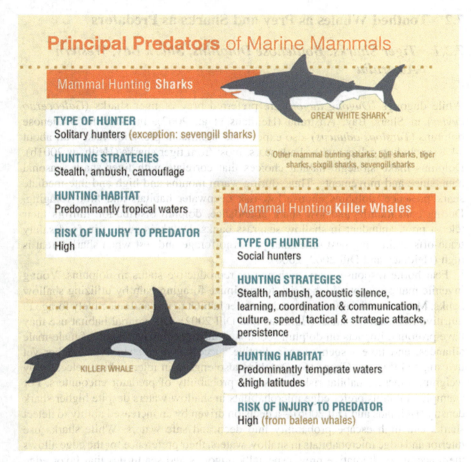

Fig. 7.1 Characteristics of principal marine mammal predators

Wirsing et al. 2008; Kiszka et al. 2015; Srinivasan et al. 2018), the marine mammal community is yet to embrace this critical dimension of behavioral ecology.

In this chapter, I present representative examples of fear responses in odontocetes and the potential effect on population dynamics, predator-prey interactions, and behavioral cascades. I highlight mainly toothed whale prey and predator perspectives and decision-making. Focal topics involving killer whales, which attack and scare a variety of marine mammals, are not restricted to toothed whale prey, and therefore other cetacean examples are included to demonstrate various killer whale strategies. But case studies involving predation risk from sharks are limited to toothed whales as prey. Further, some examples are well-documented in the literature, while others provide new avenues for analyzing risk effects and TMII. Finally, I discuss the broader impetus and relevance of predation risk theory and mechanistic models in the analysis of human "predator" impacts on marine mammals. The goal is to highlight this challenging facet of animal ecology and its widespread applications.

7.2 Toothed Whales as Prey and Sharks as Predators

7.2.1 Tiger Sharks: Bottlenose Dolphins, Shark Bay, Western Australia

While dugongs (*Dugong dugon*) are preferred prey of tiger sharks (*Galeocerdo cuvier*) in Shark Bay, Australia (Heithaus et al. 2002), Indo-Pacific bottlenose dolphins (*Tursiops aduncus*) also experience significant predation risk, with about 74% (excluding calves) with shark scars, most from tiger sharks (Heithaus 2001b). Dolphins make strategic habitat choices that correlate with tiger shark seasonal abundance and movements. Thus, during warm months and high and intermediate shark presence, dolphins choose low-risk deepwater habitats and limit foraging. During cold months and low shark abundance, dolphin distribution mirrors their teleost prey, abundant in shallow seagrass banks. Dolphins engage in food safety trade-offs optimizing best times and places to forage and rest when shark threat is high (Heithaus and Dill 2002, 2006).

Fear-borne responses vary with age or reproductive status in dolphins. Young juvenile males accept a higher risk to maximize foraging gain by utilizing shallow banks. Mothers with calves exploit deepwater habitat to rest and selectively forage in nutritive-rich shallow areas (Heithaus and Dill 2002). Differential habitat use may have profound impacts on dolphin social structure, such as formation of male-male alliances, and fitness such as reproductive success and calf survival, neither yet investigated (but see Chap. 16). Dolphins also engage in microhabitat selection by weighing inherent habitat risk against the probability of predator encounters. For example, dolphins prefer edge microhabitats in shallow waters despite higher shark density (Heithaus and Dill 2006)—a decision driven by an increased ability to detect sharks and high escape probability into deep and safe waters. While sharks use interior and edge microhabitats in shallow waters, their preference for the edge allows them access to alternative prey, especially dugongs and sea turtles that favor edge microhabitats. In summary, the Indo-Pacific bottlenose dolphin-tiger shark system is one of the best-described marine systems that provide evidence of fear-induced congruency of energy gain and reduced predator encounters in dolphins.

7.2.2 Bull Sharks: Bottlenose Dolphins, Sarasota Bay, Florida, USA

Common bottlenose dolphins (*Tursiops truncatus*) in Sarasota Bay, Florida, are year-round residents with variable patterns of habitat use (Irvine et al. 1981). They seek shallow waters often near seagrass beds in rough approximation with peak spring-summer abundance of bull sharks (*Carcharhinus leucas*), the dolphins' dominant predatory threat (Wells et al. 1980). Seagrass habitats offer a rich source of prey fish, especially in spring and summer (Barros and Wells 1998; Barros et al. 2010). Seagrass habitats also function as critical nursery areas for young dolphins, affording possible protection from bull sharks and providing improved feeding opportunities (Wells 1991). In autumn and winter, they occur more often in passes, channels, and coastal waters of the Gulf of Mexico.

Capture-release health assessment studies revealed that 90 of 246 dolphins exhibited shark bite scars, representing 36.6% of the population (Wilkinson 2014), with no sex differences in scarring rate. Further, Wilkinson (2014) documented an increased likelihood of shark bite presence in open habitats where dolphins maintain larger average group sizes (Wells et al. 1987). This system is ideal to investigate the role of sharks in influencing dolphin habitat choice relative to foraging, social, and critical environmental parameters.

7.2.3 Shark Prey Summary

Great white (*Carcharodon carcharias*) (Heithaus 2001a), Galapagos (*Carcharhinus galapagensis*) (Gobush 2010), sevengill (*Notorhynchus cepedianus*) (Lucifora et al. 2005), and Pacific sleeper sharks (*Somniosus pacificus*) (Frid et al. 2006) target marine mammal prey. Great whites, tiger sharks, and sevengills exhibit an ontogenetic diet shift, consuming marine mammal prey as they grow in size (Ebert 2002). In non-dolphin prey, there is evidence of risk-sensitive behavior as evidenced in Cape fur seals (*Arctocephalus pusillus pusillus*), which modulate habitat use and grouping and social patterns in low (Egg Island, South Africa) versus high predation risk (Seal Island, False Bay, South Africa) sites based on spatiotemporal variation in great white occurrence (de Vos et al. 2015). Moreover, habitat utilization patterns among Cape fur seals between high-risk sites show that use of landscape complexity as *refugia* can prove to be an effective antipredator strategy against seasonal shark predation (Wcisel et al. 2015).

Among dolphins, there is evidence of shark bites, occasional shark attacks, and shark consumption based on stomach contents (Heithaus 2001a). Without credible evidence of predator avoidance behaviors and understanding of prey-predator ecology, it is challenging to analyze shark predation risk effects. Based on shark predator diet and predator and dolphin behaviors, it has been speculated (Lucifora et al. 2005) that La Plata River dolphins (*Pontoporia blainvillei*) select habitats at least partially due to predation.

Predation risk from sharks is variable across species and areas. For example, snubfin dolphins, *Orcaella heinsohni* in northwestern Australia and *T. aduncus* in some regions of the southwestern Indian Ocean (Heithaus et al. 2017; Smith et al. 2018), experience high predation risk; but Atlantic spotted dolphins, *Stenella frontalis* in Bimini, Bahamas (Melillo-Sweeting et al. 2014), show relatively low shark bite scarring. There are other examples where shark predation risk has been attributed but not evaluated. Spinner dolphins (*Stenella longirostris*) in Hawai'i are believed to rest in nearshore waters during daytime to avoid predation from deepwater sharks (Norris and Dohl 1980), with abundant evidence of shark bite scarring in Hawai'i spinner dolphins (Norris et al. 1994). However, there has been no investigation to analyze risk effects arising from probable antipredator behaviors, such as dolphins engaging in

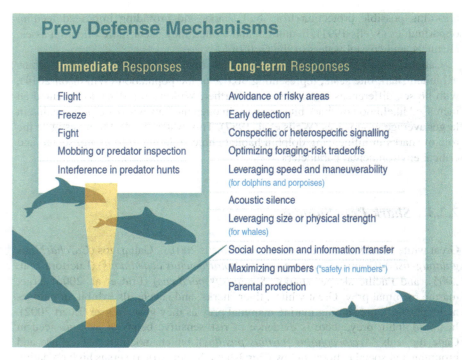

Fig. 7.2 Common antipredator mechanisms exhibited by marine mammal prey

nearshore-offshore movements or any potential energetic compromise involved in feeding due to nocturnal threats from sharks (Norris et al. 1994).

The absence of significant predation or shark bite scars does not exclude dolphins from exhibiting fear or evolving behaviors to minimize risk. Instead, these examples suggest that in many tropical systems, some large sharks can be a significant source of undocumented predation risk in dolphins (Heithaus 2001a). Thus, the construction of predator-prey models would be incomplete without the inclusion of sublethal effects on prey life history and fitness in areas where animals are agitated and responsive to shark presence or threats (Fig. 7.2).

7.3 Toothed Whales as Prey and Killer Whales as Predators

7.3.1 Killer Whales: Dusky Dolphins, Kaikoura, New Zealand

Killer whales occasionally attack dusky dolphins (*Lagenorhynchus obscurus*) off Kaikoura, New Zealand (Würsig et al. 1997; Constantine et al. 1998). Usually, dusky dolphins off Kaikoura maintain a near predictable routine of nocturnal feeding

7 Predator/Prey Decisions and the Ecology of Fear 151

(almost exclusively) on mesopelagic organisms in the productive Kaikoura Canyon waters and resting during the day in nearshore and canyon edge waters (generally, ≤200 m deep). These nearshore-offshore movements are unrestricted by social status or age/sex classes (Srinivasan et al. 2010; Markowitz 2004). Mothers with calves may also preferentially use waters <20 m deep during the day and engage in some daytime foraging (Weir 2007; Weir et al. 2008). Lifestyle decisions coincide with seasonal killer whale presence that peaks during austral spring, summer, and fall (Srinivasan and Markowitz 2010). In austral winter when killer whales are rare, dusky dolphins are distributed more broadly across the habitat with no marked depth preferences (Cipriano 1992; Markowitz 2004; Dahood 2009).

During heightened risk brought on by the immediate presence of killer whales, dolphins flee the area or move to shallow waters (<10 m deep). Dolphin evasive tactics are effective if predators are detected early and dolphins have quick access to shallow water refuges. As a more enduring strategy, dolphins have evolved solutions to improve escape probabilities and survival while attaining energy demands by developing risk-averse behaviors through habitat selection, social segregation, and other group living strategies (Markowitz 2004; Weir et al. 2008).

To measure costs and benefits associated with these choices, Srinivasan et al. (2010) used a spatially explicit individual-based model to measure different evolutionary strategies by testing various combinations of hiding time in refuge and detection distances while simultaneously modifying prey (dolphin) and predator (killer whale) behaviors pre- and post-encounter to assess the costs and benefits of evolving antipredator behaviors. Results revealed an overall 2.7% loss in foraging time associated with the evolution of a nearshore/offshore movement. Further, risk-averse strategies determined by large detection distances had approximately a 98% reduction in killer whale encounters regardless of the level of killer whale presence. But risky strategies determined by short detection distances and short hiding times resulted in the lowest foraging time lost but also maximum encounters with killer whales. In contrast, a risk-averse strategy could result in 38% lost foraging time at the highest level of killer whale presence.

Differentiated social categories and behaviors in dusky dolphins off Kaikoura lend themselves to additional questions about state-dependent risk-taking, as it relates to body condition or reproductive status and food safety trade-offs. Srinivasan et al. (2018) found that a lactating mother with calf would incur overall lower energetic costs than an adult without a calf despite pursuing more risk-averse behaviors of reduced activity, choosing shallower waters to rest and hiding longer in the refuge. Lactating dolphin mothers with calf have a heavy energetic burden to carry. Their daily energy demands could be almost twice that of an adult dolphin having to not only feed the calf but also support the calf in her slipstream, imposing hydrodynamic drag. Since dolphin calves are weaned after one year or more (Srinivasan et al. 2018), the dolphin mother has to continually optimize her decision-making to ensure her and her calf's survival. The gains the mother makes in reduced encounters with killer whales and foraging expenditure are likely to result in lost foraging time, and significant foraging calories lost relative to adults without calves. Also, Srinivasan et al. (2018) found that risk-averse behaviors can be energetically expensive for dolphins with or without a calf. Moreover, the

152   M. Srinivasan

consumption of high-quality prey does not negate the overwhelming influence and effects of elevated predation risk on energetic costs for dolphins.

Such mechanistic models allow us to explore different facets of animal decision-making and its evolutionary consequences under risk. The emergence time from a refuge post-predator encounter can be energetically expensive or increase the starvation risk for vulnerable mothers with calves reliant on patchy or unpredictable prey. As feeding alternatives diminish, animals may take more risks, be less vigilant, and thus more exposed to attacks. Further empirical evidence on seasonal food densities and variability, dolphin foraging behavior and precise daily energetic requirements, and predator behavior will help evaluate theoretical predictions. Near Kaikoura, large shark scarring evidence is rare, and sharks do not appear to have a significant effect on dusky dolphin behavioral outcomes, but this could change. The presence of multiple predators with different spatiotemporal behavior could create unexpected risk effects.

In dolphin societies, there are advantages of social dynamics such as group cohesion, communication, and defense. Social networks and information transfer can potentially enhance group vigilance and fitness against fierce and stealthy predators, such as killer whales, but these topics need further investigation (Würsig 1986; Gowans et al. 2008).

Killer whales in New Zealand appear to be mostly opportunistic foragers, but there are indications of three killer whale ecotypes specialized in taking (1) marine mammals, (2) sharks and rays, and (3) other fishes (Visser 1999). Genetic information and long-duration tracking data are lacking to determine range-wide movements, although some photo-documentation discerns individual prey preferences and behavior. Similar to other regions where killer whales hunt dolphins, mammal-hunting killer whales appear to be nomadic, using stealth to attack highly maneuverable prey.

Not all killer whale activity elicits evasive behaviors in dusky dolphins (Srinivasan and Markowitz 2010). It is likely that dolphins are responding to specific types of killer whales and measure their responses accordingly. On occasion, mom and calf nursery groups appear to leave the area well ahead of the arrival of killer whales in the system (Constantine et al. 1998). Based on anecdotal evidence from tour boat operations, dolphins may also be quieter and not vocalize before killer whales enter the area. So, it is possible that dolphin social communication networks enable them to detect nearby killer whales in advance, but dedicated studies are necessary to understand killer whale movements, hunting success, dive profiles, group size dynamics, energy needs, population structure, as well as detailed temporal patterns near Kaikoura. Without the predator piece of the puzzle, the behavioral dynamics and consequences of this clever predator-prey interaction will remain unresolved.

7.3.2 Killer Whales: Narwhals and Belugas, Eastern Canadian Arctic

In the eastern Canadian Arctic, with increasing effects of anthropogenic climate change manifested through rapid decline in sea ice, killer whales have become a

7 Predator/Prey Decisions and the Ecology of Fear

regular presence in Hudson Bay, expanding into ice-free zones and acquiring unprecedented access to marine mammal prey more times in the year (Ferguson et al. 2010; O'Corry-Crowe et al. 2016). Similar to other areas, killer whale predation events are rarely observed, and rarer still are such predation events documented from beginning to end.

Surveys suggest that killer whales target many marine mammal prey in the eastern Canadian Arctic. These include bowhead whales (*Balaena mysticetus*), narwhals (*Monodon monoceros*), beluga whales (*Delphinapterus leucas*), and phocid seals. Killer whales may attack male and female narwhals, but injury risk is probably higher when attacking tusked adult male narwhals.

Ferguson et al. (2010, 2012a) obtained a representative picture of killer whale historic and current occurrence patterns and predation behavior in the eastern Canadian Arctic, from Inuit Traditional Ecological Knowledge (TEK). Killer whale predation risk in this region is spatially varied. Narwhal-killer whale interactions were most common in Arctic Bay, northeast Baffin region of west Greenland, and Repulse Bay. For belugas, predation risk appeared to be highest in southwest Hudson Bay, and for bowhead whales, Foxe Basin is where killer whale attacks on bowhead calves, and possibly subadults, are concentrated. Killer whales hunt bowhead whales during summer and focus on belugas and narwhals during early and late ice-free seasons (Ferguson et al. 2012b).

Inuit hunters fittingly refer to Arctic marine mammal antipredator behaviors as "aarlirijuk" ("fear of killer whales") in their Inuktitut South Baffin dialect. In the eastern Canadian Arctic, marine mammal prey exhibit typical antipredator behaviors of fleeing to shallow waters, sometimes even beaching themselves. When killer whales are near, narwhals may exhibit shallow breathing, become quieter, venture into very shallow waters (<2 m) (Laidre et al. 2006), seek alternative routes to avoid killer whale encounters, and possibly detect killer whale presence a few days before their arrival. Bowheads are believed to "run" at the sight of killer whales into ice or seek shallow waters, inlets, and fjords, which deter killer whales from following. Bowhead whales also dive and stay in deep waters to avoid killer whale predation.

Belugas, besides fleeing to shorelines and shallow waters, may arrive in the region late to minimize killer whale encounters (Westdal et al. 2016). In a short-term satellite telemetry study in Hudson Bay, belugas clustered during the predation event, reducing their range size from 285 km^2 four days before the killer whale attack to about 172 km^2 on the day of the attack. In the aftermath of the predation event, belugas resumed normal home ranges, suggesting a quick recovery.

In contrast, near Admiralty Inlet, in the eastern Canadian Arctic, narwhals dispersed and doubled their home range during the killer whale attack and then resumed normal habitat use within a day of the attack. Post-attack, they also expanded their territory by moving into offshore waters (Laidre et al. 2006), where they do not regularly occur. These observations were confirmed in a similar but temporally extended study (Breed et al. 2017) in Admiralty Inlet. The authors simultaneously satellite tracked a single killer whale (representing the pod) and seven narwhals in Admiralty Inlet and found that the mere presence of killer whales disrupted long-term habitat use patterns as described above. Narwhals reacted to killer whales nearly 100 km away by moving close to shore. Breed et al. (2017)

suggest that predator effects may lead to persistent changes in habitat use and behavior, which could potentially have fitness consequences given other stressors such as resource and environmental changes.

Narwhals and belugas have evolved strategies to minimize predation risk, and their summer habitat preferences reflect this. These patterns may change with changing killer whale dynamics and the Arctic seascape. Eventually, strategic antipredator defenses could have substantial energetic and reproductive costs if they interfere with foraging habits and calving patterns. Further, change in distribution and behaviors of sentinel marine mammal species are predicted to have cascading effects sweeping through the ecosystem, including effects on subsistence harvests by Inuit Peoples (O'Corry-Crowe et al. 2016).

Killer whales in the eastern Canadian Arctic appear to specialize in hunting and consuming marine mammal prey and have adapted to a changing Arctic landscape by modifying their distribution and employing mixed feeding strategies. Acting as roaming silent apex predators, continually searching, chasing, and hunting different prey, killer whales create unpredictability. Killer whale hunting strategies probably improve hunting success and thereby fulfill their high nutritive demands (Ferguson et al. 2012b). Killer whales, like wolves (*Canis lupus*), draw strength from their tight-knit social structure to attack larger and powerful prey through active cooperation and communication, which sometimes negates the need for large group sizes and brute force (Ferguson et al. 2012a).

The power and intimidation factor of killer whales lies in their skill of using prey-specific attack and kill strategies. For bowhead whales, they appear to block the blowhole, suffocate and drown the whale, as well as remove large chunks of the body that injure and weaken the animal. For narwhals, they circle prey, ram the animal, and attempt to tire out the animal before killing it. Similarly, when hunting belugas, killer whales circle the target animal, launch several attacks to cause internal damage, and take out large chunks of body tissue. In the case of belugas, killer whales may engage in surplus killing (Ferguson et al. 2012a), like lions (*Panthera leo*), without consuming them.

Killer whales in this region also duplicate strategies commonly observed in Type B killer whales hunting seals in Antarctica (Pitman and Ensor 2003). They create coordinated waves to flush out seals from ice floes. These behaviors suggest that killer whales may have independently learned or developed evolutionarily stable strategies for attacking ice-associated seals, in the far reaches of both hemispheres. The impacts of killer whales on already climate-stressed Arctic marine mammals are a needed area for research.

7.3.3 Notes on Killer Whale—Eastern North Pacific Gray Whale Interactions

Every spring, Monterey Bay, California, is a prime place to observe transient killer whale predation on northbound migrating eastern Pacific gray whale (*Eschrichtius robustus*) mothers and calves (Weller 2009). To the extent possible, northward migrating gray whale cow-calf pairs follow the shoreline along their migratory path (Barrett-Lennard et al. 2011). However, the topography of the Monterey Canyon forces the whale pair to take a path that cuts across deep waters of the Canyon and within killer whale favored hunting grounds. Like narwhals in the eastern Canadian Arctic, gray whales engage in "snorkel behavior," surfacing quietly with nearly invisible blows and hardly breaking the water surface (Ford and Reeves 2008; Srinivasan, personal observation). The quiet surfacing and breathing patterns are harmonious in the cow-calf pair—a remarkable survival skill in the calves. Like much other killer whale prey, during an attack the whales attempt to flee to extremely shallow waters, where killer whales are unable to follow or lack the necessary water depth to separate the mother from the calf and drown the calf (Pitman et al. 2017).

Tour boat operators off Monterey maintain records of annual predation events, including data on California transient killer whales involved in gray whale cow-calf attacks (Black and Ternullo 2009). Killer whale movements indicate a preference for canyon edges, which they patrol in a back and forth manner (Black et al. 2003; Srinivasan, unpublished data), apparently to allow them to "lie in wait" for cryptically traveling cow-calf pairs.

Gray whale fear of killer whales is apparent in other parts of the migratory corridor. For example, off San Diego, Burrage (1964, p. 551) noted that "Since Gray Whales, if not threatened, usually proceed at a slow, unhurried pace, the actions of the individuals in question (i.e., frequent spouting, fast milling around, high speed dash, etc.) seem indicative of nervousness or agitation prompted by their awareness or realization of the impending presence of the Killer Whales and the potential danger of an encounter with same. There were no visible signs of any other source-no boats-for the apparent fright demonstrated by the Grays I observed."

Gray whales face another killer whale gauntlet near Unimak Island, Alaska (Barrett-Lennard et al. 2011), at the junction of the western Pacific Ocean and the Bering Sea, where mammal-hunting killer whale density is high and so is gray whale calf mortality. Gray whale mothers employ tested defenses of seeking extremely shallow waters (<3 m) to discourage pursuits, vigorously fighting the predators by tail thrashings and blocking access to the calf. As in other risky habitats, gray whales likely are highly vigilant, using crypsis and alternate routes to minimize encounters (Barrett-Lennard et al. 2011). Again, there has been a limited analysis of the risk effects of acute and chronic fear responses on gray whale energy budgets and population dynamics.

7.3.4   *Killer Whale Prey Summary*

Killer whales are powerful almost cosmopolitan predators, with a slightly skewed distribution toward temperate latitudes. Examples discussed here are only part of diverse toothed whale prey that killer whales frighten, attack, and sometimes consume (Jefferson et al. 1991; Baird 2000; Ford and Ellis 2014). For example, killer whales can attack larger toothed whales, such as sperm whales (*Physeter macrocephalus*), which are agitated by killer whale presence (Cure et al. 2013), and aggregate into a marguerite formation when attacked (Pitman et al. 2001).

There are perhaps more commonalities than differences in the various morphotypes of killer whales found around the world (de Bruyn et al. 2013). Killer whale marine mammal prey specialization and attack methods are often reflective of the prey availability, type, and habitat. Killer whales appear to be opportunistic in resource-limited environments, feeding on fish, marine mammals, and cephalopods [e.g., Hawai'i (Baird et al. 2006)], and are more specialized in prey-rich environments [e.g., eastern Aleutian Islands (Ford and Ellis 2014)]. Some of these specializations and predation techniques are probably culturally and matrilineally transmitted (see also Chap. 11), while others are influenced by the changing environment. Mammal-hunting killer whale attack strategies can include the characteristics of acoustic silence, persistence, socially mediated cooperation and coordination, guile, learning, and adaptability (Deecke et al. 2005; Reeves et al. 2006)—key skills in the killer whale repertoire.

Little is known about mammal-hunting killer whale success rates for different prey types, but see Baird (2000) and Ford et al. (1998) for preliminary analyses. A comprehensive meta-analysis of mammal-hunting killer whale effectiveness and costs may provide a thorough assessment of how killer whale hunting strategies evolved. Like terrestrial social predators, one can presume that these killer whales factor in energy demands, prey handling, injury risk, and hunting success before and after predation attempts (Mukherjee and Heithaus 2013). Studies of charismatic megafauna in marine systems usually focus on prey decision-making and ecological consequences but rarely through the lens of a predator. Like their prey, killer whales have to negotiate risks and benefits of foraging decisions (potentially leading to evolutionary consequences) in rapidly changing seascapes. Also, preying killer whales may sometimes experience aggressive reactions from potential prey or other mammals in the vicinity of a predation event. The best-described example of nontarget prey response is the interference of humpback whales (*Megaptera novaeangliae*) in killer whale predation events involving other marine mammal prey (Pitman et al. 2017). The in-depth mechanics of a predation event and the social strategies employed by mammal-hunting killer whales are important topics for future research.

Given shared sensory modalities with their prey, mammal-hunting killer whales resort to flexible and stable strategies depending on the selected target prey, including variability in the age, sex, and group size involved in the kill (Baird and Dill 1996). Once an attack is in progress, for large whale prey, killer whales rely on numbers and focus on neonates, yearlings, or subadults, collective brute force, and cooperative strategies to exhaust, weaken, and injure whales and importantly separate the prey mother from her offspring. Although not described here, killer whales also employ these strategies when pursuing humpback whale neonate calves in Western Australia (Pitman et al. 2015).

7 Predator/Prey Decisions and the Ecology of Fear

For dolphins and porpoises, the element of surprise is crucial and so also is stamina to exhaust or herd the dolphins into enclosed areas (Dahleheim and Towell 1994; Ford and Ellis 2014). Planning, persistence, and cunning can pay off when killer whales hunt pinniped prey and small dolphins. Also, killer whales may target multiple dolphins or pinniped species, leaving scarce evidence behind (Constantine et al. 1998; Ford and Ellis 1999).

In sum, the coercive effects of killer whales can transcend individual species and permeate the entire suite of vulnerable prey present in killer whale foraging grounds. There are now numerous examples in the literature of several toothed whale species displaying evasive or aggressive reactions to killer whale playback sounds, despite lack of visible predation events (Curé et al. 2011, 2015; Bowers et al. 2018), which reinforces the existence of fear-mediated behavioral responses in a range of species.

7.4 Conclusion

The ecology of fear is a not a mere concept but emphasizes the top-down forcing that fierce predators can play in shaping ecosystems by affecting multiple prey species through direct and indirect pathways (Kiszka et al. 2015). Few studies have examined how top predators can structure ecological communities through nonconsumptive effects, potentially galvanizing trophic cascades (Kiszka et al. 2015). But promising areas of research exist.

The killer whale-sea otter (*Enhydra lutris*)-sea urchin-kelp forest trophic cascade (Estes et al. 1998) describes how killer whale predation caused a severe decline in otter populations in the Aleutian archipelago, which in turn altered the kelp forest ecosystem by releasing sea urchins (sea otter prey) from predation pressure—this classic study could be revisited by exploring both TMII and DMII to measure the magnitude of killer whale lethal and sub-lethal impacts. It is possible that in the Salish Sea, northwest Pacific, increased presence of mammal-hunting killer whales (Houghton et al. 2015) has altered harbor seal feeding patterns, which in turn may explain an increase in steelhead smolts (*Oncorhynchus mykiss*) (PSSMSW 2018). Off Unimak Island, killer whales may store gray whale carcasses in shallow waters for repeated feeding on the kill (Barrett-Lennard et al. 2011), inadvertently facilitating other species such as Pacific sleeper sharks and bears (*Ursus arctos*) to scavenge and benefit from the hunt.

Overall, there is a lack of research exploring functional relationships between marine mammal antipredator behaviors and evolutionary consequences to the affected species and ecological community. One of the impediments for conducting such research is the complexity of marine systems and logistics of obtaining simultaneous prey, predator, and ecosystem profiles. The presence of multiple predators and existence of intra-guild predation (the eating and killing of potential competitor species using similar resources or habitats) can further complicate cause and effect and cost-benefit analyses of energy budgets under risk. A more obvious hurdle to overcome is acknowledging state-dependent risk effects and TMII in marine mammal systems and thinking broadly when analyzing predator-prey interactions.

An aspirational goal is to recognize the broader value of predation risk studies and models (Cresswell 2010; Schmitz et al. 2017; Gallagher et al. 2017; Moll et al. 2017) in analyzing conservation problems and assessing anthropogenic stressor effects (e.g., climate change, noise, shipping, and fisheries) (Dill 2017). Fear-released systems brought about by predator decline or removals could have unanticipated and distorted trophic cascades (Heithaus et al. 2008; Frid et al. 2008), by freeing prey and indirectly impacting fisheries resources as in terrestrial systems. One example is the systematic extirpation of wolves that left increasing elk, *Cervus canadensis* (prey), populations to overgraze and denude vegetation in Yellowstone National Park (Ripple and Beschta 2004).

Animals have evolved a versatile suite of antipredator behavioral attributes to survive and reproduce in the marine environment. These same qualities are expressed when animals are exposed to nonbiological stressors as evident from how cetaceans respond to sonar activity as they would to killer whales (Harris et al. 2018), even if less pronounced. The risk-disturbance hypothesis (Frid and Dill 2002), which states that human-caused disturbance is akin to predation risk, has motivated the redesigning of behavioral response studies to ocean noise pollution (Harris et al. 2018). Currently, new frameworks are available or in development to assess noise impacts on marine mammal vital rates by applying an expanded framework—Population Consequences of Disturbance (PCoD) model (King et al. 2015). But, there is tremendous scope to adapt and expand the use of existing predation risk models and TMII concepts to understand the effects of human-caused threats, including additive effects of predation risk and nonbiological stressors on vulnerable species.

Acknowledgment Illustrations are by Jacqui Fenner, ECS Federal, under contract to the Office of Science and Technology, National Marine Fisheries Service, National Oceanic and Atmospheric Administration.

References

Baird RW (2000) The killer whale: foraging specializations and group hunting. University of Chicago Press, Chicago

Baird RW, Dill LM (1996) Ecological and social determinants of group size in transient killer whales. Behav Ecol 7:408–416

Baird RW, Mcsweeney DJ, Bane C, Barlow J, Salden DR, Antoine LRK, Leduc RG, Webster DL (2006) Killer whales in Hawaiian waters: information on population identity and feeding habits. Pac Sci 60:523–530

Barrett-Lennard LG, Matkin CO, Durban JW, Saulitis EL, Ellifrit D (2011) Predation on gray whales and prolonged feeding on submerged carcasses by transient killer whales at Unimak Island, Alaska. Mar Ecol Prog Ser 421:229–241

Barros NB, Wells RS (1998) Prey and feeding patterns of resident bottlenose dolphins (*Tursiops truncatus*) in Sarasota Bay, Florida. J Mammal 79:1045–1059

Barros NB, Ostrom PH, Stricker CA, Wells RS (2010) Stable isotopes differentiate bottlenose dolphins off west-central Florida. Mar Mamm Sci 26:324–336

Bertram BCR (1978) Living in groups: predators and prey. Blackwell, Oxford

Black N, Ternullo R (2009) Mammal-hunting killer whales in Monterey Bay: increased trends in occurrence and predation on gray whale calves with analysis of hunting behavior, association patterns, and abundance over 22 years. In: 18th Biennial conference on the biology of marine mammals, Quebec City

Black N, Ternullo R, Schulman-Janiger A, Ellis G, Dahlheim M, Stap P (2003) Behavior and ecology of killer whales in Monterey Bay, California. In: 15th Biennial conference on the biology of marine mammals, Greensboro

Bouskila A (2010) Games played by predators and prey. In: Breed MD, Moore J (eds) Encyclopedia of animal behavior. Academic, San Diego

Bowers MT, Friedlaender AS, Janik VM, Nowacek DP, Quick NJ, Southall BL, Read AJ (2018) Selective reactions to different killer whale call categories in two delphinid species. J Exp Biol 221:1–12

Breed GA, Matthews CJ, Marcoux M, Higdon JW, Leblanc B, Petersen SD, Orr J, Reinhart NR, Ferguson SH (2017) Sustained disruption of narwhal habitat use and behavior in the presence of Arctic killer whales. Proc Natl Acad Sci USA 114:2628–2633

Brown JS (2010) Ecology of fear. In: Breed MD, Moore J (eds) Encyclopedia of animal behavior. Academic, San Diego

Brown JS, Laundre JW, Gurung M (1999) The ecology of fear: optimal foraging, game theory, and trophic interactions. J Mammal 80:385–399

Brown JS, Kotler BP, Bouskila A (2001) Ecology of fear: foraging games between predators and prey with pulsed resources. Ann Zool Fenn 38:71–87

Burrage BR (1964) An observation regarding gray whales and killer whales. Trans Kans Acad Sci 67:550–551

Cipriano F (1992) Behavior and occurrence patterns, feeding ecology, and life history of dusky dolphins (*Lagenorhynchus obscurus*) off Kaikoura, New Zealand. Dissertation, University of Arizona, Tucson, AZ

Constantine R, Visser I, Buurman D, Buurman R, Mcfadden B (1998) Killer whale (*Orcinus orca*) predation on dusky dolphins (*Lagenorhynchus obscurus*) in Kaikoura, New Zealand. Mar Mamm Sci 14:324–330

Creel S, Christianson D (2008) Relationships between direct predation and risk effects. Trends Ecol Evol 23:194–201

Cresswell WLJQJL (2010) Predator-hunting success and prey vulnerability: quantifying the spatial scale over which lethal and non-lethal effects of predation occur. J Anim Ecol 79:556–562

Curé C, Antunes R, Alves AC, Visser F, Miller P (2011) Sperm whales and pilot whales strongly react to killer whale playbacks. In: 19th Biennial conference on the biology of marine mammals, Tampa

Cure C, Antunes R, Alves AC, Visser F, Kvadsheim PH, Miller PJO (2013) Responses of male sperm whales (*Physeter macrocephalus*) to killer whale sounds: implications for anti-predator strategies. Sci Rep 3:7

Curé C, Sivle LD, Visser F, Wensveen PJ, Isojunno S, Harris CM, Kvadsheim PH, Lam FPA, Miller PJO (2015) Predator sound playbacks reveal strong avoidance responses in a fight strategist baleen whale. Mar Ecol Prog Ser 526:267–282

Dahlheim ME, Towell RG (1994) Occurrence and distribution of Pacific white-sided dolphins (*Lagenorhynchus obliquidens*) in Southeastern Alaska, with notes on an attack by killer whales (*Orcinus orca*). Mar Mamm Sci 10:458–464

Dahood A (2009) Dusky dolphin (*Lagenorhynchus obscurus*) occurrence and movements patterns near Kaikoura, New Zealand. MS, Texas A&M University

de Bruyn PJN, Tosh CA, Terauds A (2013) Killer whale ecotypes: is there a global model? Biol Rev 88:62–80 de Vos A, Justin O'Riain M, Kock AA, Meyer MA, Kotze PGH (2015) Behavior of Cape fur seals (*Arctocephalus pusillus pusillus*) in response to spatial variation in white shark (*Carcharodon carcharias*) predation risk. Mar Mamm Sci 31:1234–1251

Deecke VB, Ford JKB, Slater PJB (2005) The vocal behaviour of mammal-eating killer whales: communicating with costly calls. Anim Behav 69:395–405

Dill LM (2017) Behavioural ecology and marine conservation: a bridge over troubled water? ICES J Mar Sci 74:1514–1521

Ebert DA (2002) Ontogenetic changes in the diet of the sevengill shark (*Notorynchus cepedianus*). Mar Freshw Res 53:517–523

Eilam D (2005) Die hard: a blend of freezing and fleeing as a dynamic defense—implications for the control of defensive behavior. Neurosci Biobehav Rev 29:1181–1191

Estes JA, Tinker MT, Williams TM, Doak DF (1998) Killer whale predation on sea otters linking oceanic and nearshore ecosystems (*Orcinus orca, Enhydra lutris*). Science 282:473–476

Ferguson SH, Higdon JW, Chmelnitsky EG (2010) The rise of killer whales as a major Arctic predator. In: Ferguson SH, Lisetto LL, Mallory ML (eds) A little less Arctic: top predators in the world's largest Northern Inland Sea. Springer, Hudson Bay

Ferguson SH, Higdon JW, Westdal KH (2012a) Prey items and predation behavior of killer whales (*Orcinus orca*) in Nunavut, Canada based on Inuit hunter interviews. Aquat Biosyst 8:16

Ferguson SH, Kingsley MCS, Higdon JW (2012b) Killer whale (*Orcinus orca*) predation in a multi-prey system. Popul Ecol 54:31–41

Ford JKB, Ellis GM (1999) Transients: mammal-hunting killer whales of British Columbia, Washington, and Southeastern Alaska. University of British Columbia Press, Vancouver

Ford JKB, Ellis GM (2014) You are what you eat: foraging specializations and their influence on the social organization and behavior of killer whales. In: Yamagiwa J, Karczmarski L (eds) Primates and Cetaceans. Springer, Tokyo

Ford JKB, Reeves RR (2008) Fight or flight: antipredator strategies of baleen whales. Mamm Rev 38:50–86

Ford JKB, Ellis GM, Barrett-Lennard LG, Morton AB, Palm RS, Balcomb KC III (1998) Dietary specialization in two sympatric populations of killer whale (*Orcinus orca*) in coastal British Columbia and adjacent waters. Can J Zool 76:1456–1471

Frid A, Dill L (2002) Human-caused disturbance stimuli as a form of predation risk. Conserv Ecol 6:16

Frid A, Baker GG, Dill LM (2006) Do resource declines increase predation rates on North Pacific harbor seals? A behavior-based plausibility model. Mar Ecol Prog Ser 312:265–275

Frid A, Dill LM, Thorne RE, Blundell GM (2007) Inferring prey perception of relative danger in large-scale marine systems. Evol Ecol Res 9:635–649

Frid A, Baker GG, Dill LM (2008) Do shark declines create fear-released systems? *Oikos* 117:191–201

Gallagher AJ, Creel S, Wilson RP, Cooke SJ (2017) Energy landscapes and the landscape of fear. Trends Ecol Evol 32:88–96

Gobush KS (2010) Shark predation on Hawaiian Monk Seals: workshop II & post-workshop developments, 5–6 November 2008. US Dep. Commer., NOAA Tech. Memo., NOAA-TM-NMFS-PIFSC-21

Gowans S, Würsig B, Karczmarski L (2008) The social structure and strategies of delphinids: predictions based on an ecological framework. Adv Mar Biol 53:195–294

Harris CM, Thomas L, Falcone EA, Hildebrand J, Houser D, Kvadsheim PH, Lam F-PA, Miller PJO, Moretti DJ, Read AJ, Slabbekoorn H, Southall BL, Tyack PL, Wartzok D, Janik VM, Blanchard J (2018) Marine mammals and sonar: dose-response studies, the risk-disturbance hypothesis and the role of exposure context. J Appl Ecol 55:396–404

Heithaus MR (2001a) Predator-prey and competitive interactions between sharks (order Selachii) and dolphins (suborder Odontoceti): a review. J Zool 253:53–68

Heithaus MR (2001b) Shark attacks on bottlenose dolphins (*Tursiops aduncus*) in Shark Bay, Western Australia: attack rate, bite scar frequencies and attack seasonality. Mar Mamm Sci 17:526–539

Heithaus MR (ed) (2004) Predator-prey interactions. CRC, Boca Raton

Heithaus MR, Dill LM (2002) Food availability and tiger shark predation risk influence bottlenose dolphin habitat use. Ecology 83:480–491

Heithaus MR, Dill LM (2005) Does tiger shark predation risk influence foraging habitat use by bottlenose dolphins at multiple spatial scales? In: 16th Biennial conference on the biology of marine mammals, San Diego

Heithaus MR, Dill LM (2006) Does tiger shark predation risk influence foraging habitat use by bottlenose dolphins at multiple spatial scales? Oikos 114:257–264

Heithaus MR, Dill LM, Marshall GJ, Buhleier B (2002) Habitat use and foraging behavior of tiger sharks (*Galeocerdo cavier*) in a seagrass ecosystem. Mar Biol 140:237–248

Heithaus MR, Frid A, Wirsing AJ, Worm B (2008) Predicting ecological consequences of marine top predator declines. Trends Ecol Evol 23:202–210

Heithaus MR, Kiszka JJ, Cadinouche AL, Dulau-Drouot V, Boucaud V, Pérez-Jorge S, Webster I (2017) Spatial variation in shark-inflicted injuries to Indo-Pacific bottlenose dolphins (*Tursiops aduncus*) of the southwestern Indian Ocean. Mar Mamm Sci 33:335–341

Houghton J, Baird RW, Emmons CK, Hanson MB (2015) Changes in the occurrence and behavior of mammal-eating killer whales in Southern British Columbia and Washington State, 1987–2010. Northwest Sci 89:154–169

Hugie DM (2001) Applications of evolutionary game theory to the study of predator-prey interactions. National Library of Canada = Bibliothèque nationale du Canada

Inman AJ, Krebs J (1987) Predation and group living. Trends Ecol Evol 2:31–32

Irvine AB, Scott MD, Wells RS, Kaufmann JH (1981) Movements and activities of the Atlantic bottlenose dolphin, *Tursiops truncatus*, near Sarosota, Florida. Fish Bull 79:671–688

Jefferson TA, Stacey PJ, Baird RW (1991) A review of killer whale interactions with other marine mammals – predation to coexistence. Mamm Rev 21:151–180

King SL, Schick RS, Donovan C, Booth CG, Burgman M, Thomas L, Harwood J, Kurle C (2015) An interim framework for assessing the population consequences of disturbance. Methods Ecol Evol 6:1150–1158

Kiszka JJ, Heithaus MR, Wirsing AJ (2015) Behavioural drivers of the ecological roles and importance of marine mammals. Mar Ecol Prog Ser 523:267–281

Laidre KL, Heide-Jorgensen MP, Orr JR (2006) Reactions of narwhals, *Monodon monoceros*, to killer whale, *Orcinus orca*, attacks in the eastern Canadian Arctic. Can Field Nat 120:457–465

Lima SL (1998) Stress and decision making under the risk of predation: recent developments from behavioral, reproductive, ecological perspectives. Adv Study Behav 27:215–290

Lima SL (2002) Putting predators back into behavioral predator-prey interactions. Trends Ecol Evol 17:70–75

Lima SL, Dill LM (1990) Behavioral decisions made under the risk of predation: a review and prospectus. Can J Zool 68:619–640

Lucifora LO, Menni RC, Escalante AH (2005) Reproduction, abundance and feeding habits of the broadnose sevengill shark *Notorynchus cepedianus* in north Patagonia, Argentina. Mar Ecol Prog Ser 289:237–244

Macleod R, Macleod CD, Learmonth JA, Jepson PD, Reid RJ, Deaville R, Pierce GJ (2007) Mass-dependent predation risk and lethal dolphin porpoise interactions. Proc R Soc Lond Ser B Biol Sci 274:2587–2593

Markowitz TM (2004) Social organization of the New Zealand dusky dolphin. Doctoral dissertation, Texas A&M University

Melillo-Sweeting K, Turnbull SD, Guttridge TL (2014) Evidence of shark attacks on Atlantic spotted dolphins (*Stenella frontalis*) off Bimini, The Bahamas. Mar Mamm Sci 30:1158–1164

Moll RJ, Redilla KM, Mudumba T, Muneza AB, Gray SM, Abade L, Hayward MW, Millspaugh JJ, Montgomery RA (2017) The many faces of fear: a synthesis of the methodological variation in characterizing predation risk. J Anim Ecol 86:749–765

Mukherjee S, Heithaus MR (2013) Dangerous prey and daring predators: a review. Biol Rev 88:550–563

Norris K, Dohl T (1980) The structure and function of cetacean schools. In: Herman LH (ed) Cetacean behavior: mechanisms and functions. Wiley, New York, pp 211–261

Norris KS, Würsig B, Wells RS, Würsig M (1994) The Hawaiian spinner dolphin. University of California Press, Berkeley

O'Corry-Crowe G, Mahoney AR, Suydam R, Quakenbush L, Whiting A, Lowry L, Harwood L (2016) Genetic profiling links changing sea-ice to shifting beluga whale migration patterns. Biol Lett 12:20160404

Pitman RL, Ensor P (2003) Three forms of killer whales (*Orcinus orca*) in Antarctic waters. J Cetacean Res Manag 5:131–139

Pitman RL, Ballance LT, Mesnick SI, Chivers SJ (2001) Killer whale predation on sperm whales: observations and implications. Mar Mamm Sci 17:494–507

Pitman RL, Fearnbach H, Ballance LT, Durban JW, Totterdell JA, Kemps H (2015) Whale killers: prevalence and ecological implications of killer whale predation on humpback whale calves off Western Australia. Mar Mamm Sci 31:629–657

Pitman RL, Deecke VB, Gabriele CM, Srinivasan M, Black N, Denkinger J, Durban JW, Mathews EA, Matkin DR, Neilson JL, Schulman-Janiger A, Shearwater D, Stap P, Ternullo R (2017) Humpback whales interfering when mammal-eating killer whales attack other species: mobbing behavior and interspecific altruism? Mar Mamm Sci 33:7–58

Preisser EL, Bolnick DI, Benard MF (2005) Scared to death? The effects of intimidation and consumption in predator-prey interactions. Ecology 86:501–509

PSSMSW (2018) Salish sea marine survival project – Puget sound steelhead marine survival: 2013–2017 research findings summary. Long live the kings. Puget Sound Steelhead Marine Survival Workgroup, Seattle

Reeves RR, Berger J, Clapham PJ (2006) Killer whales as predators of large baleen whales and sperm whales. In: Estes JA, Demaster DP, Doak DF, Williams TM Jr (eds) Whales, whaling, and ocean ecosystems. University of California Press, Berkeley

Ripple WJ, Beschta RL (2004) Wolves and the ecology of fear: can predation risk structure ecosystems? Bioscience 54:755–766

Schmitz OJ, Beckerman AP, O'Brien KM (1997) Behaviorally mediated trophic cascades: effects of predation risk on food web interactions. Ecology 78:1388–1399

Schmitz OJ, Krivan V, Ovadia O (2004) Trophic cascades: the primacy of trait-mediated indirect interactions. Ecol Lett 7:153–163

Schmitz OJ, Grabowski JH, Peckarsky BL, Preisser EL, Trussell GC, Vonesh JR (2008) From individuals to ecosystem function: toward an integration of evolutionary and ecosystem ecology. Ecology 89:2436–2445

Schmitz OJ, Miller JRB, Trainor AM, Abrahms B (2017) Toward a community ecology of landscapes: predicting multiple predator-prey interactions across geographic space. Ecology 98:2281–2292

Sih A (1980) Optimal behavior – can foragers balance 2 conflicting demands. Science 210:1041–1043

Smith F, Allen SJ, Bejder L, Brown AM (2018) Shark bite injuries on three inshore dolphin species in tropical northwestern Australia. Mar Mamm Sci 34:87–99

Sorato E, Gullett PR, Griffith SC, Russell AF (2012) Effects of predation risk on foraging behaviour and group size: adaptations in a social cooperative species. Anim Behav 84:823–834

Srinivasan M, Markowitz TM (2010) Predator threats and dusky dolphin survival strategies. In: Würsig B, Würsig M (eds) The dusky dolphin. Elsevier, Amsterdam

Srinivasan M, Grant WE, Swannack TM, Rajan J (2010) Behavioral games involving a clever prey avoiding a clever predator: An individual-based model of dusky dolphins and killer whales. Ecol Model 221:2687–2598

Srinivasan M, Swannack TM, Grant WE, Rajan J, Würsig B (2018) To feed or not to feed? Bioenergetic impacts of fear-driven behaviors in lactating dolphins. Ecol Evol 8:1384–1398

Visser IN (1999) A summary of interactions between orca (*Orcinus orca*) and other cetaceans in New Zealand waters. NZ Nat Sci 24:101–112

Wcisel M, O'Riain MJ, de Vos A, Chivell W (2015) The role of refugia in reducing predation risk for Cape fur seals by white sharks. Behav Ecol Sociobiol 69:127–138

Weir JW (2007) Dusky dolphin nursery groups off Kaikoura, New Zealand. MS Texas A&M University

Weir JS, Duprey NMT, Würsig B (2008) Dusky dolphin (*Lagenorhynchus obscurus*) subgroup distribution: are shallow waters a refuge for nursery groups? Can J Zool 86:1225–1234

Weller DW (2009) Predation on marine mammals. In: Perrin WF, Würsig B, Thewissen JGM (eds) Encyclopedia of marine mammals, 2nd edn. Academic, San Diego

Wells RS (1991) The role of long-term study in understanding the social structure of a bottlenose dolphin community. In: Pryor K, Norris KS (eds) Dolphin societies – discoveries and puzzles. University of California Press, Berkeley

Wells RS, Irvine AB, Scott MD (1980) The social ecology of inshore odontocetes. Wiley, New York

Wells RS, Scott MD, Irvine AB (1987) The social structure of free-ranging bottlenose dolphins. Curr Mammal 1:247–305

Werner EE, Peacor SD (2003) A review of trait-mediated indirect interactions in ecological communities. Ecology 84:1083–1100

Westdal KH, Davies J, Macpherson A, Orr J, Ferguson SH (2016) Behavioural changes in Belugas (*Delphinapterus leucas*) during a killer whale (*Orcinus orca*) attack in southwest Hudson Bay. Can Field-Nat 130:315–319

Wilkinson KA (2014) An analysis of shark bites on resident bottlenose dolphins (*Tursiops truncatus*) in Sarasota Bay, Florida and implications for habitat use. University of Florida, Gainesville

Wirsing AJ, Heithaus MR, Dill LM (2007) Fear factor: do dugongs (*Dugong dugon*) trade food for safety from tiger sharks (*Galeocerdo cuvier*)? Oecologia 153:1031–1040

Wirsing AJ, Heithaus MR, Frid A, Dill LM (2008) Seascapes of fear: evaluating sublethal predator effects experienced and generated by marine mammals. Mar Mamm Sci 24:1–15

Würsig B (1986) Delphinid foraging strategies. In: Schusterman RJ, Thomas JA, Wood FG (eds) Dolphin cognition and behavior: a comparative approach. Lawrence Erlbaum Associates, Hillsdale

Würsig B, Cipriano F, Slooten E, Constantine R, Barr K, Yin S (1997) Dusky dolphins (*Lagenorhynchus obscurus*) of New Zealand: status of present knowledge. Reports of the Scientific Committee, International Whaling Commission

Chapter 8
Odontocete Social Strategies and Tactics Along and Inshore

Katherine McHugh

Abstract Odontocetes are social animals, and long-term studies of nearshore species have documented high levels of social complexity, cultural innovation, cooperation, and social bonding within populations. While odontocete social lives may ultimately owe their existence to the predator protection benefits of grouping, it is becoming clear that there is great variability in the nature of social relationships and fitness consequences of social behavior for whales, dolphins, and porpoises. Although much of what we know still comes from limited longitudinal studies of identified individuals from a handful of species at multiple sites, information from new populations and species highlights the flexibility and vulnerability of odontocete societies in close proximity to humans and the need for robust conservation and management plans that account for social and cultural processes.

Keywords Social complexity · Culture · Social learning · Behavioral flexibility · Cooperation · Fission–fusion dynamics · Social vulnerability

8.1 Introduction

Sociality is one of the hallmarks of the odontocete suborder, with members displaying some of the most complex societies in the animal kingdom. Sociality likely has evolved because it confers survival and reproductive advantages within varied environments. However, the strengths and complexities of some social bonds among odontocetes appear remarkable—stories of apparently grieving mothers carrying dead infants for days (Bearzi et al. 2018), entire social groups stranding alongside the sick (Odell et al. 1980), menopausal grandmothers caring for extended

K. McHugh (✉)
Chicago Zoological Society's Sarasota Dolphin Research Program, c/o Mote Marine Laboratory, Sarasota, FL, USA
e-mail: kmchugh@mote.org

© Springer Nature Switzerland AG 2019
B. Würsig (ed.), *Ethology and Behavioral Ecology of Odontocetes*, Ethology and Behavioral Ecology of Marine Mammals,
https://doi.org/10.1007/978-3-030-16663-2_8

Fig. 8.1 Resident common bottlenose dolphins, *Tursiops truncatus*, in Sarasota Bay, Florida, swim in shallow waters nearshore, where they frequently encounter human activities within their community range. Image credit: Sarasota Dolphin Research Program, taken under National Marine Fisheries Service Scientific Research Permit No. 15543

family members (Brent et al. 2015), and males forming complex lifetime alliances (Wells 2014; Connor and Krützen 2015) confound those seeking simple evolutionary explanations (see also Chap. 1).

Socioecological models link female grouping patterns to predictability and distribution of resources and potential threats from predators, with male strategies then built around distribution and availability of potentially receptive females (Gowans et al. 2008). This general mammalian "rule" of females focusing on food and fear and males monopolizing mates somewhat fits observed patterns among odontocetes, but we are still discovering the broad diversity and flexibility of social strategies among toothed whales and dolphins. Most of our understanding is based on research limited to a handful of species and almost entirely on those that live close to shore. A few long-term studies that document details of social relationships of identified individuals over multiple generations, and the potential fitness consequences of different social strategies, have shed light on richness of odontocete social lives (Wells et al. 1980; Connor et al. 1998; Mann and Karniski 2017).

Some odontocetes display a large capacity to adapt and survive in the face of rapid environmental change and increasing anthropogenic impacts. But, many odontocete populations and species are nevertheless at risk—especially those found in nearshore and inshore habitats where ranges overlap extensively with human activities (Fig. 8.1). While some nearshore populations are well documented, we may be too late in many cases to fully learn about newly acquired social patterns of which we are only starting to gain a glimpse.

8 Odontocete Social Strategies and Tactics Along and Inshore

My aim here is to introduce several themes emerging from growing knowledge about social lives of odontocetes living close to shore. This serves as an extension of Chap. 1 (by Gowans), which focused on grouping behaviors among odontocetes, and as a jumping off point for the more detailed accounts and examples presented in Part II. This chapter excludes discussions of social strategies of river dolphins (*Platanista*, *Inia*, or *Lipotes*) and porpoises (Phocoenidae) because detailed social information of these is generally lacking (summarized in chapters by Sutaria et al. and Teilmann and Sveegard, respectively).

8.2 Costs and Benefits of Grouping

For dolphins and whales in an aquatic environment with few places to hide, the primary benefit of group living appears to be predator protection from sharks and killer whales, along with perhaps some degree of protection from conspecific harassment or aggression (e.g., Wells et al. 1980; Connor 2000, 2002; Gowans et al. 2008; Möller 2012; Würsig and Pearson 2015). Cooperation in resource acquisition may also favor group formation. Major costs of group living are competition for limited and generally non-defensible food resources, mate competition, and the potential for disease transmission through social networks. Connor (2000, 2002) argued that low transport costs and large range sizes in aquatic environments help to reduce costs of grouping and philopatry for dolphins and whales, in comparison with terrestrial mammals. In addition, perhaps as a means to balance the competing costs and benefits of group formation, dynamic fission–fusion societies are a dominant social organization among nearshore odontocetes but with more stable groupings in several species and specific environments (Connor 2000; Gowans et al. 2008). Sexual segregation outside of breeding and female-biased social and/or geographic philopatry also appear repeatedly among odontocetes close to shore, again with exceptions for those where opposite sex kin are frequent associates and in a few bisexually bonded populations (Möller 2012).

Yet, despite decades of research on group dynamics and social strategies, long-term data remain limited to several species with multiple sites or populations. While long-term studies were few early on (Wells et al. 1980, Connor et al. 1998; Chap. 16), by 2017, researchers had greatly expanded the breadth of species represented by intensive, long-term studies—with 30 of 75 odontocete species now having some long-term study of at least 10 years, 19 of which have longitudinal data tracking identified individuals over time (Mann and Karniski 2017). However, there still remain only six spp. (including the two *Tursiops*) for which there are long-term longitudinal studies at multiple locations to allow for understanding of intra-specific variation in social and grouping patterns (Table 8.1).

168 K. McHugh

Table 8.1 Long-term, longitudinal studies (LTS) of odontocetes providing insight into social strategies

Family	No. of species[a]	No. of species with LTS[b]	No. of study locations[b]	IUCN status[c]
Offshore				
Physeteridae (sperm whale)	1	1	5	VU
Kogiidae (dwarf/ pygmy sperm whale)	2	–	–	DD
Ziphiidae (beaked whales)	22	3	1 (each spp.)	DD (20), LC (2)
Inshore/rivers				
Platanistidae (S. Asian river dolphin)	1	–	–	EN
Iniidae (Amazon river dolphin)	1	1	2	DD
Lipotidae (Yangtze river dolphin)	1	–	–	CR
Pontoporiidae (Franciscana)	1	–	–	VU
Arctic				
Monodontidae (narwhal, beluga)	2	1	1	LC
Cosmopolitan				
Delphinidae (dolphins)	37	13	1 (9 spp.) 2 (*S. guianensis*) 4 (*T. aduncus*) 9 (*T. truncatus*) 11 (*O. orca*)	DD (15), LC (14), NT (1), VU (3), EN (3), CR (1)
Phocoenidae (porpoises)	7	–	–	DD (2), LC (2), VU (1), EN (1), CR (1)
Total	75	19	46 (6 spp. *with multiple LTS sites*)	40 (DD), 8 (EN/CR), 7 (NT/VU), 20 (LC)

[a]Committee on taxonomy. 2017. List of marine mammal species and subspecies. Society for Marine Mammalogy, www.marinemammalscience.org, consulted on August 7, 2018
[b]Mann and Karniski 2017. LTS = includes only continuous, systematic long-term ($\geq$10 years), longitudinal studies (tracking identified individuals). No. of study locations also only includes LTS
[c]IUCN Red List of Threatened Species, www.iucnredlist.org, consulted on August 8, 2018. *DD* Data Deficient, *LC* Least Concern, *NT* Near Threatened, *VU* Vulnerable, *EN* Endangered, *CR* Critically Endangered

8.3 Social Complexity

Nearshore odontocete societies vary in group size and stability of bonds. There are short-term associations thought to be characteristic of porpoises, stable matrilineal pods of resident killer whales (*Orcinus orca*), and multitiered alliance networks of

Fig. 8.2 Male common bottlenose dolphins, *Tursiops truncatus*, engage in highly active sociosexual behavior during the juvenile period as they build relationships that may become important alliances later in life. Image credit: Sarasota Dolphin Research Program, taken under National Marine Fisheries Service Scientific Research Permit No. 20455

bottlenose dolphin (*Tursiops* sp.) males operating within a fluid fission–fusion society (Connor 2000). There is even limited evidence for socially monogamous (or at least longer term) relationships between unrelated male and female franciscana dolphins (*Pontoporia blainvillei*; Wells et al. 2013).

The majority of species live in fluid societies with a high degree of fission–fusion dynamics and complex interaction, communication, and cognitive requirements (Aureli et al. 2008). Dyadic relationships vary in strength and stability and may depend on behavioral context (e.g., Gero et al. 2005; Gazda et al. 2015). Most species display long-term bonds with kin and nonkin, often mediated by complex affiliative, aggressive, and sociosexual behaviors (Fig. 8.2). Dolphins and whales use vocal communication, touch, and synchrony to maintain affiliative bonds and display physical and acoustic aggression intra- and intersexually (Connor 2002). Cooperation between relatives and nonkin is frequent in odontocete social lives, from simple group hunting to complex cooperative foraging techniques requiring division of labor (Gazda et al. 2005), unrelated males forming lifetime cooperative alliances (Connor and Krützen 2015), and even species where menopausal grandmothers within stable matrilineal groups care for family past their own reproductive lives (e.g., Brent et al. 2015). Cooperative behaviors may provide direct or inclusive fitness benefits to those involved (or would result in fitness consequences to individuals who refused to participate).

Some larger-bodied odontocetes display stable matrilineal groups, such as sperm whales (*Physeter macrocephalus*; Whitehead 2003), killer whales (Bigg et al. 1990; Ford et al. 2000), long- and short-finned pilot whales (*Globicephala* spp., Amos et al. 1993; Kasuya and Marsh 1984), false killer whales (*Pseudorca crassidens*; Baird et al. 2008), and pygmy killer whales (*Feresa attenuata*; McSweeney et al. 2009). Strong lifetime associations are formed between family members, with daughters (and in some cases sons) staying with mothers through their own adulthood, and matriarchs likely serve pivotal leadership roles (Brent et al. 2015). Substantial allocare of young by other group members occurs (including allonursing, e.g., sperm whales, Gero et al. 2009), as does widespread food sharing (e.g., killer whales, Ford and Ellis 2006). Some matrilineal species also display an extremely rare feature among animals—extended postreproductive life spans of females who continue caring for family after ceasing their own reproduction. They are killer whales, short-finned pilot whales, false killer whales (Photopoulou et al. 2017), beluga whales (*Delphinapterus leucas*), and narwhals (*Monodon monoceros*; Ellis et al. 2018).

Allocare, food-sharing, and postreproductive care are forms of cooperative behavior that may have energetic or fitness consequences to the provider but which appear to remain viable social strategies because potential consequences are overridden by affiliative and inclusive fitness benefits of cooperating with and caring for close family members within tightly bonded kin groups. Killer whales are the only nearshore species with stable matrilineal groups and also the only species for which bisexual social and geographic philopatry is common among resident pods. Postreproductive females are more likely to lead foraging groups (Brent et al. 2015), particularly when food is scarce and for their sons, thus serving as repositories of ecological knowledge while maximizing their own inclusive fitness. Intergenerational reproductive conflict and costs to older mothers co-breeding with daughters, coupled with increases in local group relatedness with age, may have favored evolution of early reproductive senescence among resident killer whales (Croft et al. 2017).

Strong cooperative male alliances among bottlenose dolphins (*Tursiops* spp.) are a striking feature of societies that otherwise exhibit high levels of fission–fusion dynamics, with frequent changes in group composition and membership. In Sarasota Bay, Florida, unrelated males form stable long-term alliances, coordinating with a single partner to obtain mating opportunities and fend off predators and rivals, and these paired males enjoy higher reproductive success than males without a partner (Wells 2014). Elsewhere, alliances are more flexible in size, with larger alliances conferring greater fitness benefits (e.g., Wiszniewski et al. 2012). Male Indo-Pacific bottlenose dolphins in a large and open social network in Shark Bay, Australia, form multitiered alliances seeking access to receptive females (Connor and Krützen 2015; Chap. 16), possibly a unique mammalian social structure (Randić et al. 2012). While these complex alliances are currently only well understood among bottlenose dolphins studied long term, it is possible that other species or populations may display strategies that would hold more clues to development of complex social organizations and associated cognitive and communicative abilities.

Odontocete social complexity likely requires significant cognitive capacity for individual recognition and communication while tracking relationships with (and perhaps between) other community members. Social demands are probably not the sole driver of large brain size and cognitive ability in dolphins and whales, but high encephalization is thought to have developed in humans, primates, and others due primarily to challenges associated with managing a dynamic social world, including cooperation and coordination, while also linked to breadth of cultural behaviors or ecological factors (Fox et al. 2017; also Chap. 18). Although odontocetes provide an important comparative group for considering selection pressures favoring large brains (Connor 2007), their social intelligence cannot (and should not) be understood apart from special factors of their ecology that separate them from other groups like primates (Barrett and Würsig 2014). For example, odontocetes are social predators that live in an underwater environment favoring sensory and communication systems built around sound, which differs greatly from the ecological context of terrestrial counterparts. Nevertheless, some species display cognitive and communicative features indicative of the critical importance of individual recognition and long-term tracking of social relationships, such as the development and referential use of individually distinctive signature whistles in bottlenose dolphins (Sayigh et al. 1998; King et al. 2013) coupled with long-term social memory for these signals (Bruck 2013). Recent efforts to understand proximate and ultimate links between social complexity and communicative complexity in birds and other animals (e.g., Freeberg et al. 2012; Freeberg and Krams 2015), and model broader connections between cooperation, acoustic communication, and cognition in predator species (e.g., Kershenbaum and Blumstein 2017), provide new directions for understanding odontocete social lives.

8.4 Behavioral Flexibility and Cultural Diversity

Odontocete societies close to shore exhibit intraspecific variation indicative of both a high degree of behavioral flexibility and cultural diversity within and among populations. For example, killer whale ecotypes display differing behavioral patterns in feeding specializations, seasonal habitat use, group size, social organization, long-term bonds, and acoustic communication patterns (Ford and Ellis 2014; see also Chap. 11). Resident killer whales, which specialize on salmon, have stable bisexually philopatric matrilineal pods (Bigg et al. 1990). Within these pods lie the core matrilines, where both sons and daughters remain in their natal group for life, with up to four generations of maternally related individuals forming strong and stable long-term bonds and sharing stereotyped vocal calls (Ford and Ellis 2014). In contrast, transient killer whales in the same waters feed on marine mammals and generally occur in smaller groups that maximize their foraging efficiency but do not share the same strict matriline structure of residents (Baird and Whitehead 2000). Female dispersal of transients is commonly observed at sexual maturity, with sons also dispersing occasionally, and strong long-term bonds form almost exclusively between mothers and sons. Associations between matrilines are much more dynamic, without consistent long-term groupings into hierarchical pods or vocal clans, as in residents (Ford and Ellis 2014). Additional culturally transmitted ecological specializations occur in killer whale populations offshore and in other geographic areas, making killer whale ecotypes one of the strongest examples of cultural evolution in the animal kingdom, providing a better understanding of dynamics and consequences of culturally driven specialization and reproductive isolation (Whitehead and Ford 2018).

Some odontocete populations exhibit differing patterns of social associations and bonding dependent upon local ecological characteristics and geographic isolation. While bottlenose dolphins in most locations usually exhibit fluid fission–fusion societies with some stronger sex-specific bonds, a population in the fjords of Doubtful Sound, New Zealand, instead forms large mixed-sex groups displaying close, long-term associations and stable community structure with no dispersal (Lusseau et al. 2003). This population lives in a harsh, low-productivity environment that may require higher cooperation and group stability, whereby population isolation and ecological constraints may drive social behavior differences from bottlenose dolphins elsewhere (Lusseau et al. 2003). Similarly, Indo-Pacific humpback dolphins (*Sousa chinensis*) also form fluid fission–fusion communities. However, the Taiwan Strait population exhibits stronger cohesion and long-term stability in social network structure, built primarily around presumed mom–calf pairs. This unique pattern may be a response to local conditions making long-term cooperative behavior, specifically calf care, more advantageous to buffer against effects of small population size, isolation, and other stressors (Dungan et al. 2015). Roughly similar scenarios may affect sociality in facultative freshwater dolphins, with differences in social patterns observed in different environments. For example, Irrawaddy dolphins (*Orcaella brevirostris*) live in coastal waters throughout Asia as well as in river systems and inland lakes, where many populations are in severe decline. In Indonesia, the open ocean coastal population is less social with interactions focused on feeding, whereas riverine dolphins display more long-term intensive associations across behavioral contexts (Kreb 2004). This intraspecific pattern is in contrast to the "true" river dolphins whose social behavior is less well known but who appear to display less sociability and lower group sizes than coastal species (Smith and Reeves 2012; Gomez-Salazar et al. 2012).

Both intraspecific and intrapopulation differences in social behavior occur in semipelagic delphinids that switch from pelagic to nearshore waters on diurnal or seasonal bases. Two "habitat switching" species with flexibility in social behavior based on environment are (1) island-associated versus atoll-living spinner dolphins (*Stenella longirostris*) off Hawaii and the Northwestern Hawaiian Islands (Karczmarski et al. 2005) and (2) dusky dolphins (*Lagenorhynchus obscurus*) off New Zealand that show both diel and seasonal habitat shifts (Würsig and Pearson 2015). While spinner dolphins also occur offshore, the focus here is on populations that generally stay close to shore during the day and shift to deeper waters to forage at night. Off Kona in the Main Hawaiian Islands, spinner dolphins are a diel fission–fusion society, resting in small bays during the day in groups of 20–100 with flexible

8 Odontocete Social Strategies and Tactics Along and Inshore

membership that then coalesce at night in groups of hundreds to forage and socialize in deep waters—where large numbers are likely important both for predator protection and mating (Norris et al. 1994). However, off the more isolated Midway Atoll, spinner dolphins lack this diel fission–fusion pattern, instead forming stable, bisexually bonded groups that use the same atoll each day for resting and move offshore together each night to forage (Karczmarski et al. 2005). Occasional "macro fission–fusion" events bring groups of animals from one atoll to another, where they gradually integrate into the existing social group, but otherwise atoll spinner dolphins exhibit little fluidity in social associations (Karczmarski et al. 2005).

In New Zealand, dusky dolphins display behavioral flexibility in disparate habitats. They occur seasonally in coastal shallows in the Marlborough Sounds and year-round in waters off Kaikoura, where a deep canyon comes close to shore. Near Kaikoura, dusky dolphins spend days resting and socializing in shallows within a large school or smaller nursery groups and then move offshore to feed on fish and squid from the deep scattering layer at night (Würsig and Pearson 2015). They exhibit low fission–fusion dynamics, splitting up mostly during foraging, and their grouping patterns are indicative of fairly high predation risk from killer whales (Chap. 7). In contrast, a subset of the population changes foraging and social strategies as they travel to a seasonal habitat in Admiralty Bay. This winter habitat is a smaller extensive shallow area with relatively low predation risk, where dusky dolphins cooperatively herd schooling fishes in small groups during the day and rest in the same shallow nearshore waters at night (Würsig and Pearson 2015). Group size is variable, but generally smaller than the large school observed in Kaikoura, and duskies display high social fluidity with some behaviorally specific and longer-term preferential associations (Pearson et al. 2017). Although the driver of this unique seasonal "mode-switching" behavior is unknown, some bonds observed in Admiralty Bay persist across years, suggesting that seasonal migratory behavior could be a cultural tradition in dusky dolphins (Whitehead et al. 2004; Pearson et al. 2017; Chap. 18).

Arctic odontocetes—narwhals and beluga whales—have poorly understood social strategies but also display intrapopulation changes in social behavior along with seasonal habitat shifts related to ice cover. Narwhals are deep divers that congregate nearshore in bays and fjords in summer. Fidelity to nearshore summering areas appears matrilineally driven, and subpopulation structure is based on these predictable and likely culturally inherited migration patterns (Heide-Jorgensen et al. 2015). Narwhals in summer tend to travel in large groups containing many smaller, typically sexually segregated, clusters (Marcoux et al. 2009). Despite hypotheses of a matrilineal social structure and recent evidence indicating that narwhal females may exhibit postreproductivity (Ellis et al. 2018), genetic and fatty acid analysis of ice-entrapped groups suggests that relatives may not forage together, perhaps indicating instead that narwhals could display a fission–fusion social structure (Watt et al. 2015). Winter observations in the offshore pack ice are limited, but in Baffin Bay they form large aggregations of thousands of animals at remarkably high densities (on average 77 animals per km; Laidre and Heide-Jorgensen 2011), indicating a shift in sociality with season.

Belugas also display a seasonal shift in social behavior, where animals that use discrete coastal summering areas converge on common wintering grounds (O'Corry-Crowe et al. 2018). Belugas group based on relatedness during migration, indicating that relatives, especially females, maintain a migratory social structure that likely facilitates learning of different routes (Colbeck et al. 2013). Outside of migration, summer groups appear segregated by age, sex, and reproductive status (Loseto et al. 2006). Recent genetic analyses confirm natal philopatry to summer aggregation sites and migratory routes, with large numbers of closely related belugas together at these coastal sites, providing evidence that migratory culture and kinship have promoted the formation of demographically distinct subunits that nevertheless overlap spatially and temporally (O'Corry-Crowe et al. 2018). The most basic social units in belugas may be immediate matrilines, which could explain multiple observations of allonursing and allocare for individuals in human care (e.g., Leung et al. 2010; Hill and Campbell 2014) and the potential existence of menopause and postreproductive periods (Ellis et al. 2018). Older beluga females presumably could play similar roles to killer whale matriarchs serving as repositories of ecological knowledge, in this case perhaps how to avoid ice entrapment during migration. If they play a leadership role in migratory groups, as appears likely, this role may improve survival of kin during vulnerable periods.

Intrapopulation differences in social patterns in response to environmental disturbance and anthropogenic interactions also occur. Resident bottlenose dolphins in Sarasota Bay, Florida, have temporary but substantial changes in grouping and social network dynamics in response to severe harmful algal blooms, whereby dolphins associate in larger groups and with more individuals during red tide events, and the community network becomes significantly more connected and compacted (McHugh et al. 2011). It is uncertain whether these social shifts reflect an adaptive response facilitating information transfer or capitalization of different prey resources or are simply a consequence of individuals aggregating in relatively less impacted areas, but they have potential consequences in terms of heightened disease transfer during ecologically stressful times (McHugh et al. 2011). On a longer time scale, larger social restructuring occurred in response to two major hurricanes in the Bahamas, when about one-third of resident bottlenose and spotted dolphins (*Stenella frontalis*) disappeared. Here, bottlenose dolphins that initially interacted in one larger fluid fission–fusion community received an influx of immigrants and then split into two distinct units that rarely interacted though both new social clusters contained previous residents and immigrants (Elliser and Herzing 2011). In contrast, spotted dolphins showed no increase in immigration post-hurricane, but their social patterns also shifted, displaying increases in social differentiation and cohesion within previously established clusters and a change in male alliance structure, where adult alliances simplified but juveniles began to make alliance-level associations not seen previously (Elliser and Herzing 2014). Lastly, interspecific association between these two sympatric species also changed; while bottlenose and spotted dolphins still regularly interacted post-hurricane, there was a reduction in sexual-aggressive behavior and aggressive encounters (Elliser and Herzing 2016). Again, the specific reasons for these shifts are unclear, but they show a level of social flexibility in odontocetes that allows for adjustments in the face of environmental challenges.

An increasing number of studies have also documented examples of within-population differences in social patterns or social segregation based on anthropogenic interactions, especially human-centered foraging specializations utilized by only a subset of animals. For example, some bottlenose dolphins in Laguna, Brazil, participate in a cooperative foraging interaction with local beach-casting fishermen (Pryor and Lindbergh 1990), and these cooperative dolphins have stronger within-class associations and are divided into separate social clusters from noncooperative dolphins in the overall community network (Daura-Jorge et al. 2012). This social partitioning is not based solely on differences in space-use patterns but is focused around use of this specific foraging tactic, which is likely transmitted via social learning. Similar patterns occur in bottlenose dolphin populations where a subset of individuals interacts with commercial trawlers. Social differentiation and clustering within the shared community range of resident bottlenose dolphins in Savannah, Georgia, are based on human-related foraging associated with trawlers (a socially learned behavior) but not foraging via begging for food (possibly individually learned) that is distributed among clusters (Kovacs et al. 2017). In Moreton Bay, Australia, another bottlenose dolphin society changed due to interactions with fishermen, with humans driving social segregation (Chilvers and Corkeron 2001). Initially, there were two distinct social communities based on trawler-associated foraging rather than ranging differences, whereby community core areas overlapped, but individuals differed in feeding modes, habitat preferences, and group size, with trawler dolphins associating in larger groups than those that did not feed in association with humans. These social patterns nearly vanished after a reduction in commercial trawling, with the community now displaying less differentiation and more compaction in their social network, with more and stronger associations between individuals and a loss of prior partitioning, indicating that social structure can adapt quickly and may be resilient to disturbance (Ansmann et al. 2012). Lastly, killer whale social structure in the Strait of Gibraltar is being shaped by fishery interactions in a location where a subset of pods interact with a local tuna fishery (Esteban et al. 2016). Originally only one pod interacted, which then fissioned into two, with stronger within-pod interactions than seen in others, and this emerging situation may be an active example where social spreading of a novel foraging behavior begins to drive population fragmentation (Esteban et al. 2016). Thus, odontocete sociality and foraging behavior are both flexible and intertwined within local communities, with social learning of foraging specializations possibly driving social segregation.

Observed variation in foraging traditions or specializations provides some of the best examples of cultural diversity in odontocetes. Cultural processes, with social learning as a key driver, have resulted in an impressive variety of vocal and behavioral "cultures" observed among the few whales and dolphins that have been studied extensively (Whitehead and Rendell 2015). While attributions of culture to whales and dolphins have been somewhat contentious (e.g., Rendell and Whitehead 2001; Laland and Janik 2006; McGrew 2015), remarkably diverse foraging strategies and differing patterns of foraging traditions observed among and within multiple bottlenose dolphin communities most likely result from cultural transmission (see Chap. 15; Wells 2003; Whitehead and Rendell 2015). Foraging specializations are passed on via social learning among associates, and many tactics either involve social coordination (such as cooperative mud ring and driver–barrier feeding styles) or social consequences for the individuals or communities where they are practiced [such as the human-centered foraging behaviors discussed above and the difficult and time-consuming sponge tool-use of Indo-Pacific bottlenose dolphins (Krützen et al. 2005, 2014; Mann et al. 2012)]. Thus, cultural diversity and social strategies in odontocetes may be highly interdependent, with culturally transmitted ecological specializations potentially driving social patterns in more species than currently documented.

8.5 Vulnerability

Odontocete societies alongshore cannot escape the persistent influences of human activities on their health, behavior, and critical habitats. Nearshore and inshore odontocete species' ranges overlap substantially with humans, and local populations often face severe impacts from anthropogenic noise, fishing pressure, industrial and recreational activity, directed and incidental takes, coastal pollution, and habitat loss and degradation. Although many odontocete species are still relatively unknown (over half are "data deficient," IUCN 2018), 13 of the 14 species listed as Vulnerable to Endangered on the IUCN Red List are found inshore or nearshore (Table 8.1), as are all three species listed as Critically Endangered. In addition, all 11 Critically Endangered odontocete subpopulations or subspecies come from inshore or nearshore habitats, which may be isolated from other suitable areas or members of their species.

While human impacts are clearly the major cause of declines, odontocetes appear perhaps less resilient in the face of conservation challenges than their mysticete cousins, many of which have demonstrated clear signs of population recovery after depletion from past hunting (e.g., Wade et al. 2012). Although life history traits constraining rapid population growth, such as late ages of first reproduction and low calving rates, explain some of this difference, social and behavioral factors are also likely at play, whereby the obligate sociality of many odontocetes means that their survival and reproductive success depends on maintaining connections to others (Wade et al. 2012). Certain aspects of odontocete societies, including high levels of mutual dependence, social cohesion, and reliance on social groups for predator defense and care of young, as well as intergenerational transfer of knowledge and leadership of older individuals, may be especially important during times of scarce prey or in avoiding high-risk circumstances (Wade et al. 2012). Species for which these factors are relevant are especially vulnerable to impacts from lost individuals, which may result in disruption or fragmentation of social bonds/units such that removal of only a few individuals can have disproportionate impacts on survival or birth rates. The endangered southern resident killer whales are an example of this phenomenon, with shifts in social cohesion resulting from recent population declines (Parsons et al. 2009), and the potential that removal of very few key individuals could be devastating (Williams and Lusseau 2006). Disproportionate impacts occur with losses of any matriarchal repositories of information whose removal leads to reduced survival of daughters and sons (Foster et al. 2012) or of juvenile females who may act as social brokers maintaining cohesion between groups within a larger pod (Williams and Lusseau 2006).

Geographically isolated inshore odontocete populations also appear particularly vulnerable. There are now multiple examples where critically endangered subpopulations display different social dynamics—especially stronger/longer-term bonding and cooperative behavior—than observed elsewhere due in part to their isolation from others but likely also because remaining social bonds have become especially important to survival in these settings [e.g., Fiordland bottlenose dolphins (Lusseau et al. 2003), Taiwanese humpback dolphin (Dungan et al. 2015), Irrawaddy dolphins in rivers (Kreb 2004)]. Dungan et al. (2015) suggest that small populations subject to ecological stress from anthropogenic disturbance and other factors may develop these more stable social patterns to dampen impacts to survivorship and reproductive success, regardless of their isolation from others. However, Smith and Reeves (2012) stress the particular vulnerability of riverine cetaceans to effects of habitat degradation and population fragmentation due to human activities including dam construction, whereby further isolation continues to drive declines of small populations (see also Chap. 19). While the particular challenges in each situation may differ, the likelihood of heightened social reliance within struggling populations necessitates a focus on minimizing impacts that disrupt social dynamics in conservation planning, because—again—the loss of very few individuals may exert much larger population-level impacts than their numbers would otherwise suggest.

Strong, long-term bonds among odontocetes can lead to heartbreaking, and perhaps dangerous, responses to dead companions. Nurturant responses to dead conspecifics are much more commonly observed in odontocetes than mysticetes, and primarily among the Delphinidae, with probable mothers typically attending to calves or juveniles (Bearzi et al. 2018; Reggente et al. 2016). While these behaviors may begin as initial attempts to revive or protect, in many cases what could be adaptive caring behavior turns maladaptive, with long-term carrying of or standing by the dead (Bearzi et al. 2018). These behaviors may represent the strong attachment of mother–calf bonds or grief and mourning of lost long-term companions but either way likely result in at least short-term energetic consequences to attending animals and possibly longer-term fitness consequences in cases where individuals actually strand alongside others (e.g., Odell et al. 1980). In populations facing increasing stressors due to anthropogenic and environmental impacts, it is possible that what are now infrequent instances of potentially maladaptive caring for dead may become more common and pose greater risks within small, strongly bonded populations.

In some cases, social strategies that benefit individuals in other contexts may also become maladaptive in the face of anthropogenic impacts or when some individuals adopt risky behaviors. Cultural processes can both aid and constrain potential responses to environmental change (Keith and Bull 2017; Whitehead 2010; Whitehead et al. 2004), and so understanding aspects of behavioral ecology, including extent of behavioral plasticity and specialization, capacity for social learning, and patterns of individual behavioral variation, become increasingly important in conservation efforts for odontocetes the world over (Brakes and Dall 2016; Chap. 10).

While complex social interactions allow opportunities for social learning and cultural transmission that could provide resilience in the face of anthropogenic stressors and environmental change, these dynamic and widespread social connections can also leave individuals vulnerable to disease transmission or spread of maladaptive behaviors. Although potentially helpful in adapting quickly to changing environments, behavioral flexibility itself—especially in a foraging context—can be maladaptive. Foraging tactics connected to human activities, including commercial and recreational fisheries, are becoming more common among odontocete populations, and behaviors such as depredation and begging from or being provisioned by humans may pose risks to participating individuals, including entanglement in or ingestion of fishing gear or vessel strikes. While typically only a few individuals begin utilizing these strategies, they then can spread to others via social learning, potentially endangering larger proportions of vulnerable local communities (e.g., Donaldson et al. 2012a, b; Christiansen et al. 2016; see also Chap. 10).

8.6 Conclusion

While this chapter only scratches the surface of nearshore and inshore odontocete social lives, it sheds light on a few key features of toothed whale societies: social complexity, behavioral flexibility, cultural diversity, and vulnerability to anthropogenic impacts. Despite difficulties of observing social interactions underwater, long-term research is unraveling aspects of odontocete sociality that contribute to survival and reproductive success within complex societies and varied ways in which nearshore odontocetes depend upon each other. More work is needed on understudied species and geographic areas to develop a comprehensive understanding of social strategies and variability within and among taxa and provide behavioral and social integrity information that is critical for effective conservation of declining populations.

References

Amos BC, Schlötterer C, Tautz D (1993) Social structure of pilot whales revealed by analytical DNA profiling. Science 260:670–672

Ansmann IC, Parra GJ, Chilvers BL, Lanyon JM (2012) Dolphins restructure social system after reduction in commercial fisheries. Anim Behav 84:575–581

Aureli F, Schaffner CM, Boesch C, Bearder SK, Call J, Chapman CA, Connor R, Di Fiore A, Dunbar RIM, Henzi SP, Holekamp K, Korstjens AH, Layton R, Lee P, Lehmann J, Manson JH, Ramos-Fernandez G, Strier KB, van Schaik CP (2008) Fission-fusion dynamics: new research frameworks. Curr Anthropol 49:627–654

Baird RW, Whitehead H (2000) Social organization of mammal-eating killer whales: group stability and dispersal patterns. Can J Zool 78:2096–2105

Baird RW, Gorgone AM, McSweeney DJ, Webster DL, Salden DR, Deakos MH, Ligon AD, Schorr GS, Barlow J, Mahaffy SD (2008) False killer whales (*Pseudorca crassidens*) around the main Hawaiian Islands: long-term site fidelity, inter-island movements, and association patterns. Mar Mamm Sci 24:591–612

Barrett L, Würsig B (2014) Why dolphins are not aquatic apes. Anim Behav Cogn 1:1–18

Bearzi G, Kerem D, Furey NB, Pitman RL, Rendell L, Reeves RR (2018) Whale and dolphin behavioural responses to dead conspecifics. Zoology 128:1–15

Bigg MA, Olesiuk PF, Ellis GM, Ford JKB, Balcomb KC (1990) Social organization and genealogy of resident killer whales (*Orcinus orca*) in the coastal waters of British Columbia and Washington State. Rep Int Whaling Commission Spec Iss 12:383–405

Brakes P, Dall SRX (2016) Marine mammal behavior: a review of conservation implications. Front Mar Sci 3:87

Brent LJN, Franks DW, Foster EA, Balcomb KC, Cant MA, Croft DP (2015) Ecological knowledge, leadership, and the evolution of menopause in killer whales. Curr Biol 25:746–750

Bruck JN (2013) Decades-long social memory in bottlenose dolphins. Proc R Soc B 280:20131726

Chilvers BL, Corkeron PJ (2001) Trawling and bottlenose dolphins' social structure. Proc R Soc B 268:1901–1905

Christiansen F, McHugh KA, Bejder L, Siegal EM, Lusseau D, McCabe EB, Lovewell G, Wells RS (2016) Food provisioning increases the risk of injury in a long-lived marine top predator. R Soc Open Sci 3:160560

Colbeck GJ, Duchesne P, Postma LD, Lesage V, Hammill MO, Turgeon J (2013) Groups of related belugas (*Delphinapterus leucas*) travel together during their seasonal migrations in and around Hudson Bay. Proc R Soc B 280:20122552

Connor RC (2000) Group living in whales and dolphins. In: Mann J, Connor RC, Tyack PL, Whitehead H (eds) Cetacean societies: field studies of dolphins and whales. University of Chicago Press, Chicago, pp 199–218

Connor RC (2002) Ecology of group living and social behavior. In: Hoelzel AR (ed) Marine mammal biology: an evolutionary approach. Blackwell, Oxford, pp 353–370

Connor RC (2007) Dolphin social intelligence: complex alliance relationships in bottlenose dolphins and a consideration of selective environments for extreme brain size evolution in mammals. Philos Trans R Soc B 362:587–602

Connor RC, Krützen M (2015) Male dolphin alliances in Shark Bay: changing perspectives in a 30-year study. Anim Behav 103:223–235

Connor RC, Mann J, Tyack PL, Whitehead H (1998) Social evolution in toothed whales. Trends Ecol Evol 13:228–232

Croft DP, Johnstone RA, Ellis S, Nattrass S, Franks DW, Brent LJN, Mazzi S, Balcomb KC, Ford JKB, Cant MA (2017) Reproductive conflict and the evolution of menopause in killer whales. Curr Biol 27(2):298–304. https://doi.org/10.1016/j.cub.2016.12.015

Daura-Jorge FG, Cantor M, Ingram SN, Lusseau D, Simoes-Lopes PC (2012) The structure of a bottlenose dolphin society is coupled to a unique foraging cooperation with artisanal fishermen. Biol Lett 8:702–705

Donaldson R, Finn H, Bejder L, Lusseau D, Calver M (2012a) The social side of human-wildlife interaction: wildlife can learn harmful behaviours from each other. Anim Conserv 15:427–435

Donaldson R, Finn H, Bejder L, Lusseau D, Calver M (2012b) Social learning of risky behavior: importance for impact assessments, conservation and management of human-wildlife interactions. Anim Conserv 15:442–444

Dungan SZ, Wang JY, Araújo CC, Yang SC, White BN (2015) Social structure in a critically endangered Indo-Pacific humpback dolphin (*Sousa chinensis*) population. Aquat Conserv Mar Freshw Ecosyst. https://doi.org/10.1002/aqc.2562

Ellis S, Franks DW, Nattrass S, Currie TE, Cant MA, Giles D, Balcomb KC, Croft DP (2018) Analyses of ovarian activity reveal repeated evolution of post-reproductive lifespans in toothed whales. Sci Rep 8:12833

Elliser CR, Herzing DL (2011) Replacement dolphins? Social restructuring of a resident pod of Atlantic bottlenose dolphins, *Tursiops truncatus*, after two major hurricanes. Mar Mamm Sci 27:39–59

Elliser CR, Herzing DL (2014) Social structure of Atlantic spotted dolphins, *Stenella frontalis*, following environmental disturbance and demographic changes. Mar Mamm Sci 30:329–347

Elliser CR, Herzing DL (2016) Changes in interspecies association patterns of Atlantic bottlenose dolphins, *Tursiops truncatus*, and Atlantic spotted dolphins, *Stenella frontalis*, after demographic changes related to environmental disturbance. Mar Mamm Sci 32:602–628

Esteban R, Verborgh P, Gauffier P, Giménez J, Foote AD, de Stephanis R (2016) Maternal kinship and fisheries interaction influence killer whale social structure. Behav Ecol Sociobiol 70:111–122

Ford JKB, Ellis GM (2006) Selective foraging by fish-eating killer whales *Orcinus orca* in British Columbia. Mar Ecol Prog Ser 316:185–199

Ford JKB, Ellis GM (2014) You are what you eat: foraging specializations and their influence on the social organization and behavior of killer whales. In: Yamagiwa J, Karczmarski L (eds) Primates and Cetaceans: field research and conservation of complex mammalian societies. Springer, Tokyo

Ford JKB, Ellis GM, Balcomb KC (2000) Killer whales: the natural history and genealogy of *Orcinus orca* in the waters of British Columbia and Washington. University of British Columbia Press, Vancouver

Foster EA, Franks DW, Mazzi S, Darden SK, Balcomb KC, Ford JKB, Croft DP (2012) Adaptive prolonged postreproductive life span in killer whales. Science 337:1313

Fox KCR, Muthukrishna M, Shultz S (2017) The social and cultural roots of whale and dolphin brains. Nat Ecol Evol. https://doi.org/10.1038/s41559-017-0336-y

Freeberg TM, Krams I (2015) Does social complexity link vocal complexity and cooperation? J Ornithol. https://doi.org/10.1007/s10336-015-1233-2

Freeberg TM, Dunbar RIM, Ord TJ (2012) Social complexity as a proximate and ultimate factor in communicative complexity. Philos Trans R Soc B 367:1785–1801

Gazda SK, Connor RC, Edgar RK, Cox F (2005) A division of labour with role specialization in group-hunting bottlenose dolphins (*Tursiops truncatus*) off Cedar Key, Florida. Proc R Soc B 272:135–140

Gazda S, Iyer S, Killingback T, Connor R, Brault S (2015) The importance of delineating networks by activity type in bottlenose dolphins (*Tursiops truncatus*) in Cedar Key, Florida. R Soc Open Sci 2:140263

Gero S, Bejder L, Whitehead H, Mann J, Connor R (2005) Behaviorally specific preferred associations in bottlenose dolphins, *Tursiops* spp. Can J Zool 83:1566–1573

Gero S, Engelhaupt D, Rendell L, Whitehead H (2009) Who cares? Between-group variation in alloparental caregiving in sperm whales. Behav Ecol 20:838–843

Gomez-Salazar C, Trujillo F, Whitehead H (2012) Ecological factors influencing group sizes of river dolphins (*Inia geoffrensis* and *Sotalia fluviatilis*). Mar Mamm Sci 28:E124–E142

Gowans S, Würsig B, Karczmarski L (2008) The social structure and strategies of delphinids: predictions based on an ecological framework. Adv Mar Biol 53:195–294

Heide-Jorgensen MP, Nielsen NH, Hansen RG, Schmidt HC, Blackwell SB, Jorgensen OA (2015) The predictable narwhal: satellite tracking shows behavioural similarities between isolated subpopulations. J Zool 297:54–65

Hill HM, Campbell CA (2014) The frequency and nature of allocare by a group of belugas (*Delphinapterus leucas*) in human care. Int J Comp Psychol 27:501–514

Karczmarski L, Würsig B, Gailey G, Larson KW, Vanderlip C (2005) Spinner dolphins in a remote Hawaiian atoll: social grouping and population structure. Behav Ecol 16:675–685

Kasuya T, Marsh H (1984) Life history and reproductive biology of the short-finned pilot whale, *Globicephala macrorhynchus*, off the Pacific Coast of Japan. Rep Int Whaling Commission (Spec Iss) 6:259–310

Keith SA, Bull JW (2017) Animal culture impacts species' capacity to realise climate-driven range shifts. Ecography 40:296–304

Kershenbaum A, Blumstein DT (2017) Introduction to the special column: communication, cooperation, and cognition in predators. Curr Zool 63:295–299

King SL, Sayigh LS, Wells RS, Fellner W, Janik VM (2013) Vocal copying of individually distinctive signature whistles in bottlenose dolphins. Proc R Soc B 280:20130053

Kovacs CJ, Perrtree RM, Cox TM (2017) Social differentiation in common bottlenose dolphins (*Tursiops truncatus*) that engage in human-related foraging behaviors. PLoS One 12:e0170151

Kreb D (2004) Facultative river dolphins: conservation and social ecology of freshwater and coastal Irrawaddy dolphins in Indonesia. PhD Thesis, University of Amsterdam

Krützen M, Mann J, Heithaus MR, Connor RC, Bejder L, Sherwin WB (2005) Cultural transmission of tool use in bottlenose dolphins. Proc Natl Acad Sci 102:8939–8943

Krützen M, Kriecker S, MacLeod CD, Learmonth J, Kopps AM, Walsham P, Allen SJ (2014) Cultural transmission of tool use by Indo-Pacific bottlenose dolphins (*Tursiops* sp.) provides access to a novel foraging niche. Proc R Soc B 281:201440374

Laidre KL, Heide-Jorgensen MP (2011) Life in the lead: extreme densities of narwhals *Monodon monoceros* in the offshore pack ice. Mar Ecol Prog Ser 423:269–278

Laland KN, Janik VM (2006) The animal cultures debate. Trends Ecol Evol 21:542–547

Leung ES, Vergara V, Barrett-Lennard LG (2010) Allonursing in captive belugas (*Delphinapterus leucas*). Zoo Biol 29:1–5

Loseto LL, Richard P, Stern GA, Orr J, Ferguson SH (2006) Segregation of Beaufort Sea beluga whales during the open-water season. Can J Zool 84:1743–1751

Lusseau D, Schneider K, Boisseau OJ, Haase P, Slooten E, Dawson SM (2003) The bottlenose dolphin community of Doubtful Sound features a large proportion of long-lasting associations—can geographic isolation explain this unique trait? Behav Ecol Sociobiol 54:396–405

Mann J, Karniski C (2017) Diving beneath the surface: long-term studies of dolphins and whales. J Mammal 98:621–630

Mann J, Stanton MA, Patterson EM, Beinenstock EJ, Singh LO (2012) Social networks reveal cultural behavior in tool-using dolphins. Nat Commun 3:980

Marcoux M, Auger-Methe M, Humphries MM (2009) Encounter frequencies and grouping patterns of narwhals in Koluktoo Bay, Baffin Island. Polar Biol 32:1705–1716

McGrew WC (2015) Cetaceans in the culture club? Curr Anthropol 56:927–928

McHugh KA, Allen JB, Barleycorn AA, Wells RS (2011) Severe *Karenia brevis* red tides influence juvenile bottlenose dolphin (*Tursiops truncatus*) behavior in Sarasota Bay, Florida. Mar Mamm Sci 27:622–643

McSweeney DJ, Baird RW, Mahaffy SD, Webster DL, Schorr GS (2009) Site fidelity and association patterns of a rare species: Pygmy killer whales (*Feresa attenuata*) in the main Hawaiian Islands. Mar Mamm Sci 25:557–572

Möller LM (2012) Sociogenetic structure, kin associations and bonding in delphinids. Mol Ecol 21:745–764

Norris KS, Würsig B, Wells RS, Würsig M (1994) The Hawaiian spinner dolphin. University of California Press, Berkeley

O'Corry-Crowe G, Suydam R, Quakenbush L, Potgieter B, Harwood L, Litovka D, Ferrer T, Citta J, Burkanov V, Frost K, Mahoney B (2018) Migratory culture, population structure and stock identity in North Pacific beluga whales (*Delphinapterus leucas*). PLoS One 13:e0194201

Odell DK, Asper ED, Baucom J, Cornell LH (1980) A recurrent mass stranding of the false killer whale, *Pseudorca crassidens*, in Florida. Fish Bull 78:171–177

Parsons KM, Balcomb KC, Ford JKB, Durban JW (2009) The social dynamics of southern resident killer whales and conservation implications for this endangered population. Anim Behav 77:963–971

Pearson HC, Markowitz TM, Weir JS, Würsig B (2017) Dusky dolphin (*Lagenorhynchus obscurus*) social structure characterized by social fluidity and preferred companions. Mar Mamm Sci 33:251–276

Photopoulou T, Ferreira IM, Best PB, Kasuya T, Marsh H (2017) Evidence for a postreproductive phase in female false killer whales *Pseudorca crassidens*. Front Zool 14:30

Pryor K, Lindbergh J (1990) A dolphin-human fishing cooperative in Brazil. Mar Mamm Sci 6:77–82

Randić S, Connor RC, Sherwin WB, Krützen M (2012) A novel mammalian social structure in Indo-Pacific bottlenose dolphins (*Tursiops* sp): complex male alliances in an open social network. Proc R Soc B 279:3083–3090

Reggente MAL, Alves F, Nicolau C, Freitas L, Cagnazzi D, Baird RW, Galli P (2016) Nurturant behavior toward dead conspecifics in free-ranging mammals. J Mammal 97:1428–1434

Rendell L, Whitehead H (2001) Culture in whales and dolphins. Behav Brain Sci 24:309–382

Sayigh LS, Tyack PT, Wells RS, Solow AR, Scott MD, Irvine AB (1998) Individual recognition in wild bottlenose dolphins: a field test using playback experiments. Anim Behav 57:41–50

Smith BD, Reeves RR (2012) River cetaceans and habitat change: generalist resilience or specialist vulnerability. J Mar Biol 201:718935

Wade PR, Reeves RR, Mesnick SL (2012) Social and behavioural factors in cetacean responses to overexploitation: are odontocetes less "resilient" than mysticetes? J Mar Biol 2012:567276

Watt CA, Petersen SD, Ferguson SH (2015) Genetics and fatty acids assist in deciphering narwhal (*Monodon monoceros*) social groupings. Polar Biol 38:1971–1981

Wells RS (2003) Dolphin social complexity: lessons from long-term study and life history. In: de Waal FBM, Tyack PL (eds) Animal social complexity: intelligence, culture, and individualized societies. Harvard University Press, Cambridge, pp 32–56

Wells RS (2014) Social structure and life history of bottlenose dolphins near Sarasota Bay, Florida: insights from four decades and five generations. In: Yamagiwa J, Karczmarski L (eds) Primates and Cetaceans: field research and conservation of complex mammalian societies. Springer, Tokyo, pp 149–172

Wells RS, Irvine AB, Scott MD (1980) The social ecology of inshore odontocetes. In: Herman LM (ed) Cetacean behavior: mechanisms and processes. Wiley, New York, pp 263–317

Wells RS, Bordino P, Douglas DC (2013) Patterns of social association in the franciscana, *Pontoporia blainvillei*. Mar Mamm Sci 29:E520–E528

Whitehead H (2003) Sperm whales: social evolution in the ocean. University of Chicago Press, Chicago

Whitehead H (2010) Conserving and managing animals that learn socially and share cultures. Learn Behav 38:329–336

Whitehead H, Ford JKB (2018) Consequences of culturally-driven ecological specialization: killer whales and beyond. J Theor Biol 456:279–294

Whitehead H, Rendell L (2015) The cultural lives of whales and dolphins. The University of Chicago Press, Chicago

Whitehead H, Rendell L, Osborne RW, Würsig B (2004) Culture and conservation of non-humans with reference to whales and dolphins: review and new directions. Biol Conserv 120:427–437

Williams R, Lusseau D (2006) A killer whale social network is vulnerable to targeted removals. Biol Lett 2(4):497–500

Wiszniewski J, Corrigan S, Beheregaray LB, Möller LM (2012) Male reproductive success increases with alliance size in Indo-Pacific bottlenose dolphins (*Tursiops aduncus*). J Anim Ecol 81:423–431

Würsig B, Pearson HC (2015) Dolphin societies: structure and function. In: Herzing DL, Johnson CM (eds) Dolphin communication and cognition: past, present and future. MIT Press, Cambridge, pp 77–105

Chapter 9
Oceanic Dolphin Societies: Diversity, Complexity, and Conservation

Sarah L. Mesnick, Lisa T. Ballance, Paul R. Wade, Karen Pryor, and Randall R. Reeves

Abstract Sociality—collective living—confers multiple advantages to oceanic dolphins, including enhanced foraging, predator avoidance, and alloparental care and may be particularly important in oceanic environments where prey is patchy and refuge nonexistent. This chapter covers broad aspects of the social lives of the delphinid community that inhabits the vast eastern tropical Pacific Ocean (ETP). Our approach is socio-ecological: the chapter ties dolphin social structure and mating systems to environmental factors, including oceanographic patterns, distribution of prey, and risk of predation that shape behavior. By merging a top-down look at schools distributed over a variable environment, with a bottom-up look from the perspective of subgroups that comprise schools, a picture of fission–fusion

---

The original version of this chapter was revised. A correction to this chapter can be found at
https://doi.org/10.1007/978-3-030-16663-2_24

S. L. Mesnick (✉)
Southwest Fisheries Science Center, National Marine Fisheries Service, National Oceanic and Atmospheric Administration, La Jolla, CA, USA

Scripps Institution of Oceanography, UC San Diego, La Jolla, CA, USA
e-mail: sarah.mesnick@noaa.gov

L. T. Ballance
Southwest Fisheries Science Center, National Marine Fisheries Service, National Oceanic and Atmospheric Administration, La Jolla, CA, USA

Scripps Institution of Oceanography, UC San Diego, La Jolla, CA, USA

Marine Mammal Institute, Oregon State University, Newport, OR, USA

P. R. Wade
Alaska Fisheries Science Center, National Marine Fisheries Service, National Oceanic and Atmospheric Administration, Sand Point, WA, USA

K. Pryor
Charlestown, MA, USA

R. R. Reeves
Okapi Wildlife Associates, Hudson, QC, Canada

© Springer Nature Switzerland AG 2019
B. Würsig (ed.), *Ethology and Behavioral Ecology of Odontocetes*, Ethology and Behavioral Ecology of Marine Mammals,
https://doi.org/10.1007/978-3-030-16663-2_9

societies emerges. We also consider impacts of the tuna purse seine fishery on the socio-ecology of affected dolphins and discuss likely effects on behavior, learning, social bonds, and population dynamics.

ETP dolphin societies are diverse, spatially and compositionally fluid (pure or mixed species), yet socially complex and structured. They have distinct schooling, reproductive, and sexual characteristics, different patterns of association with other species, and differing degrees of interaction with the tuna purse seine fishery. Individuals may have distinct roles (older, experienced, and post-reproductive females), form stable or at least semi-stable subgroups (female/young, adult male, juvenile), and leave or join the company of others in response to a variety of social and ecological factors, including distribution of prey and risk of predation. In some taxa, individuals school with a small number of companions who may be related and recognize one another (common bottlenose, *Tursiops truncatus*; Risso's, *Grampus griseus*; rough-toothed, *Steno bredanensis*; and striped dolphins, *Stenella coeruleoalba*), while in other species school size is larger, membership is fluid, and unrelated individuals abound (pantropical spotted, *Stenella attenuata*; spinner, *Stenella longirostris*; and common dolphins, *Delphinus delphis*). Mating systems are variable among species and sometimes within species, likely reflecting differences in habitat productivity. In some taxa, e.g., eastern spinners (*S. l. orientalis*), a few sexually mature males may be responsible for most mating, while in other taxa, e.g., "whitebelly" spinners, large relative testes suggest a more "open" mating system where many males in the school engage in copulation.

For pantropical spotted and spinner dolphins in the ETP, the behavior of schooling with large tuna that has led to their ecological success and abundance has also led to their depletion by making them a target of purse seiners. Schooling and sociality, normally adaptive traits, have caused ETP dolphins to become collateral damage in the tuna fishery. Yet dolphins have learned some things from their experiences with purse seiners. Some individuals know how to evade capture or, alternatively, how to await a lowering of the net ("backdown") to escape. But, behavior that helps to avoid capture can cause high stress, exertion, or social separation and disruption, and these could be factors slowing or inhibiting population recovery. Survival and reproductive success of oceanic dolphins likely depends largely on social and behavioral factors that may also help determine their ability to recover from severe depletion caused by human activities.

Keywords Oceanic dolphins · Eastern tropical Pacific Ocean · Pantropical spotted dolphin · Spinner dolphin · Social structure · Social organization · Mating systems · Social disruption · Resilience · Tuna purse seine fishery

9.1 Introduction

Sociality—collective living—defines odontocete cetaceans, including oceanic dolphins. Group living enhances foraging, improves predator avoidance, and allows alloparental care, and it may be particularly advantageous in unpredictable oceanic environments that vary over space and time (Norris and Schilt 1988; Whitehead 2007; Acevedo-Gutierrez 2018). Survival and reproductive success largely depend

9 Oceanic Dolphin Societies: Diversity, Complexity, and Conservation

Fig. 9.1 Dolphins of the eastern tropical Pacific (ETP). Top: Central American spinner dolphin, "whitebelly" spinner dolphin; Middle: striped dolphin, coastal pantropical spotted dolphin; Bottom: offshore pantropical spotted dolphin, short-beaked common dolphin. Latin names are in Appendix 1. The species have overlapping ranges but distinct habitat preferences and different school sizes, social structure, and behavior (see text). Photo credit: Southwest Fisheries Science Center, NOAA Fisheries

on social and behavioral factors, which may help determine the ability to recover from severe depletion caused by human activities (Wade et al. 2012).

This chapter covers broad aspects of the behavioral ecology of dolphins in the vast eastern tropical Pacific Ocean (ETP) (Fig. 9.1). Our approach is socio-ecological: we tie dolphin sociality to the environmental factors that help to shape behavior, including oceanographic patterns, distribution of prey, and risk of predation. We focus on pantropical spotted (*Stenella attenuata*) and spinner (*S. longirostris*) dolphins, the most intensely studied oceanic dolphins because of their interactions with the tuna purse seine fishery. We review social structure and sexual systems and consider behavioral responses to environmental variation and anthropogenic disturbance.

This chapter relies heavily on the knowledge of pioneers who devoted themselves to understanding behavior of oceanic dolphins so that animals captured in purse seine nets could be released alive (Perrin 1969; Norris et al. 1978; Pryor and Kang 1980), observations from dedicated surveys onboard research and fishing vessels (Au and Pitman 1986; Scott and Cattanach 1998; Gerrodette and Forcada 2005; Wade et al. 2007; Gerrodette et al. 2008; Cramer et al. 2008), data from specimens collected from the fishery (Perrin 1975; Perrin et al. 1976; Perrin and Reilly 1984; Hohn et al. 1985; Myrick et al. 1986; Scott et al. 2012), and a comparative approach.

9.1.1 ETP Delphinids: Diversity and a Brief Overview of Factors Related to Sociality

The ETP supports a diverse and abundant community of odontocetes, including 15 species of delphinids (Appendix 1; we use common names in the text, Latin names, subspecies and management stocks are listed in the appendix; Ballance et al. 2006). It is the only tropical ocean with multiple endemic subspecies of spinner and pantropicalspotted dolphins and several species are divided into stocks for management purposes. ETP delphinids occupy overlapping but different habitats within the region, and have overlapping but varying prey preferences (Au and Pitman 1986; Reilly and Fiedler 1994; Forney et al. 2012; Scott et al. 2012). They have distinct schooling, reproductive, and sexual characteristics, different patterns of association with other species, and differing degrees of interaction with the tuna purse seine fishery (Perrin and Reilly 1984; Perrin 1975; Perrin et al. 1976, 1991; Gerrodette et al. 2008). They provide an excellent opportunity for examining the nature of sociality in oceanic habitats and the impacts of human activities on these social systems.

ETP dolphins share a number of life history characteristics that contribute to their sociality. They have overlapping generations, long life spans (25–40+ years for many species; Perrin and Reilly 1984), large brains (Würsig 2018), and protracted periods of maternal investment (Perrin and Reilly 1984). The period of maternal care typically ranges from ca. 2 to 4 years and may be as high as 5+ years (Perrin and Reilly 1984; Cramer et al. 2008). Some species have post-reproductive females but numbers are small (Perrin and Reilly 1984; Myrick et al. 1986). Little is known about the role that reproductively senescent females may have in ETP dolphins, but the idea that older, more experienced individuals might help raise young or be repositories of socio-ecological knowledge, such as where to find prey or how to avoid risky situations, is thought to be important in other social odontocetes (Whitehead 2003; Foster et al. 2012; Kasuya 2017; Wade et al. 2012; Photopoulou et al. 2017; Ellis et al. 2018).

ETP dolphins are highly mobile, covering large distances in search of food that tends to be patchy in tropical oceans (Ballance et al. 2006). Prey includes epi- and

9 Oceanic Dolphin Societies: Diversity, Complexity, and Conservation 187

mesopelagic small schooling fishes and squids, larger fishes, and, for certain larger dolphin species, other cetaceans (Pitman and Stinchcomb 2002; Scott et al. 2012; Jefferson and LeDuc 2018). ETP dolphins may have overlapping diets but also exhibit distinct foraging specializations. Pantropical spotted and spinner dolphins, for example, are largely nocturnal foragers on the deep-scattering layer yet also show evidence of prey species specialization (Scott et al. 2012).

ETP dolphins contend with other predatory cetaceans, especially killer whales (*Orcinus orca*), and with large sharks (Au 1991; Scott et al. 2012). Travelling in groups presents more sensory input for locating prey and also provides more protection from predation than travelling alone in a habitat with nowhere to hide other than within the group (the cover-seeking concept; Williams 1964). The advantage can extend to multispecies aggregations where the combined number of individuals of all species may decrease risk of predation to individuals (an extension of the selfish herd hypothesis; Hamilton 1971) and increase sensory capabilities to detect predators more efficiently than one species alone could provide (Norris and Dohl 1980; Scott et al. 2012).

To coordinate travel, foraging, and predator defense, ETP dolphins communicate over large distances. On a more intimate scale, they also need to communicate to coordinate social interactions such as mating, alloparental care, affiliation, and aggression. Dolphin communication involves at least three basic modalities: acoustic, visual, and tactile. Water transmits sound efficiently, providing an effective medium for long-distance communication. Dolphins produce a diversity of acoustic signals that play different roles in communication and convey a broad range of intention cues (Lammers et al. 2003) (see Chap. 2). In the ETP, maximum detection distances of over 11 km have been recorded (Rankin et al. 2008), and many species can be discriminated acoustically (Oswald et al. 2003). Visual communication is limited to perhaps 10s of meters. Coordinated movements may be facilitated by face and body coloration, which provides signs of direction and orientation of school members. Distinct pigment patterns and sexually dimorphic features such as the brilliant white rostrum tip (pantropical spotted dolphin) and enlarged post-anal keel (eastern spinner dolphin) can provide clear visual information about age and sex of individuals swimming together (Würsig et al. 1990). Lastly, the most intimate and shortest-distance communication occurs in the tactile domain, which can be used to convey a wide range of emotional or alertness states. Importantly, oceanic dolphins integrate the joint sensory capabilities of many animals sensing many stimuli and in many directions to transmit important environmental information within the school (Norris and Dohl 1980).

9.1.2 The Eastern Tropical Pacific Is a Unique Habitat for Oceanic Dolphins

The ETP as used here refers to a 21 million km^2 open-ocean region bounded to the east by the coasts of North, Central, and South America from Southern California to Northern Peru and extending westward to the central Pacific. The ETP is primarily oceanic (seaward of the continental shelf) and includes three water masses, three

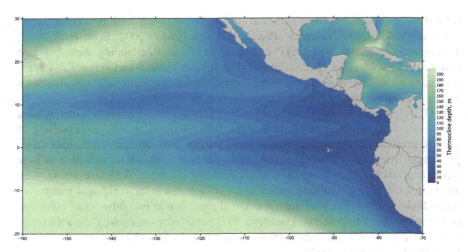

Fig. 9.2 Thermocline depth in the eastern tropical Pacific (ETP). Background shading is annual mean thermocline depth (m, 1980–2015); adapted from Fiedler et al. (2017)

major surface currents, and a number of smaller-scale semipermanent surface features including the Costa Rica Dome and Equatorial Front (Appendix 2; Fiedler and Talley 2006). Far from homogeneous, the physical environment is spatially heterogeneous and temporally variable on seasonal, interannual, and multi-year scales. This heterogeneity is also characteristic of the water column. Of particular importance biologically is the thermocline, sharp and shallow in the east and along the 10°N latitude thermocline ridge, where it can be just 10s of meters below the surface (Fig. 9.2; Au and Perryman 1985; Ballance et al. 2006; Fiedler and Lavin 2017).

A hallmark of the ETP is the prevalence of multispecies aggregations of pantropical spotted and spinner dolphins, yellowfin tuna (*Thunnus albacares*), and large, speciose flocks of seabirds (Au and Perryman 1985). The shallow, typically sharp thermocline of the ETP is thought to be a critical factor accounting for the association of tuna and dolphins (Au and Perryman 1985). Although the dolphins and tuna are pantropical, the association is more common in the ETP than in other tropical oceans, and in the ETP, it is more prevalent where the thermocline is shallow and sharp (in the east and along the 10°N thermocline ridge; Fig. 9.2). Several dolphin taxa are involved, most often the offshore pantropical spotted dolphin and the eastern spinner dolphin. The association between dolphins and tuna is a strong bond that appears to be based on reduced risk of predation for both species (Scott et al. 2012), whereas association of seabirds with dolphins and tunas is based on access to prey (Au and Pitman 1986; Ballance et al. 1997). The dolphin species that associate with tuna are abundant and ecologically successful compared to other non-tuna-associated dolphin species (Au and Pitman 1986).

Surfacing dolphins and flocks of seabirds provide a clear visual signal of the presence of tuna schools, and by the 1950s, the fishing industry was using "dolphin sets," encircling schools of dolphins with purse seine nets to catch large schools of large-bodied tuna (Perrin 1969), forming the basis for one of the world's largest tuna

9 Oceanic Dolphin Societies: Diversity, Complexity, and Conservation 189

fisheries. When "setting on dolphins," schools are chased and corralled by speed-boats until the purse seine can be wrapped around them and co-schooling tuna. In the early years of the fishery, incidental mortality of dolphins was high. It is estimated that more than six million dolphins were killed in dolphin sets, four million by 1972. This incidental mortality was a major driver of passage of the US Marine Mammal Protection Act (MMPA) of 1972 (Lo and Smith 1986; Wade 1995; Gosliner 1999). The decline in numbers of individuals in three dolphin stocks was so significant that these stocks have been considered depleted under the MMPA (Wade et al. 2007). By 1990, changes in fishing gear and practices had significantly reduced incidental mortality of dolphins in the fishery, creating an expectation that depleted dolphin stocks would recover. However, interactions with the fishery still occur frequently. Every year there are ca. 11,000 sets on dolphin schools and millions of animals are chased, captured, and released annually (Archer et al. 2010a; IATTC 2018). The most recent data from fishery-independent assessment surveys have indicated that dolphin stocks are not showing clear signs of recovery and a number of hypotheses have been posited to explain the findings (Gerrodette and Forcada 2005; Wade et al. 2007; Gerrodette et al. 2008).

In addition to hypotheses related to the demographic effects of mortality and injury caused by the fishery and the effects of ecosystem change on the dolphins, Wade et al. (2012) proposed that the purse seine fishery could have affected, and could be continuing to affect, the dolphin populations in other ways. Those authors suggested that disruption of the social structure of dolphin schools and of key social bonds among individuals could be a factor inhibiting, or at least slowing, population recovery. Wade et al. (2012) cited the ETP dolphins as one of several odontocete examples in support of the hypothesis that populations of social odontocetes are less "resilient" to heavy exploitation than populations of mysticetes due mainly to social and behavioral factors.

The remainder of this chapter focuses on the social structure and mating systems of ETP dolphins as we consider their behavioral responses to natural variation and anthropogenic disturbance. The world is increasingly human-dominated, and the question of the resilience of human-disturbed animal populations is likely to arise more frequently in the future. Our experiences may inform this question in other systems.

9.2 Schooling

Schooling has a number of important functions that likely contribute to survival and reproductive success including enhanced foraging, predator avoidance and communal defense, sensory integration, and social facilitation. Such functions may be particularly important in oceanic environments where prey is patchy and physical refuge nonexistent (Norris and Dohl 1980; Norris and Schilt 1988; Norris and Johnson 1994; Whitehead 2007; Acevedo-Gutierrez 2018). Wade et al. (2012) highlighted the importance of several social and behavioral factors for highly social odontocetes, specifically (1) social cohesion and social organization, (2) cooperation

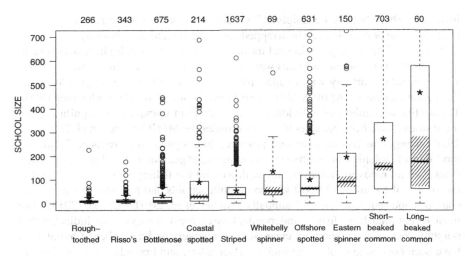

Fig. 9.3 School size distributions for 10 dolphin species, subspecies and stocks from all ETP surveys 1986–2006, pure single species) schools only. Calibration factors as described in Gerrodette et al. (2002) have been applied. Sample sizes are shown along the top. Means (*), medians (heavy horizontal lines), 95% confidence intervals on the medians (hatched boxes), interquartile ranges (open boxes), standard spans (dashed lines), and outliers (circles) are shown for sightings used in abundance estimation. Some outliers are not shown. See text for comparison of group size in pure *versus* mixed schools

in avoiding predators, (3) allomaternal care such as "babysitting" and communal nursing, (4) opportunities for social learning, (5) and perhaps leadership by older or more experienced individuals that know where and when to find scarce prey resources and how to avoid predators or respond appropriately to human activities.

The benefits of schooling can also accrue to mixed (two or more) species aggregations, especially when dolphins have similar requirements and abilities, feed on similar prey, swim at comparable speeds, and perhaps share acoustic and social signaling. Although some species rarely if ever swim with different species, others occur frequently in mixed-species groups. In the ETP, common bottlenose dolphins (*Tursiops truncatus*) regularly associate with a number of other cetacean species, while striped dolphins (*Stenella coeruleoalba*) rarely do (although they regularly occur with common dolphins (*Delphinus delphis*) in the California Current). Fraser's dolphins (*Lagenodelphis hosei*) and melon-headed whales (*Peponocephala electra*) often swim together, and pantropical spotted and spinner dolphins are regularly found in mixed-species schools, although they do so less commonly in other areas.

School size differs among dolphin species and, geographically, within species. In the ETP, school size is highly variable and strongly skewed, ranging from a dozen individuals to several thousand (Fig. 9.3). Common bottlenose, Risso's (*Grampus griseus*), rough-toothed (*Steno bredanensis*), and striped dolphins all have smaller median group sizes, while pantropical spotted, spinner, and particularly common dolphins occur in larger and sometimes extremely large schools. The diversity of school sizes among species suggests different ways of group living in the open ocean. At one end of the continuum are species in which membership and

9 Oceanic Dolphin Societies: Diversity, Complexity, and Conservation

knowledge are shared among a small number of companions who may all be related and recognize one another. At the other end are large schools where membership is fluid and unrelated individuals abound. It follows that membership may be more stable in the former and more dynamic in the latter, but this remains to be determined. Although school size is variable within a species, individual species seem to have characteristic average school sizes. For example, a comparison of survey results from the tropical Indian Ocean, ETP, and Gulf of Mexico shows that, on average, pantropical spotted, spinner, and common dolphins consistently occur in larger schools, and striped and rough-toothed dolphins occur in smaller schools (Ballance and Pitman 1998). The factors that determine optimal school size for different species are unknown. Ecological factors are likely important, including predation pressure and the distribution and abundance of prey.

Within a species, in addition to ecological factors, school size varies due to other factors such as time of day, social composition (sex and age), type of activity (feeding, traveling, socializing), and whether schools are pure (monospecific) or mixed species (Scott and Cattanach 1998; Ballance and Pitman 1998; Gerrodette and Forcada 2005; Gerrodette et al. 2008). For example, in the ETP, median school size differs between the two subspecies of pantropical spotted dolphins: 29.5 (95% CI = 11–94) for coastal offshore pantropical spotted and 62 (95% CI = 31–118) for offshore pantropical spotted dolphins. Common bottlenose dolphins also differ in school size and the degree of mixing throughout their range. Scott and Chivers (1990) found that school size (mean and range) was considerably larger offshore than in coastal waters or around islands, and the percentage of mixed-species schools increased offshore. However, median school size was similar among coastal, island, offshore, and far-offshore habitats (10–12 individuals), indicating that small group size is the norm across habitat types.

The degree of interspecific mixing in ETP dolphin schools is highly variable and species- or subspecies-specific. For example, the two subspecies of pantropical spotted dolphins differ not only in school size but also in their degree of mixing with other species: 7.8% of coastal pantropical spotted dolphin schools are mixed species ($n = 232$ schools), while 57% of offshore pantropical spotted dolphin schools are mixed species ($n = 1467$), usually with local stocks of spinner dolphins. The association between pantropical spotted and spinner dolphins is common (Fig. 9.4), yet not obligate, and the frequency differs between the two species. Compared to offshore pantropical spotted dolphins (57%), eastern spinners are more likely to be found in mixed-species schools 76% of the time ($n = 1002$; see also Cramer et al. 2008). Mixing appears to be more important to spinner dolphins, and perhaps particularly mothers and calves (Cramer et al. 2008). Mixed-species schools are larger than pure schools. The median school size of mixed pantropical spotted and spinner schools is 200 (95% CI = 112–334) as compared to 54.0 (95% CI = 23–113) for pure schools of pantropical spotted dolphins (both subspecies combined) and 77.0 (95% CI = 37–185.5) for pure schools of spinner dolphins. Analysis of aerial photographs shows that when schooling together, the two species do not mix (Wayne Perryman, pers. comm., 26 Nov. 2018). The numerically dominant species tended to be located in the front and periphery of the school (Cramer et al. 2008), consistent with the idea that eastern spinners might use the schools of the

Fig. 9.4 Photo of a portion of a mixed school of eastern spinner and offshore spotted dolphins in the eastern tropical Pacific, with masked boobies, *Sula dactylatra*, overhead. Spinner dolphins can be identified by the triangular forward-canted dorsal fin and extremely large post-anal keel of adult males. Spotted dolphins are mainly on the far side of the image. The school is photographed from a research vessel. Evasive behavior may be effective in avoiding capture within tuna nets but may also lead to social separation, which may slow or inhibit population recovery. Photo by Robert Pitman/Southwest Fisheries Science Center, NOAA Fisheries. Photo credit: US government image; no copyright needed

larger-bodied offshore pantropical spotted dolphins to gain protection or as refugia during diurnal quiescent resting periods (Norris and Dohl 1980; Cramer et al. 2008). Different mechanisms may be operating when small groups of offshore pantropical spotted dolphins are found in schools of eastern spinner dolphins (see below).

School size also varies on different temporal scales (e.g., daily, Scott and Cattanach 1998; interannually, Ballance et al. 2006; Gerrodette et al. 2008). Pantropical spotted, spinner, and common dolphin school sizes increase during the mid- to late-morning hours, reaching a peak in the afternoon, and then decrease toward evening (Scott and Cattanach 1998). Scott and Cattanach (1998) and Scott and Chivers (2009) suggested that during the day, oceanic dolphins minimize predation risk and maximize social interactions by aggregating in large schools but then at dusk disperse to feed on vertically migrating prey at night and to minimize competition for food. Daytime aggregation provides additional advantages, such as more efficient searching for prey patches and increased opportunities for mating and information transfer. Similar patterns of daytime aggregation and nighttime dispersion are observed in spinner dolphins in Hawaii (Würsig et al. 1994; Chap. 17).

9 Oceanic Dolphin Societies: Diversity, Complexity, and Conservation

9.3 Social Behavior and Structure of ETP Pantropical Spotted Dolphins

The study of association patterns of oceanic dolphins is challenging because of logistical difficulties of conducting long-term field studies of known individuals, which are needed to reveal relationships. We review what is known of the social structure and social behavior of offshore pantropical spotted dolphins, arguably the best studied of any oceanic species. Information is from aerial and ship-based observations (Cramer et al. 2008; Gerrodette and Forcada 2005; Gerrodette et al. 2008; Scott et al. 2012), tagging and tracking studies designed to examine the nature of the tuna-dolphin bond and movement patterns of dolphins and tuna (Perrin et al. 1979; Scott and Chivers 2009), life history studies (Perrin 1975; Perrin et al. 1976; Perrin and Reilly 1984; Hohn et al. 1985; Myrick et al. 1986), and particularly from two research cruises aboard tuna purse seine vessels designed to examine interactions among individuals confined in the nets (Norris et al. 1978; Pryor and Kang 1980). We focus on the structure and behavior of subgroups, the smaller and likely more stable clusters of individuals within the larger, more ephemeral schools. By merging this bottom-up look at subgroups with a top-down look at schools, the picture of a fission–fusion society emerges, in which individuals join and leave the company of others to form aggregations that can number from a few individuals to hundreds or thousands (Wade et al. 2012).

The various sources of data on the sex and age structure of spotted dolphins reveal two general patterns. First, while most schools contain all classes of individuals (Perrin et al. 1976; Pryor and Kang 1980), there is some evidence of school segregation (Perrin and Hohn 1994, Archer and Chivers 2002). Second, the ratio of males to females appears to decline with age (Perrin et al. 1976). Tagging and tracking studies (Perrin et al. 1979; Forney et al. 2002; Scott and Chivers 2009) show that schools are highly fluid (animals tagged together are rarely recaptured together) and variable (an individual dolphin may travel in schools of vastly different sizes) but also comprising smaller subgroups whose members remain together over time. For example, a group of four adult male offshore pantropical spotted dolphins from a school of over 2000 were tagged and tracked for 3 days. The males remained in close association with each other over the course of several resightings within schools ranging in size from 16 to 700 dolphins. Similar observations were made of two adult females, one with a calf, that were found together consistently over a period of at least 2 days (Forney et al. 2002; Forney pers. comm., Dec 2018).

Consistent with observations from aerial photographs, spotted and spinner dolphins within the confines of the net did not mix. In a set holding approximately 500 spotted dolphins, for example, the animals positioned themselves in three separate, monospecific groups, "as if they were enclosed in three separate transparent plastic bags" (Pryor and Shallenberger 1991, p. 177). In the net, groups of spotted dolphins were observed in subgroups, typically <20 animals, which swam and surfaced in unison and maintained physical contact with one another (Pryor and Shallenberger 1991). Typical behavior within subgroups was "affiliative" and essentially defined the subgroups: unison breathing, unison swimming, close inter-animal distances, and much body contact. In captivity, spotted dolphins that engage in such activities almost invariably have established relationships and it is likely that those in the wild have as well (Pryor and Kang 1980; Norris and Dohl 1980). Separate subgroups representative of all classes of animals (males and females, young adults, juveniles, and mothers and young) were present in the nets. These subgroups swam near one another and interacted frequently, yet maintained their distance from other subgroups. The researchers hypothesized that associations among subgroups may also be long term and that they are, in effect, members of a "tribe". Large schools possibly represent aggregations of different tribes (Pryor and Kang 1980).

In the observed sets, subgroups accounted for about one-half the animals, most of the remaining individuals within the net moved in single loose aggregation. A few animals moved about individually but rarely for long. Pryor and Kang (1980) and Pryor and Shallenberger (1991) identified seven kinds of pantropical spotted dolphin associations, usually consisting of 2–10 animals: female/calf pairs, female/young subgroups, mother–calf–adult triads, juvenile subgroups, young adult subgroups, courting pairs, and dominant adult male "squads". The following descriptions are paraphrased from Pryor and Kang (1980) and Pryor and Shallenberger (1991). The same subgroups were identified in spinner dolphins captured within the nets although they mingled more freely than the spotted dolphins.

Female-Calf Pairs and Female-Young Subgroups Mother-calf pairs were the most obvious and identifiable groupings. Females also accompanied older juveniles up to young adult size. Because young become independent of their mothers at 3–7 years of age, older calves can be quite large (Hohn et al. 1985; Cramer et al. 2008). Mother-calf pairs exhibited a close-knit association, physical proximity, extensive play behavior, much body rubbing, and little or no aggressive behavior. Except for the "lost baby" phenomenon (see below), the researchers observed no animals smaller than ¾-adult-size juveniles without at least one adult female in close and constant association. Female-calf pairs occurred together with other similar pairs in female-young ("nursery") subgroups that changed membership frequently (Fig. 9.5). Spotted dolphin schools sometimes contained a very small number of post-reproductive females (Myrick et al. 1986). Thus, older females may remain in association with their daughters and daughters' offspring or at least with other females in the school. These observations suggest that allomaternal care is a role shared by the adult females within a school.

Adult Male Subgroups Adult male subgroups comprised three to eight large adult males, all with prominent post-anal keels, heavily spotted bodies ("fused" color pattern of reproductively mature males), and striking facial coloration—black facial masks and conspicuous white-tipped rostrums (the latter perhaps to accentuate open-mouth threat displays; Fig. 9.6). The typical behavior of these adult male "squads" was coordinated cruising at a constant but slow rate. The males swam shoulder-to-shoulder with flippers overlapping, moving with extreme precision and without altering course or speed, while other individuals moved aside. These male "squads"

9 Oceanic Dolphin Societies: Diversity, Complexity, and Conservation

Fig. 9.5 A female-young subgroup of spotted dolphins in a tuna net. These groups consisted of adult females, each accompanied by a calf or young juvenile. Photo credit: Karen Pryor

Fig. 9.6 Adult male subgroup of spotted dolphins in a tuna net. An individual from another adult male subgroup rising from below gapes his jaws in a threat gesture at the group passing over him. Photo credit: Karen Pryor

made investigative excursions in the nets and carried out the majority of threat displays observed, sometimes clashing vigorously with other male subgroups. We know little about the function of adult male "squads," yet observations suggest they have a role in social ordering (Johnson and Norris 1994) and possibly in reproduction (Pryor and Shallenberger 1991).

Juvenile Subgroups Juvenile subgroups were comprised of three to six small animals, together in near-perfect synchrony while breathing and swimming, although these subgroups were not present in every set (Fig. 9.7). The most commonly observed behaviors exhibited by these subgroups were swimming and rafting. It was unclear why some juveniles formed these closely associated groups while others, of the same size, remained paired with females, but the difference may be related to dispersal of young males.

In addition to the subgroups recognized above, Pryor and Kang (1980) and Pryor and Shallenberger (1991) also observed triads and courting pairs. In mother–young–adult triads, the adult females exhibited protective behavior toward the calf, often travelling with the small calf sandwiched between them. Adult male–female pairs swimming closely together and interacting were considered courting pairs. Courting

Fig. 9.7 A juvenile subgroup of spotted dolphins "rafting head-down". Juveniles were the only subgroup observed to exhibit this behavior. Photo credit: Karen Pryor

pairs were observed in some but not all sets, and displayed two categories of behavior toward one another: aggressive and nonaggressive. When another adult male in the school tried to join a nonaggressive courting pair, one or both animals in the pair threatened the intruding male until he left. The rest of the school, the individuals that were not in the subgroups mentioned above, generally consisted of large juveniles and young adults. These young adult subgroups fluctuated constantly in size and composition. These subgroups varied in size up to 50 individuals but were similar to other subgroups with small inter-animal distances and abundant affiliative displays.

Pryor and Kang (1980) and Pryor and Shallenberger (1991) also described the behavioral repertoires of spotted dolphins within purse seine nets and compared it with their extensive observations of captive spotted dolphins in Hawaii (Pryor 1975, 2004). This intimate look at behavior sheds additional light on the nature and function of social bonds among individuals as animals engaged with one another in affiliative, aggressive, sexual, protective, nurturing, and play behavior. The researchers also recorded behavior in the net that resembled known captive behavior such as agitation, disorientation, and panic.

The following two brief summaries highlight the nature of social bonds among individuals. The researchers observed that individuals exhibited a wide range of behavior in response to stressful or frightening situations in the net and showed various signs of agitation. When a captive or wild dolphin smacked the water surface with a fluke or made an abrupt leap, the entire school rapidly dove out of sight. Inside the net, individuals aggregated when startled. When speedboat engines started up, animals increased swim speed, and mother-calf pairs and young adult subgroups joined dominant males near them and for a brief time increased to one group swimming in unison. The researchers also observed that in some sets, there appeared to be "lost babies"—small, months-old calves calling repeatedly as they swam alone within the school. Adult females would investigate but not stay with these calves. More often, the researchers saw individuals that were the same size as young mothers sinking passively in the net, tail first, behavior they identified as "grieving." Similar behavior in captivity was interpreted as "giving up." Such behavior is

9 Oceanic Dolphin Societies: Diversity, Complexity, and Conservation

consistent with the observation that chase and encirclement can separate mothers and calves (see below). Recent work has shown that behavior interpreted as grieving is also found in other cetacean species, and the idea is garnering new ways of thinking about emotional states and social bonds, especially in delphinids (Bearzi et al. 2018).

9.4 Mating Systems of ETP Dolphins

Little is known about mating systems of oceanic dolphins. Sexually receptive females are dispersed and highly mobile, making it difficult for males to monopolize access to females. Yet, across taxa, testis size (relative to body size), pattern of sexual dimorphism, and mating system are linked, and data on these features can be used to make predictions about mating strategies of the sexes (Dines et al. 2015). When females routinely mate with more than one male, sexual selection has favored the evolution of large testes capable of delivering greater numbers of sperm per ejaculation (sperm competition; Parker 1970). Greater relative testis size is correlated with a more open, i.e., polygynandrous (multi-mate, or "promiscuous") mating system, as compared to polygynous or monogamous mating. In fact, the relationship between relative testis size and mating system is so strong that testis size can be used as an indicator of mating system (Gomendio et al. 1998).

Testis size ranges widely among ETP species. Maximum weights (mass of right testis and epidiymis) ranged from 421 g (central American spinner dolphin) to 548 (striped dolphin), 843 g (eastern spinner dolphin), 1354 g (whitebelly spinner dolphin), 1385 g (coastal spotted dolphin), 2029 g (common dolphin), and 3254 g (offshore pantropical spotted dolphin). These data suggest variable mating systems, both among and within species. In a phylogenetic analysis of residual testes weights (maximum testes mass regressed onto maximum body lengths), Dines et al. (2015) identified species with testes masses higher or lower than expected based on body size. ETP species or related populations included striped dolphin (lower than expected) and offshore pantropical spotted and common dolphins (higher than expected). These results suggest a more structured, polygynous mating system in the former and a more open, polygynandrous system in the latter. The spinner dolphin is a special case, exhibiting an unusually wide range of intraspecific variation in testis size and sexual dimorphism.

Perrin and Mesnick (2006) found a strong correlation between the striking sexual dimorphism in eastern spinner dolphins, in which adult males have enormous post-anal keels and forward-canted dorsal fins (Fig. 9.1) and reduced testis size (to 843 g), and the opposite pattern in the corresponding characteristics of whitebelly spinners, with reduced sexual dimorphism and larger relative testis size (to 1354 g). Only 4 of 699 (0.6%) eastern spinner males examined had testis plus epididymis weighing more than 700 g, the level at which all epididymides contained sperm. This is roughly a 25-fold reduction in the number of reproductive males compared to the number for whitebelly spinners. These numbers and the fact that increased dimorphism and decreased testis size are indicative of increased polygyny and pre-copulatory sexual competition in other mammals suggest that the mating system varies geographically in the spinner dolphin, with a gradient from a more open, polygynandrous mating system in the whitebelly spinner to a more polygynous mating system in the eastern spinner. The differences may be due to habitat. Eastern spinners are endemic to the far-eastern, core area of the ETP, where the thermocline is shallowest and productivity highest and where spinner dolphins regularly associate with spotted dolphins and yellowfin tuna. As in other mammals, the more favorable habitat may enable females to acquire benefits by residing in groups where bonds between individuals are more stable and males to spend more time and energy competing for access to receptive females.

Except for the relatively extreme case of eastern spinner dolphins, sexual dimorphism is slight in ETP delphinids, with small differences in body size (ca. 3–7%; Dines et al. 2015), cranial features, and coloration (Perrin 1975). The most striking sexually dimorphic character shared across multiple species is the post-anal keel, the development and function of which are not well known, although it may serve as a visual signal that makes adult males easily recognizable (Mesnick and Ralls 2018a). The post-anal keel is most extreme in eastern spinner dolphins but is also noticeable in offshore pantropical spotted dolphins and yet relatively inconspicuous in common dolphins. When sexual size dimorphism is slight, males may gain access to females by forming alliances with other males: what they lack in size, they make up in numbers. "Dimorphism by association" may provide mating advantages to cooperating males who collectively herd females (Tolley et al. 1995; Connor and Krützen 2015; see also Chap. 16). As noted above, coordinated subgroups of adult males occur in pantropical spotted dolphins and are also observed in spinner dolphins in Hawaii (Norris and Johnson 1994), but whether these males cooperate during mating is unknown.

Even less is known about female mating strategies. Female ETP dolphins are diffusely seasonal breeders (Perrin and Reilly 1984). With a 2–3-year interbirth interval on average (and more for some species), females may maximize their reproductive success by being good mothers and may also enhance fitness by choosing males or mating with multiple males. Multiple matings may function to ensure insemination or to confuse paternity and reduce potential aggression. Multiple mating may also enable females to promote sperm competition and exert cryptic female choice (Orbach et al. 2017; Mesnick and Ralls 2018b).

9.5 ETP Dolphins and Interactions with the Tuna Purse Seine Fishery

Since its inception in the 1950s, the purse seine fishery has been a significant factor in the lives of ETP dolphins and an ongoing focus of research to understand direct and indirect impacts of the fishery on dolphins, tuna, and the ecosystem (Perrin 1969; Reilly et al. 2005; Gerrodette and Forcada 2005; Wade et al. 2007; Archer et al. 2010a, b; Gerrodette et al. 2012). Dolphins are long-lived, large-brained animals, and long-term exposure to the fishery represents considerable opportunity for them to learn from experience. Since the late 1960s and early 1970s, the purse seine

9 Oceanic Dolphin Societies: Diversity, Complexity, and Conservation

fishery has conducted ca. 11,000 sets on dolphins per year (IATTC 2018), extending over generations of dolphins of several species, subspecies, and management stocks. Individual offshore spotted dolphins interact with the fishery (i.e. are chased by the speedboats) between 2 and 50 times per year, depending on size of the school (Edwards and Perkins 1998; see also Archer et al. 2010a). Although currently, >99% of the dolphins are reportedly released alive, it is clear that most offshore pantropical spotted and eastern spinner dolphins alive today have a history of repeated, long-term interaction with the fishery. While we know that exposure to the fishery has impacts on dolphins (Curry 1999; Archer et al. 2001, 2004; Forney et al. 2002; Reilly et al. 2005; Edwards 2006; Noren and Edwards 2007; Cramer et al. 2008; Kellar et al. 2013), we do not know whether intrinsic growth rate of the dolphin populations is great enough to offset or nullify these impacts, that is, whether the populations are recovering.

The two main dolphin populations targeted by the fishery, northeastern offshore pantropical spotted dolphins and eastern spinner dolphins, are considered to be depleted under the MMPA. Wade et al. (2007) estimated that this spotted dolphin population was at 20% of its pre-exploitation level of 3.6 million and the eastern spinner at 29% of its pre-exploitation level of 1.8 million dolphins, based on data collected through 2002. The expectation is that these populations, in the absence of substantial human-caused mortality, should be growing. Yet, despite significant reductions in reported bycatch mortality since the early 1990s (from hundreds of thousands killed per year to fewer than 1000 per year, a rate of removal that should be sustainable), as of the last fishery-independent abundance survey (2006), there was still uncertainty whether either population was showing clear signs of recovery (Gerrodette et al. 2008). The most recent point abundance estimates of both stocks are higher than previous estimates, but the variance around those estimates is too large to rule out the possibility of non-recovery (95% confidence intervals on rates of change from 1998 to 2006 included zero; Fig. 18, Gerrodette et al. 2008). Previous studies using data through 2000 (Lennert-Cody et al. 2001; Gerrodette and Forcada 2005) concluded that neither of the two targeted dolphin populations was recovering at a rate consistent with its depleted status and low reported bycatch. How the populations have been faring since the last survey in 2006 is unknown, so the question of recovery remains open.

One of the indicators of the impact of the fishery on dolphins has been the relationship between dolphin evasiveness and fishery exposure, documented in both short- and long-term studies (Orbach 1977; Schramm Urrutia 1997; Sevenbergen 1997; Heckel et al. 2000; Mesnick et al. 2002; Lennert-Cody and Scott 2005; Archer et al. 2010b). Since the early days of the fishery, tuna fishermen and scientists have observed that in high-density fishing areas, dolphins appear more adept at evading chase and encirclement: they tend to be wary of purse seiners, require longer chases and more chase boats to round them up, and are more difficult to encircle. Behavioral changes in swimming and schooling dynamics occur during all stages of the fishing operation as the school coalesces, runs away from the seiner, and splits to avoid the chase boats or to evade capture during the deployment of the net, and again after release. Some spotted dolphin schools in more heavily fished areas "explode," the school breaking into multiple smaller groups and sometimes forcing the set to be aborted—fishermen refer to these groups as the "untouchables" (Orbach 1977).

Once inside the net, the dolphins have been trained by decades of fishing to wait for the "backdown" procedure (Norris et al. 1978; Pryor and Kang 1980; Forney et al. 2002; Santurtún and Galindo 2002). After the school is encircled and the net closed ("pursed") at the bottom, the seiner is skillfully maneuvered so that the corkline sinks and dolphins can escape over a specialized portion of the net designed to prevent accidental entanglements. Backdown effectively releases the vast majority of dolphins while retaining the tuna that swim lower in the net. Observations of passive behavior during encirclement and active behavior during backdown indicate that dolphins have learned the standard sequence of fishing operations and have developed adaptive behaviors (Pryor and Kang 1980; Santurtún and Galindo 2002), which may reduce stress as a result of learning from repeated capture (St. Aubin et al. 1996). However, changes in fishing operations can cause changes in dolphin behavior inside the net (Forney et al. 2002).

Wade et al. (2012) raised the question of whether the continued chase and encirclement of dolphins impairs the ability of affected populations to recover from depletion, a lack of resilience to exploitation of this kind and scale. There is also the issue of whether substantial "cryptic" mortality is occurring and contributing to the lack of recovery (if indeed recovery is not underway). Setting on dolphins can cause separation of individuals in the school during high-speed chases (due to different swimming capabilities, which is a particular concern for mothers with young calves, or due to schools splitting to evade capture), during encirclement with the deployed purse seine (due to individuals ending up inside or outside the net), and during release of the dolphins (due to confusion and flight responses as the dolphins flee from the net during backdown). While the mother-calf bond is strong, mothers and calves can become separated during purse seine operations (Archer et al. 2001, 2004) because calves probably cannot maintain the relatively high speeds associated with flight and evasion during and after sets (Edwards 2006, 2007; Noren and Edwards 2007). Archer et al. (2004) estimated that 75–95% of lactating females killed did not appear to be accompanied by a calf when killed and assumed that their orphaned calves would not have survived. This could represent a substantial level of cryptic mortality.

The tuna fishery may also have an effect on spinner dolphins through disruption of their mating system. The more structured, polygynous mating system of eastern spinners (as compared to that of whitebelly spinners), and the small number of males with fully developed testes, leads to the prediction that only a tiny fraction of the male population is responsible for much of the reproduction (Perrin and Mesnick 2006). If few males participate in mating, social separation or removal of dominant males could suppress reproduction. With many other adult males present, yet not reproductively (and perhaps socially) competent, the duration of the impact is unknown.

The proportion of females pregnant, calf production, and the length of time before disassociation (a proxy for the age of weaning) decrease when exposure to purse seining increases (Cramer et al. 2008; Kellar et al. 2013), and calf production was declining since at least 1987 for both eastern spinner and northeastern offshore spotted dolphins (Cramer et al. 2008). This can be interpreted as suggesting that

9 Oceanic Dolphin Societies: Diversity, Complexity, and Conservation

fishing intensity (the frequency of sets on individual dolphins) causes reduced pregnancy rates and also possibly lower neonatal calf survival, and therefore that the fishery has population-level impacts beyond the reported direct kill. Cramer et al. (2008) concluded the decline in reproductive output (at least through 2003) was the proximate cause (or one of the proximate causes) of the failure of the dolphin populations to recover at expected rates in the years following the significant reduction in bycatch levels.

Further investigation of the various known or suspected direct and indirect impacts of the fishery on dolphins is important. While the effects of chasing and encirclement have been cited to help explain the possible lack of recovery, the consequences of repeated and ongoing exposure to the fishery, and the mechanisms leading to those consequences, are not yet well understood. Additional field surveys should help determine whether there has been any population recovery and if and how dolphin behavior has changed in response to both environmental variability and human activities, particularly the purse seine fishery.

9.6 Synthesis and Future Directions

ETP dolphins are social mammals living in one of the largest open spaces on the planet—marine space the size of the continent of Africa. Individuals are rarely if ever alone. They live in groups like some primates, forage in the open like many ungulates, and respond to potential predation like schooling fish (see Norris and Dohl 1980; Norris 1994). Group living confers multiple advantages to individuals. It requires behavioral integration and social synchrony, made possible by a multimodality communication system. ETP dolphin schools are spatially, temporally, and compositionally fluid, yet socially complex and structured, possibly on multiple levels. Individuals may have distinct roles, form stable or at least semi-stable subgroups, and leave or join the company of others in response to a variety of social and ecological cues. Mating systems are variable among species and sometimes within species, likely reflecting differences in local ocean productivity. In at least one taxon—the eastern spinner dolphin—a few sexually mature males may be responsible for a disproportionate portion of the mating, while in other taxa, e.g., whitebelly spinner, large testes suggest a more "open" mating system where many males in the school engage in copulation.

Much about the nature of the societies of oceanic dolphins remains a mystery. Little is known about the stability or duration of bonds among individuals, subgroups, schools, and aggregations, and assessing these characteristics, even if only qualitatively, is logistically challenging. How recurrent are relationships and at what scales? Is there longer-term structure above the subgroup? Biologging has provided tremendous insights, and we hope there will be additional opportunities to apply longer-duration tags and to increase sample sizes for different age and sex classes. Little is also understood about the factors determining optimal school size. Spatial analysis of oceanographic, ecological, and social predictors may provide some clues about the mechanisms underlying the differences in school size among and within species.

Members of social species regularly make decisions to coordinate activities, including where to forage, when to move, and how to respond to risks. Yet, neither fish nor dolphin schools have been shown to have leaders for short-term movements or evasive actions; the responsiveness of the school appears to be mediated by sensory integration (Norris and Dohl 1980; Parrish et al. 2002). Long-term movements, however, may be different. Do older, more experienced individuals carry and share socio-ecological knowledge? While evolutionary theory predicts that shared decision-making is the most efficient, in many species leaders seem to emerge (Smith et al. 2016). The concept of leadership is important for understanding long-term movement, collective foraging, and social interactions, such as conflict resolution. It is our hope that future work will examine these questions in social odontocetes and particularly in societies where we see evidence of the presence of older, post-reproductive females (see Chap. 11; Smith et al. 2018).

For spotted and spinner dolphins in the ETP, the same behavior of schooling with large tuna that has led to their ecological success and abundance has also led to their depletion by making them a target of purse seiners. Schooling and sociality, normally adaptive traits, have caused ETP dolphins to become collateral damage of the tuna fishery. Yet dolphins have learned some things from their experience with purse seiners. Some individuals know how to evade capture or, alternatively, how to await the backdown procedure. While much of the potential for recovery may reside in the experienced animals alive today, the very behavior that helps them avoid capture may require or lead to levels of stress, exertion, and social separation and disruption that may inhibit, or slow, population recovery.

Group living has many benefits for social animals. Social behavior, group cohesion, and fitness benefits are interrelated and tied to overall likelihood of population persistence in a broad range of terrestrial and marine species (Allee 1931; Ehrenfeld 1970; Gilpin and Soulé 1986; Dobson and Poole 1998; Gosling and Sutherland 2000; Wade et al. 2012; UNEP/CMS 2018; among others). However, being highly social can make some species more vulnerable to human exploitation. In a recent review of odontocete cetaceans that have experienced significant exploitation but show little to no signs of recovery, including ETP dolphins, Wade et al. (2012) used the concept of a species' or population's resilience, that is, its ability to recover from depletion caused by human activities, and they suggested that highly social odontocetes are more vulnerable to exploitation than baleen whales, which do not usually form long-term social bonds beyond the period of calf dependency, which is relatively short. They highlighted several aspects of sociality that have received relatively little attention in assessments of oceanic dolphin populations: social complexity, sexual behavior, alloparental care, social learning, and leadership. It is our hope that the information presented here provides a better understanding of oceanic dolphin societies and the role of social and behavioral factors in determining whether and how a species or population is able to respond and recover from a disturbance caused by human activities.

Acknowledgments The authors extend their gratitude to the many scientists who have devoted themselves to the oceanic dolphins of the ETP and the "tuna-dolphin issue." We especially thank Bill Perrin for his extensive knowledge (history, taxonomy, morphology, school structure, reproductive strategies) and Bob Pitman (ecology, evolution) for generously sharing their experience and

9 Oceanic Dolphin Societies: Diversity, Complexity, and Conservation

insights. Both Bill and Bob provided thoughtful review and valuable comments on the manuscript. Thanks to Karin Forney (tagging and tracking), Katie Cramer and Wayne Perryman (school geometry), and Tim Gerrodette (abundance and trends) for sharing their knowledge and unique insights. Thanks to Tim Gerrodette for compiling and analyzing the data and creating the figure on school size. Paul Fiedler generated the beautiful maps of thermocline depth and temperature in the ETP. Paula Olson and Bob Pitman provided photographs of ETP dolphins. We also thank command and crew and fellow observers of the many NOAA and tuna fishing vessels that provided access to these remarkable animals and their oceanic habitat.

Appendices

Appendix 1

Delphinid species recorded in the eastern tropical Pacific. Adapted from Ballance et al. (2006). Endemic taxa are noted. Naming convention follows the Committee on Taxonomy (2018) (taxonomy) and Dizon et al. (1994) (management stocks).

Delphinus delphis — Common dolphin, saddleback dolphin

> *D. d. delphis* — Common dolphin
>
>> Recognized ETP management stocks:
>>> Northern short-beaked common dolphin
>>> Central short-beaked common dolphin
>>> Southern short-beaked common dolphin
>
> *D. d. bairdii* — Eastern North Pacific long-beaked common dolphin

Feresa attenuata — Pygmy killer whale
Globicephala macrorhynchus — Short-finned pilot whale
Grampus griseus — Risso's dolphin, *grampus*
Lagenodelphis hosei — Fraser's dolphin
Lagenorhynchus obliquidens — Pacific white-sided dolphin
Lagenorhynchus obscurus — Dusky dolphin
Orcinus orca — Killer whale, orca
Peponocephala electra — Melon-headed whale, Electra dolphin
Pseudorca crassidens — False killer whale
Stenella attenuata — Pantropical spotted dolphin

> *S. a. attenuata* — Offshore pantropical spotted dolphin
>
>> Recognized ETP management stocks:
>>> Northeastern offshore pantropical dolphin
>>> Western-southern offshore pantropical dolphin
>
> *S. a. graffmani* — Coastal pantropical spotted dolphin — Endemic

Stenella coeruleoalba — Striped dolphin

Stenella longirostris — Spinner dolphin

    S. l. centroamericana — Central American spinner dolphin — Endemic
    S. l. longirostris — Gray's spinner dolphin

        Whitebelly spinner (*S. l. orientalis* × *S. l. longirostris*)

    S. l. orientalis — Eastern spinner dolphin — Endemic

Steno bredanensis — Rough-toothed dolphin
Tursiops truncatus — Common bottlenose dolphin
T. t. truncatus — Common bottlenose dolphin

Appendix 2: Basic Oceanography of the Eastern Tropical Pacific

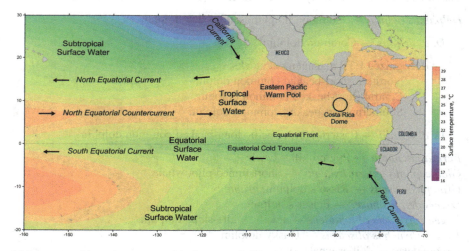

Major geographic and oceanographic features of the eastern tropical Pacific (ETP), including surface currents and water masses. Background shading is annual mean surface temperature (°C, 1980–2015); adapted from Fiedler et al. (2017)

References

Acevedo-Gutierrez A (2018) Group behavior. In: Würsig B, Thewissen JGM, Kovacs KM (eds) Encyclopedia of marine mammals, 3rd edn. Academic/Elsevier, London, pp 428–435
Allee WC (1931) Animal aggregations: a study in general sociology. University of Chicago Press, Chicago

9 Oceanic Dolphin Societies: Diversity, Complexity, and Conservation

Archer FI, Chivers SJ (2002) Age structure of the northeastern spotted dolphin incidental kill by year for 1971–1990 and 1996–2000. SWFSC Administrative Report LJ-02-12. NMFS, Southwest Fisheries Science Center, La Jolla, CA

Archer F, Gerrodette T, Dizon A, Abella K, Southern ŠÁ (2001) Unobserved kill of nursing dolphin calves in a tuna purse-seine fishery. Mar Mamm Sci 17:540–554

Archer F, Gerrodette T, Chivers S, Jackson A (2004) Annual estimates of the unobserved incidental kill of pantropical spotted dolphin (*Stenella attenuata attenuata*) calves in the tuna purse-seine fishery of the eastern tropical Pacific. Fish Bull 102:233–244

Archer FI, Redfern JV, Gerrodette T, Chivers SJ, Perrin WF (2010a) Estimation of relative exposure of dolphins to fishery activity. Mar Ecol Prog Ser 410:245–255

Archer FI, Mesnick SL, Allen AC (2010b) Variation and predictors of vessel-response behavior in a tropical dolphin community. NOAA Technical Memorandum NMFS, NOAA-TM-NMFS-SWFSC-457, pp 1–60

Au DWK (1991) Polyspecific nature of tuna schools: shark, dolphin and seabird associates. Fish Bull 89:343–354

Au DWK, Perryman WL (1985) Dolphin habitats in the eastern tropical Pacific. Fish Bull US 83: 623–643

Au DK, Pitman RL (1986) Seabird interactions with dolphins and tuna in the eastern tropical Pacific. Condor 88:304–317

Ballance LT, Pitman RL (1998) Cetaceans of the western tropical Indian Ocean: distribution, relative abundance, and comparisons with cetacean communities of two other tropical eco-systems. Mar Mamm Sci 14:429–459

Ballance LT, Pitman RL, Reilly SB (1997) Seabird community structure along a productivity gradient: importance of competition and energetic constraint. Ecology 78:1502–1518

Ballance LT, Pitman RL, Fiedler PC (2006) Oceanographic influences on seabirds and cetaceans of the eastern tropical Pacific: a review. Prog Oceanogr 69:360–390

Bearzi G, Kerem D, Furey NB, Pitman RL, Rendell L, Reeves Randall RR (2018) Whale and dolphin behavioural responses to dead conspecifics. Zoology 128:1–15

Committee on Taxonomy (2018) List of marine mammal species and subspecies. Society for Marine Mammalogy. www.marinemammalscience.org. Accessed 26 Nov 2018

Connor RC, Krützen M (2015) Male dolphin alliances in Shark Bay: changing perspectives in a 30-year study. Anim Behav 103:223–235

Cramer KL, Perryman WL, Gerrodette T (2008) Declines in reproductive output in two dolphin populations depleted by the yellowfin tuna purse-seine fishery. Mar Ecol Prog Ser 369:273–285

Curry B (1999) Stress in mammals: the potential influence of fishery-induced stress on dolphin in the eastern tropical Pacific. U.S. Department of Commerce, NOAA Technical Memorandum NMFS SWFSC-260, 121 pp

Dines JP, Mesnick SL, Ralls K, May-Collado L, Agnarsson I, Dean MD (2015) A trade-off between precopulatory and postcopulatory trait investment in male cetaceans. Evolution 69:1560–1572

Dizon AE, Perrin WF, Akin PA (1994) Stocks of dolphins (Stenella spp. and *Delphinus delphis*) in the eastern tropical Pacific: a phylogeographic classification. U.S. Department of Commerce, NOAA Technical Memorandum NMFS SWFSC-119

Dobson A, Poole J (1998) Conspecific aggregation and conservation biology. In: Caro T (ed) Behavioral ecology and conservation biology. Oxford University Press, New York, pp 193–208

Edwards EF (2006) Duration of unassisted swimming activity for spotted dolphin (*Stenella attenuata*) calves: implications for mother-calf separation during tuna purse-seine sets. Fish Bull 104:125–135

Edwards EF (2007) Fishery effects on dolphins targeted by tuna purse-seiners in the eastern tropical Pacific Ocean. Int J Comp Psychol 20:217–227

Edwards EF, Perkins PC (1998) Estimated tuna discard from dolphin, school, and log sets in the eastern tropical Pacific Ocean, 1989–1992. Fish Bull 96:210–222

Ehrenfeld DW (1970) Biological conservation. Holt, Rinehart, and Winston, New York

Ellis S, Franks DW, Nattrass S, Currie TE, Cant MA, Giles D, Balcomb KC, Croft DP (2018) Analyses of ovarian activity reveal repeated evolution of post-reproductive lifespans in toothed whales. Sci Rep 8:12833

Fiedler PC, Lavin MF (2017) Oceanographic conditions of the eastern tropical Pacific. In: Glynn PW, Manzello DP, Enochs IC (eds) Coral reefs of the eastern tropical Pacific: persistence and loss in a dynamic environment. Springer, Dordrecht, pp 59–83

Fiedler PC, Talley LD (2006) Hydrography of the eastern tropical Pacific: a review. Prog Oceanogr 69:143–180

Fiedler PC, Redfern JV, Ballance LT (2017) Oceanography and cetaceans of the Costa Rica Dome region. U.S. Department of Commerce, NOAA Technical Memorandum NMFS SWFSC-590

Forney KA, St. Aubin DJ, Chivers SJ (2002) Chase encirclement stress studies on dolphins involved in eastern tropical Pacific Ocean purse-seine operations during 2001. Administrative Report LJ-02-32. NMFS, Southwest Fisheries Science Center, La Jolla, CA

Forney KA, Ferguson MC, Becker EA, Fiedler PC, Redfern JV, Barlow J, Vilchis IL, Ballance LT (2012) Habitat-based spatial models of cetacean density in the eastern Pacific Ocean. Endanger Species Res 16:113–133

Foster EA, Franks DW, Mazzi S, Darden SK, Balcomb KC, Ford JK, Croft DP (2012) Adaptive prolonged postreproductive life span in killer whales. Science 337:1313

Gerrodette T, Forcada J (2005) Non-recovery of two spotted and spinner dolphin populations in the eastern tropical Pacific Ocean. Mar Ecol Prog Ser 291:1–21

Gerrodette T, Perryman W, Barlow J (2002) Calibrating group size estimates of dolphins in the eastern tropical Pacific Ocean. Southwest Fisheries Science Center Administrative Report LJ-02-08. National Marine Fisheries Service. National Oceanic and Atmospheric Administration

Gerrodette T, Watters G, Perryman W, Ballance L (2008) Estimates of 2006 dolphin abundance in the eastern tropical Pacific, with revised estimates from 1986–2003. U.S. Department of Commerce, NOAA Technical Memorandum NMFS SWFSC-422

Gerrodette TIM, Olson R, Reilly S, Watters G, Perrin W (2012) Ecological metrics of biomass removed by three methods of purse-seine fishing for tunas in the eastern tropical Pacific Ocean. Conserv Biol 26(2):248–256

Gilpin ME, Soulé ME (1986) Minimum viable populations: processes of species extinction. In: Soulé ME (ed) Conservation biology: the science of scarcity and diversity. Sinauer Associates, Sunderland, pp 19–34

Gomendio M, Harcourt AH, Roldán ERS (1998) Sperm competition in mammals. In: Birkhead TR, Møller AP (eds) Sperm competition and sexual selection. Academic, San Diego, pp 667–756

Gosliner ML (1999) The tuna-dolphin controversy. In: Twiss JR Jr, Reeves RR (eds) Conservation and management of marine mammals. Smithsonian Institution, Washington, pp 120–155

Gosling LM, Sutherland WJ (eds) (2000) Behaviour and conservation. Cambridge University Press, Cambridge

Hamilton WD (1971) Geometry for the selfish herd. J Theor Biol 31:295–311

Heckel G, Murphy KE, Compeán Jiménez GA (2000) Evasive behavior of spotted and spinner dolphins (*Stenella attenuata* and *S. longirostris*) during fishing for yellowfin tuna (*Thunnus albacares*) in the eastern Pacific Ocean. Fish Bull 98:692–703

Hohn AA, Chivers SJ, Barlow J (1985) Reproductive maturity and seasonality of male spotted dolphins, *Stenella attenuata*, in the eastern tropical Pacific. Mar Mamm Sci 1:273–293

IATTC (2018) Tunas, billfishes and other pelagic species in the eastern Pacific Ocean in 2017. Fisheries Status Reports, No 16-2018. Reports No 1-16. www.iattc.org/PublicationsENG.htm

Jefferson TA, LeDuc R (2018) Delphinids, overview. In: Würsig B, Thewissen JGM, Kovacs KM (eds) Encyclopedia of marine mammals, 3rd edn. Academic/Elsevier, San Diego, pp 242–246

Johnson CM, Norris KS (1994) Social behavior. In: Norris KS, Würsig B, Wells RS, Würsig M (eds) The Hawaiian spinner dolphin. University of California Press, Berkeley, pp 243–286

Kasuya T (2017) Small cetaceans of Japan. Exploitation and biology. CRC, Boca Raton

Kellar NM, Trego ML, Chivers SJ, Archer FI (2013) Pregnancy patterns of pantropical spotted dolphins (*Stenella attenuata*) in the eastern tropical Pacific determined from hormonal analysis of blubber biopsies and correlations with the purse-seine tuna fishery. Mar Biol 160:3113–3124

Lammers MO, Au WW, Herzing DL (2003) The broadband social acoustic signaling behavior of spinner and spotted dolphins. J Acoust Soc Am 114:1629–1639

Lennert-Cody CE, Scott MD (2005) Spotted dolphin evasive response in relation to fishing effort. Mar Mamm Sci 21:13–28

Lennert-Cody CE, Buckland ST, Marques FF (2001) Trends in dolphin abundance estimated from fisheries data: a cautionary note. J Cetacean Res Manage 3:305–320

Lo NCH, Smith TD (1986) Incidental mortality of dolphins in the eastern tropical Pacific, 1959-72. Fish Bull 84:27–34

Mesnick SL, Ralls K (2018a) Sexual dimorphism. In: Würsig B, Thewissen JGM, Kovacs KM (eds) Encyclopedia of marine mammals, 3rd edn. Academic/Elsevier, San Diego, pp 848–853

Mesnick SL, Ralls K (2018b) Mating systems. In: Würsig B, Thewissen JGM, Kovacs KM (eds) Encyclopedia of marine mammals, 3rd edn. Academic/Elsevier, San Diego, pp 586–592

Mesnick SL, Archer FI, Allen AC, Dizon AE (2002) Evasive behavior of eastern tropical Pacific dolphins relative to effort by the tuna purse seine fishery. Administrative Report LJ-02-30. NMFS, Southwest Fisheries Science Center, La Jolla, CA

Myrick AC Jr, Hohn AA, Barlow J, Sloan PA (1986) Reproductive biology of female spotted dolphins, *Stenella attenuata*, from the eastern tropical Pacific. Fish Bull 84:247–259

Noren SR, Edwards E (2007) Physiological and behavioral development in delphinid calves: implications for calf separation and mortality due to tuna purse-seine sets. Mar Mamm Sci 23: 15–29

Norris KS (1994) Comparative view of cetacean social ecology, culture and evolution. In: Norris KS, Würsig B, Wells RS, Würsig M (eds) The Hawaiian spinner dolphin. University of California Press, Berkeley, pp 301–349

Norris KS, Dohl TP (1980) The structure and functions of cetacean schools. In: Hermann LM (ed) Cetacean behavior: mechanisms and functions. Wiley, New York, pp 211–261

Norris KS, Johnson CM (1994) Schools and schooling. In: Norris KS, Würsig B, Wells RS, Würsig M (eds) The Hawaiian spinner dolphin. University of California Press, Berkeley, pp 232–242

Norris KS, Schilt CR (1988) Cooperative societies in three-dimensional space: on the origins of aggregations, flocks, and schools, with special reference to dolphins and fish. Ethol Sociobiol 9:149–179

Norris KS, Stuntz WE, Rogers W (1978) The behavior of porpoises and tuna in the eastern tropical Pacific yellowfin tuna fishery – preliminary studies. Final report, U.S. Marine Mammal Commission Contract MM6AC022 to Center for Coastal Marine Studies, University of California, Santa Cruz. NTIS PB 283 970

Orbach MK (1977) Hunters, seamen, and entrepreneurs: the tuna seinermen of San Diego. University of California Press, Berkeley

Orbach DN, Marshall CD, Mesnick SL, Würsig B (2017) Patterns of cetacean vaginal folds yield insights into functionality. PLoS One 12:e0175037

Oswald JN, Barlow J, Norris TF (2003) Acoustic identification of nine delphinid species in the eastern tropical Pacific Ocean. Mar Mamm Sci 19:20–37

Parker GA (1970) Sperm competition and its evolutionary consequences in the insects. Biol Rev 45:525–567

Parrish JK, Viscido SV, Grünbaum D (2002) Self-organized fish schools: an examination of emergent properties. Biol Bull 202:296–305

Perrin WF (1969) Using porpoise to catch tuna. World Fish 18:42

Perrin WF (1975) Variation of spotted and spinner porpoise (genus *Stenella*) in the eastern Pacific and Hawaii. Bulletin of the Scripps Institution of Oceanography, University of California

Perrin WF, Hohn AA (1994) Pantropical spotted dolphin *Stenella attenuata*. In: Ridgway SH, Harrison R (eds) Handbook of marine mammals: the first book of dolphins, vol 5. Academic Press, London, pp 71–98

Perrin WF, Mesnick SL (2006) Sexual ecology of the spinner dolphin, *Stenella longirostris*, geographic variation in mating system. Mar Mamm Sci 19:462–483

Perrin WF, Reilly SB (1984) Reproductive parameters of dolphins and small whales of the family Delphinidae. In: Perrin WF, Brownell RL Jr, DeMaster DP (eds) Reproduction in whales, dolphins and porpoises. Report of the International Whaling Commission (Special Issue 6). Cambridge, pp 97–133

Perrin WF, Coe JM, Zweifel JR (1976) Growth and reproduction of the spotted porpoise, *Stenella attenuata*, in the offshore eastern tropical Pacific. Fish Bull 74:229–269

Perrin WF, Evans WE, Holts DB (1979) Movements of pelagic dolphins (*Stenella* spp.) in the eastern tropical Pacific as indicated by results of tagging, with summary of tagging operations. U.S. Department of Commerce, NOAA Technical Report NMFS SSRF-737

Perrin WF, Akin PA, Kashiwada JV (1991) Geographic variation in external morphology of the spinner dolphin *Stenella longirostris* in the eastern Pacific and implications for conservation. Fish Bull 89:411–428

Photopoulou T, Ferreira IM, Best PB, Kasuya T, Marsh H (2017) Evidence for a post-reproductive phase in female false killer whales *Pseudorca crassidens*. Front Zool 14:30

Pitman RL, Stinchcomb C (2002) Rough-toothed dolphins (*Steno bredanensis*) as predators of mahimahi (*Coryphaena hippurus*). Pac Sci 56:447–450

Pryor K (1975) Lads before the wind. Harper and Row, Oxford

Pryor K (2004) On behavior. Sunshine Books, Waltham

Pryor K, Kang I (1980) Social behavior and school structure in pelagic porpoises (*Stenella attenuata* and *S. longirostris*) during purse seining for tuna. Southwest Fisheries Center Administrative Report LJ-80-11C. NMFS, Southwest Fisheries Science Center, La Jolla, CA

Pryor K, Shallenberger IK (1991) Social structure in spotted dolphins (*Stenella attenuata*) in the tuna purse seine fishery in the eastern tropical Pacific. In: Pryor K, Norris KS (eds) Dolphin societies: discoveries and puzzles. University of California, Berkeley

Rankin S, Barlow J, Oswald J, Ballance L (2008) Acoustic studies of marine mammals during seven years of combined visual and acoustic line-transect surveys for cetaceans in the eastern and central Pacific Ocean. U.S. Department of Commerce, NOAA Technical Memorandum NMFS SWFSC-429

Reilly SB, Fiedler PC (1994) Interannual variability of dolphin habitats in the eastern tropical Pacific. I: Research vessel surveys, 1986–1990. Fish Bull 92:434–450

Reilly SB, Donahue MA, Gerrodette T, Forney K, Wade P, Ballance L, Forcada J, Fiedler P, Dizon A, Perryman W, Archer FI, Edwards E (2005) Report of the Scientific Research Program under the International Dolphin Conservation Program Act. U.S. Department of Commerce, NOAA Technical Memorandum NMFS SWFSC-372. https://swfsc.noaa.gov/publications/TM/SWFSC/NOAA-TM-NMFS-SWFSC-372.PDF

Santurtún E, Galindo F (2002) Coping behaviors of spotted dolphins during fishing sets. Administrative Report LJ-02-36C. NMFS, Southwest Fisheries Science Center, La Jolla, CA

Schramm Urrutia Y (1997) Activity level of offshore spotted (*Stenella attenuata*) and eastern spinner dolphins (*S. longirostris*), during tuna purse seining in the eastern Pacific Ocean. MS Thesis, Universidad Autónoma de Baja California, Ensenada

Scott MD, Cattanach KL (1998) Diel patterns in aggregations of pelagic dolphins and tunas in the eastern Pacific. Mar Mamm Sci 14:401–422

Scott MS, Chivers SJ (1990) Distribution and herd structure of bottlenose dolphins in the eastern tropical Pacific Ocean. In: Leatherwood S, Reeves RR (eds) The bottlenose dolphin. Academic, San Diego, pp 387–402

Scott MD, Chivers SJ (2009) Movements and diving behavior of pelagic spotted dolphins. Mar Mamm Sci 25:137–160

9 Oceanic Dolphin Societies: Diversity, Complexity, and Conservation

Scott MD, Chivers SJ, Olson RJ, Fiedler PC, Holland K (2012) Pelagic predator associations: tuna and dolphins in the eastern tropical Pacific Ocean. Mar Ecol Prog Ser 458:283–302

Sevenbergen K (1997) Abnormal behavior in dolphins chased and captured in the ETP tuna fishery: learned or stress-induced behavior? M.S. Thesis, University of California, San Diego

Smith JE, Gavrilets S, Mulder MB, Hooper PL, El Mouden C, Nettle D, Hauert C, Hill K, Perry S, Pusey AE, van Vugt M (2016) Leadership in mammalian societies: emergence, distribution, power, and payoff. Trends Ecol Evol 31:54–66

Smith JE, Ortiz CA, Buhbe MT, van Vugt M (2018) Obstacles and opportunities for female leadership in mammalian societies: a comparative perspective. Leadersh Q. https://doi.org/10.1016/j.leaqua.2018.09.005

St. Aubin DJ, Ridgway SH, Wells RS, Rhinehart H (1996) Dolphin thyroid and adrenal hormones: circulating levels in wild and semi-domesticated *Tursiops truncatus*, and influence of sex, age, and season. Mar Mamm Sci 12:1–13

Tolley KA, Read AJ, Wells RS, Urian KW, Scott MD, Irvine AB, Hohn AA (1995) Sexual dimorphism in wild bottlenose dolphins (*Tursiops truncatus*) from Sarasota, Florida. J Mammal 76:1190–1198

UNEP/CMS (2018) Report on the CMS workshop on conservation implications of animal culture and social complexity. ScC-SC3/Inf.8. Convention on the conservation of migratory species of wild animals, 3rd meeting of the sessional committee of the CMS Scientific Council, Bonn, 29 May–1 June 2018

Wade PR (1995) Revised estimates of incidental of dolphins (Delphinidae) by the purse-seine tuna fishery in the eastern tropical Pacific, 1959–1972. Fish Bull 93:345–354

Wade PR, Watters GM, Gerrodette T, Reilly SB (2007) Depletion of spotted and spinner dolphins in the eastern tropical Pacific: modeling hypotheses for their lack of recovery. Mar Ecol Prog Ser 343:1–14

Wade PR, Reeves RR, Mesnick SL (2012) Social and behavioural factors in cetacean responses to overexploitation: are odontocetes less "resilient" than mysticetes? J Mar Biol 2012:567276

Whitehead H (2003) Sperm whale societies; social evolution in the ocean. University of Chicago Press, Chicago

Whitehead H (2007) Learning, climate and the evolution of cultural capacity. J Theor Biol 245:341–350

Williams GC (1964) Measurement of consocation among fishes and comments on the evolution of schooling. Publ Mus Mich State Univ Biol Ser 2:349–384

Würsig B (2018) Intelligence. In: Würsig B, Thewissen JGM, Kovacs KM (eds) Encyclopedia of marine mammals, 3rd edn. Academic/Elsevier, San Diego, pp 512–517

Würsig B, Kieckhefer TR, Jefferson TA (1990) Visual displays for communication in cetaceans. In: Thomas J, Kastelein R (eds) Sensory abilities of cetaceans. Plenum, New York, pp 545–559

Würsig B, Wells RS, Norris KS, Würsig M (1994) In: Norris KS, Würsig B, Wells RS, Würsig M (eds) The Hawaiian spinner dolphin. University of California Press, Berkeley, pp 65–102

Chapter 10
Odontocete Adaptations to Human Impact and Vice Versa

Giovanni Bearzi, Sarah Piwetz, and Randall R. Reeves

Abstract Some mammalian species that have not succumbed to pervasive human impacts and encroachments have managed to adapt to certain types of human activities. Several odontocetes have modified their behavior to persist, and in some cases even prosper, in human-altered riverine, coastal, and oceanic habitat. Examples include cooperation with fishers to catch fish, depredation on fishing gear, scavenging, and other kinds of opportunistic foraging (e.g., behind trawlers, around fish farms, or near built structures such as dams and offshore platforms). Some populations have adapted to life in human-made channels and waterways. We review information on the variety of odontocete adaptations to human encroachment, highlight some of the risks and benefits, and try to single out factors that may trigger or contribute to adaptation. Adaptation often brings wildlife into close contact with humans, which leads to conflict. We discuss the challenges of coexistence and contend that we humans, too, need to adjust our behavior and change how we perceive and value wildlife for coexistence to be possible. In addition to good management and conservation action, tolerance on our part will be key for allowing wildlife—odontocetes included—to persist. We advocate cultural and even spiritual shifts that can foster tolerance, nurture the social change that leads to appreciation for wildlife, and create more opportunities to preserve nature.

Keywords Toothed whales · Dolphins · Porpoises · Behavior · Human impact · Fisheries · Depredation · Scavenging · Adaptation · Coexistence

---

G. Bearzi (✉)
Dolphin Biology and Conservation, Cordenons PN, Italy

OceanCare, Wädenswil, Switzerland

S. Piwetz
Dolphin Biology and Conservation, Cordenons PN, Italy

Texas Marine Mammal Stranding Network, Galveston, TX, USA

R. R. Reeves
Okapi Wildlife Associates, Hudson, QC, Canada

© Springer Nature Switzerland AG 2019
B. Würsig (ed.), *Ethology and Behavioral Ecology of Odontocetes*, Ethology and
Behavioral Ecology of Marine Mammals,
https://doi.org/10.1007/978-3-030-16663-2_10

10.1 Introduction

Human populations are dramatically modifying and depleting ecosystems, with losses of biodiversity and increasing rates of extinction that presage a bleak and impoverished future for life on Earth (Schipper et al. 2008; Wilson 2010; Pimm et al. 2014; Ceballos et al. 2015, 2017; Lebreton et al. 2018; Steffen et al. 2018). Planet-wide negative effects of our activities are often referred to as human "impact" (Halpern et al. 2008; Goudie 2018) or "ecological footprint" (Wackernagel and Rees 1998; Sanderson et al. 2002). Human "encroachment" has been used in a similar context (e.g., Watson et al. 2015), often to emphasize our use of, and intrusion into, natural landscapes and seascapes. Faced with pervasive impacts and encroachments, some animal species that have not succumbed or vanished altogether have learned to respond and adapt to certain types of human activities. On land, mammalian carnivores that have shown remarkable adaptations and comebacks in recent times include gray wolves, coyotes, red foxes, spotted and striped hyenas, and brown and black bears (Bateman and Fleming 2012; Carter and Linnell 2016). Some large carnivores persist in human-dominated agricultural landscapes, while others have gradually recolonized semi-wild territories, where they take advantage of human activities occasionally or systematically (Athreya et al. 2013; Chapron et al. 2014).

As human populations increased and colonized riverine, coastal, and marine habitats, several odontocete populations came into contact with humans and were faced with the challenges of adapting to artificially modified habitat. At best, they adapted to human activities, for instance, by modifying behavior to take advantage of foraging opportunities, primarily in relation to fishing, which led to forms of commensalism, mutualism, or depredation (Leatherwood 1975; Powell and Wells 2011; Bonizzoni et al. 2014; Tixier et al. 2018). Some of these adaptations date back centuries or even millennia, as documented by historical reports of dolphins cooperating with coastal fishers to the advantage of both parties (Busnel 1973; Hall 1984), in ways that may still be observed today (Pryor and Lindbergh 1990; Simões-Lopes et al. 1998; Neil 2002; Smith et al. 2009).

Here, we illustrate the range of odontocete behavioral responses to human encroachment. Our focus is on substantial changes in behavior that have developed in response to new opportunities, or unnatural environments, with emphasis on cases of more than mere tolerance, where the adaptive behavior has either benefited the cetaceans or simply allowed for their persistence. One important caveat applies: apparent advantages gained by the cetaceans may be partially or entirely offset by disadvantages that come with human encroachment, whether direct or indirect. Therefore, a presumed benefit or advantage may be only compensatory, with no net gain realized by the cetaceans. For example, odontocetes that depredate on fishing gear (advantage: enhanced foraging) often incur risks of entanglement or hooking or enter into competition with fishers (disadvantages: incidental death or injury, deliberate killing in retaliation). Even in cases involving noncompetitive and lower-risk behavioral adaptation (e.g., to life in human-made channels), any short-term benefit to individual animals may be outweighed by costs at the population level.

Caveat notwithstanding, at least some odontocete populations have managed to survive and persist in environments that are far from pristine, with individuals having learned to benefit from certain aspects of human encroachment. We attempt to identify some of the mechanisms underpinning adaptation, and, perhaps more importantly, we contend that wildlife adaptations to human encroachment give us the opportunity to preserve and coexist with some of nature, odontocetes included. To avail ourselves of that opportunity, we must be prepared to adapt our own behavior toward wildlife and ensure that interactions are inspired by tolerance, some degree of conflict is accepted, and the benefits of coexistence are recognized and emphasized.

10.2 The Ways and Challenges of Adaptation

Defining changes in behavior and associating such changes with one particular stimulus or a set of stimuli can be difficult. For example, the common bottlenose dolphin *Tursiops truncatus* is by nature an opportunistic forager. Its *modus operandi* for finding, identifying, sampling, and becoming at least temporarily fixated on a given food source appears to be that of a prospector. This is its normal behavior, so one would expect diet to change over time just as the spectrum of possibilities, or opportunities, changes. On that premise it can be argued that the changed behavior, or behavioral adaptation, of bottlenose dolphins is really nothing of the sort. The animals simply continue to behave opportunistically, whether that means rooting in the mud or sand to startle flatfish or gobies, chase and herd mullet against a sloping shoreline, or patrol behind trawl nets to capture stunned or discarded prey. The intrinsic behavioral aptitude and character of bottlenose dolphins, as quintessential opportunists, doesn't change; it is the seascape and the possibilities it offers that changes.

Behavioral adaptation to altered conditions represents the degree of plasticity that is intrinsic to the animals' normal behavioral repertoire. Behavioral responses, however, may be maladaptive, or the plasticity of behavior may be insufficient to offset the loss in welfare, even when the behavioral response appears beneficial (Wong and Candolin 2015). Adaptation to human encroachment is a double-edged sword: it can bring benefits or at least compensate for some of the harm but at the same time involves risks to health or to life itself. Assessing whether benefits gained through adaptation are outweighed by the harm caused by human encroachment is often next to impossible, largely due to the lack of baseline data and the difficulty of determining cause-effect linkages, as well as the difficulty of resolving timescale mismatches between short- and long-term gains and losses.

Odontocetes exhibit remarkable resilience, and many populations have managed to persist, and in a few cases even appear to thrive, within areas exposed to heavy human encroachment. For example, (1) common bottlenose dolphins occur at especially high densities in the semi-enclosed, highly eutrophic and degraded Amvrakikos Gulf, Greece (Bearzi et al. 2008; Gonzalvo et al. 2016); (2) common bottlenose dolphins in the Northern Evoikos Gulf, Greece, forage around fish farms within a bay polluted by a large ferronickel plant (Bonizzoni et al. 2014); (3) common bottlenose dolphins regularly occur in the Galveston Ship Channel, USA, a high-traffic, heavy-industry, urbanized area (Piwetz 2019; Fig. 10.1); (4) Indo-Pacific bottlenose dolphins *Tursiops aduncus* frequently stay in Fremantle Inner Harbour, Western Australia, part of a busy, noisy industrial port exposed to high levels of vessel traffic (up to 56 vessels per hour; Marley et al. 2017); (5) Indo-Pacific humpback dolphins *Sousa chinensis* in Hong Kong have long been exposed to massive and constant disturbance from shoreline development and land reclamation (Jefferson and Karczmarski 2001; Jefferson and Curry 2015; Fig. 10.1); (6) common dolphins *Delphinus delphis* reside in Port Phillip Bay, a highly urbanized, shallow, semi-enclosed embayment in SE Australia (Mason et al. 2016); and (7) South Asian river dolphins *Platanista gangetica* persist in the Buriganga River, Bangladesh, a polluted river that flows through Dhaka "with huge industrial, household discharges, encroachments and vehicle loads and sometimes regarded as ecologically dead" (Alam et al. 2015). In some of these cases, the apparent resilience could in fact reflect nothing more than the lack of options, i.e., the animals simply have "nowhere to go" (Forney et al. 2017). In a few other cases, however, the animals could certainly move away but they appear to remain deliberately, or they return to degraded habitat (as is the case of the Amvrakikos Gulf; Bearzi et al. 2011b), often because prey is more abundant compared to surrounding areas.

In Table 10.1 we provide examples of published cases of behavioral adaptation or changed behavior in response to human encroachment. Recognizing that many reviews of odontocete "interactions" with human activities have already been published, including those involving opportunistic depredation of fishing gear, scavenging, and facilitated predation (Fertl and Leatherwood 1997; Northridge and Hofman 1999; Plagányi and Butterworth 2005; Gilman et al. 2006; Kock et al. 2006; Read 2008; Hamer et al. 2012; Werner et al. 2015; Northridge 2018), we have sought to offer a quick overview rather than a comprehensive literature survey. The examples included in Table 10.1 encompass cases of adaptation of odontocete species to fisheries and aquaculture, as well as occurrence in human-made channels and canals, near dams, near oil and gas platforms, or close to urbanized areas. We are aware that several other types of adaptation exist, which we did not consider here. For instance, we did not consider (1) deliberate food provisioning of habituated dolphins and the potential benefits or negative consequences that might derive from "friendly" interactions of various kinds with humans (Lockyer 1990; Orams 2002; Durden 2005; Bejder et al. 2006; Cunningham-Smith et al. 2006; Smith et al. 2008; de Sá Alves et al. 2011); (2) potential indirect benefits to some odontocetes resulting from man-driven climate and ecosystem changes, regime shifts, and food web modifications (e.g., decline of predators; Myers and Worm 2003; Ferretti et al. 2008); and (3) bow-riding (Würsig 2018) and wave-riding (Williams et al. 1992) as a way of potentially reducing the energetic costs of movement.

What are the main factors triggering adaptive changes in behavior, which may ultimately allow for survival under human encroachment? Predictable occurrence of prey that is more concentrated and/or easier to catch in and around fishing gear has promoted behavioral adaptation and specialization in a variety of odontocete species, worldwide (Table 10.1). Fish and other organisms caught in fishing nets, hooked on

10 Odontocete Adaptations to Human Impact and Vice Versa

Fig. 10.1 Examples of odontocete adaptations to human impact. (**a**) Common bottlenose dolphins, *Tursiops truncatus*, in the Galveston Ship Channel, USA, have adapted to living in a high-traffic, heavy-industry, urbanized area (photo by Giovanni Bearzi, Dolphin Biology and Conservation). (**b**) Indo-Pacific humpback dolphin, *Sousa chinensis*, in Hong Kong, where the population is exposed to massive and constant disturbance from shoreline development and land reclamation (photo courtesy of Hong Kong Cetacean Research Project). (**c**) A common bottlenose dolphin patrolling near a coastal fish farm in the Gulf of Corinth, Greece, in search of wild prey that concentrate around the cages (photo by Silvia Bonizzoni, Dolphin Biology and Conservation). (**d**) Common bottlenose dolphins foraging behind midwater pair trawlers in the northern Adriatic Sea off Veneto, Italy (photo by Silvia Bonizzoni, Dolphin Biology and Conservation). (**e**) An Amazon River dolphin, *Inia geoffrensis*, catching a large piranha (possibly a redeye piranha *Serrasalmus rhombeus*) along a dam wall in Brazil's Tocantins River (photo by Claryana C. Araújo). (**f**) A South Asian river dolphin, *Platanista gangetica*, in the waters immediately below a barrage in the Sapta Koshi River,

fishing lines, or used as bait cannot escape and can be extracted or bitten as an easy though risky meal. After nets have been hauled, organisms discarded at sea as bycatch may unintentionally provision the surrounding odontocetes and reinforce scavenging (Norris and Prescott 1961; Leatherwood 1975; Wassenberg and Hill 1990; Couperus 1994, 1997; Svane 2005). Begging for food can also occur, e.g., in association with depredation and scavenging around recreational fishing boats (Powell and Wells 2011). Around fish farms, artificial substrate and infrastructure, combined with input of nutrients or manufactured fish feed, result in concentration of wild prey, while the infrastructure itself may facilitate prey capture (Díaz López 2006; Piroddi et al. 2011; Bonizzoni et al. 2014). Shellfish aquaculture also may offer enriched habitat where dolphins can forage more effectively on wild fish that aggregate around farm buoys and rafts (Díaz López and Methion 2017), though some studies suggest avoidance of farm areas (Markowitz et al. 2004; Watson-Capps and Mann 2005; Pearson et al. 2012). Finally, rare cases of symbiosis between dolphins and fishers may bring the desired mutual benefits of concentration and relative ease of catching prey (Pryor and Lindbergh 1990; Simões-Lopes et al. 1998; Neil 2002; Smith et al. 1997, 2009).

Depredation on fishing gear is arguably the most common and best-documented type of odontocete adaptation and one that often results in mortality and conflict with humans. Depredation, scavenging, and foraging in the proximity of fishing gear expose cetaceans to greater risk of becoming hooked or entangled (Couperus 1997; Friedlaender et al. 2001; Baird et al. 2015; Stepanuk et al. 2018), and incidental mortality from encounters with fishing gear has become the main direct and immediate threat to many odontocete populations around the world (Read et al. 2006; Reeves et al. 2013; Taylor et al. 2017). Also, as depredation becomes established and widespread within a geographic region, the risk of retaliation by irate fishers tends to increase. This can lead to deliberate harm and killing (Loch et al. 2009) or even culling campaigns (Kasuya 1985; Bearzi et al. 2004). Whalers retaliated forcefully against killer whales that regularly scavenged whale carcasses during the industrial whaling era (Robertson 1954, p. 159; Whitehead and Reeves 2005).

The economic impacts of depredation and food web competition have often been overestimated (Lavigne 2003; Plagányi and Butterworth 2005). In some cases, the perception of reduced landings caused by dolphin depredation of nets is much greater than the actual economic damage, and reports of such damage may be inflated, either deliberately or due to misperceptions (Bearzi et al. 2011a; Rechimont

---

Fig. 10.1 (continued) Nepal (photo by Grant Abel). (**g**) Northern bottlenose whales, *Hyperoodon ampullatus*, feeding on fish discarded from a deep-set gillnet vessel fishing for Greenland halibut *Reinhardtius hippoglossoides*, and (**h**) following a shrimp trawler off Labrador, eastern Canada (photos courtesy of Jack Lawson, Fisheries and Oceans Canada). (**i**) A sperm whale, *Physeter macrocephalus*, surfacing between dives to remove sablefish *Anoplopoma fimbria* from demersal longline gear deployed in the eastern Gulf of Alaska (photo by Paul Norwood, courtesy of Jan Straley, SEASWAP/NOAA permit #14122). (**j**) Killer whales, *Orcinus orca*, selectively removing Patagonian toothfish *Dissostichus eleginoides* from a longline deployed by the Sapmer fleet off Île aux Cochons, Crozet Archipelago (photo by Bertrand Loyer, Saint Thomas Productions)

10 Odontocete Adaptations to Human Impact and Vice Versa

Table 10.1 Examples of cooperation, depredation, scavenging, opportunistic foraging, and other odontocete adaptations to human impact

Adaptation	Species	Area	References
Fishing: cooperative	*Orcaella brevirostris*	Myanmar: Ayeyarwady River	Smith et al. (1997, 2009)
	Sousa teuszii	W Africa: Mauritania	Busnel (1973)
	Tursiops truncatus	W Africa: Mauritania	Busnel (1973)
		E Australia	Hall (1984), Neil (2002)
		Brazil: Laguna	Pryor and Lindbergh (1990), Cantor et al. (2018)
		Brazil: Tramandaí	Simões-Lopes et al. (1998), Zappes et al. (2011)
Fishing: longlines	*Physeter macrocephalus*	S Atlantic Ocean: South Georgia	Purves et al. (2004), Towers et al. (2018)
		Argentina and Malvinas	Goetz et al. (2011)
		S Chile	Hucke-Gaete et al. (2004)
		Indian Ocean: Crozet Islands	Guinet et al. (2015)
		USA: Alaska	Mathias et al. (2009), Peterson and Hanselman (2017)
	Hyperoodon ampullatus	E Canada	COSEWIC (2011), J. Lawson, pers. comm.
	Orcinus orca	S Atlantic Ocean: South Georgia	Purves et al. (2004), Towers et al. (2018)
		Australia	Bell et al. (2006), Tixier et al. (2018)
		S Brazil	Dalla Rosa and Secchi (2007)
		S Chile	Hucke-Gaete et al. (2004)
		Indian Ocean: Crozet Islands	Guinet et al. (2015)
		Indian Ocean: St. Paul/ Amsterdam Islands	Tixier et al. (2018)
		New Zealand: Three Kings Islands	Visser (2000)
		USA: Alaska	Peterson and Hanselman (2017)
	Pseudorca crassidens	Atlantic, Indian, and Pacific Ocean	Ramos-Cartelle and Mejuto (2008)
		Australia	Bell et al. (2006)
		Portugal: Azores	Hernandez-Milian et al. (2008)
		S Brazil	Dalla Rosa and Secchi (2007)
		N Pacific Ocean	Forney et al. (2011)
	Globicephala macrorhynchus	W Atlantic Ocean	Garrison (2007)

(continued)

218 G. Bearzi et al.

Table 10.1 (continued)

Adaptation	Species	Area	References
	Globicephala melas	Mediterranean: S Spain	López et al. (2012)
	Grampus griseus	W Atlantic Ocean	Garrison (2007)
		S Brazil	Dalla Rosa and Secchi (2007)
		Mediterranean: S Spain	López et al. (2012)
	Delphinus delphis	Mediterranean: S Spain	López et al. (2012)
	Stenella coeruleoalba	Mediterranean: S Spain	López et al. (2012)
	Steno bredanensis	USA: Hawaii	Schlais (1984), West et al. (2011)
Fishing: troll	*Tursiops truncatus*	USA: E Florida	Zollett and Read (2006)
Fishing: hand-jig	*Grampus griseus*	Portugal: Azores	Cruz et al. (2014)
Fishing: pole and line	*Delphinus delphis*	Portugal: Azores	Silva et al. (2002), Cruz et al. (2016)
	Stenella frontalis	Portugal: Azores	Silva et al. (2002), Cruz et al. (2016)
	Tursiops truncatus	Portugal: Azores	Silva et al. (2002), Cruz et al. (2016)
Fishing: recreational angling	*Steno bredanensis*	Angola	Weir and Nicolson (2014)
	Tursiops truncatus	USA: W Florida	Powell and Wells (2011)
Fishing: trammel and gill nets	*Tursiops truncatus*	Mediterranean: France, Corsica	Rocklin et al. (2009)
		Mediterranean: Italy, Sardinia	Lauriano et al. (2004)
		Mediterranean: Italy, Aeolian archipelago	Blasi et al. (2015)
		Mediterranean: Italy, Favignana Island	Buscaino et al. (2009)
		Mediterranean: Spain, Balearic Islands	Brotons et al. (2008); Gazo et al. (2008)
		Mediterranean: Tunisia, Kerkennah Islands	Ayadi et al. (2013)
		Mediterranean: Turkey	Gönener and Özdemir (2012)
		Mexico: Veracruz	Rechimont et al. (2018)
		USA: North Carolina	Cox et al. (2003)
Fishing: crab pots	*Tursiops truncatus*	USA: E Florida	Noke and Odell (2002)
Fishing: stake nets	*Orcaella brevirostris*	E India: Chilika lagoon	D'Lima et al. (2013)
	Sotalia guianensis	S Brazil: Cananéia	Monteiro-Filho (1995)

(continued)

10 Odontocete Adaptations to Human Impact and Vice Versa

Table 10.1 (continued)

Adaptation	Species	Area	References
Fishing: trawl nets	*Physeter macrocephalus*	E Canada: Grand Banks	Karpouzli and Leaper (2004), J. Lawson, pers. comm.
	Hyperoodon ampullatus	E Canada: Davis Strait/ Grand Banks	COSEWIC (2011), J. Lawson, pers. comm.
	Orcinus orca	North Sea	Couperus (1994), Luque et al. (2006)
	Cephalorhynchus hectori	New Zealand: Banks Peninsula	Rayment and Webster (2009)
	Lagenorhynchus acutus	NE Atlantic Ocean: Celtic Sea	Couperus (1997), Morizur et al. (1999)
	Sousa chinensis	Hong Kong	Jefferson (2000)
	Sousa plumbea	Mediterranean: Turkey	Ozbilgin et al. (2018)
	Sousa sahulensis	E Australia: Cleveland Bay	Parra (2006)
	Tursiops aduncus	E Australia: Moreton Bay	Chilvers and Corkeron (2001)
		E Australia: Moreton Bay	Corkeron et al. (1990), Wassenberg and Hill (1990)
		E Australia: New South Wales	Broadhurst (1998)
	Tursiops truncatus	NW Australia: Pilbara	Jaiteh et al. (2013), Allen et al. (2014, 2017)
		Mediterranean: Adriatic Sea	Fortuna et al. (2010)
		Mediterranean: Croatia	Bearzi et al. (1999)
		Mediterranean: Spain, Balearic Islands	Gonzalvo et al. (2008)
		Gulf of Mexico	Leatherwood (1975)
		Gulf of California	Leatherwood (1975)
		USA: Georgia	Kovacs and Cox (2014), Kovacs et al. (2017)
		USA: Texas, Galveston Ship Channel	Fertl (1994), Piwetz (2019)
	Phocoena phocoena	Denmark: North Sea and Baltic Sea	Clausen and Andersen (1988)
Fishing: fish farms	*Tursiops truncatus*	Mediterranean: Greece, Eastern Ionian Sea	Piroddi et al. (2011)
		Mediterranean: Greece, Northern Evoikos Gulf	Bonizzoni et al. (2014)
		Mediterranean: Greece, Gulf of Corinth	Bearzi et al. (2016)
		Mediterranean: Italy, Sardinia	Díaz López (2006, 2017)

(continued)

Table 10.1 (continued)

Adaptation	Species	Area	References
	Phocoena phocoena	Canada: Bay of Fundy	Haarr et al. (2009)
Fishing: mussel farms	*Tursiops truncatus*	N Spain: Galicia	Díaz López and Methion (2017)
Whaling industry	*Orcinus orca*	–	Whitehead and Reeves (2005)
Oil and gas platforms	*Tursiops truncatus*	Mediterranean: Adriatic Sea	Triossi et al. (2013)
		S Brazil	Cremer et al. (2009)
	Phocoena phocoena	North Sea: Dogger Bank	Todd et al. (2009)
Dams	*Platanista gangetica*	Nepal: Sapta Koshi River	Grant Abel, pers. comm. (see main text)
	Inia geoffrensis	Brazil: Tocantins River	Araújo and Wang (2014)
Channels and canals	*Sousa plumbea*	Mediterranean through Red Sea	Frantzis (2018)
	Sousa sahulensis	E Australia: Cleveland Bay	Parra (2006)
	Tursiops truncatus	USA: W Florida	Allen et al. (2001)
Ice channels	*Orcinus orca*	McMurdo Sound, Ross Sea, Antarctica	Andrews et al. (2008)
Urbanized areas	*Delphinus delphis*	SE Australia: Port Phillip Bay	Mason et al. (2016)
	Tursiops aduncus	W Australia: Fremantle Inner Harbour	Marley et al. (2017)

et al. 2018). Odontocete adaptations to fishing are certainly not limited to depredation, nor are they always to the detriment of the fishery. For instance, some species use nets as barriers to corral fish, sometimes resulting in increased fishery landings (Rocklin et al. 2009; D'Lima et al. 2013).

As noted earlier, another major factor unrelated to fishing that can trigger changes in odontocete behavior is habitat modification including channels, canals, dams, and other artificial structures (Table 10.1). Artificial channels may provide suitable habitat for dolphins and their prey. For example, common bottlenose dolphins spend considerable time, or even their entire lives, in the Intracoastal Waterway of the eastern and southern USA, particularly in North Carolina (Haviland-Howell et al. 2007) and Florida (Odell and Asper 1980), and in the deep-dredged Galveston Ship Channel, an important foraging area for dolphins (Piwetz 2019). Leatherwood and Reeves (1983) noted that dolphins in Texas "use intensively and regularly the artificially maintained passes at Port Mansfield and Port Isabel, as well as their associated ship channels." Those authors further surmised, "The creation and maintenance of navigable passes and the resultant reduction in lagoon salinity may have created 'new' habitat for dolphins," possibly modifying their distribution and/or increasing their numbers.

At least one odontocete species is known to have taken advantage of a human-made canal to expand its range. Humpback dolphins observed in the Mediterranean waters of Israel, Turkey, and Greece are believed to be Indian Ocean humpback dolphins *Sousa plumbea* that have passed from the Red Sea through the Suez Canal or at least descendants of dolphins that used this route to colonize an entirely different ocean basin (Frantzis 2018). We are not aware of any other cetaceans using an artificial canal system to switch ocean basins. Similar to canals on land, channels opened through sea ice by icebreaking ships allow cetaceans to move into otherwise inaccessible areas, where fast-adapting species can gain the benefit of new foraging opportunities. Killer whales *Orcinus orca* have learned to take advantage of this by using icebreaker channels through the fast ice in McMurdo Sound (western Ross Sea, Antarctica) (Pitman and Ensor 2003; Andrews et al. 2008). Concern has been expressed that narwhals and belugas entering the artificial channels created by icebreakers in the Arctic put themselves at greater risk of ice-entrapment, which is known to cause genuinely accidental mass mortality of those two odontocete species even without human involvement (Stirling 1980; Stirling et al. 1981).

Dams built in waterways for power generation, flood control, or irrigation are, for the most part, extremely harmful to freshwater cetaceans—fragmenting populations, disrupting natural flow patterns and sediment transport processes, blocking fish migrations, and reducing the quantity of fresh water in lakes and river channels by diverting it for urban, industrial, and agricultural uses (Smith and Reeves 2012). There are, however, instances in which river dolphins appear to gain some benefit from these structures. For example, in Brazil's Tocantins River, botos *Inia geoffrensis* apparently take advantage of the prey concentrations that form immediately downstream of a dam (Araújo and Wang 2014; Fig. 10.1). It has also been suggested that the dolphins "may be exposed to less human disturbance downstream of dams because of restriction zones in which human activities are prohibited for safety concerns, which may be beneficial [to the dolphins] in the short-term" (Araújo and Wang 2014). Similarly, Grant Abel observed and photographed South Asian river dolphins *Platanista gangetica* in Nepal aggregated (along with gillnet fishers) and foraging in the turbulent waters immediately below the Sapta Koshi River barrage (Fig. 10.1). In both of these examples, any benefits to the dolphins need to be weighed against the losses and risks to them associated with damming, as well as gillnetting in the Nepal case, but it is nevertheless interesting that both of these ancient, highly specialized odontocete species (Cassens et al. 2000) show such adaptability to this type of human encroachment on the natural environment.

10.3   The Most Adaptable Odontocetes

Adaptations to human encroachment are often expressions of the behavioral repertoire of species that are naturally inclined to respond creatively to new stimuli and opportunities. Identification of those qualities—physiological, behavioral, or cognitive—that make one species more capable and inclined than another to adapt to and benefit from human encroachment represents a valuable research avenue (Barrett et al. 2019). A better understanding of these "survival skills" could help to reshape our perception of odontocetes as something other than opportunistic freeloaders and evil pests (Bearzi et al. 2004, 2010) or new-age creatures possessing superior intelligence and supernatural abilities (Lilly 1967; Wyllie 1993).

Assessing the degree of adaptation to human impact across cetacean taxa would require relatively complex analyses that take into account perception and observer bias (possibly through the use of culturomics or other weighting tools, to account inter alia for the fact that some odontocetes are studied more than others; Fox et al. 2017; Bearzi et al. 2018). Pending a rigorous analytical approach, our overview (Table 10.1) suggests that the most adaptable odontocetes are largely coastal delphinids, particularly the behaviorally flexible and opportunistic genus *Tursiops* along with several other coastal and riverine species. These species are not only more likely to be exposed to human impact (due to their proximity to land) but also naturally inclined to respond to environmental variation. For instance, some coastal and riverine odontocetes can display high levels of specialization and innovation, including the use of tools (Smolker et al. 1997; Krützen et al. 2005; Martin et al. 2008; Allen et al. 2011; Patterson and Mann 2011; Araújo and Wang 2012; Barber 2016) and creative foraging tactics (Lopez and Lopez 1985; Silber and Fertl 1995; Sargeant et al. 2005; Torres and Read 2009).

Local or ecologically specialized populations of a number of oceanic and continental slope species, including sperm whales *Physeter macrocephalus*, offshore killer whales, false killer whales *Pseudorca crassidens*, Risso's dolphins *Grampus griseus*, and rough-toothed dolphins *Steno bredanensis*, also show evidence of "adapting to," in the sense of taking advantage of, longline fishing (Table 10.1). Among beaked whales, there are reports of frequent scavenging and depredation by northern bottlenose whales *Hyperoodon ampullatus* on the longline fishery off Labrador and in western Davis Strait, Canada (COSEWIC 2011; J. Lawson, pers. comm.). Adaptability to encroachment shown by most offshore, deepwater species (obviously less exposed to human impact than coastal species) seems largely limited to depredation on longlines. Sperm whales and northern bottlenose whales, however, are known to feed in association with trawlers off eastern Canada (Karpouzli and Leaper 2004; COSEWIC 2011; J. Lawson, pers. comm.). Some odontocetes incidentally caught in trawl nets (but not included in Table 10.1) might have been actively extracting prey from nets or scavenging in a trawler's wake (e.g., common dolphins *Delphinus delphis*; Svane 2005; Spitz et al. 2013; Thompson et al. 2013). Therefore, the geographic and taxonomic extent of odontocete adaptations to fishing may be greater than outlined here.

Porpoises (family Phocoenidae) seem comparatively less likely than delphinids (family Delphinidae) to adapt to and gain any benefit from human activities, especially if one considers their occurrence in coastal waters that are often exposed to high levels of human encroachment. While harbor porpoises *Phocoena phocoena* were reported to occur near a fish farm in Canada and around offshore gas installations in the North Sea (Todd et al. 2009; Haarr et al. 2009; Table 10.1), it is unclear whether these represent true adaptations. Oil and gas platforms may provide substrate, shelter, or enhanced foraging opportunities for demersal and other fishes, and some studies suggest they also attract common bottlenose dolphins (Triossi et al. 2013; Cremer et al. 2009). The actual attraction potential, however, would need to be confirmed by weighted studies conducted near and away from the platforms, rather than primarily or exclusively in their proximity.

What other attributes besides an opportunistic bent and behavioral flexibility make some odontocete species more adaptable to human encroachment than others? Because several of the relatively adaptable odontocete species rank high in terms of encephalization and brain complexity (Fox et al. 2017; Ridgway et al. 2017), the hypothesis that encephalization predicts adaptation to human impact is worth investigating. While some of the animal taxa most adapted to human encroachment do not have large brains (e.g., domiciliary cockroaches, order Blattodea; Lihoreau et al. 2012), encephalization is generally considered an important factor promoting behavioral adaptation in vertebrates. As Jerison (1985) put it: "What behavior or dimensions of behavior evolved when encephalization evolved? The answer: the relatively unusual behaviors that require increased neural information processing capacity, beyond that attributable to differences among species in body size." Large brains can confer advantages in the form of behavioral flexibility, carrying fitness benefits to individuals facing novel or altered environmental conditions (Reader and Laland 2002; Marino 2005; Sol et al. 2008). Mammalian species with large brains relative to their body mass generally tend to be more successful than species with relatively small brains in dealing with environmental change, when the response demands behavioral flexibility in the form of learning and innovation. In other words, "enlarged brains can provide a survival advantage in novel environments" (Sol et al. 2008). Among cetaceans, encephalization has been linked to the breadth of social and cultural behavior across species (Connor 2007; Fox et al. 2017; Bearzi et al. 2018) but, so far, not specifically to adaptation to human encroachment. A tentative hypothesis, calling for validation by future research, is that the odontocete species most adaptable to human impact have evolved to respond to changing environments, possess a naturally flexible behavioral repertoire predisposing them to innovation (including through fast-paced cultural transmission of information; Whitehead and Ford 2018), have large brains relative to their body mass, and of course have significant chances of entering into contact with human encroachment. As noted above, an alternative possibility is that what may look like resilience and adaptation is, at least in some cases, merely the result of a lack of options (Forney et al. 2017).

10.4 From Conflict to Coexistence

Conflict between people and wildlife dates back to the dawn of mankind (Woodroffe et al. 2005). Growing human encroachment on even the most remote areas has resulted in massive declines of large terrestrial and marine carnivores (Myers et al. 2007; Ripple et al. 2014). However, several large mammalian species have survived and even come back from being nearly eradicated (Chapron et al. 2014), including in urban areas (Gehrt et al. 2010; Bateman and Fleming 2012). Recent studies have documented surprising adaptations to encroachment, such as increased nighttime activity (Gaynor et al. 2018). Finding ways to mitigate conflict and allow the surviving mammals to share habitat with humans is now a pressing conservation concern worldwide (Treves et al. 2006; Carter and Linnell 2016; Chapron and López-Bao 2016). In the case of some large carnivores on land, and of sharks at sea, the conflict includes occasional attacks on humans that can generate significant resistance to conservation efforts (Löe and Röskaft 2004; Neff and Hueter 2013; Treves and Bruskotter 2014; Penteriani et al. 2016). In the case of cetaceans, the risks to humans are almost entirely related to resource competition rather than to human health and safety. Human responses to conflict with wildlife have historically entailed the use of lethal methods. However, in some cases we have learned to adapt through nonlethal approaches, mitigating conflict and avoiding risky situations through better understanding of animal behavior.

Coexistence with large carnivores has been described as a "dynamic but sustainable state" in which both the people and the large, often dangerous animals adapt to living in shared landscapes, with institutional arrangements that "ensure long-term carnivore population persistence, social legitimacy, and tolerable levels of risk" (Carter and Linnell 2016). Achieving coexistence requires a considerable degree of tolerance as well as risk management, e.g., through protective legislation, supportive public opinion, and a variety of thoughtfully designed tools and practices (Messmer 2000; Chapron et al. 2014; Ripple et al. 2016). Successful coexistence is not only contingent on the economic and material aspects of interaction with wildlife but is also strongly related to social, cultural, cognitive, and emotional factors (Treves and Bruskotter 2014). For instance, despite livestock depredations by snow leopards *Panthera uncia*, Tibetan Buddhist monasteries protect the leopards and their habitat because of cultural and religious values (Li et al. 2014). In Libya, where dolphins are regarded as "holy" and special animals, it is said "if you kill a dolphin you don't catch fish anymore" (Bearzi 2006). Superstitions, myths, and cultural taboos, however, can also motivate dolphin killing (Loch et al. 2009). As conservation scientists, we should not promote ignorance, which forms the basis for the belief systems that underlie superstition, myth, and taboo. Rather, we should support consolidated cultural values that have proven to be relevant and effective for nature conservation. Monetary incentives can sometimes promote tolerance, but several studies of land carnivores have shown that conservation success must rely strongly on lasting social and cultural change rather than on purely economic rewards (Hazzah et al. 2014; Treves and Bruskotter 2014).

When tolerated and allowed to coexist, wildlife can sometimes persist and recolonize areas with moderate or even high levels of human encroachment. The alternative to a coexistence model is a "separation model," whereby it is assumed that large predators can survive only within protected areas. The separation model, however, has been described as "a consequence of former policy goals to exterminate these species" (Chapron et al. 2014). While the importance of protected areas for cetacean conservation is undeniable (Hoyt 2011), such areas are often too small, fragmented, and poorly managed (Guidetti et al. 2008) to support viable populations of large and highly mobile marine predators. In addition to increasing the size and number of protected areas to meaningful levels (e.g., from the current $<1\%$ of the global oceans to the 10–30% targets suggested by international agreements; Cullis-Suzuki and Pauly 2010), identifying and protecting travel corridors among geographically proximate protected areas, and ensuring that appropriate regulations are enforced within them (Edgar et al. 2014), managers (and scientists) need to endorse and promote wildlife values for society (Messmer 2000), fostering tolerance of damage or risk, and appreciation of animals and biodiversity. Because the benefits and liabilities are distributed unevenly among different stakeholders, and the burden of conflicts with odontocetes falls mainly on the shoulders (and bank accounts) of fishers, they should be involved in problem solving, encouraged to tolerate some loss, compensated when justified, and helped in finding suitable alternative means of gaining their livelihoods.

10.5 Can We Adapt to the Odontocetes that Have Adapted to Us?

Cetacean species that had been severely depleted by whaling (including odontocetes such as the sperm whale; Whitehead 2002) recovered substantially after large-scale commercial exploitation ceased. Likewise, populations of large carnivores on land increased after the killing stopped, and favorable legislation was introduced, at times irrespective of human population density (Linnell et al. 2001). Regulating human behavior and resource exploitation can allow odontocete populations to survive and recover even within areas of heavy human encroachment. Studies of human-wildlife conflict suggest that coexistence is possible: the long-term viability of wild fauna in habitat shared with people relies on good management practice and tolerance, the latter sometimes being even more important than the former (Li et al. 2014).

Learning more about ways to instill tolerance and nurture the social change that leads to appreciation for wildlife is crucially important. How capable are we humans of adapting and changing our perceptions so that we regard at least some of the damage caused by odontocetes as an understandable response to stresses for which we are partly or entirely responsible? Beyond merely calling for management measures to protect endangered species, can we come to view wild creatures with the wonder and respect they deserve and seek ways to share with them the privilege of living on this fragile and finite planet?

These kinds of questions shift the emphasis away from the realm of science and open the door to philosophical or even spiritual speculation. E. O. Wilson argued that we have a natural affinity for life ("biophilia") that constitutes the essence of our humanity and binds us to all other living species (Wilson 1984; Kellert and Wilson 1993). His aptly titled book, *The Creation: An Appeal to Save Life on Earth*, attempts to bridge science and religion, contending that "the Creation, whether you believe it was placed on this planet by a single act of God or accept the scientific evidence that it evolved autonomously during billions of years, is the greatest heritage, other than the reasoning mind itself, ever provided to humanity" (Wilson

2010). If one accepts and embraces the concept of biophilia, the focus turns to learning how to coexist, cultivate tolerance, and accept that some of the human-wildlife conflict is inevitable and ultimately natural, even necessary for our own well-being and survival (Monbiot 2014; Svenning et al. 2016).

Acknowledgments We are grateful to Peter J. Corkeron for contributing valuable thoughts and information and to Silvia Bonizzoni for insightful comments. Grant Abel provided a report with photos of his observations of South Asian river dolphins in the Sapta Koshi River, Nepal, and Jack Lawson provided a summary of interactions between odontocetes and fisheries on the Grand Banks, Newfoundland, Canada. Grant Abel, Claryana C. Araújo, Silvia Bonizzoni, Samuel Hung, Thomas A. Jefferson, Jack Lawson, Bertrand Loyer, and Jan Straley generously contributed photos displaying cases of odontocete adaptation to human impact.

References

Alam SMI, Hossain MM, Baki MA, Bhouiyan NA (2015) Status of ganges dolphin, *Platanista gangetica* (Roxburgh, 1801) in the river Buriganga, Dhaka. Bangladesh J Zool 43:109–120

Allen MC, Read AJ, Gaudet J, Sayigh LS (2001) Fine-scale habitat selection of foraging bottlenose dolphins *Tursiops truncatus* near Clearwater, Florida. Mar Ecol Prog Ser 222:253–264

Allen SJ, Bejder L, Krützen M (2011) Why do Indo-Pacific bottlenose dolphins (*Tursiops* sp.) carry conch shells (*Turbinella* sp.) in Shark Bay, Western Australia? Mar Mamm Sci 27:449–454

Allen SJ, Tyne JA, Kobryn HT, Bejder L, Pollock KH, Loneragan NR (2014) Patterns of dolphin bycatch in a north-western Australian trawl fishery. PLoS One 9:e93178

Allen SJ, Pollock KH, Bouchet PJ, Kobryn HT, McElligott DB, Nicholson KE, Smith JN, Loneragan NR (2017) Preliminary estimates of the abundance and fidelity of dolphins associating with a demersal trawl fishery. Sci Rep 7:4995

Andrews RD, Pitman RL, Ballance LT (2008) Satellite tracking reveals distinct movement patterns for type B and type C killer whales in the southern Ross Sea, Antarctica. Polar Biol 31:1461–1468

Araújo CC, Wang JY (2012) Botos (*Inia geoffrensis*) in the upper reaches of the Tocantins river (Central Brazil) with observations of unusual behavior, including object carrying. Aquat Mamm 38:435–440

Araújo CC, Wang JY (2014) The dammed river dolphins of Brazil: impacts and conservation. Oryx 49:17–24

Athreya V, Odden M, Linnell JD, Krishnaswamy J, Karanth U (2013) Big cats in our backyards: persistence of large carnivores in a human dominated landscape in India. PLoS One 8:e57872

Ayadi A, Bradai MN, Ghorbel M, M. (2013) Les pingers comme moyen d'atténuation des interactions négatives du grand dauphin avec les filets maillants aux Iles Kerkennah (Tunisie). Rapport Commission Intérnationale pour l'Exploration Scientifique de la Mer Méditerranée 40:786

Baird RW, Mahaffy SD, Gorgone AM, Cullins T, McSweeney DJ, Oleson EM, Bradford AL, Barlow J, Webster DL (2015) False killer whales and fisheries interactions in Hawaiian waters: evidence for sex bias and variation among populations and social groups. Mar Mamm Sci 31:579–590

Barber TM (2016) Variety and use of objects carried by provisioned wild Australian humpback dolphins (*Sousa sahulensis*) in Tin Can Bay, Queensland, Australia. Int J Comp Psychol 29:21

Barrett LP, Stanton LA, Benson-Amran S (2019) The cognition of "nuisance" species. Anim Behav 147:167–177

Bateman PW, Fleming PA (2012) Big city life: carnivores in urban environments. J Zool 287:1–23

10 Odontocete Adaptations to Human Impact and Vice Versa

Bearzi G (2006) Action plan for the conservation of cetaceans in Libya. In: Regional activity Centre for Specially Protected Areas (RAC/SPA). Environment General Authority and Marine Biology Research Center, Libya, p 50

Bearzi G, Politi E, Notarbartolo di Sciara G (1999) Diurnal behavior of free-ranging bottlenose dolphins in the Kvarneric (northern Adriatic Sea). Mar Mamm Sci 15:1065–1097

Bearzi G, Holcer D, Notarbartolo di Sciara G (2004) The role of historical dolphin takes and habitat degradation in shaping the present status of northern Adriatic cetaceans. Aquat Conserv Mar Freshwat Ecosyst 14:363–379

Bearzi G, Agazzi S, Bonizzoni S, Costa M, Azzellino A (2008) Dolphins in a bottle: abundance, residency patterns and conservation of bottlenose dolphins *Tursiops truncatus* in the semiclosed eutrophic Amvrakikos gulf, Greece. Aquat Conserv Mar Freshwat Ecosyst 18:130–146

Bearzi G, Pierantonio N, Bonizzoni S, Notarbartolo di Sciara G, Demma M (2010) Perception of a cetacean mass stranding in Italy: the emergence of compassion. Aquat Conserv Mar Freshwat Ecosyst 20:644–654

Bearzi G, Bonizzoni S, Gonzalvo J (2011a) Dolphins and coastal fisheries within a marine protected area: mismatch between dolphin occurrence and reported depredation. Aquat Conserv Mar Freshwat Ecosyst 21:261–267

Bearzi G, Bonizzoni S, Gonzalvo J (2011b) Mid-distance movements of common bottlenose dolphins in the coastal waters of Greece. J Ethol 29:369–374

Bearzi G, Bonizzoni S, Santostasi NL, Furey NB, Eddy L, Valavanis VD, Gimenez O (2016) Dolphins in a scaled-down Mediterranean: the Gulf of Corinth's odontocetes. In: di Sciara GN, Podestà M, Curry BE (eds) Mediterranean marine mammal ecology and conservation, Advances in marine biology, vol 75. Academic, Oxford, pp 297–331

Bearzi G, Kerem D, Furey NB, Pitman RL, Rendell L, Reeves RR (2018) Whale and dolphin behavioural responses to dead conspecifics. Zoology 128:1–15

Bejder L, Samuels A, Whitehead H, Gales N, Mann J, Connor R, Heithaus M, Watson-Capps J, Flaherty C, Krützen M (2006) Decline in relative abundance of bottlenose dolphins exposed to long-term disturbance. Conserv Biol 20:1791–1798

Bell C, Shaughnessy P, Morrice M, Stanley B (2006) Marine mammals and Japanese long-line fishing vessels in Australian waters: operational interactions and sightings. Pac Conserv Biol 12:31–39

Blasi MF, Giuliani A, Boitani L (2015) Influence of trammel nets on the behaviour and spatial distribution of bottlenose dolphins (*Tursiops truncatus*) in the Aeolian archipelago, Southern Italy. Aquat Mamm 41:295–310

Bonizzoni S, Furey NB, Pirotta E, Valavanis VD, Würsig B, Bearzi G (2014) Fish farming and its appeal to common bottlenose dolphins: modelling habitat use in a Mediterranean embayment. Aquat Conserv Mar Freshwat Ecosyst 24:696–711

Broadhurst MK (1998) Bottlenose dolphins, *Tursiops truncatus*, removing by-catch from prawntrawl codends during fishing in New South Wales, Australia. Mar Fish Rev 60:9–14

Brotons JM, Grau AM, Rendell L (2008) Estimating the impact of interactions between bottlenose dolphins and artisanal fisheries around the Balearic Islands. Mar Mamm Sci 24:112–127

Buscaino G, Buffa G, Sarà G, Bellante A, Tonello AJ, Hardt FAS, Cremer MJ, Bonanno A, Cuttitta A, Mazzola S (2009) Pinger affects fish catch efficiency and damage to bottom gill nets related to bottlenose dolphins. Fish Sci 75:537–544

Busnel RG (1973) Symbiotic relationship between man and dolphins. Trans N Y Acad Sci 35:112–131

Cantor M, Simões-Lopes PC, Daura-Jorge FG (2018) Spatial consequences for dolphins specialized in foraging with fishermen. Anim Behav 139:19–27

Carter NH, Linnell JD (2016) Co-adaptation is key to coexisting with large carnivores. Trends Ecol Evol 31:575–578

Cassens I et al (2000) Independent adaptation to riverine habitats allowed survival of ancient cetacean lineages. PNAS 97:11343–11347

Ceballos G, Ehrlich PR, Barnosky AD, García A, Pringle RM, Palmer TM (2015) Accelerated modern human-induced species losses: entering the sixth mass extinction. Sci Adv 1:e1400253

Ceballos G, Ehrlich PR, Dirzo R (2017) Biological annihilation via the ongoing sixth mass extinction signaled by vertebrate population losses and declines. Proc Natl Acad Sci 114: E6089–E6096

Chapron G, López-Bao JV (2016) Coexistence with large carnivores informed by community ecology. Trends Ecol Evol 31:578–580

Chapron G et al (2014) Recovery of large carnivores in Europe's modern human-dominated landscapes. Science 346:1518–1519

Chilvers BL, Corkeron PJ (2001) Trawling and bottlenose dolphins' social structure. Proc R Soc Lond B Biol Sci 268:1901–1905

Clausen B, Andersen S (1988) Evaluation of bycatch and health status of the harbour porpoise (*Phocoena phocoena*) in Danish waters. Danish Rev Game Biol 13:1–20

Connor RC (2007) Dolphin social intelligence: complex alliance relationships in bottlenose dolphins and a consideration of selective environments for extreme brain size evolution in mammals. Philos Trans R Soc Lond B 362:587–602

Corkeron PJ, Bryden MM, Hedstrom KE (1990) Feeding by bottlenose dolphins in association with trawling operations in Moreton Bay, Australia. In: Leatherwood S, Reeves RR (eds) The bottlenose dolphin. Academic, London, pp 329–336

COSEWIC (2011) COSEWIC assessment and status report on the northern bottlenose whale *Hyperoodon ampullatus* in Canada. Committee on the Status of Endangered Wildlife in Canada, Ottawa. xii + 31 pp. http://www.sararegistry.gc.ca/default.asp?lang=En&n=3BF95D10-1#_Toc295995736

Couperus AS (1994) Killer whales (*Orcinus orca*) scavenging on discards of freezer trawlers north east of the Shetland islands. Aquat Mamm 20:47–51

Couperus AS (1997) Interactions between Dutch midwater-trawl and Atlantic white-sided dolphins (*Lagenorhynchus acutus*) southwest of Ireland. J Northwest Atlantic Fish Sci 22:209–218

Cox TM, Read AJ, Swanner D, Urian K, Waples D (2003) Behavioral responses of bottlenose dolphins, *Tursiops truncatus*, to gillnets and acoustic alarms. Biol Conserv 115:203–212

Cremer MJ, Barreto AS, Hardt FAS, Júnior AJT, Mounayer R (2009) Cetacean occurrence near an offshore oil platform in southern Brazil. Biotemas 22:247–251

Cruz MJ, Jordao VL, Pereira JG, Santos RS, Silva MA (2014) Risso's dolphin depredation in the Azorean hand-jig squid fishery: assessing the impacts and evaluating effectiveness of acoustic deterrents. ICES J Mar Sci 71:2608–2620

Cruz MJ, Menezes G, Machete M, Silva MA (2016) Predicting interactions between common dolphins and the pole-and-line tuna fishery in the Azores. PLoS One 11:e0164107

Cullis-Suzuki S, Pauly D (2010) Marine protected area costs as "beneficial" fisheries subsidies: a global evaluation. Coast Manag 38:113–121

Cunningham-Smith P, Colbert DE, Wells RS, Speakman T (2006) Evaluation of human interactions with a provisioned wild bottlenose dolphin (*Tursiops truncatus*) near Sarasota Bay, Florida, and efforts to curtail the interactions. Aquat Mamm 32:346–356

D'Lima C, Marsh H, Hamann M, Sinha A, Arthur R (2013) Positive interactions between Irrawaddy dolphins and artisanal fishers in the Chilika lagoon of eastern India are driven by ecology, socioeconomics, and culture. Ambio 43:614–624

Dalla Rosa L, Secchi ER (2007) Killer whale (*Orcinus orca*) interactions with the tuna and swordfish longline fishery off southern and South-Eastern Brazil: a comparison with shark interactions. J Mar Biol Assoc UK 87:135–140

de Sá Alves LCP, Andriolo A, Orams MB, de Freitas Azevedo A (2011) The growth of "botos feeding tourism", a new tourism industry based on the boto (Amazon river dolphin) *Inia geoffrensis* in the Amazonas state, Brazil. Sitientibus Série Ciências Biológicas 11:8–15

Díaz López B (2006) Bottlenose dolphin (*Tursiops truncatus*) predation on a marine fin fish farm: some underwater observations. Aquat Mamm 32:305–310

Díaz López B (2017) Temporal variability in predator presence around a fin fish farm in the Northwestern Mediterranean Sea. Mar Ecol 38:e12378

Díaz López B, Methion S (2017) The impact of shellfish farming on common bottlenose dolphins' use of habitat. Mar Biol 164:83

Durden WN (2005) The harmful effects of inadvertently conditioning a wild bottlenose dolphin (*Tursiops truncatus*) to interact with fishing vessels in the Indian River Lagoon, Florida, USA. Aquat Mamm 31:413–419

Edgar GJ et al (2014) Global conservation outcomes depend on marine protected areas with five key features. Nature 506:216–220

Ferretti F, Myers RA, Serena F, Lotze HK (2008) Loss of large predatory sharks from the Mediterranean Sea. Conserv Biol 22:952–964

Fertl D (1994) Occurrence patterns and behavior of bottlenose dolphins (*Tursiops truncatus*) in the Galveston Ship Channel, Texas. Tex J Sci 46:299–318

Fertl D, Leatherwood S (1997) Cetacean interactions with trawls: a preliminary review. J Northwest Atl Fish Sci 22:219–248

Forney KA, Kobayashi DR, Johnston DW, Marchetti JA, Marsik MG (2011) What's the catch? Patterns of cetacean bycatch and depredation in Hawaii-based pelagic longline fisheries. Mar Ecol 32:380–391

Forney KA, Southall BL, Slooten E, Dawson S, Read AJ, Baird RW, Brownell RL Jr (2017) Nowhere to go: noise impact assessments for marine mammal populations with high site fidelity. Endanger Species Res 32:391–413

Fortuna CM et al (2010) By-catch of cetaceans and other species of conservation concern during pair trawl fishing operations in the Adriatic Sea (Italy). Chem Ecol 26:65–76

Fox KC, Muthukrishna M, Shultz S (2017) The social and cultural roots of whale and dolphin brains. Nat Ecol Evol 1:1699–1705

Frantzis A (2018) A long and deep step in range expansion of an alien marine mammal in the Mediterranean: first record of the Indian Ocean humpback dolphin *Sousa plumbea* (G. Cuvier, 1829) in the Greek Seas. BioInvas Records 7:83–87

Friedlaender AS, McLellan WA, Pabst DA (2001) Characterising an interaction between coastal bottlenose dolphins (*Tursiops truncatus*) and the spot gillnet fishery in southeastern North Carolina, USA. J Cetacean Res Manag 3:293–304

Garrison LP (2007) Interactions between marine mammals and pelagic longline fishing gear in the US Atlantic Ocean between 1992 and 2004. Fish Bull 105:408–417

Gaynor KM, Hojnowski CE, Carter NH, Brashares JS (2018) The influence of human disturbance on wildlife nocturnality. Science 360:1232–1235

Gazo M, Gonzalvo J, Aguilar A (2008) Pingers as deterrents of bottlenose dolphins interacting with trammel nets. Fish Res 92:70–75

Gehrt SD, Riley SPD, Cypher BL (2010) Urban carnivores: ecology, conflict, and conservation. The Johns Hopkins University Press, Baltimore

Gilman E, Brothers N, McPherson G, Dalzell P (2006) A review of cetacean interactions with longline gear. J Cetacean Res Manag 8:215–223

Goetz S, Laporta M, Martínez Portela J, Santos MB, Pierce GJ (2011) Experimental fishing with an "umbrella-and-stones" system to reduce interactions of sperm whales (*Physeter macrocephalus*) and seabirds with bottom-set longlines for Patagonian toothfish (*Dissostichus eleginoides*) in the Southwest Atlantic. ICES J Mar Sci 68:228–238

Gönener S, Özdemir S (2012) Investigation of the interaction between bottom gillnet fishery (Sinop, Black Sea) and bottlenose dolphins (*Tursiops truncatus*) in terms of economy. Turk J Fish Aquat Sci 12:115–126

Gonzalvo J, Valls M, Cardona L, Aguilar A (2008) Factors determining the interaction between common bottlenose dolphins and bottom trawlers off the Balearic archipelago (western Mediterranean Sea). J Exp Mar Biol Ecol 367:47–52

Gonzalvo J, Lauriano G, Hammond PS, Viaud-Martinez KA, Fossi MC, Natoli A, Marsili L (2016) The Gulf of Ambracia's common bottlenose dolphins, *Tursiops truncatus*: a highly dense and yet threatened population. In: di Sciara GN, Podestà M, Curry BE (eds) Mediterranean marine mammal ecology and conservation, Advances in marine biology, vol 75. Academic Press, Oxford, pp 259–296

Goudie AS (2018) Human impact on the natural environment. Wiley, Hoboken

Guidetti P et al (2008) Italian marine reserve effectiveness: does enforcement matter? Biol Conserv 141:699–709

Guinet C, Tixier P, Gasco N, Duhamel G (2015) Long-term studies of Crozet Island killer whales are fundamental to understanding the economic and demographic consequences of their depredation behaviour on the Patagonian toothfish fishery. ICES J Mar Sci 72:1587–1597

Haarr ML, Charlton LD, Terhune JM, Trippel EA (2009) Harbour porpoise (*Phocoena phocoena*) presence patterns at an aquaculture cage site in the Bay of Fundy, Canada. Aquat Mamm 35:203–211

Hall HJ (1984) Fishing with dolphins? Affirming a traditional aboriginal fishing story in Moreton Bay, SE Queensland. In: Coleman RJ, Covacecich J, Davie P (eds) Focus on Stradbroke. Boolarong, Brisbane, pp 16–22

Halpern BS et al (2008) A global map of human impact on marine ecosystems. Science 319:948–952

Hamer DJ, Childerhouse SJ, Gales NJ (2012) Odontocete bycatch and depredation in longline fisheries: a review of available literature and of potential solutions. Mar Mamm Sci 28:E345–E374

Haviland-Howell G, Frankel AS, Powell CM, Bocconcelli A, Herman RL, Sayigh LS (2007) Recreational boating traffic: a chronic source of anthropogenic noise in the Wilmington, North Carolina Intracoastal Waterway. J Acoust Soc Am 122:151–160

Hazzah L, Dolrenry S, Naughton L, Edwards CT, Mwebi O, Kearney F, Frank L (2014) Efficacy of two lion conservation programs in Maasailand, Kenya. Conserv Biol 28:851–860

Hernandez-Milian G et al (2008) Results of a short study of interactions of cetaceans and longline fisheries in Atlantic waters: environmental correlates of catches and depredation events. Hydrobiologia 612:251–268

Hoyt E (2011) Marine protected areas for whales, dolphins and porpoises: a world handbook for cetacean habitat conservation and planning. Earthscan, London

Hucke-Gaete R, Moreno CA, Arata J (2004) Operational interactions of sperm whales and killer whales with the Patagonian toothfish industrial fishery off southern Chile. CCAMLR Sci 11:127–140

Jaiteh VF, Allen SJ, Meeuwig JJ, Loneragan NR (2013) Subsurface behavior of bottlenose dolphins (*Tursiops truncatus*) interacting with fish trawl nets in northwestern Australia: implications for bycatch mitigation. Mar Mamm Sci 29:E266–E281

Jefferson TA (2000) Population biology of the Indo-Pacific hump-backed dolphin in Hong Kong waters. Wildl Monogr 144:1–65

Jefferson TA, Curry BE (2015) Humpback dolphins: a brief introduction to the genus *Sousa*. In: Jefferson TA, Curry BE (eds) Humpback dolphins (*Sousa* spp.): current status and conservation, part 1, Advances in marine biology, vol 72. Academic Press, Oxford, pp 1–16

Jefferson TA, Karczmarski L (2001) Sousa chinensis. Mamm Species 655:1–9

Jerison HJ (1985) Animal intelligence as encephalization. Philos Trans R Soc Lond B 308:21–35

Karpouzli E, Leaper R (2004) Opportunistic observations of interactions between sperm whales and deep-water trawlers based on sightings from fisheries observers in the Northwest Atlantic. Aquat Conserv Mar Freshwat Ecosyst 14:95–103

Kasuya T (1985) Fishery-dolphin conflict in the Iki Island area of Japan. In: Beddington JR, Beverton RJH, Lavigne DM (eds) Marine mammals and fisheries. George Allen & Unwin, London, pp 253–272

Kellert SR, Wilson EO (1993) The biophilia hypothesis. Island Press, Washington

Kock KH, Purves MG, Duhamel G (2006) Interactions between cetacean and fisheries in the Southern Ocean. Polar Biol 29:379–388

Kovacs C, Cox T (2014) Quantification of interactions between common bottlenose dolphins (*Tursiops truncatus*) and a commercial shrimp trawler near Savannah, Georgia. Aquat Mamm 40:81–94

Kovacs CJ, Perrtree RM, Cox TM (2017) Social differentiation in common bottlenose dolphins (*Tursiops truncatus*) that engage in human-related foraging behaviors. PLoS One 12:e0170151

Krützen M, Mann J, Heithaus MR, Connor RC, Bejder L, Sherwin WB (2005) Cultural transmission of tool use in bottlenose dolphins. Proc Natl Acad Sci USA 102:8939–8943

Lauriano G, Fortuna CM, Moltedo G, Notarbartolo di Sciara G (2004) Interactions between bottlenose dolphins (*Tursiops truncatus*) and the artisanal fishery in Asinara Island National Park (Sardinia): assessment of catch damage and economic loss. J Cetacean Res Manag 6:165–173

Lavigne DM (2003) Marine mammals and fisheries: the role of science in the culling debate. In: Gales N, Hindell M, Kirkwood R (eds) Marine mammals: fisheries, tourism and management issues. CSIRO, Victoria, pp 31–47

Leatherwood S (1975) Some observations of feeding behavior of bottle-nosed dolphins (*Tursiops truncatus*) in the northern Gulf of Mexico and (*Tursiops* cf. *T. gilli*) off southern California, Baja California, and Nayarit, Mexico. Mar Fish Rev 37(9):10–16

Leatherwood S, Reeves RR (1983) Abundance of bottlenose dolphins in Corpus Christi Bay and coastal southern Texas. Contrib Mar Sci 26:179–199

Lebreton L et al (2018) Evidence that the Great Pacific Garbage Patch is rapidly accumulating plastic. Sci Rep 8:4666

Li J et al (2014) Role of Tibetan Buddhist monasteries in snow leopard conservation. Conserv Biol 28:87–94

Lihoreau M, Costa JT, Rivault C (2012) The social biology of domiciliary cockroaches: colony structure, kin recognition and collective decisions. Insect Soc 59:445–452

Lilly JC (1967) The mind of the dolphin: a nonhuman intelligence. Doubleday, New York

Linnell JD, Swenson JE, Anderson R (2001) Predators and people: conservation of large carnivores is possible at high human densities if management policy is favourable. Anim Conserv 4:345–349

Loch C, Marmontel M, Simões-Lopes PC (2009) Conflicts with fisheries and intentional killing of freshwater dolphins (Cetacea: Odontoceti) in the Western Brazilian Amazon. Biodivers Conserv 18:3979–3988

Lockyer C (1990) Review of incidents involving wild, sociable dolphins, worldwide. In: Leatherwood S, Reeves RR (eds) The bottlenose dolphin. Academic Press, London, pp 337–353

Löe J, Röskaft E (2004) Large carnivores and human safety: a review. Ambio 33:283–288

Lopez JC, Lopez D (1985) Killer whales (*Orcinus orca*) of Patagonia, and their behavior of intentional stranding while hunting nearshore. J Mammal 66:181–183

López DM, Barcelona SG, Báez JC, De la Serna JM, de Urbina JMO (2012) Marine mammal bycatch in Spanish Mediterranean large pelagic longline fisheries, with a focus on Risso's dolphin (*Grampus griseus*). Aquat Living Resour 25:321–331

Luque PL, Davis CG, Reid DG, Wang J, Pierce GJ (2006) Opportunistic sightings of killer whales from Scottish pelagic trawlers fishing for mackerel and herring off North Scotland (UK) between 2000 and 2006. Aquat Living Resour 19:403–410

Marino L (2005) Big brains do matter in new environments. Proc Natl Acad Sci 102:5306–5307

Markowitz TM, Harlin AD, Würsig B, McFadden CJ (2004) Dusky dolphin foraging habitat: overlap with aquaculture in New Zealand. Aquat Conserv Mar Freshwat Ecosyst 14:133–149

Marley SA, Salgado Kent CP, Erbe C, Parnum IM (2017) Effects of vessel traffic and underwater noise on the movement, behaviour and vocalisations of bottlenose dolphins in an urbanised estuary. Sci Rep 7:13437

Martin AR, da Silva VMF, Rothery P (2008) Object carrying as sociosexual display in an aquatic mammal. Biol Lett 4:243–245

Mason S, Salgado Kent C, Donnelly D, Weir J, Bilgmann K (2016) Atypical residency of short-beaked common dolphins (*Delphinus delphis*) to a shallow, urbanized embayment in South-Eastern Australia. R Soc Open Sci 3:160478

Mathias D, Thode A, Straley J, Folkert K (2009) Relationship between sperm whale (*Physeter macrocephalus*) click structure and size derived from videocamera images of a depredating whale (sperm whale prey acquisition). J Acoust Soc Am 125:3444–3453

Messmer TA (2000) The emergence of human-wildlife conflict management: turning challenges into opportunities. Int Biodeter Biodegr 45:97–102

Monbiot G (2014) Feral: Rewilding the land, the sea, and human life. The University of Chicago Press, Chicago

Monteiro-Filho ELA (1995) Pesca interativa entre o golfinho *Sotalia fluviatilis guianensis* e a comunidade pesqueira da região de Cananéia. Bol Inst Pesca 22:15–23

Morizur Y, Berrow SD, Tregenza NJC, Couperus AS, Pouvreau S (1999) Incidental catches of marine-mammals in pelagic trawl fisheries of the Northeast Atlantic. Fish Res 41:297–307

Myers RA, Worm B (2003) Rapid worldwide depletion of predatory fish communities. Nature 423:280–283

Myers RA, Baum JK, Shepherd TD, Powers SP, Peterson CH (2007) Cascading effects of the loss of apex predatory sharks from a coastal ocean. Science 315:1846–1850

Neff C, Hueter R (2013) Science, policy, and the public discourse of shark "attack": a proposal for reclassifying human-shark interactions. J Environ Stud Sci 3:65–73

Neil DT (2002) Cooperative fishing interactions between aboriginal Australians and dolphins in eastern Australia. Anthrozoös 15:3–18

Noke WD, Odell DK (2002) Interactions between the Indian River Lagoon blue crab fishery and the bottlenose dolphin, *Tursiops truncatus*. Mar Mamm Sci 18:819–832

Norris KS, Prescott HH (1961) Observations of Pacific cetaceans of California and Mexican waters. Univ Calif Publ Zool 63:291–402

Northridge S (2018) Fisheries interactions. In: Würsig B, Thewissen JGM, Kovacs KM (eds) Encyclopedia of marine mammals, 3rd edn. Elsevier, San Diego, pp 375–383

Northridge SP, Hofman RJ (1999) Marine mammal interactions with fisheries. In: Twiss JR, Reeves RR (eds) Conservation and management of marine mammals. Smithsonian Institution Press, Washington, pp 99–119

Odell DK, Asper ED (1980) Distribution and movements of freeze-branded bottlenose dolphins in the Indian and Banana rivers, Florida. In: Leatherwood S, Reeves RR (eds) The bottlenose dolphin. Academic Press, London, pp 515–540

Orams MB (2002) Feeding wildlife as a tourism attraction: a review of issues and impacts. Tour Manag 23:281–293

Ozbilgin YD, Kalecik E, Gücü AC (2018) First record of humpback dolphins in Mersin Bay, the eastern Mediterranean, Turkey. Turk J Fish Aquat Sci 18:187–190

Parra GJ (2006) Resource partitioning in sympatric delphinids: space use and habitat preferences of Australian snubfin and indo-Pacific humpback dolphins. J Anim Ecol 75:862–874

Patterson EM, Mann J (2011) The ecological conditions that favor tool use and innovation in wild bottlenose dolphins (*Tursiops* sp.). PLoS One 6:e22243

Pearson HC, Vaughn-Hirshorn RL, Srinivasan M, Würsig B (2012) Avoidance of mussel farms by dusky dolphins (*Lagenorhynchus obscurus*) in New Zealand. NZ J Mar Freshw Res 46:567–574

Penteriani V et al (2016) Human behaviour can trigger large carnivore attacks in developed countries. Sci Rep 6:20552

Peterson MJ, Hanselman D (2017) Sablefish mortality associated with whale depredation in Alaska. ICES J Mar Sci 74:1382–1394

Pimm SL, Jenkins CN, Abell R, Brooks TM, Gittleman JL, Joppa LN, Raven PH, Roberts CM, Sexton JO (2014) The biodiversity of species and their rates of extinction, distribution, and protection. Science 344:1246752

Piroddi C, Bearzi G, Christensen V (2011) Marine open cage aquaculture in the eastern Mediterranean Sea: a new trophic resource for bottlenose dolphins. Mar Ecol Prog Ser 440:255–266

Pitman RL, Ensor P (2003) Three forms of killer whales (*Orcinus orca*) in Antarctic waters. J Cetacean Res Manag 5:131–139

Piwetz S (2019) Common bottlenose dolphin (*Tursiops truncatus*) behavior in an active narrow seaport. PLoS One 14:e0211971

Plagányi EE, Butterworth DS (2005) Indirect fishery interactions. In: Reynolds JE III, Perrin WF, Reeves RR, Montgomery S, Ragen TJ (eds) Marine mammal research: conservation beyond crisis. The Johns Hopkins University Press, Baltimore, pp 19–45

Powell JR, Wells RS (2011) Recreational fishing depredation and associated behaviors involving common bottlenose dolphins (*Tursiops truncatus*) in Sarasota Bay, Florida. Mar Mamm Sci 27:111–129

Pryor K, Lindbergh J (1990) A dolphin-human fishing cooperative in Brazil. Mar Mamm Sci 6:77–82

Purves MG, Agnew DJ, Balguerias E, Moreno CA, Watkins B (2004) Killer whale (*Orcinus orca*) and sperm whale (*Physeter macrocephalus*) interactions with longline vessels in the Patagonian toothfish fishery at South Georgia, South Atlantic. CCAMLR Sci 11:111–126

Ramos-Cartelle A, Mejuto J (2008) Interaction of the false killer whale (*Pseudorca crassidens*) and the depredation on the swordfish catches of the Spanish surface longline fleet in the Atlantic, Indian and Pacific Oceans. Report, International Commission for the Conservation of Atlantic tunas (ICCAT). Collective Volume of Scientific Papers 62:1721–1783

Rayment W, Webster T (2009) Observations of Hector's dolphins (*Cephalorhynchus hectori*) associating with inshore fishing trawlers at Banks Peninsula, New Zealand. NZ J Mar Freshw Res 43:911–916

Read AJ (2008) The looming crisis: interactions between marine mammals and fisheries. J Mammal 89:541–548

Read AJ, Drinker P, Northridge S (2006) Bycatch of marine mammals in US and global fisheries. Conserv Biol 20:163–169

Reader SM, Laland KN (2002) Social intelligence, innovation, and enhanced brain size in primates. Proc Natl Acad Sci 99:4436–4441

Rechimont ME, Lara-Domínguez AL, Morteo E, Martínez-Serrano I, Equihua M (2018) Depredation by coastal bottlenose dolphins (*Tursiops truncatus*) in the southwestern Gulf of Mexico in relation to fishing techniques. Aquat Mamm 44:469–481

Reeves RR, McClellan K, Werner TB (2013) Marine mammal bycatch in gillnet and other entangling net fisheries, 1990 to 2011. Endanger Species Res 20:71–97

Ridgway SH, Carlin KP, Van Alstyne KR, Hanson AC, Tarpley RJ (2017) Comparison of dolphins' body and brain measurements with four other groups of cetaceans reveals great diversity. Brain Behav Evol 88:235–257

Ripple WJ et al (2014) Status and ecological effects of the world's largest carnivores. Science 343:1241484

Ripple WJ et al (2016) Saving the world's terrestrial megafauna. Bioscience 66:807–812

Robertson RB (1954) Of whales & men. Simon & Shuster, New York

Rocklin D, Santoni MC, Culioli JM, Tomasini JA, Pelletier D, Mouillot D (2009) Changes in the catch composition of artisanal fisheries attributable to dolphin depredation in a Mediterranean marine reserve. ICES J Mar Sci 66:699–707

Sanderson EW, Jaiteh M, Levy MA, Redford KH, Wannebo AV, Woolmer G (2002) The human footprint and the last of the wild. Bioscience 52:891–904

Sargeant BL, Mann J, Berggren P, Krützen M (2005) Specialization and development of beach hunting, a rare foraging behavior, by wild bottlenose dolphins (*Tursiops* sp.). Can J Zool 83:1400–1410

Schipper J et al (2008) The status of the world's land and marine mammals: diversity, threat, and knowledge. Science 322:225–230

Schlais JF (1984) Thieving dolphins: a growing problem in Hawaii's fisheries. Sea Frontiers 30:293–298

Silber GK, Fertl D (1995) Intentional beaching by bottlenose dolphins (*Tursiops truncatus*) in the Colorado River Delta, Mexico. Aquat Mamm 21:183–186

Silva MA, Feio R, Prieto R, Gonçalves JM, Santos RS (2002) Interactions between cetaceans and the tuna fishery in the Azores. Mar Mamm Sci 18:893–901

Simões-Lopes PC, Fabián ME, Menegheti JO (1998) Dolphin interactions with the mullet artisanal fishing on southern Brazil: a qualitative and quantitative approach. Revista Brasileira de Zoologia 15:709–726

Smith BD, Reeves RR (2012) River cetaceans and habitat change: generalist resilience or specialist vulnerability? J Mar Biol 718935

Smith BD, Thant UH, Lwin JM, Shaw CD (1997) Investigation of cetaceans in the Ayeyarwady River and northern coastal waters of Myanmar. Asian Mar Biol 14:173–194

Smith H, Samuels A, Bradley S (2008) Reducing risky interactions between tourists and free-ranging dolphins (*Tursiops* sp.) in an artificial feeding program at Monkey Mia, Western Australia. Tour Manag 29:994–1001

Smith BD, Tun MT, Chit AM, Win H, Moe T (2009) Catch composition and conservation management of a human-dolphin cooperative cast-net fishery in the Ayeyarwady River, Myanmar. Biol Conserv 142:1042–1049

Smolker R, Richards A, Connor R, Mann J, Berggren P (1997) Sponge carrying by dolphins (Delphinidae, *Tursiops* sp.): a foraging specialization involving tool use? Ethology 103:454–465

Sol D, Bacher S, Reader SM, Lefebvre L (2008) Brain size predicts the success of mammal species introduced into novel environments. Am Nat 172:S63–S71

Spitz J, Chouvelon T, Cardinaud M, Kostecki C, Lorance P (2013) Prey preferences of adult sea bass *Dicentrarchus labrax* in the northeastern Atlantic: implications for bycatch of common dolphin *Delphinus delphis*. ICES J Mar Sci 70:452–461

Steffen W et al (2018) Trajectories of the Earth system in the Anthropocene. Proc Natl Acad Sci 201810141

Stepanuk JE, Read AJ, Baird RW, Webster DL, Thorne LH (2018) Spatiotemporal patterns of overlap between short-finned pilot whales and the US pelagic longline fishery in the Mid-Atlantic Bight: an assessment to inform the management of fisheries bycatch. Fish Res 208:309–320

Stirling I (1980) The biological importance of polynyas in the Canadian Arctic. Arctic 33:303–315

Stirling I, Cleator H, Smith TG (1981) Marine mammals. In: Stirling I, Cleator H (eds) Polynyas in the Canadian Arctic. Occasional paper 45. Canadian Wildlife Service, Environment Canada, Ottawa, pp 45–58

Svane I (2005) Occurrence of dolphins and seabirds and their consumption of by-catch during prawn trawling in Spencer Gulf, South Australia. Fish Res 76:317–327

Svenning JC et al (2016) Science for a wilder Anthropocene: synthesis and future directions for trophic rewilding research. Proc Natl Acad Sci 113:898–906

Taylor BL et al (2017) Extinction is imminent for Mexico's endemic porpoise unless fishery bycatch is eliminated. Conserv Lett 10:588–595

Thompson FN, Abraham ER, Berkenbusch K (2013) Common dolphin (*Delphinus delphis*) bycatch in New Zealand commercial trawl fisheries. PLoS One 8(5):e64438

Tixier P, Lea MA, Hindell MA, Guinet C, Gasco N, Duhamel G, Arnould JP (2018) Killer whale (*Orcinus orca*) interactions with blue-eye trevalla (*Hyperoglyphe antarctica*) longline fisheries. PeerJ 6:e5306

Todd VL, Pearse WD, Tregenza NC, Lepper PA, Todd IB (2009) Diel echolocation activity of harbour porpoises (*Phocoena phocoena*) around North Sea offshore gas installations. ICES J Mar Sci 66:734–745

Torres LG, Read AJ (2009) Where to catch a fish? The influence of foraging tactics on the ecology of bottlenose dolphins (*Tursiops truncatus*) in Florida Bay, Florida. Mar Mamm Sci 25:797–815

Towers JR, Tixier P, Ross KA, Bennett J, Arnould JP, Pitman RL, Durban JW (2018) Movements and dive behaviour of a toothfish-depredating killer and sperm whale. ICES J Mar Sci 76:1–14

Treves A, Bruskotter JT (2014) Tolerance for predatory wildlife. Science 344:476–477

Treves A, Wallace RB, Naughton-Treves L, Morales A (2006) Co-managing human-wildlife conflicts: a review. Hum Dimens Wildl 11:383–396

Triossi F, Willis TJ, Pace DS (2013) Occurrence of bottlenose dolphins *Tursiops truncatus* in natural gas fields of the northwestern Adriatic Sea. Mar Ecol 34:373–379

Visser IN (2000) Killer whale (*Orcinus orca*) interactions with longline fisheries in New Zealand waters. Aquat Mamm 26:241–252

Wackernagel M, Rees W (1998) Our ecological footprint: reducing human impact on the earth. New Society Publishers, Gabriola Island

Wassenberg TJ, Hill BJ (1990) Partitioning of material discarded from prawn trawlers in Morton Bay. Mar Freshw Res 41:27–36

Watson FG, Becker MS, Milanzi J, Nyirenda M (2015) Human encroachment into protected area networks in Zambia: implications for large carnivore conservation. Reg Environ Chang 15:415–429

Watson-Capps JJ, Mann J (2005) The effects of aquaculture on bottlenose dolphin (*Tursiops* sp.) ranging in Shark Bay, Western Australia. Biol Conserv 124:519–526

Weir CR, Nicolson I (2014) Depredation of a sport fishing tournament by rough-toothed dolphins (*Steno bredanensis*) off Angola. Aquat Mamm 40:297–304

Werner TB, Northridge S, Press KM, Young N (2015) Mitigating bycatch and depredation of marine mammals in longline fisheries. ICES J Mar Sci 72:1576–1586

West KL, Mead JG, White W (2011) *Steno bredanensis* (Cetacea: Delphinidae). Mamm Species 43:177–189

Whitehead H (2002) Estimates of the current global population size and historical trajectory for sperm whales. Mar Ecol Prog Ser 242:295–304

Whitehead H, Ford JKB (2018) Consequences of culturally-driven ecological specialization: killer whales and beyond. J Theor Biol 456:279–294

Whitehead H, Reeves RR (2005) Killer whales and whaling: the scavenging hypothesis. Biol Lett 1:415–418

Williams TM, Friedl WA, Fong ML, Yamada RM, Sedivy P, Haun JE (1992) Travel at low energetic cost by swimming and wave-riding bottlenose dolphins. Nature 355:821–823

Wilson EO (1984) Biophilia: the human bond with other species. Harvard University Press, Cambridge, MA

Wilson EO (2010) The creation: an appeal to save life on earth. W.W. Norton, New York

Wong B, Candolin U (2015) Behavioral responses to changing environments. Behav Ecol 26:665–673

Woodroffe R, Thirgood S, Rabinowitz A (2005) People and wildlife: conflict or coexistence? Cambridge University Press, Cambridge

Würsig B (2018) Bow-riding. In: Würsig B, Thewissen JGM, Kovacs KM (eds) Encyclopedia of marine mammals, 3rd edn. Elsevier, San Diego, pp 135–137

Wyllie T (1993) Dolphins, telepathy and underwater birthing. Bear, Santa Fe

Zappes CA, Andriolo A, Simões-Lopes PC, Di Beneditto APM (2011) "Human-dolphin (*Tursiops truncatus* Montagu, 1821) cooperative fishery" and its influence on cast net fishing activities in Barra de Imbé/Tramandaí, Southern Brazil. Ocean Coast Manag 54:427–432

Zollett EA, Read AJ (2006) Depredation of catch by bottlenose dolphins (*Tursiops truncatus*) in the Florida king mackerel (*Scomberomorus cavalla*) troll fishery. Fish Bull 104:343–349

Part II
Examples of Odontocete Ethology and Behavioral Ecology: Present Knowledge and Ways Forward

Chapter 11
Killer Whales: Behavior, Social Organization, and Ecology of the Oceans' Apex Predators

John K. B. Ford

Abstract The killer whale—the largest of the dolphins and the top marine predator—has a cosmopolitan distribution throughout the world's oceans. Although globally it could be considered a generalist predator with a diverse diet, it is deeply divided into ecotypes, many of which have distinct foraging strategies involving only a narrow range of prey species. These ecotypes, which often exist in sympatry, are believed to arise from culturally driven dietary specializations that develop within matrilineal social groups and are transmitted among matriline members and across generations by social learning. Specializations are maintained by behavioral conformity and social insularity of lineages, which result in reproductive isolation and, ultimately, genetic divergence of ecotypes. Ecotypes have distinct patterns of seasonal distribution, group size, social organization, foraging behavior, and acoustic activity that are related to the type of prey being sought. Sophisticated cooperative foraging tactics have evolved in some ecotypes, and prey sharing within matrilineal social groups is common. Remarkable behavioral and demographic attributes have been documented in one well-studied ecotype, including lifelong natal philopatry without dispersal of either sex from the social group, vocal dialects that encode genealogical relatedness within lineages, and multi-decade long post-reproductive periods of females. Cultural traditions of killer whales, including foraging specializations, can be deeply rooted and resistant to change, which may limit the ability of ecotypes to adapt to sudden environmental variability.

Keywords *Orcinus orca* · Orca · Ecological specialization · Cultural traditions · Matrilineal society · Foraging tactics · Dialects · Menopause

---

J. K. B. Ford (✉)
Department of Zoology, University of British Columbia, Vancouver, BC, Canada

Pacific Biological Station, Fisheries and Oceans Canada, Nanaimo, BC, Canada
e-mail: john.ford@ubc.ca

© Springer Nature Switzerland AG 2019
B. Würsig (ed.), *Ethology and Behavioral Ecology of Odontocetes*, Ethology and Behavioral Ecology of Marine Mammals,
https://doi.org/10.1007/978-3-030-16663-2_11

11.1 Introduction

The killer whale, or orca—with its striking black and white coloration, wide range in the world's oceans, and fame at oceanariums—is one of the most widely recognized and familiar marine mammals. But until quite recently, scientific understanding of *Orcinus orca* was minimal, and knowledge of the animal was based almost entirely on anecdotal and scattered opportunistic observations. From dramatic and fanciful accounts of early mariners witnessing the predatory nature of the species, it acquired a reputation of almost mythical proportion as a savage, bloodthirsty demon, malevolent to both marine animals and humans alike. In his tome *Naturalis Historiæ*, written almost 2000 years ago, Roman naturalist and philosopher Pliny the Elder described the orca as "a mightie masse and lumpe of flesh without all fashion, armed with the most terrible, sharpe, and cutting teeth" that showed no mercy in its vicious attacks on female baleen whales and their calves. More recently, nineteenth-century whaling captain and naturalist Charles Scammon concluded that "in whatever quarter of the world the Orcas are found, they seem always intent upon seeking something to destroy or devour" (Scammon 1874). It is only in recent decades that scientific field studies have shed much needed light on the true behavior and ecology of killer whales, and it has become clear that this is indeed a most remarkable and, in many respects, unique social predator.

11.2 Distribution and Population Structure

Killer whales—the largest of the dolphins (family Delphinidae)—are the most widely distributed marine mammals with a cosmopolitan range in all the world's oceans and most seas. Despite the widespread occurrence of killer whales, it is not an abundant species. They are rare in many regions, especially the tropics, and reach highest densities in cool, productive, high-latitude waters. They do not undertake long-range migrations between feeding and breeding areas, but distribution shifts associated with seasonal occurrence of prey aggregations are common (e.g., Similä et al. 1996; Ford et al. 2000). Killer whales have successfully exploited a diversity of marine habitats. They occur pelagically in deep oceanic waters but much more commonly in continental shelf and nearshore waters, where they can be found moving through confined passes and channels and up narrow inlets and fjords. Notable concentrations occur along the northwestern coast of North America, around Iceland, along the coast of northern Norway, and in the Southern Ocean around Antarctica. In the Antarctic, killer whales are commonly found up to the pack ice edge in many areas and may extend well into ice-covered waters.

Although only a single species is currently recognized, there is an increasingly compelling body of evidence that multiple species of killer whales likely exist (Morin et al. 2010). Discrete populations with distinctive genetic, morphological, ecological, and cultural attributes—variously described as types, forms, races, lineages, or, most commonly, ecotypes—have been identified in several regions. These ecotypes often exist in sympatry, sharing the same waters yet maintaining social and reproductive isolation from each other and specializing on different types of prey. Recent genetic studies of killer whales from multiple global regions suggest that ancestral killer whales diverged rapidly within the past 250,000–350,000 years through various processes of population expansion, dispersal, and colonization, with founder groups radiating into diverse ecological niches through dietary specialization and subsequent cultural and reproductive isolation followed by genetic drift or adaptation (Morin et al. 2015; Moura et al. 2015; Foote et al. 2016; Hoelzel and Moura 2016). Ecological and genetic divergence may have taken place both in allopatry and in sympatry (Hoelzel et al. 2007; Riesch et al. 2012; Hoelzel and Moura 2016).

11.3 Sympatric Ecotypes: Foraging Specializations as Cultural Traditions

Globally, the killer whale could be considered a generalist predator since an extremely diverse array of prey types has been documented, either through direct observation of predation or from identification of stomach contents of stranded animals. Overall, almost 200 species of marine vertebrates and invertebrates have been recorded as prey, including 37 species of cetaceans, 20 of pinnipeds, 44 of bony fishes, 29 of sharks and rays, 27 of seabirds, 29 of squid and octopus, 2 of sea turtles, and even 2 species of terrestrial mammals (Ford 2014, unpubl.). However, ecotypes of killer whales can have remarkably specialized diets and forage selectively for only a very small subset of the prey species that this versatile predator is capable of consuming. It seems that ecological specializations reflect cultural traditions that have evolved over millennia, passing from one generation to the next by social learning. The matrilineal social structure typical of killer whales, as described later, is well suited to the inter-generational cultural transmission of dietary habits and the specialized knowledge of when, where, and how to acquire their prey.

Sympatric ecotypes of killer whales are best known from long-term field studies in coastal waters of the eastern North Pacific, from California north to the Aleutian Islands, Alaska. Three ecotypes with largely distinct diets share these waters but do not mix—*Residents*, *Transients* (also known as *Bigg's killer whales*), and *Offshores* (Bigg 1982; Ford et al. 2000; Ford and Ellis 2014). Each ecotype is distinct genetically and differs subtly in appearance, but perhaps most striking are the numerous differences in patterns of movement and habitat use, social structure, and behavior that are related to their particular foraging specialization. The Resident ecotype in the northeastern Pacific feeds predominantly on fish, particularly Pacific salmon. Although all six species of salmon in northeastern Pacific waters are eaten by Resident killer whales, Chinook salmon (*Oncorhynchus tshawytscha*) is by far their preferred prey despite being one of the least abundant salmon species (Ford and

Ellis 2006). This is likely due to the Chinook's large size, high fat content, and year-round availability in the coastal range of Residents. Pink salmon (*O. gorbuscha*) and sockeye salmon (*O. nerka*), considerably smaller in size but vastly more abundant during their summer spawning migrations through coastal waters, are surprisingly not significant prey items. Chum salmon (*O. keta*), second in size to Chinook, is an important prey species during fall. Resident killer whales also feed occasionally on bottom-dwelling fishes and squid but have not been seen to prey on marine mammals in over 45 years of observation.

In striking contrast to the diet of Residents, the Transient ecotype feeds on small marine mammals and does not consume fish. They prey on pinnipeds (mostly harbor seals (*Phoca vitulina*) and Steller sea lions (*Eumetopias jubatus*)) and small cetaceans (mostly harbor porpoises (*Phocoena phocoena*) and Dall's porpoises (*Phocoenoides dalli*)) (Ford et al. 1998; Saulitis et al. 2000). Common minke whales (*Balaenoptera acutorostrata*) are occasionally attacked by Transient killer whales, and gray whale (*Eschrichtius robustus*) calves and juveniles are hunted during their northward spring migration (Ternullo and Black, 2002; Barrett-Lennard et al. 2011). Transients rarely attack mature large baleen whales, likely because they are too energetically costly or risky to routinely pursue (Ford and Reeves 2008). Transients have not been observed to eat any species of fish, but they do feed on a variety of squids, including the Robust clubhook squid (*Onykia robusta*), with a mantle length of more than 1 m and a weight of more than 13 kg (Ford 2014). They also have been observed killing a variety of seabird species, but only a minority of these are actually consumed.

Details of the foraging habits of Offshore killer whales are not as well known, but evidence is now compelling that this is a shark-specialist ecotype. Offshores have been observed feeding on Pacific sleeper sharks (*Somniosus pacificus*) on numerous occasions, as well as on blue sharks (*Prionace glauca*), spiny dogfish (*Squalus suckleyi*), and salmon sharks (*Lamna ditropis*). It appears that these whales feed so extensively on sharks that their teeth become severely worn due to the abrasive quality of shark skin, which is roughened by embedded dermal denticles (Ford et al. 2011; Ford 2014). The teeth of all adult stranded Offshore killer whales observed to date have been worn flat to the gums, a pattern that has not been observed in even very old Residents or Transients. The Offshores appear to eat just the liver of sharks, which is rich in lipids, and likely wear their teeth while ripping the body open to access it. The diet of Offshores is not exclusively elasmobranchs, however, since predation on some teleost fishes has also been documented (Ford et al. 2011; Ford 2014).

Five distinctive ecotypes have been described in waters of the Southern Ocean surrounding Antarctica—these are referred to as types A, B1, B2, C, and D. Antarctic Type A killer whales are open-water marine mammal predators that may specialize on Antarctic minke whales (*Balaenoptera bonaerensis*), especially during austral summer (Pitman and Ensor 2003). Type B1, also known as pack ice killer whales, have a circumpolar distribution and feed on pinnipeds in loose pack ice. They forage selectively for Weddell seals (*Leptonychotes weddellii*), typically ignoring the much more abundant crabeater seals (*Lobodon carcinophaga*) and leopard seals (*Hydrurga leptonyx*) occupying the same waters (Pitman and Durban 2012). Type B2, also called Gerlache Strait killer whales, have a similar appearance to the sympatric pack ice killer whales but are almost one-quarter smaller in size (Durban et al. 2017). They are known only from the Antarctic Peninsula, where they appear to feed mostly on penguins. Type C are also known as Ross Sea killer whales, which is where they are mostly found. This ecotype, which is even smaller than Type B2 whales, appears to feed only on fish, including Antarctic toothfish (*Dissostichus mawsoni*) and potentially Antarctic silverfish (*Pleuragramma antarcticum*) and other smaller fish species (Pitman et al. 2018). Finally, Type D, also known as subantarctic killer whales, have a circumpolar distribution in deep, oceanic waters between 40°S and 60°S (Pitman et al. 2011). It is very distinctive in appearance and highly divergent genetically from other ecotypes (Foote et al. 2013). Little is yet known about its foraging ecology, but it has been observed depredating fisheries targeting Patagonian toothfish (*Dissostichus eleginoides*), suggesting it may naturally be a fish feeder.

In the eastern North Atlantic, there appear to be at least three ecotypes, although the picture is still unclear. Type 1 is primarily a fish-feeding ecotype, targeting herring and mackerel around Norway, Iceland, and Scotland, although some groups are reported to occasionally prey on seals as well (Foote et al. 2009; Jourdain et al. 2017). Type 2 is partly sympatric with Type 1 and is a marine mammal specialist ecotype, preying on both cetaceans and pinnipeds. A third unnamed ecotype is found in the Strait of Gibraltar, where the whales hunt migrating Atlantic bluefin tuna (*Thunnus thynnus*) (Esteban et al. 2016).

It is likely that specialized killer whale ecotypes exist wherever sufficiently abundant and predictable prey resources are available to sustain them year-round. Many aspects of the biology of ecotypes, such as seasonal distribution, group size, social organization, foraging tactics, and acoustic activity, are strongly influenced by their dietary specialization, as will be explored in greater detail in the following sections. It should be noted, however, that more generalist foraging strategies may be expected of killer whales in some regions, particularly in low-productivity tropical and oceanic waters such as around Hawaii (Baird et al. 2006).

11.4 Social Organization

Like most delphinids, killer whales are highly social animals, typically living in groups of a few individuals to 20 or more. However, in contrast to the considerable social fluidity of many coastal dolphin species (Connor et al. 2000), killer whale groups are often highly stable and cohesive over time. In most areas where killer whales have been studied for long periods using individual photographic identification, their primary social units consist of strongly bonded kin related by matrilineal descent (Bigg et al. 1990; Ford and Ellis 1999; Similä and Ugarte 1999; Iñíguez et al. 2005; Tosh et al. 2008; Tixier et al. 2014). Temporal persistence of these bonds is a primary variable determining group sizes and structure in different ecotypes.

Fig. 11.1 A matrilineal group of Resident killer whales, *Orcinus orca*, with three generations. Shown is the ca. 63-year-old matriarch A30 at left in the image, in the back, her ca. 41-year-old son A38 to her left, and her 27-year-old daughter A50 to his left. A50's two young offspring, A84 (6 years old, at the rear) and A99 (newborn, surfacing), swim alongside their mother. Photo by Jared Towers

The best known society is that of Resident killer whales, which have been studied continuously for over 45 years in coastal British Columbia and Washington State (Bigg et al. 1990; Ford et al. 2000; Towers et al. 2015) and for over 30 years in southern Alaska (Matkin et al. 2014). Long-term studies have also been conducted on Resident-type killer whales off the east coast of Kamchatka (Ivkovich et al. 2010). Resident killer whales live in complex societies based on matrilineal genealogy, social association, and patterns of vocalizations. The basic social unit—the *matriline*—is a highly stable kin group that consists of a matriarch, her sons and daughters, and the daughters' offspring. Because the lifespan of females can reach 80 years and females have their first calf at 12–14 years (Olesiuk et al. 2005), matrilines may contain four and occasionally five generations of maternally related individuals. Remarkably, both males and females stay with their close kin for life—no individual whales have been observed to leave their matriline and join another on a long-term basis. Lifetime natal philopatry with the complete absence of dispersal is exceedingly rare in mammals (Wright et al. 2016). As a result, the matrilines of Resident killer whales are perhaps the most enduring social groupings of any mammal (Fig. 11.1).

Members of Resident matrilines travel together and seldom separate by more than a few kilometers or for more than a few hours. Contact is maintained among matriline members by the exchange of stereotyped underwater calls that form a dialect unique to the group (see Acoustic Behavior, below) (Ford 1991; Miller et al. 2004). Matrilines frequently travel in the company of certain other matrilines that are

11 Killer Whales: Behavior, Social Organization, and Ecology of the... 245

closely related, based on high degrees of dialect similarity, and these likely shared a common matrilineal ancestor in the recent past. Matrilines that spend the majority of their time together are designated as *pods* (Bigg et al. 1990). Pods are less stable than matrilines, and member matrilines may spend days or weeks apart. However, matrilines still spend more time with others from their pod than with those from different pods. In British Columbia, Resident pods are on average composed of three matrilines (range = 1–11), with a mean total size of 18 whales (range = 2–49; Ford et al. 2000). Residents may form large temporary aggregations involving multiple matrilines and pods, especially at times when salmon densities are high—these are often referred to as "superpods."

Resident killer whale society is not just matrilineal in structure, but matrifocal, with old females playing a key role as matriarchs of the kin group. Resident killer whales are extremely unusual among mammals in that females undergo menopause at around 40 years of age, then live for another decade on average and sometimes for another 30–40 years (Olesiuk et al. 2005). By this time, post-reproductive females are grandmothers, great-grandmothers, or even great-great-grandmothers and are central to the social cohesion and day-to-day functioning of the group. They are the oldest members of the group and no doubt have the greatest experience and ecological knowledge of when and where to locate Chinook salmon, their preferred prey. The bond between matriarchs and their adult sons is particularly strong—sons spend more time near their mother than with others in the group (Bigg et al. 1990; Brent et al. 2015) and have a higher probability of dying compared to daughters, following the death of their post-reproductive mother (Foster et al. 2012). The death of a matriarch can also increase the probability of fission of the group along maternal lines—for example, two adult sisters together with their respective offspring may spend increasing time apart after their mother dies (Stredulinsky 2016).

A level of social structure above the Resident pod is the *clan*, which is defined solely by the vocal repertoire of pods. All pods within a clan have vocal dialects that include shared call types, indicating a common ancestral matrilineal heritage. The dialects of different clans have no calls in common. Pods from different clans often travel together and mix despite their completely distinct call repertoires. Finally, above the clan is the *community*, which consists of pods that regularly associate with one another. Thus, unlike matrilines and clans, the community is defined solely by association patterns. Even though the ranges of communities may overlap, pods from one community avoid contact with those from another.

Mammal-hunting Transient killer whales in coastal waters of the northeastern Pacific have a less structured and not so strictly matrilineal society than Residents. They usually travel in groups of about 3–6 individuals, much smaller than the typical size of Resident matrilines and pods. Although the core members of the social unit are a mother and her descendants, offspring may disperse from the natal matriline for extended periods or permanently (Ford and Ellis 1999; Baird and Whitehead 2000). Female offspring usually leave their natal group around the time of sexual maturity and travel with other Transient matrilines. These young females usually give birth to their first calf shortly after dispersing. Once dispersed, females may rejoin their natal matriline occasionally, but generally only for brief periods after they have calves of their own. Male dispersal also takes place, but less predictably. Adult males that have lost their mothers through mortality often travel alone or associate with a variety of different Transient matrilines, but rarely with other lone males. The associations of Transient matrilines are very dynamic, and they do not form consistent groupings equivalent to Resident pods. Also, in contrast to Residents, Transient populations do not seem to be acoustically subdivided into clans. Instead, all transients in a population share a distinctive set of calls, although some additional calls or variants of shared calls may be specific to subregions or portions of the population (Deecke et al. 2005).

The typically small size of Transient groups is likely a result of the foraging strategy of this ecotype. Transients generally hunt other marine mammals with stealth: they swim quietly to avoid detection by their acoustically sensitive prey and attack using the element of surprise (Barrett-Lennard et al. 1996; Deecke et al. 2005). This strategy no doubt constrains group size, as larger groups such as those of Residents would increase the probability of the predators being detected by their prey. Small groups may also be most energetically efficient for Transients when hunting smaller marine mammals such as harbor seals (Baird and Dill 1996).

Most killer whale societies studied in other ocean regions are matrilineal in structure, but none has yet been found to have the lifelong natal philopatry seen in Residents. Killer whales that feed on migrating bluefin tuna in the Strait of Gibraltar form highly stable matrilineal groups (Esteban et al. 2016), although it is not clear if some dispersal may take place. Associations within a group of seal-feeding killer whales in Norway have persisted for at least 30 years (Jourdain et al. 2017). Killer whales in the subantarctic Crozet Islands (Guinet 1991) and Prince Edward Islands (Tosh et al. 2008), which prey mostly on seals and penguins, live in small groups that appear similar in structure to those of Transients with strong maternal bonds but some fluidity in group composition. A possible exception to the pattern of structured societies with enduring kinship bonds occurs in a population of herring-eating killer whales in Iceland. These animals form large groups, but association analyses have so far failed to identify any clear matrilineally based hierarchical structure, perhaps beyond a mother and her immature offspring. Instead, many associations in social clusters appear weak and temporary, resulting in a complex, dynamic sociality (Tavares et al. 2017).

11.5 Acoustic Behavior and Dialects

As with most delphinids, killer whales have a well-developed acoustic system, producing pulsed signals and whistles for social communication and echolocation clicks for orientation and discrimination of objects in their surroundings. The most characteristic signals of killer whales globally are strident burst-pulsed calls about 1–2 s long. These loud, structurally complex calls are audible at ranges of over 15 km in quiet conditions. Individual whales have stable repertoires of several to ten or more stereotyped calls that are unique at various levels of social structure. Call repertoires appear to be learned by individuals and thus are passed across generations by cultural transmission (Ford 1991; Filatova et al. 2015). In coastal waters of the northeastern Pacific, the three sympatric ecotypes—Residents, Transients, and Offshores—produce repertoires of call types that are entirely distinct from each other. Distinctive variants of call repertoires exist within each ecotype as well. Dialects at the level of the social group are very rare in vertebrates (Ford 1991; Filatova et al. 2015).

Fish-eating Resident killer whales have call repertoires that are based on matrilineal genealogy—within a community, different acoustic clans can be found, each composed of a number of pods that share a portion of their call repertoires. Structural variants of shared calls that are pod-specific or matriline-specific are typical, forming a system of related dialects within each clan. These related dialects likely reflect a common matrilineal heritage of clan members, having evolved from the common call repertoire of an ancestral group that diverged acoustically when the matriline went through a process of growth and fission. Those pods with many shared features in their dialects are probably more closely related, and have split more recently, than those with more divergent features. Newborn whales presumably learn their natal group's dialect by mimicking their mother and other close kin, and dialects are retained within the matriline. Dialects are maintained even though groups from different clans regularly travel together, each using their own distinctive calls. Dialects that encode the matrilineal identity of individuals are no doubt very effective at maintaining group contact, cohesion, and stability. They also appear to serve an important role in inbreeding prevention. Because there is no dispersal of individuals from the natal group, Resident killer whales would seemingly have a high risk of mating with immediate kin. Genetic studies in the Northern Resident community in British Columbia have revealed that Resident males tend to mate outside their pod or clan with females that have unrelated or distantly related dialects (Barrett-Lennard 2000). Multiple clans with associated group-specific dialects have been documented within Resident killer whale populations off the coasts of eastern Kamchatka (Filatova et al. 2015) and southern Alaska (Yurk et al. 2002), as well as British Columbia (Ford 1991). A variety of stereotyped calls have been described from herring-feeding populations of killer whales in Norway and Iceland, at least some of which appear to be group-specific (Moore et al. 1988; Strager 1995; Filatova et al. 2015).

While fish-eating ecotypes of killer whales tend to be highly vocal while foraging, mammal-eating ecotypes typically hunt in near silence (Guinet 1992; Barrett-Lennard et al. 1996; Deecke et al. 2005, 2011; Jourdain et al. 2017). They rarely exchange stereotyped calls and use of echolocation clicks is much reduced. This appears to be a foraging tactic—pinnipeds and cetaceans have excellent underwater hearing and would be more difficult to catch if they heard the approaching predators. Mammal-hunting whales seem to rely on passive listening and vision rather than active echolocation to detect potential prey. However, once prey is detected and captured, the whales can become quite vocal. As with fish-eating killer whales, mammal-hunting killer whales have repertoires of distinctive stereotyped call types which may vary regionally but do not appear to have dialect variation at the level of the social group, at least within the Transient killer whale ecotype off the North American west coast. As there is dispersal from the natal matriline in this ecotype and more fluid associations, group-specific calls would not be expected. Also, dispersal reduces the risk of inbreeding, so the requirement for an acoustic outbreeding mechanism may be reduced in Transients.

11.6 Foraging Specializations and Tactics

Killer whales are highly versatile social predators that forage cooperatively for their preferred prey using a variety of predatory behaviors. Across the world's oceans, there are broad similarities in social structure, group size, and acoustic activity of ecotypes that feed primarily on fish versus those that mostly hunt marine mammals. The often highly specialized tactics used to catch their prey, however, can vary from place to place and across prey species. These local specializations likely result from the innovation of novel foraging methods that, when successful, are passed on by cultural transmission within lineages.

11.6.1 Predation on Fish

Killer whales that specialize on fish tend to have relatively large group sizes, generally 10–30 or more. This is the case with salmon-eating Residents and shark-eating Offshores in the northeastern Pacific (Ford and Ellis 2014), herring-eating populations in Norway and Iceland, (Similä and Ugarte 1999; Tavares et al. 2017), and fish-eating Ross Sea (Type C) killer whales in Antarctica (Pitman and Ensor 2003). Groups of fish-eating killer whales forage in a coordinated manner, sometimes with close cooperation among individuals, especially immediate kin. Resident killer whale pods disperse widely when foraging for Chinook salmon, often spread over areas of 10 km^2 or more. Resident pods do not herd or corral fish cooperatively—rather, individuals and small maternal groups (e.g., a mother and her youngest offspring) search for and catch fish independently. The great majority of salmon captured by adult females and subadults are brought to the surface where they are broken up and shared with others in the matriline (Fig. 11.2; Wright et al. 2016). The widespread prevalence of prey sharing is remarkable given that even large Chinook salmon could be easily consumed by an individual alone. All age–sex classes share their prey with their matrilineal kin but rarely with individuals outside the matriline, even when multiple matrilines forage together. This would be expected if prey sharing is driven primarily by kin selection, whereby individuals receive the greatest inclusive fitness gains by provisioning close relatives. Prey sharing also serves to reduce competition among kin and promotes the maintenance of matrilineal cohesion. Reproductive females share mostly with their offspring but also occasionally with their siblings, mother, or other more distant kin. Post-reproductive females

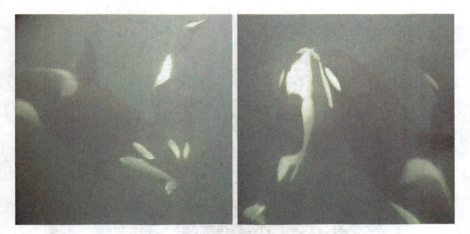

Fig. 11.2 An adult female Resident killer whale, *Orcinus orca*, sharing a salmon with her 1-year-old calf. Taken from video recording made by Cetacean Research Program, Pacific Biological Station, Canada

share their prey more often with their adult sons than adult daughters, perhaps because their older sons have far greater reproductive potential, and enhancing their survival through provisioning would increase her inclusive fitness. Females only have about five surviving offspring in their lifetime (Olesiuk et al. 2005), while adult males can potentially have many offspring by mating with multiple females outside the matriline. Genetic studies in one Resident community have revealed that just two adult males jointly sired at least 36 progeny, or roughly half of all sampled offspring born in the population over a 25-year period (Ford et al. 2018).

In coastal fjords of Norway, killer whales feed on herring that congregate in high densities during fall and winter. To do so, the whales employ a cooperative foraging tactic known as "carousel feeding" to capture these small schooling fishes: this involves a group of whales encircling and herding a school of herring into a tight ball close to the surface (Fig. 11.3). Once the school is concentrated, individuals dive under the school and strike it with their tail flukes, stunning herring in the process. They then pick up and eat the debilitated herring floating in the water column (Similä and Ugarte 1993). Herring-feeding killer whales in Iceland also use tail slaps to debilitate the fish, but they first appear to manipulate the herring into a dense school by emitting a particular kind of pulsed call rich in energy at low frequencies to which herring are most sensitive (Simon et al. 2006). It is interesting that fish-eating Resident killer whales in the northeast Pacific do not target herring, which are extremely abundant in their range during spring spawning—perhaps the whales in this area lack any effective tactic for exploiting these small, schooling fish.

Fish-eating killer whales in New Zealand target several species of large rays that live on or in muddy seabeds. They sometimes dig into the soft substrate to dislodge the fish, surfacing with mud on their rostrum. The captured rays, which can reach many 10s of kg, are generally shared among group members (Visser 2000). Feeding

Fig. 11.3 Aerial view of killer whales, *Orcinus orca*, cooperatively encircling and foraging on a school of herring, northern Norway. Photo courtesy of Norwegian Orca Survey

on rays can be risky since they have sharp venomous spines that can cause injury or mortality to the whales (Duignan et al. 2000), but it appears to be a successful foraging strategy in New Zealand waters.

In the Strait of Gibraltar, a small population of killer whales specializes on predation of bluefin tuna as the fish enter and exit the Mediterranean Sea during their breeding migration (Esteban et al. 2016). To catch these swift tuna, the whales employ an endurance-exhaustion technique involving chases of individual fish that may last 30–40 min at swimming speeds of 12–14 km/h (Guinet et al. 2007). Killer whales can sustain sufficient swimming speeds necessary to catch small to medium (0.8–1.5 m) tuna using this technique but appear unable to match the swimming ability of larger fish.

A foraging technique that has developed among numerous fish-eating killer whale populations is depredation of human fisheries. In many different locations around the world, including the Pacific, Atlantic, and Southern Oceans and the Mediterranean and Bering Seas, killer whales have learned to remove fish caught on longline fishing gear as it is being hauled in. In most cases, depredation involves fish species that are naturally targeted by the whales, but in others the whales have learned to take and consume fish that may not be part of their regular diet. The whales can be highly selective in the fish they remove from the longlines—they typically take the largest individuals of species that are rich in oil, such as Patagonian toothfish and sablefish

11 Killer Whales: Behavior, Social Organization, and Ecology of the... 251

(*Anoplopoma fimbria*), while ignoring small individuals or less desirable species with low oil content (Yano and Dahlheim 1995; Guinet et al. 2014). Killer whales can learn to depredate soon after a fishery begins within their range (Guinet et al. 2014), or the behavior can begin abruptly in fisheries that have existed for years, with the innovation of the technique by a single group, which can then spread to others in the population (Matkin and Saulitis 1994). Once the behavior is established, the whales can be extremely difficult to dissuade from the practice, and depredations may persist despite defensive actions such as shootings and underwater detonations. See Chap. 10 for a description of killer whale interactions with humans.

11.6.2 Predation on Marine Mammals

Killer whales that specialize on marine mammal prey tend to have two behavioral traits in common—they have small group sizes (typically <10 individuals) and they forage in silence to avoid alerting their acoustically sensitive prey and to facilitate passive listening to detect potential prey. These features have been noted in the northeastern Pacific Transient ecotype (Deecke et al. 2005), Norway (Jourdain et al. 2017), Shetland (Deecke et al. 2011), Argentina (Iñíguez et al. 2005; J. Ford, unpubl. obs.), and the Crozet Islands (Guinet 1992). Small group sizes likely allow the stealthy hunting of the whales' vigilant prey by reducing the need to exchange vocalizations for coordination and by making the whales less conspicuous visually. In coastal British Columbia and southern Alaska, Transient killer whales forage in small groups averaging 3–4 whales for harbor seals, the most abundant marine mammal in the region and most common prey (Ford et al. 1998). Harbor seals are easy prey for these whales, and the great majority are captured and consumed quickly following detection. Other prey species of Transients are more difficult quarry. Fast-swimming Dall's porpoise require prolonged high-speed pursuits and escape in about half of observed chases (Ford et al. 1998). If their attempt to flee is unsuccessful, the porpoises are dispatched by ramming from below (Fig. 11.4). Steller sea lions, which may weigh in excess of 1000 kg, are a challenging and dangerous prey species as they can mount an aggressive defense. The whales attack by circling a sea lion in open water and taking turns ramming the animal or striking it with tail flukes while avoiding being bitten. By repeatedly pummeling the sea lion in this manner, which may go on for 1.5–2 h, it is finally debilitated enough to safely grasp and drown underwater, whereupon it is broken up and shared by the group.

Killer whales in Argentine Patagonia and the subantarctic Crozet Islands employ a novel and risky behavior to hunt young southern elephant seals (*Mirounga leonina*) and, in Patagonia, South American sea lions (*Otaria flavescens*). This hunting method involves intentional stranding on shallow, sloping beaches to catch naïve pups as they make their first forays into the water (Lopez and Lopez 1985; Guinet 1992). Considerable skill and agility are needed to successfully execute this tactic without becoming permanently stranded, and it is most commonly undertaken by adults. Young whales are provisioned by their mothers until they learn the technique

Fig. 11.4 A Dall's porpoise, *Phocoenoides dalli*, rammed at high speed by a mammal-hunting Transient (or Bigg's) killer whale, *Orcinus orca*. Photo by Jared Towers

and begin to attempt intentional strandings on their own at 5–6 years of age (Hoelzel 1991; Guinet and Bouvier 1995). During observations at the Crozet Islands, juvenile whales became temporarily stranded on two occasions and likely would have been unable to get off the beach without assistance from their mother on one occasion and human observers on the other (Guinet and Bouvier 1995).

Perhaps the most sophisticated cooperative hunting technique of mammal-eating killer whales is that of pack ice (Type B1) killer whales in Antarctica. Groups of these whales visually seek out a seal hauled out on a small ice floe by spy-hopping and then attempt to wash the targeted seal off the ice with a wave created by rushing toward the floe in a tightly coordinated manner just beneath the surface (Fig. 11.5; Visser et al. 2008; Pitman and Durban 2012). This wave-wash procedure may be repeated multiple times over periods of 30 min or more before the seal is dislodged or the attack is abandoned (Pitman and Durban 2012).

11.7 Conclusions

Killer whales are multifaceted social predators that have evolved to successfully occupy a variety of ecological niches in the world's oceans. They live in matrilineal groups of close kin with enduring bonds that may persist for many years or, in some

Fig. 11.5 Pack ice (Type B1) killer whales, *Orcinus orca*, creating a wave to wash a Weddell seal, *Leptonychotes weddellii*, off an ice floe, Antarctic Peninsula. Photo by John Durban

populations, for life. They have patterns of seasonal movements, group size, social organization, foraging behavior, and acoustic activity that are finely tuned to each ecotype's particular dietary habits. These adaptations appear to be acquired by social learning and are maintained by cultural transmission across generations in the matriline. Killer whales can be highly unconventional in their life history and behavior. The well-studied salmon-eating Resident killer whales of coastal northeastern Pacific waters have several features that are extremely unusual in mammals—they live in multi-generation matrilines from which there is no dispersal of either sex, matrilines have discrete group-specific dialects acquired by vocal learning, there is extensive nonreciprocal food sharing among all age/sex classes within matrilines, and females have prolonged post-reproductive periods that can last for decades. No doubt there are more unusual features in behavior and life history yet to be discovered in other ecotypes and regions.

Ecological specializations such as those seen in killer whales can be a mixed blessing. On the one hand, specialists may have a competitive advantage over generalists in foraging efficiency (as in the old adage, "jack-of-all-trades, master of none") (Kassen 2002). Through innovation and social learning, culturally driven specialization may lead to foraging strategies and tactics that are highly refined for a particular prey type. On the other hand, a high degree of specialization on a particular resource may lead to a reduced ability to efficiently exploit other kinds of prey. The culturally acquired knowledge and behaviors that make certain ecotypes adept at intercepting and catching seasonally migrating fishes such as salmon or herring would be ineffective for hunting marine mammals. Similarly, mammal-hunting specialists would be unlikely to have the acquired skills needed to be an adept fish predator. A specialist foraging strategy may be successful if abundance of the targeted resource is stable and predictable, but should its availability suddenly decline, specialists may be incapable of switching to alternative prey without a considerable reduction in foraging efficiency. Indeed, recent modelling of hypothetical specialist and generalist killer whale populations has shown that cultural specialization is adaptive in the short term and leads to increased fitness, whereas generalization is rarely adaptive. However, cultural evolution of specialization can lead to increased rates of group extirpation (Whitehead and Ford 2018).

It might be expected that a lifestyle based on learning and innovation would give killer whales the flexibility to adapt to fluctuating resource availability. But the cultural traditions of killer whales are deeply rooted and can be very resistant to change. An example is the vocal dialects of Resident killer whales. Clan-specific call repertoires in use today are, in most cases, little changed from those recorded as far back as the late 1950s (Ford 1991, unpubl. data). Despite decades of regular associations of pods from different clans, no diffusion or transfer of calls between them has taken place. The dietary specializations of ecotypes are also firmly fixed traditions. For example, three mammal-eating Transient killer whales that were live-captured and held temporarily in an ocean net pen refused to eat any fish offered to them daily for 2.5 months, by which time they were starving. On day 75, one animal died of malnutrition, but on the 79th day, the two remaining whales finally began to eat salmon that was provided to them and continued to do so until they were released back to the wild several months later, whereupon they resumed their mammal-hunting habits (Ford and Ellis 1999). In contrast, salmon-eating Resident killer whales readily ate fish offered to them within a few days of capture (the existence of ecotypes with different diets was not known to the captors at the time).

Sympatric salmon-eating Resident and mammal-eating Transient killer whale populations off the west coast of Canada have both been affected by fluctuations in the abundance of their prey (Ford et al. 2010). Resident killer whale survival and calving rates are correlated with availability of Chinook salmon, their primary prey species. Mortality rates increased sharply and reproduction decreased during several consecutive years of unusually low range-wide Chinook abundance in the 1990s. It appears that the whales either did not switch to alternative prey during this period or, if they did, it was insufficient to offset nutritional stress and its consequences. One population—the Southern Residents—remains critically endangered, and limited Chinook salmon availability is considered one of the main factors preventing recovery (Lacy et al. 2017). Transient killer whales were rare in coastal British Columbia when field studies began in the 1970s, likely because decades of predator control measures had seriously depleted their primary prey, harbor seals. It is likely that the whales either left the area or suffered decreased survival and reproduction in response to the removal of their food supply, as seen in Residents. Since then, harbor seal abundance has rebounded to historical levels, and Transients have experienced substantial population growth (Ford et al. 2007; Ford 2014).

11 Killer Whales: Behavior, Social Organization, and Ecology of the... 255

Culturally driven foraging specialization appears to have resulted in the adaptive radiation of killer whales into a variety of ecological niches within the past 250,000 years (Riesch et al. 2012; Foote et al. 2016). However, cultural conservatism may limit the ability of these specialized ecotypes to adapt quickly to perturbations in their food supply over short time frames. But in some cases killer whales do seem to have the capacity for innovation in order to take advantage of a novel food supply—the abrupt onset of depredation behavior to take fish from a long-existing fishery is one such case. Hopefully, highly specialized killer whale ecotypes will have sufficient flexibility and resilience to withstand future changes in their ecosystem, especially where they arise from conflict with humans for the same food resources.

Acknowledgment Many thanks to John Durban, Eve Jourdain (Norwegian Orca Survey) and Jared Towers for kindly allowing use of their photographs.

References

Baird RW, Dill LM (1996) Ecological and social determinants of group size in transient killer whales. Behav Ecol 7:408–416
Baird RW, Whitehead H (2000) Social organization of mammal-eating killer whales: group stability and dispersal patterns. Can J Zool 78:2096–2105
Baird RW, McSweeney DJ, Bane C, Barlow J, Salden DR, Antoine LRK, LeDuc RG, Webster DL (2006) Killer whales in Hawaiian waters: information on population identity and feeding habits. Pac Sci 60:523–530
Barrett-Lennard LG (2000) Population structure and mating systems of northeastern Pacific killer whales. Ph.D. dissertation. University of British Columbia, Vancouver
Barrett-Lennard LG, Ford JKB, Heise KA (1996) The mixed blessing of echolocation: differences in sonar use by fish-eating and mammal-eating killer whales. Anim Behav 51:553–565
Barrett-Lennard LG, Matkin CO, Durban JW, Saulitis EL, Ellifrit D (2011) Predation on gray whales and prolonged feeding on submerged carcasses by transient killer whales at Unimak Island, Alaska. Mar Ecol Prog Ser 421:229–241
Bigg MA (1982) An assessment of killer whale stocks off Vancouver Island, British Columbia. Report of the International Whaling Commission 32:655–666
Bigg MA, Olesiuk PF, Ellis GM, Ford JKB, Balcomb KC (1990) Social organization and genealogy of resident killer whales (*Orcinus orca*) in the coastal waters of British Columbia and Washington state. Report of the International Whaling Commission 12:383–405
Brent LJ, Franks DW, Foster EA, Balcomb KC, Cant MA, Croft DP (2015) Ecological knowledge, leadership, and the evolution of menopause in killer whales. Curr Biol 25:746–750
Connor RC, Wells R, Mann J, Read A (2000) The bottlenose dolphin: social relationships in a fission-fusion society. In: Mann J, Connor RC, Tyack P, Whitehead H (eds) Cetacean societies: field studies of whales and dolphins. University of Chicago Press, Chicago, pp 91–126
Deecke VB, Ford JKB, Slater PJ (2005) The vocal behaviour of mammal-eating killer whales: communicating with costly calls. Anim Behav 69:395–405
Deecke VB, Nykänen M, Foote AD, Janik VM (2011) Vocal behaviour and feeding ecology of killer whales *Orcinus orca* around Shetland, UK. Aquat Biol 13:79–88
Duignan PJ, Hunter JE, Visser IN, Jones GW, Nutman A (2000) Stingray spines: a potential cause of killer whale mortality in New Zealand. Aquat Mamm 26:143–147

Durban JW, Fearnbach H, Burrows DG, Ylitalo GM, Pitman RL (2017) Morphological and ecological evidence for two sympatric forms of type B killer whale around the Antarctic Peninsula. Polar Biol 40:231–236

Esteban R, Verborgh P, Gauffier P, Giménez J, Foote AD, de Stephanis R (2016) Maternal kinship and fisheries interaction influence killer whale social structure. Behav Ecol Sociobiol 70:111–122

Filatova OA, Samarra FI, Deecke VB, Ford JK, Miller PJ, Yurk H (2015) Cultural evolution of killer whale calls: background, mechanisms and consequences. Behaviour 152:2001–2038

Foote AD, Newton J, Piertney SB, Willerslev E, Gilbert MTP (2009) Ecological, morphological and genetic divergence of sympatric North Atlantic killer whale populations. Mol Ecol 18:5207–5217

Foote AD, Morin PA, Pitman RL, Ávila-Arcos MC, Durban JW, van Helden A, Sinding MHS, Gilbert MTP (2013) Mitogenomic insights into a recently described and rarely observed killer whale morphotype. Polar Biol 36:1519–1523

Foote AD, Vijay N, Ávila-Arcos MC, Baird RW, Durban JW, Fumagalli M, Gibbs RA, Hanson MB, Korneliussen TS, Martin MD, Robertson KM (2016) Genome-culture coevolution promotes rapid divergence of killer whale ecotypes. Nat Commun 7:11693

Ford JKB (1991) Vocal traditions among resident killer whales (*Orcinus orca*) in coastal waters of British Columbia. Can J Zool 69:1454–1483

Ford JKB (2014) Marine mammals of British Columbia. Royal BC Museum handbook, mammals of BC, vol 6. Royal BC Museum, Victoria, p 460

Ford JKB, Ellis GM (1999) Transients: mammal-hunting killer whales of British Columbia, Washington, and Southeastern Alaska. UBC Press, Vancouver, p 96

Ford JKB, Ellis GM (2006) Selective foraging by fish-eating killer whales *Orcinus orca* in British Columbia. Mar Ecol Prog Ser 316:185–199

Ford JKB, Ellis GM (2014) You are what you eat: ecological specializations and their influence on the social organization and behaviour of killer whales. In: Yamagiwa J, Karczmarski L (eds) Primates and cetaceans: field research and conservation of complex mammalian societies. Springer, New York, NY, pp 75–98

Ford JKB, Ellis GM, Barrett-Lennard LG, Morton AB, Palm RS, Balcomb KC III (1998) Dietary specialization in two sympatric populations of killer whales (*Orcinus orca*) in coastal British Columbia and adjacent waters. Can J Zool 76:1456–1471

Ford JKB, Ellis GM, Balcomb KC (2000) Killer whales: the natural history and genealogy of *Orcinus orca* in British Columbia and Washington, vol 102, 2nd edn. UBC Press, Vancouver

Ford JKB, Ellis GM, Durban JW (2007) An assessment of the potential for recovery of west coast transient killer whales using coastal waters of British Columbia. Research document 2007/088, Fisheries and Oceans Canada. Canadian Science Advisory Secretariat, Ottawa

Ford JKB, Ellis GM, Olesiuk PF, Balcomb KC (2010) Linking killer whale survival and prey abundance: food limitation in the oceans' apex predator? Biol Lett 6:139–142

Ford JKB, Ellis GM, Matkin CO, Wetklo MH, Barrett-Lennard LG, Withler RE (2011) Shark predation and tooth wear in a population of northeastern Pacific killer whales. Aquat Biol 11:213–224

Ford JKB, Reeves RR (2008) Fight or flight: antipredator strategies of baleen whales. Mammal Rev 38(1):50–86

Ford MJ, Parsons KM, Ward EJ, Hempelmann JA, Emmons CK, Hanson MB, Balcomb KC, Park LK (2018) Inbreeding in an endangered killer whale population. Anim Conserv. https://doi.org/10.1111/acv.12413

Foster EA, Franks DW, Mazzi S, Darden SK, Balcomb KC, Ford JKB, Croft DP (2012) Adaptive prolonged postreproductive life span in killer whales. Science 337:1313–1313

Guinet C (1991) L'orque (*Orcinus orca*) autour de l'Archipel Crozet comparaison avec d'autres localités. Rev Ecol (Terre Vie) 46:1991

Guinet C (1992) Comportement de chasse des orques *Orcinus orca* autour des Îles Crozet. Can J Zool 70:1656–1667

Guinet C, Bouvier J (1995) Development of intentional stranding hunting techniques in killer whale (*Orcinus orca*) calves at Crozet Archipelago. Can J Zool 73(1):27–33

Guinet C, Domenici P, De Stephanis R, Barrett-Lennard L, Ford JKB, Verborgh P (2007) Killer whale predation on bluefin tuna: exploring the hypothesis of the endurance-exhaustion technique. Mar Ecol Prog Ser 347:111–119

Guinet C, Tixier P, Gasco N, Duhamel G (2014) Long-term studies of Crozet Island killer whales are fundamental to understanding the economic and demographic consequences of their depredation behaviour on the Patagonian toothfish fishery. ICES J Mar Sci 72:1587–1597

Hoelzel AR (1991) Killer whale predation on marine mammals at Punta Norte, Argentina; food sharing, provisioning and foraging strategy. Behav Ecol Sociobiol 29:197–204

Hoelzel AR, Moura AE (2016) Killer whales differentiating in geographic sympatry facilitated by divergent behavioural traditions. Heredity 117:481–482

Hoelzel AR, Hey J, Dahlheim ME, Nicholson C, Burkanov V, Black N (2007) Evolution of population structure in a highly social top predator, the killer whale. Mol Biol Evol 24:1407–1415

Iñíguez, M., Tossenberger, V.P., and Gasparrou, C. 2005. Socioecology of killer whales (*Orcinus orca*) in northern Patagonia, Argentina. Unpublished paper to the IWC Scientific Committee, 9 p. Ulsan, Korea, June 2005. (SC/57/SM5)

Ivkovich T, Filatova OA, Burdin AM, Sato H, Hoyt E (2010) The social organization of resident-type killer whales (*Orcinus orca*) in Avacha Gulf, Northwest Pacific, as revealed through association patterns and acoustic similarities. Mammalian Biology-Zeitschrift für Säugetierkunde 75:198–210

Jourdain E, Vongraven D, Bisther A, Karoliussen R (2017) First longitudinal study of seal-feeding killer whales (*Orcinus orca*) in Norwegian coastal waters. PLoS One 12(6):e0180099

Kassen R (2002) The experimental evolution of specialists, generalists, and the maintenance of diversity. J Evol Biol 15:173–190

Lacy RC, Williams R, Ashe E, Balcomb KC III, Brent LJ, Clark CW, Croft DP, Giles DA, MacDuffee M, Paquet PC (2017) Evaluating anthropogenic threats to endangered killer whales to inform effective recovery plans. Sci Rep 7:14119

Lopez JC, Lopez D (1985) Killer whales (*Orcinus orca*) of Patagonia, and their behavior of intentional stranding while hunting nearshore. J Mammal 66:181–183

Matkin CO, Saulitis E (1994) Killer whale (*Orcinus orca*): biology and management in Alaska. Prepared for US Marine Mammal Commission by North Gulf Oceanic Society, Homer, AK

Matkin CO, Testa JW, Ellis GM, Saulitis EL (2014) Life history and population dynamics of southern Alaska resident killer whales (*Orcinus orca*). Mar Mamm Sci 30:460–479

Miller PJO, Shapiro AD, Tyack PL, Solow AR (2004) Call-type matching in vocal exchanges of free-ranging resident killer whales, *Orcinus orca*. Anim Behav 67:1099–1107

Moore SE, Francine JK, Bowles AE, Ford JKB (1988) Analysis of calls of killer whales, *Orcinus orca*, from Iceland and Norway. Rit Fiskideildar 11:225–250

Morin PA, Archer FI, Foote AD, Vilstrup J, Allen EE, Wade P, Durban J, Parsons K, Pitman R, Li L, Bouffard P, Abel Nielsen SC, Rasmussen M, Willerslev E, Gilbert MTP, Harkins T (2010) Complete mitochondrial genome phylogeographic analysis of killer whales (*Orcinus orca*) indicates multiple species. Genome Res 20(7):908–916

Morin PA, Parsons KM, Archer FI, Ávila-Arcos MC, Barrett-Lennard LG, Dalla Rosa L, Duchêne S, Durban JW, Ellis GM, Ferguson SH, Ford JKB, Ford MJ, Garilao C, Gilbert MTP, Kaschner K, Matkin CO, Peterson SD, Robertson KM, Visser IN, Wade PR, Ho SYW, Foote AD (2015) Geographic and temporal dynamics of a global radiation and diversification in the killer whale. Mol Ecol 24:3964–3979

Moura AE, Kenny JG, Chaudhuri RR, Hughes MA, Reisinger RR, De Bruyn PJN, Dahlheim ME, Hall N, Hoelzel AR (2015) Phylogenomics of the killer whale indicates ecotype divergence in sympatry. Heredity 114:48

Olesiuk PF, Ellis GM, Ford JKB (2005) Life history and population dynamics of northern resident killer whales (*Orcinus orca*) in British Columbia. Research Document 2005/045. Canadian Science Advisory Secretariat, Fisheries & Oceans Canada, Ottawa, ON

Pitman RL, Durban JW (2012) Cooperative hunting behavior, prey selectivity and prey handling by pack ice killer whales (*Orcinus orca*), type B, in Antarctic Peninsula waters. Mar Mamm Sci 28:16–36

Pitman RL, Durban JW, Greenfelder M, Guinet C, Jorgensen M, Olson PA, Plana J, Tixier P, Towers JR (2011) Observations of a distinctive morphotype of killer whale (*Orcinus orca*), type D, from subantarctic waters. Polar Biol 34(2):303–306

Pitman RL, Ensor P (2003) Three forms of killer whales (*Orcinus orca*) in Antarctic waters. J Cetacean Res Manag 5(2):131–140

Pitman RL, Fearnbach H, Durban JW (2018) Abundance and population status of Ross Sea killer whales (*Orcinus orca*, type C) in McMurdo Sound, Antarctica: evidence for impact by commercial fishing? Polar Biol 41:781–792

Riesch R, Barrett-Lennard LG, Ellis GM, Ford JKB, Deecke VB (2012) Cultural traditions and the evolution of reproductive isolation: ecological speciation in killer whales? Biol J Linn Soc 106:1–17

Saulitis E, Matkin C, Barrett-Lennard L, Heise K, Ellis G (2000) Foraging strategies of sympatric killer whale (*Orcinus orca*) populations in Prince William Sound, Alaska. Mar Mamm Sci 16:94–109

Scammon CM (1874) The marine mammals of the north-western coast of North America: described and illustrated; together with an account of the American whale-fishery. JH Carmany, San Francisco

Similä T, Ugarte F (1993) Surface and underwater observations of cooperatively feeding killer whales in northern Norway. Can J Zool 71:1494–1499

Similä T, Ugarte F (1999) Patterns in social organisation and occurrence among killer whales photo-identified in northern Norway. European Research on Cetaceans 12:220. In: Evan PGH, Parsons ECM (eds) Proceedings of the twelfth annual conference of the European Cetacean Society, Monaco, 20–24 Jan 1998. Artes Graficas Soler, Valencia

Similä T, Holst JC, Christensen I (1996) Occurrence and diet of killer whales in northern Norway: seasonal patterns relative to the distribution and abundance of Norwegian spring-spawning herring. Can J Fish Aquat Sci 53:769–779

Simon M, Ugarte F, Wahlberg M, Miller L (2006) Icelandic killer whales *Orcinus orca* use a pulsed call suitable for manipulating the schooling behaviour of herring *Clupea harengus*. Bioacoustics 16:57–74

Strager H (1995) Pod-specific call repertoires and compound calls of killer whales, *Orcinus orca* Linnaeus, 1758, in the waters of northern Norway. Can J Zool 73:1037–1047

Stredulinsky EH (2016) Determinants of group splitting: an examination of environmental, demographic, genealogical and state-dependent factors of matrilineal fission in a threatened population of fish-eating killer whales (*Orcinus orca*). MSc thesis. University of Victoria, BC

Tavares SB, Samarra FI, Miller PJ (2017) A multilevel society of herring-eating killer whales indicates adaptation to prey characteristics. Behav Ecol 28:500–514

Ternullo R, Black N (2002) Predation behavior of transient killer whales in Monterey Bay, California. In: Fourth international Orca symposium and workshop. CEBC-CNRS, Villiers en Bois, pp 156–159

Tixier P, Gasco N, Guinet C (2014) Killer whales of the Crozet Islands: photoidentification catalogue 2014. Villiers en Bois: Centre d'Etudes Biologiques de Chizé-CNRS, vol 10, p m9. https://www.researchgate.net/publication/268978107_Tixier_et_al_CROZET_KILLER_WHALES_PHOTO_ID_CATALOGUE_2014_POSTER

Tosh CA, De Bruyn PJ, Bester MN (2008) Preliminary analysis of the social structure of killer whales, *Orcinus orca*, at subantarctic Marion Island. Mar Mamm Sci 24:929–940

Towers JR, Ellis GM, Ford JKB (2015) Photo-identification catalogue and status of the northern resident killer whale population in 2014. Canadian Technical Report of Fisheries and Aquatic Science 3139:vi + 75

Visser IN (2000) Orca (*Orcinus orca*) in New Zealand waters. Ph.D. dissertation, University of Auckland, New Zealand

Visser IN, Smith TG, Bullock ID, Green GD, Carlsson OG, Imberti S (2008) Antarctic Peninsula killer whales (*Orcinus orca*) hunt seals and a penguin on floating ice. Mar Mamm Sci 24:225–234

Whitehead H, Ford JKB (2018) Consequences of culturally-driven ecological specialization: killer whales and beyond. J Theor Biol 456:279–294

Wright BM, Stredulinsky EH, Ellis GM, Ford JKB (2016) Kin-directed food sharing promotes lifetime natal philopatry of both sexes in a population of fish-eating killer whales, *Orcinus orca*. Anim Behav 115:81–95

Yano K, Dahlheim ME (1995) Behavior of killer whales *Orcinus orca* during longline fishery interactions in the southeastern Bering Sea and adjacent waters. Fish Sci 61:584–589

Yurk H, Barrett-Lennard L, Ford JKB, Matkin CO (2002) Cultural transmission within maternal lineages: vocal clans in resident killer whales in southern Alaska. Anim Behav 63:1103–1119

Chapter 12
Sperm Whale: The Largest Toothed Creature on Earth

Mauricio Cantor, Shane Gero, Hal Whitehead, and Luke Rendell

Abstract Among large variations in size, habitat use, trophic niche, and social systems of toothed whales, one species—the sperm whale—stands out as an animal of extremes. The world's largest biological sonar operated by the largest brain on Earth shapes much of sperm whales' lives as efficient predators, exploiting massive biological resources at great depths. They are nomads with home ranges spanning thousands of kilometers horizontally and more than a kilometer vertically. These three-dimensional movements and extremely low reproductive rates place a premium on cooperative calf care, making it central to the tight matrilineal social units of female sperm whales in tropical and subtropical waters. The social units themselves are elements of sympatric cultural clans with distinctive behaviors and vocal dialects. Males leave their maternal units in their teens, gradually moving to higher latitudes and becoming less social until, when very much larger than the females, they make periodic forays to warmer waters for mating. New technology is beginning to give us insight into the behaviors of this extraordinary animal, but its long life span means that long-term studies using simple methods are still immensely valuable.

Mauricio Cantor and Shane Gero are co-first authors for this chapter.

M. Cantor (✉)
Departamento de Ecologia e Zoologia, Universidade Federal de Santa Catarina, Florianópolis, Brazil

Centro de Estudos do Mar, Universidade Federal do Paraná, Pontal do Paraná, Brazil

S. Gero
Department of Bioscience, Aarhus University, Aarhus, Denmark

H. Whitehead
Department of Biology, Dalhousie University, Halifax, Canada

L. Rendell
Sea Mammal Research Unit, School of Biology, University of St. Andrews, St. Andrews, UK

© Springer Nature Switzerland AG 2019
B. Würsig (ed.), *Ethology and Behavioral Ecology of Odontocetes*, Ethology and Behavioral Ecology of Marine Mammals,
https://doi.org/10.1007/978-3-030-16663-2_12

Keywords Foraging · Traveling · Socializing · Acoustic communication · Social structure · Culture

12.1 Introduction

Sperm whales are odd creatures. Their distinctiveness is molded by their surroundings—the deep ocean and their rich social lives. In offshore waters, sperm whales make a living diving to deep strata peppered by unpredictable food patches in an otherwise dark desert. But, for a sperm whale, this desert is a social place.

For most people, they are the archetype of a whale—ask a child to draw a whale and you will likely get a big-headed, out-of-the-ordinary line drawing resembling a sperm whale (Fig. 12.1a). Despite living in an environment unfamiliar to us, sperm whales are engrained in our culture, from starring in one of the most-read novels (Melville 1851) to lubricating the machinery of the industrial revolution (e.g., Whitehead 2002). Thanks to a modern appreciation of the natural world and to increasing research effort, our image of sperm whales has shifted from brutish leviathan to docile giant.

Sperm whales are large-bodied animals—the largest of any toothed predators. Perhaps more arresting than size is the sexual dimorphism in body size and weight, the most marked among cetaceans (Fig. 12.1b). Mature males average about 16 m in length and up to about 45,000 kg compared to the average 11 m, 15,000 kg females. They also have the largest brain in absolute mass (~7.8 kg), but not the largest brain relative to their immense body sizes (see Whitehead and Rendell 2014). Large brains are associated with high cognitive capacities, and sperm whales' rich social lives and complex communication system make excellent use of such abilities.

A glance at sperm whale external morphology reveals odd adaptations for a deep-sea life (Fig. 12.1a). Their blow does not come from the top of their heads, but is left-skewed, coming from a single asymmetric blowhole. About one quarter to one third of the animal's body makes up the barrel-shaped head, which is largely composed of the spermaceti organ (Clarke 1970). This organ is packed with spermaceti oil that has physical properties of wax, from which the whales get their name. That massive head houses the world's largest biological sound generator. At both ends are air sacs, and air from distal sacs at the tip of their noses is forced through the cartilaginous phonic lips (or *museau du singe*) to produce a click. The click is directed backward to the frontal sac placed in front of the skull and then reverberates forward, through a mass of connective tissue and oil termed the junk that lies underneath the spermaceti and serves to direct sounds from the head into the environment. The resulting bursts of high-pressure sounds—the clicks—are loud (236 dB *re*: 1 µPa rms; Møhl et al. 2003) and forward-directional (Møhl et al. 2000).

Loud clicks allow sperm whales to "see" much further than the end of their noses. In this three-dimensional aquatic world, sound travels about 4.5 times the speed that it does in air. Because sound does not attenuate quickly in the dense medium of

Fig. 12.1 Morphology, behavior, and methods for studying sperm whales, *Physeter macrocephalus*. (**a**) A close encounter highlights the distinctive massive head and its powerful biosonar (©Amanda Cotton). (**b**) A traveling social unit of females in the company of a mature male highlights the sexual dimorphism in body sizes (©Marina Milligan). (**c**) A whale with squid in mouth in a rare glimpse of prey brought to the surface (©Robyn and Wade Hughes). (**d**) While resting and socializing, females are relatively active at the surface with social unit members (©Patrick Dykstra). (**e**) A suckling calf illustrates how maternal and allomaternal care of young is a strong force within social units (©Patrick Dykstra). (**f**) Methods for studying sperm whale behavioral ecology include DTags to record dive profiles (©Keri Wilk), (**g**) arrays of omni- and directional hydrophones (©Marina Milligan), and (**h**) individual photo identification via natural marks in longitudinal studies (©Dominica Sperm Whale Project)

water—while light travels only a few hundred meters at most—it is especially efficient for sensing the physical environment while scanning for prey, predators, and peers (Madsen et al. 2002). From echolocation to communication, theirs is a world of sound.

Sperm whales are ecologically important throughout much of the deep oceans. They are distributed circumglobally (Jaquet 1996) and eat a worldwide biomass comparable to that of all human fisheries combined, roughly 100 million tons per year (Whitehead 2003). This enormous consumption helps to regulate mesopelagic food webs via top-down control of mesopredator numbers directly and of producers and detritivores indirectly. It also acts as a "biological pump" that counteracts the downward flow of carbon through migration and sinking of organic matter. Sperm whales contribute to nutrient cycling by feeding at the typically dark but nutrient-rich depths and defecating at the sea surface where nutrients may be in short supply. The fecal plumes rich in nitrogen and iron content promote plankton growth and primary productivity, when brought to the euphotic zone where the sun penetrates, especially in the iron-depleted waters in the southern hemisphere, contributing to an estimated removal of 200,000 tons more carbon from the atmosphere than they produce through respiration (Lavery et al. 2010).

By occupying such a niche, where few other animals venture, they largely escaped competition, except from each other. Sperm whales live long, grow slowly, and mature late; females produce few energy-expensive offspring; high calf survival rates are vital to population growth (Whitehead 2003). By being large-bodied animals that behave cooperatively, they also largely escape predation, except from killer whales (*Orcinus orca*) and humans. Much of this ecological success is mediated by their acoustic sensory capabilities that drive their foraging skills and social lives.

Here, we revisit the classic and recent findings on the behavioral ecology of sperm whales to illustrate the lives of deep-diving odontocetes and discuss how well general principles of behavioral ecology hold up in this species. Our goal is to offer an overview of sperm whales as important predators and as a cooperative, social, and cultural species. We dive into their adaptations to mesopelagic habitats, assessing their foraging and movement behaviors; then we surface to assess their social lives and learned traditions. We close the chapter by raising some challenging questions we hope the next technological advances will help to answer.

12.2 Foraging

Sperm whales are the largest toothed predators, but it is not their teeth that make them efficient hunters. Their mouths are impressive for the nearly 90-degree wide gape and large jaws; but they have only vestigial maxillary teeth and no proper oral cavity, lips, or cheek (Werth 2004; Fig. 12.1a). The mandibular teeth are fully exposed along the lower jaw, but may not even be used to capture or handle prey (Werth 2004). What makes sperm whales efficient predators are their noses. Scientists

12 Sperm Whale: The Largest Toothed Creature on Earth 265

have come up with a variety of theories for the evolution of the sperm whale head, from buoyancy control (Clarke 1970) to battering ram (Carrier et al. 2002), but the only explanation that does not sink under the weight of its own contradictions is that it is uniquely adapted to produce highly focused echolocation clicks from the largest biosonar on Earth.

Foraging sperm whales dive after a range of large food items that dwell in deeper waters—mainly deep-sea squids, but also fishes (Kawakami 1980). Stomach contents suggest that preferred prey are gelatinous cephalopods of the Histioteuthidae family, but those of at least other five families, including the giant (*Architeuthis* spp.) and jumbo squids (*Dosidicus gigas*), are also consumed. To find prey in dark deep waters, sperm whales echolocate. The challenge of directly observing foraging has led to many hypotheses about how whales capture and consume prey. For instance, their whitish outer mouth lining—in sharp contrast with their otherwise largely dark bodies—suggests that sperm whales could lure bioluminescent squid (Beale 1839). Stomach contents reveal that prey are captured with few bite marks and suggest suction feeding (Werth 2004). It is now clear that sperm whales do not acoustically stun their prey with intense clicks (Fais et al. 2016), as previously suggested (Norris and Møhl 1983). However, it was not until the advent of animal-borne technologies that could dive along with the whales that we began to understand the finer mechanics of their deep foraging behaviors.

The overarching challenge for predators is maximizing net energy, that is, accruing higher intake via prey capture than the energy spent searching for and handling them. For air-breathing predators foraging with a restricted oxygen supply, underwater time must be spaced out by breaks at the surface. Sperm whales perform deep and long dives of ~400–1200 m for ~40–50 min and split up with short breaks of ~9 min at the surface to load up their blood and muscles with oxygen (Watwood et al. 2006). Typical sperm whale dives are "U-shaped" (Fig. 12.2a). When foraging, sperm whales use their long-range biosonar through ~80% of the dives (Watwood et al. 2006). They produce predictable and characteristic long series of regularly spaced clicks at 0.5–2 s intervals (Fig. 12.2b). When homing in on and capturing prey, sperm whales produce "buzzes" (or "creaks"; Fig. 12.2c) that are rapid accelerating sequences of clicks (Fais et al. 2016; Miller et al. 2004).

There are some sex differences in foraging behavior. Females focus on cephalopods about 0.1–1.0 kg (Fig. 12.1c), but males show greater variation in foraging strategies, in part due to their wider ranging habits. In colder high-latitude waters, males hunt the colossal squids (*Mesonychoteuthis hamiltoni*) and a selection of mesopelagic and demersal fish in Antarctic/Arctic waters (Teloni et al. 2008; Hanselman et al. 2018). This variation in prey and habitats enables at least two distinct types of foraging dives: shallower dives with fewer and longer buzzes suggesting evasive and larger prey and deeper dives with relatively more but shorter buzzes suggesting denser and less evasive prey (Teloni et al. 2008; Fais et al. 2015).

There is still much to learn about sperm whale foraging. Very little is known about the ontogeny of diving, foraging, and echolocation; but calves can make long (up to 44 min) and deep (down to 662 m) dives prior to age 1, producing echolocation clicks and, rarely, buzzes at this young age (Tønnesen et al. 2018). The role of

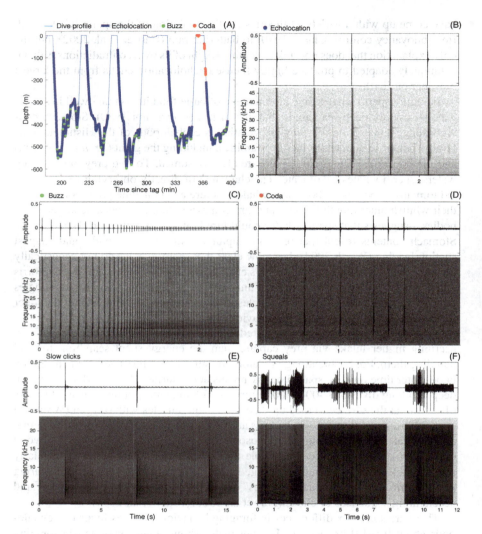

Fig. 12.2 The diverse acoustic behavior of sperm whales, *Physeter macrocephalus*. (**a**) Dive profile of a tagged sperm whale showing the typical "U shape". Foraging sounds, such as (**b**) echolocation clicks and (**c**) buzzes, are produced in the deeper phases of the dive. Social sounds are mostly produced in the shallower phases, here illustrated by (**d**) the 1 + 1 + 3 coda type recorded in Dominica and typically heard in the Caribbean; (**e**) slow clicks recorded from males in the Gulf of Alaska (note the energy emphasis at 2–4 kHz and the reverberations; courtesy L. Wild/SEASWAP); and three "squeals" (**f**) recorded in the southeast Pacific (note that they overlap other social sounds, both codas and chirrups). Waveforms and spectrograms were produced with 1024 point FFT, 50% overlap, Hanning window

12 Sperm Whale: The Largest Toothed Creature on Earth

learning in the development and tuning of diving behaviors remains poorly understood. We are largely prevented from studying optimal foraging in sperm whales because mapping prey distribution at such depths is currently prohibitively expensive. Behavioral decisions on habitat use and with whom to forage are also important factors in foraging success. On this topic, we know much more.

12.3 Traveling

Sperm whales are nomadic, with home ranges that can span thousands of kilometers. In the mesopelagic world, food resources are patchy, usually short-lived and unpredictable in space and time. So, sperm whales can be almost anywhere that there are deep ice-free waters, and they are usually on a constant search.

There are marked differences in distribution between the sexes. Females, often with young, concentrate in lower-latitude waters, generally less than 40° north or south (Whitehead et al. 2008; Mizroch and Rice 2013). Females roam more widely in the Pacific than they do in the Atlantic, where only males have been documented traveling long distances (Gero et al. 2007). Males in general show a wider range of movements than females, sometimes traveling thousands of kilometers across ocean basins and at other times staying resident in rather small coastal areas for months or years (Mizroch and Rice 2013; Rødland and Bjørge 2015). There is little evidence that sperm whales follow regular migration routes or clear seasonal agendas, as do most baleen whales.

Sperm whales likely use information on relevant environmental conditions, gleaned directly or through conspecifics (e.g., eavesdropping on the click patterns of other sperm whale groups), to update knowledge about habitat and food patch quality. Their movements are typically a relatively straight track at about 4 km/h (Fig. 12.1b), which means they may cover about 90 km, horizontally, in a day. When food abounds, however, paths are more convoluted, so that displacements are restricted to about 10–20 km (Whitehead 2003). Over larger spatiotemporal scales, we know much less about the decisions sperm whales make about movements, but we suspect that habitat knowledge handed down over generations plays an important role.

In social animals such as sperm whales, a consensus must be reached on movement decisions if a group is to remain together. Foraging groups of female sperm whales often make long and seemingly disorganized turns that may be indicative of shared decision-making (Whitehead 2016). In the Pacific, a typical group containing 30+ individuals can take 1 h or more to complete a turn, and within this period there is much individual variation in heading (Whitehead 2016). The precise mechanics of female group decisions remain poorly understood, and we do not know the extent to which an individual plays a leading role, perhaps a matriarch, with a long life span of experience to draw upon (as in killer whales; Brent et al. 2015). Female sperm whales can be slow and seemingly disorganized when deciding where to go; but the same is not true about deciding with whom they travel.

12.4 The Social Sperm Whale

The social behavior of sperm whales follows their sexual divergences in lifestyles. Adult males are generally solitary. In their teens, males depart from their natal groups to move poleward. After leaving their female relatives, they live quasi-solitary lives in polar or near-polar waters where they feed, grow, and mature, only returning to the tropical and subtropical breeding grounds in their late 20s to search for receptive females (Best 1979). In warmer waters, the now large males roam around, apparently avoiding one another while visiting groupings of females for periods of minutes to hours (Fig. 12.1b). But males can engage in social gatherings among themselves. In low latitudes, younger and smaller males may form loose aggregations—bachelor groups (Best 1979); in high latitudes, they occasionally cluster (Curé et al. 2013). The drivers of male clustering are not understood—perhaps protection against larger predators or feeding/mating competitors, perhaps simply a relict sociality from their nursery years.

Females are deeply social beings, always in the company of others. Together with calves and juveniles, they spend most of their lives in stable, nearly matrilineal familial units (Fig. 12.1d). Such social units range in membership from 3 to over 20 close relatives as well as unrelated individuals (Lyrholm et al. 1999; Konrad et al. 2018). Social units make up the primary tier of female sperm whale societies (Fig. 12.3), within which individuals move and feed together and share knowledge. However, there are social preferences, partially driven by kinship such that strict matrilineal lines appear to structure relationships among unit members (Gero et al. 2013; Konrad et al. 2018). Females cooperatively raise and defend their calves; they may also find food together by eavesdropping on each other or sharing information about food patches (Whitehead 2003). The absence of territoriality and within-unit mating suggests reduced competition among unit members (Christal and Whitehead 2001). More strikingly, within the social units (and in rare cases between), females can suckle each other's calves (Gero et al. 2009; Fig. 12.1e).

While sperm whale calves may be physiologically capable of following their mothers during long deep dives (Tønnesen et al. 2018), calves usually remain near the surface, following other adults of the social unit when the adults emerge to breathe (Whitehead 2003). A sperm whale calf represents a significant energy investment in terms of gestation, lactation, and long periods of care. Maturation is slow. Females mature sexually at about 9 years old and reproduce until their 40s, while males attain sexual maturity later (the prolonged puberty can last until their 20s) and grow for longer (into their 30s to reach full physical maturity near their 50s). Throughout their lives, females give birth to a single calf at a time, following a gestation of 14–16 months, at about 5-year intervals (Best 1979). Thus, each calf matters, and the structure of female sperm whale society revolves around them (Gero et al. 2013).

Female social units often associate temporarily with other such units, to form groups—the reasons behind this behavior are poorly understood, but may relate to increased protection from predators. Where studied well, such as off Dominica in the

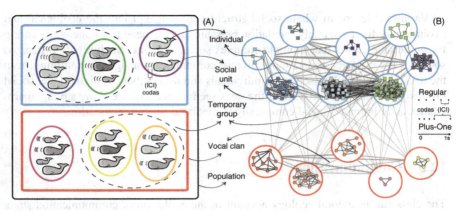

Fig. 12.3 Social patterns and culture in sperm whales, *Physeter macrocephalus*. (**a**) Schematic of sperm whale multilevel social structure. Females and young live in nearly permanent social units of tens of members (colored ellipses) for many years. Social units form temporary groups (dashed ellipses) that last from a few hours to a few days; these groups are preferentially formed among social units that share coda repertoires, forming vocal clans (colored rectangles) within the same population (black rectangle). (**b**) Empirical data from sperm whales in the Eastern Pacific. In the social network, modules of individual females (small colored nodes) are connected by social relationships (black links whose thicknesses are proportional to the time that individuals were seen together) and define social units (open colored circles). Note that in some occasions members of different social units occur together, representing temporary groups. In the overlaid acoustic network, modules of social units connected by acoustic behavioral similarity (gray links whose thicknesses are proportional to the similarity of their coda repertoires) represent vocal clans. Black dots represent clicks, and spaces represent inter-click intervals (ICI), illustrating the rhythm of typical codas of each clan. Figure adapted from Cantor et al. 2015

Caribbean, there is evidence for "bond groups," pairs of units that show social preferences for each other enduring across decades, possibly for life (Gero et al. 2015). Wherever studied, these temporary groups are composed of social units that share similar vocal repertoires (Fig. 12.3), which we return to later. Thus, we currently understand sperm whale social structure as complex and hierarchical.

Social structure is also apparently flexible, varying considerably from place to place. For example, estimates of both group and social unit sizes (excluding calves) are larger in the Pacific (groups, 25–50; units, around 12) than in the Atlantic (groups, 5–18; units, 5–6). Calves are more abundant per group in the Atlantic than in the Pacific, and this could lead to, or perhaps result from, differences of intensity and strategies of within-unit maternal and allomaternal care (Whitehead et al. 2012). The matrilineal structure of the social units tends to be less marked in the Pacific than in the Atlantic (Whitehead et al. 2012; Konrad et al. 2018). In the Pacific, groups with different vocal dialects are sympatric (Rendell and Whitehead 2003), which appears to be the case in the Atlantic as well (Gero et al. 2016b). The causes of such large-scale variation in social structure are, again, poorly understood. It is possible that elevated predation pressure by killer whales in the Pacific has selected larger group sizes (Whitehead et al. 2012).

Variations in sperm whale social structure give insight into the ecological and evolutionary forces, namely, predation pressure and the need to forage while caring for vulnerable calves that promote complex social structure. Such highly organized societies are expected to use more diverse communication signals, stemming from the need to identify themselves as individuals and as members of higher level social groups (Freeberg et al. 2012). In the next section, we explore how, among sperm whales as in humans, social complexity is mediated by communication.

12.5   Communication

The classical behavioral ecology account of the evolution of communication starts with *cues*, by-products of some activity that nonetheless provide reliable information for others to use. In sperm whales, the obvious example would be the production of echolocation buzzes indicating to a listener that the producer has encountered prey. If the use of that information by others then has selective consequences for the cue producer, the behavior can evolve into a *signal*, selected because its production changes the behavior of receivers. The sperm whale click generator evolved to form a crucial element of sperm whale sensory biology, but its inadvertent role as a producer of cues has led to it being pressed into service as a producer of communication signals.

While sperm whales could plausibly communicate using a range of modes—chemical, visual, and so forth—in the marine habitat, the acoustic modality has several advantages. Sound can propagate over hundreds of kilometers in the right conditions. By comparison, chemical communication is slow and short range, and we do not have evidence of its use among sperm whales. While visual communication is also limited underwater, sperm whales often come within visual range of each other. Such proximity associates with intense social behavior, so it is possible that body posture and movements such as jaw opening could have communicative function. Sperm whales also display aerial behaviors such as breaching and lobtailing (striking the water with the tail flukes) that result in visual and acoustic cues that appear to play a role in communication (Waters and Whitehead 1990).

Nonetheless, it is vocal communication that has been best studied and that we focus on here. Despite tens of thousands of vocalizations having been analyzed, there has been no experimental demonstration of responses in other sperm whales, through the types of playback studies that are the gold standard for research into animal communication (e.g., King and Janik 2013). Thus, while our level of knowledge about the nature and function of such interactions is below that of other species on which experiments have been performed, we still have good evidence that sperm whales pay attention to each others' vocal production.

The nasal click generator produces a diverse range of acoustic signals, including "slow clicks," "squeals," and "codas." Slow clicks (Fig. 12.2e) are intense slow-repetition clicks associated with mature males. They are more than just slow echolocation clicks, showing reverberations that give them a "ringing" quality, hence the alternative term "clang." First recorded when mature males were consorting with female groups on breeding grounds (Whitehead 2003), hypotheses about their function have revolved around mating competition. The spermaceti organs of mature males comprise up to one third of their body length compared to around one quarter in females and thus represent an exponentially greater physiological investment. Arguments have been made that this morphological and behavioral sexual dimorphism results from sexual selection "on a grand scale" (Cranford 1999). But slow clicks are also produced by males at high latitudes, where females are rare (Jaquet et al. 2001; Oliveira et al. 2013). In these contexts, slow clicks do not overlap with foraging; they tend to be produced at or near the surface (Jaquet et al. 2001, Oliveira et al. 2013), features that suggest communicative function. Thus, perhaps male-male competition cannot completely explain slow click production, and it may be that while the sexual selection hypothesis is broadly correct, the signals have also been co-opted for communication at high latitudes. By contrast, "squeals" are short bursts of clicks produced at such high repetition rates that they take on a low tonal mewing quality (Fig. 12.2f). Their production is relatively rare and strongly associated with social behavior or high arousal behavioral states (Weir et al. 2007). If squeals are a signal, they must function in particularly close range interactions. While rare, squeals could turn out to be a significant part of sperm whale communication repertoire. Squeals and slow clicks are not, however, the entire story for sperm whale acoustic communication.

Codas are stereotyped patterns of clicks that can be grouped into recognizable "types." (Coda types are distinguished based on the temporal patterning of the clicks they contain; e.g., five regularly spaced clicks are termed a "five-regular," or 5R, type; the 4 + 1 type contains four regularly spaced clicks followed by a longer pause before the final click, while a "1 + 1 + 3" coda contains longer gaps between the first two clicks followed by three clicks in quick succession.) Codas were first reported in the scientific literature by Watkins and Schevill (1977), who noted stereotyped patterns of clicks produced at the end of dives—thereby forming "codas" to the dive in the musical sense of the term.

Codas are also produced at the beginning, not just the end, of dives (e.g., Schulz et al. 2011) and most prolifically during periods of social behavior at the surface (Whitehead and Weilgart 1991; Fig. 12.2d). In tropical waters, codas are produced almost exclusively by animals in the size range consistent with mature females and immature males, rather than mature males (Marcoux et al. 2006), although males can produce them at high latitudes (Curé et al. 2013). The clicks that make up codas are also markedly different to those typically used in echolocation, as they are less powerful and more omnidirectional than echolocation clicks (Madsen et al. 2002); this is expected of a signal selected for communication as opposed to echolocation. These changes could result from inflation of air sacs within the spermaceti complex (Madsen et al. 2002). This, along with clear separation in time and space between foraging and coda production (Watwood et al. 2006)—and the prolific production during periods of close-quarter surface interactions (Weilgart and Whitehead 1993)—is why codas are generally considered to be a principal form of acoustic communication in sperm whales.

More insight into individual coda production came when acoustic size measurement methods linked the production of coda types to specific individuals (Schulz et al. 2011). The whales in one Caribbean social unit share production of the single coda type that dominates their unit's repertoire, the 1 + 1 + 3 type. Most units in this population share the 1 + 1 + 3 type, and neither individuals nor units can be identified by the way they produce this coda type (Gero et al. 2016b). Thus, the coda repertoire of a sperm whale social unit is a shared group-level signal, with sharing extending beyond the immediate social unit. This supports the notion that they are learned signals, given the presence of both closely related and unrelated individuals in these units, with learning of the 1 + 1 + 3 type influenced by individuals other than the mother (Gero et al. 2008).

There is another layer of complexity in one coda type, the 5R. This type is common around the world—it is the second most common type heard in the Eastern Caribbean (Gero et al. 2016b) and widely used in the Pacific (Weilgart and Whitehead 1993; Amano et al. 2014) and the Azores (Oliveira et al. 2016). Furthermore, it appears to be produced at the start of sequences of codas more often than expected by chance (Weilgart and Whitehead 1993). In multiple social units in the Eastern Caribbean (Gero et al. 2016b), and in different groups encountered around the Azores (Oliveira et al. 2016), whales produce this 5R coda in individually distinctive ways—with subtle variations in rhythm and tempo.

Studies of how coda vocalization varies between oceans, regions, social units, and individuals have given us new windows into sperm whale social structure at spatial scales that are difficult to observe otherwise. "Vocal clans" of sperm whales are collections of social units that share a part of the repertoire that forms a large proportion of their coda production and is readily distinguishable from the repertoires of units from other clans (Rendell and Whitehead 2003; Fig. 12.3b). The sharing of coda repertoires—belonging to a vocal clan—appears to be a significant structuring factor in sperm whale society worldwide. Coda repertoires may therefore serve as vocal markers of clan membership (Rendell and Whitehead 2003). Two of the largest clans in the Eastern Tropical Pacific were those characterized by the production of "+1" codas (all with a longer final interval at the end), and hence termed the "*Plus-One*" clan, and the production of codas in which clicks are regularly spaced, hence termed the "*Regular*" clan (Rendell and Whitehead 2003; Cantor et al. 2016). In the Eastern Caribbean, a single coda type is sufficient to discriminate between the two sympatric clans—the *EC1* clan produces primarily 1 + 1 + 3 codas, while the *EC2* does not and produces predominantly 5R3 codas instead (Gero et al. 2016a). Similarly, a single coda type, the 3 + 1, dominates the coda repertoire in the Mediterranean sperm whale population (Pavan et al. 2000). Most significantly, when social units form temporary associations with other units in the Pacific, they do so with units of the same vocal clan, even though the clans are sympatric (Rendell and Whitehead 2003). This pattern is reflected in the Atlantic Ocean, where sympatric clans have been confirmed in the Eastern Caribbean (Gero et al. 2016a). But, there is variation—contrasting evidence from the North Pacific suggests that clans off Japan are more consistent with geographic rather than sympatric dialect variation (Amano et al. 2014).

Clans are not static. There were dramatic changes in population composition off the Galápagos Islands from the 1980s to the 2010s, with extensive turnover of individuals. By the end of the twentieth century, sperm whales had all but disappeared from these waters, for reasons that are not understood. After a period of absence, they came back in the early 2010s—but these were not the same sperm whales. The incomers had the vocal dialects of two other clans—one that had previously only been heard once before around the Galápagos and another that was previously known only from Chilean waters—suggesting a large-scale displacement of clans (Cantor et al. 2016). By contrast, the same coda dialect has been recorded in the Eastern Caribbean over the past 30 years, indicating that at least one of the clans present in those waters has remained in that area for the same 30-year period (Gero et al. 2016a, b). By the nature of nomadic behavior of sperm whales, clan structure can be spatially flexible; but its dialect and membership remain stable over time. While individuals, and units, may roam widely; social identity does not change.

Sperm whale codas are rich in information about the producer's identity. We understand some diversity and patterns of variation in coda production but do not understand the communication function of codas. The most direct evidence of communication may come from a study of coda exchanges, matching, and overlapping in two social units—one from the Pacific and one from the Caribbean (Schulz et al. 2008). These exchanges form duet-like sequences, in which whales synchronize timing of coda production (Schulz et al. 2008). In some birds, such overlapping is thought to be an aggressive signal (e.g., Dabelsteen et al. 1997). This does not seem to be the case here, because there were no other signs of aggressive interactions, and overlapping occurred between individuals from the same social unit that we would expect to also engage in cooperative caring of calves. Instead, these vocal exchanges could function to affirm the shared unit and/or clan membership of the whales involved (Schulz et al. 2008), in a similar manner as affiliative signals in birds (Kelley et al. 2008). Rapidly and synchronously matching a shared coda type effectively signals a shared repertoire and hence probable social affiliation (Schulz et al. 2008). There is a potential parallel between the sense of connection and belonging provoked by shared rhythmic behavior in humans (drumming, clapping, dancing) and the notion that shared rhythmic codas in sperm whales may communicate the message "we belong."

12.6 Social Learning

Rendell and Whitehead (2003) suggested that variations of sharing codas represented learned and culturally transmitted group-level vocal signatures. However, evidence for vocal learning is difficult to test in sperm whales. There are weak relationships between maternal lineage (as indicated by mitochondrial DNA that is passed down the maternal line) and coda repertoires (Whitehead et al. 1998), but these are not strong enough to explain vocal variation as a product of genetic differences. Furthermore, this correlation disappears when genetic sequence divergence is measured (Rendell et al. 2012). Therefore, the most likely explanation for coda dialect variation between groups is that coda repertoires are learned by young sperm whales, generally from their mother, who is also the source of their mitochondrial DNA. However, off Dominica, fine-scale similarity of coda repertoire is not correlated within kinship within or between social units, suggesting that if a mother's codas are the prime dialect model for the young sperm whale, then her influence is diluted by unit- or clan-level conformity (Konrad et al. 2018).

More direct evidence in favor of learning comes from the acoustic repertoire of calves, which often produces a greater diversity of, and low consistency within, coda types (Schulz et al. 2008; Gero et al. 2016a, b). This is perhaps indicative of "babbling" as in humans and other species (Gero et al. 2016a). Among the community off Dominica, calves take at least 2 years to learn to produce the stereotyped patterns of the coda types in their natal dialect (Gero et al. 2016a). Therefore, over many generations, repeated learning could have introduced small copying errors that accumulated down lineages, giving rise to the patterns of dialect diversity of today.

If divisions appear so critical to sperm whale society, how did they come to be? It is impossible to investigate historical processes empirically, but recent modeling work used populations of virtual whales with characteristics informed by known parameters of sperm whale life history. The models showed that clans with different dialects, resembling real-world ones, emerged when virtual whales learned from each other in a specific manner, conforming to the most similar individuals around them (Cantor et al. 2015). Similar processes explain vocal dialect diversity in birds and humans, and such repeated episodes of social learning (from others as opposed to individual trial and error) can give rise to a second cultural inheritance system based on information held in brains rather than genes (Whiten 2017). So, we argue, it is appropriate to describe sperm whale vocal dialects as cultural traditions.

Vocal dialects do not appear to be the only traditions in sperm whale societies, and vocal learning is not the only form of sperm whale social learning. The best evidence for nonvocal learning comes not from female groups, but from the rise of a new tradition among males foraging at high latitudes. This is the taking of fish from longlines set by fishers in the waters off south Alaska that began in the late 1990s in one place (West Yakutat). This behavior subsequently spread east and west along almost the entire south Alaska seaboard by 2010, in a pattern that fits to "wave-of-advance" models originally formulated for describing the dynamics of prehistoric human groups (Schakner et al. 2014).

While female-based clans have distinctive repertoires of coda vocalizations, they also show variation in other behavioral patterns. Groups of the most well-known clans of the Pacific—*Regular* and *Plus-One*—consistently vary in movement patterns and use of habitat and also forage and socialize differently (Whitehead and Rendell 2004; Marcoux et al. 2007a; Cantor and Whitehead 2015). Likewise, social behavior varies between clans: individuals of the *Regular* clan dive more synchronously and show briefer associations that are more evenly distributed across individuals than the members of the *Plus-One* clan (Cantor and Whitehead 2015). Thus, clan membership has implications for more than vocal dialect; clans are repositories of multiple traditions, incorporating ways of communicating, knowledge about

12 Sperm Whale: The Largest Toothed Creature on Earth

habitats and their features, and knowledge about how to manage social relationships. Multiple traditions make a culture. It therefore seems reasonable to describe clans as cultural groups. Those populations where multiple clans occur in sympatry may be termed a multicultural society.

Cultural variation can have implications for the fitness of clan members (Marcoux et al. 2007b). Since social units are driven by care of young (Gero et al. 2013), differences in social relationship quality and duration among social unit members—and surface time between foraging dives—could affect the quality of allomaternal care, and this could lead to differences in calf survival between clans (Cantor and Whitehead 2015). Likewise, differences in diet, movements, and foraging styles result in some clans having higher feeding success than others but also in being affected differently by ecological shifts, such as El Niño-Southern Oscillation events (Whitehead and Rendell 2004). Thus, cultural variations between clans could represent alternative strategies for dealing with the challenges of survival in a changing ocean.

The behavioral diversity embedded in sperm whale clans highlights an important point. If conservation has as its focus the preservation of diversity, a substantial amount of the behavioral diversity of sperm whales, and the knowledge that underpins it, might be held among the culturally defined clans, as opposed to genetic populations or arbitrary management stocks. The maintenance of such diversity could be important in the long-term resilience of sperm whale populations. These cultural traditions, honed by the environment in which they were innovated and passed on across generations of mothers, may be a significant reason why sperm whale societies survive. Thus, our long-term conservation plans should take sperm whale cultural factors into account (Whitehead 2010; Brakes et al. 2019).

We return to the notion of belonging. Belonging to a clan is crucial for a sperm whale—whether in a region inhabited by just one, or in one of the multicultural zones, belonging defines much of how sperm whales live their lives. Clans may contain tens of thousands of individuals, thus featuring among the largest mammalian cultural groups outside humans (Whitehead and Rendell 2014). The question remains unanswered of why some populations contain multiple sympatric clans and some do not. When we eventually answer this question, we may understand something quite profound about the societal essence of being a sperm whale.

12.7 Outstanding Questions

Some behavior of sperm whales fits predictions from behavioral ecology. That they stay when feeding is good and move when it is poor (Whitehead 2003) is classical optimal foraging. The fission of particularly large social units of females (Christal et al. 1998) and the fusion of small ones (Konrad et al. 2018) similarly fit theories of optimal group size. Male dispersal is the norm among mammals, and male sperm whales disperse. However, other elements of sperm whale life do not fit so neatly with the expectations of a behavioral ecologist. Why do males head for the far ends of the Earth, thousands of kilometers from the nearest female? And why are they so big? Extreme male-biased sexual dimorphism in mammals is presumed to be the result of important intermale competition for, or defense of, resources or females or young. But it is hard to envisage serious competition for, or defense of, resources by male sperm whales. There is no evidence of them defending females or young, and physical competition seems rare. So why are male sperm whales so massive? Could it be female choice?

For females, cooperative, long-term groupings seem to make sense in the difficult and dangerous world of animals always on the move in the pelagic zone. Females of many other mammalian species form cooperative groups, helping each other survive and reproduce (Clutton-Brock 2016; Lukas and Clutton-Brock 2018). The nature of cooperative groups of female mammals can be arranged on a continuum indexed by female relatedness within groups, and the position along this continuum predicts many social attributes (Lukas and Clutton-Brock 2018). With little relatedness among group members, females tend to be arranged in a dominance hierarchy, mediated by aggression. Conversely, when most group members are close kin, there tends to be division of labor, reproductive suppression, infanticide by females, and considerable alloparental care. Sperm whales, with a mean within-social-unit relatedness of about 0.14 (Konrad et al. 2018), are at the lower end of the continuum and, as with other mammals at this general level of within-group kinship, do not seem to have female infanticide, reproductive suppression, or division of labor. But they may be unusual in their apparent lack of female-female aggression and in the importance of alloparental care. Why?

The study of sperm whale behavior is undergoing a technological revolution, affording us data-rich perspectives on what they do over short timescales. Drones document social behavior at the surface, as well as measuring size, health, and collecting biological samples; tags detail each movement and each sound underwater (Fig. 12.1f); acoustic arrays (including those on animal-borne tags) capture their world of sound (Fig. 12.1g) and infer deepwater foraging behavior, communication interchanges, and maybe even prey fields; high-resolution molecular genetics allow detailed measures of kinship and pedigree; and artificial intelligence processes such as machine learning and computer vision are enabling computers to recognize individual whales from photographs or clans from acoustic recordings. These new methods allow us to address the problems of vast geographic scales and to penetrate the depths of the oceans to discern what sperm whales do below the surface, where direct observation is not currently possible, but where they spend most of their time. Methods are improving in reliability, output, ease of use, and reduced cost.

Despite new technologies, however, long-term observational studies (Fig. 12.1h) continue to provide crucial insights. Much of what we know about the lives of sperm whales come from data painstakingly collected in often very simple ways of observing, sound recording with simple hydrophones, and photographing, data amassed over years of expeditions (e.g., Whitehead 2003; Gero et al. 2016a, b; Cantor et al. 2016). These efforts plus the new ones are essential for interpreting increasingly detailed and powerful analytical methods of processing new data, with computer tools for realistically simulating the behavior of the whales.

12.8 Closing Remarks

Today we can appreciate that sperm whales, with their familial ties, learned traditions, and social organization, draw some parallels with our own societal and cultural lives. Today's researchers still face unanswered questions about the behavior of sperm whales. Today's oceans are different from those in which sperm whale foraging strategies, roaming habits, social structures, and cultures arose. Today's sperm whales, the relicts of two massive hunts in the past two centuries (Whitehead 2002), are faced with increasing human presence, are confronted with entanglement in fishing gear, are run down or harassed by shipping fleets, encounter plastics masquerading as prey, and are subject to the changes brought about by a warming, acidified ocean.

Tomorrow's ecology will be different, but tomorrow's sperm whales will still be the largest toothed creature on Earth. How much will their ways of life change because of our behavior? Tomorrow's researcher will not only be charged with figuring out what it means to be a sperm whale but also, and perhaps more importantly, what it is like to live as our neighbors on this shared planet. Armed with knowledge born of necessity, tomorrow's researchers will need to find collaborative solutions to enable sperm whale survival so that this generation's whales can pass on their cultural inheritance to their calves in a healthy ocean.

Acknowledgments We thank the researchers and volunteers involved in 30+ years of collaboration on land and at sea. MC received a PMP/BS postdoctoral fellowship (UFPR/UNIVALI 46/2016). SG is supported by a Villum Foundation technical and scientific research grant.

References

Amano M, Kourogi A, Aoki K, Yoshioka M, Mori K (2014) Differences in sperm whale codas between two waters off Japan: possible geographic separation of vocal clans. J Mammal 95:169–175

Beale T (1839) The natural history of the sperm whale. John Van Voorst, London

Best PB (1979) Social organization in sperm whales, *Physeter macrocephalus*. In: Behavior of marine animals. Springer, Boston, pp 227–289

Brakes P, Dall SRX, Aplin LM, Bearhop S, Carroll EL, Ciucci P, Fishlock V, Ford JKB, Garland EC, Keith SA, McGregor PK, Mesnick SL, Noad MJ, Notarbartolo di Sciara G, Robbins MM, Simmonds MP, Spina F, Thornton A, Wade PR, Whiting MJ, Williams J, Rendell L, Whitehead H, Whiten A, Rutz C (2019) Animal cultures matter for conservation. Science 363 (6431):1032–1034

Brent LJ, Franks DW, Foster EA, Balcomb KC, Cant MA, Croft DP (2015) Ecological knowledge, leadership, and the evolution of menopause in killer whales. Curr Biol 25:746–750

Cantor M, Whitehead H (2015) How does social behavior differ among sperm whale clans? Mar Mam Sci 31:1275–1290

Cantor M, Shoemaker LG, Cabral RB, Flores CO, Varga M, Whitehead H (2015) Multilevel animal societies can emerge from cultural transmission. Nat Commun 6:8091

Cantor M, Whitehead H, Gero S, Rendell L (2016) Cultural turnover among Galápagos sperm whales. R Soc Open Sci 3:160615

Carrier DR, Deban SM, Otterstrom J (2002) The face that sank the Essex: potential function of the spermaceti organ in aggression. J Exp Biol 205:1755–1763

Christal J, Whitehead H (2001) Social affiliations within sperm whale (*Physeter macrocephalus*) groups. Ethology 107:323–340

Christal J, Whitehead H, Lettevall E (1998) Sperm whale social units: variation and change. Can J Zool 76:1431–1440

Clarke MR (1970) Function of the spermaceti organ of the sperm whale. Nature 228:873

Clutton-Brock T (2016) Mammal societies. Wiley, Hoboken, NJ

Cranford TW (1999) The sperm whale's nose: sexual selection on a grand scale? Mar Mam Sci 15:1133–1157

Curé C, Antunes R, Alves AC, Visser F, Kvadsheim PH, Miller PJ (2013) Responses of male sperm whales (*Physeter macrocephalus*) to killer whale sounds: implications for anti-predator strategies. Sci Rep 3:1579

Dabelsteen T, McGregor PK, Holland JO, Tobias JA, Pedersen SB (1997) The signal function of overlapping singing in male robins. Anim Behav 53:249–256

Fais A, Aguilar Soto N, Johnson M et al (2015) Sperm whale echolocation behaviour reveals a directed, prior-based search strategy informed by prey distribution. Behav Ecol Sociobiol 69:663–674

Fais A, Johnson M, Wilson M et al (2016) Sperm whale predator-prey interactions involve chasing and buzzing, but no acoustic stunning. Sci Rep 6:1–13

Freeberg TM, Dunbar RI, Ord TJ (2012) Social complexity as a proximate and ultimate factor in communicative complexity. Philos Trans B 367:1785–1801

Gero S, Gordon JCD, Carlson C et al (2007) Population estimate and inter-island movement of sperm whales, *Physeter macrocephalus*, in the Eastern Caribbean. J Cetacean Res Manag 9:143–150

Gero S, Engelhaupt D, Whitehead H (2008) Heterogeneous social associations within a sperm whale, *Physeter macrocephalus*, unit reflect pairwise relatedness. Behav Ecol Sociobiol 63:143–151

Gero S, Engelhaupt D, Rendell L, Whitehead H (2009) Who cares? Between-group variation in alloparental caregiving in sperm whales. Behav Ecol 20:838–843

Gero S, Gordon J, Whitehead H (2013) Calves as social hubs: dynamics of the social network within sperm whale units. Proc R Soc B 280:20131113

Gero S, Gordon JCD, Whitehead H (2015) Individualized social preferences and long-term social fidelity between social units of sperm whales. Anim Behav 102:15–23

Gero S, Whitehead H, Rendell L (2016a) Individual, unit, and vocal clan level identity cues in sperm whale codas. R Soc Open Sci 3:150372

Gero S, Bøttcher A, Whitehead H, Madsen PT (2016b) Socially segregated, sympatric sperm whale clans in the Atlantic Ocean. R Soc Open Sci 3:160061

Hanselman DH, Pyper BJ, Peterson MJ (2018) Sperm whale depredation on longline surveys and implications for the assessment of Alaska sablefish. Fish Res 200:75–83

Jaquet N (1996) How spatial and temporal scales influence understanding of sperm whale distribution: a review. Mammal Rev 26:51–65

Jaquet N, Dawson S, Douglas L (2001) Vocal behavior of male sperm whales: why do they click? J Acoust Soc Am 109:2254–2259

Kawakami T (1980) A review of sperm whale food. Sci Rep Whales Res Inst 32:199–218

Kelley LA, Coe RL, Madden JR, Healy SD (2008) Vocal mimicry in songbirds. Anim Behav 76:521–528

King SL, Janik VM (2013) Bottlenose dolphins can use learned vocal labels to address each other. Proc Natl Acad Sci 110:13216–13221

Konrad C, Gero S, Frasier T, Whitehead H (2018) Kinship influences sperm whale social organization within, but generally not among, social units. R Soc Open Sci 5:180914

Lavery TJ, Roudnew B, Gill P, Seymour J, Seuront L, Johnson G, Mitchell JG, Smetacek V (2010) Iron defecation by sperm whales stimulates carbon export in the Southern Ocean. Proc Soc R B 277(1699):3527–3531

Lukas D, Clutton-Brock T (2018) Social complexity and kinship in animal societies. Ecol Lett 21:1129–1134

Lyrholm T, Leimar O, Johanneson B, Gyllensten U (1999) Sex-biased dispersal in sperm whales: contrasting mitochondrial and nuclear genetic structure of global populations. Proc Soc R B 266:347–354

Madsen PT, Wahlberg M, Møhl B (2002) Male sperm whale (Physeter macrocephalus) acoustics in a high-latitude habitat: implications for echolocation and communication. Behav Ecol Sociobiol 53:31–41

Marcoux M, Whitehead H, Rendell L (2006) Coda vocalizations recorded in breeding areas are almost entirely produced by mature female sperm whales (Physeter macrocephalus). Can J Zool 84:609–614

Marcoux M, Whitehead H, Rendell L (2007a) Sperm whale feeding variation by location, year, social group and clan: evidence from stable isotopes. Mar Ecol Prog Ser 333:309–314

Marcoux M, Rendell L, Whitehead H (2007b) Indications of fitness differences among vocal clans of sperm whales. Behav Ecol Sociobiol 61:1093–1098

Melville H (1851) Moby-Dick; or, the whale. Harper Brothers Press, London, UK

Miller PJ, Johnson MP, Tyack PL (2004) Sperm whale behaviour indicates the use of echolocation click buzzes 'creaks' in prey capture. Proc R Soc B 271:2239–2247

Mizroch SA, Rice DW (2013) Ocean nomads: distribution and movements of sperm whales in the North Pacific shown by whaling data and discovery marks. Mar Mamm Sci 29:E136–E165

Møhl B, Wahlberg M, Madsen PT et al (2000) Sperm whale clicks: directionality and source level revisited. J Acoust Soc Am 107:638–648

Møhl B, Wahlberg M, Madsen PT, Heerfordt A, Lund A (2003) The monopulsed nature of sperm whale clicks. J Acoust Soc Am 114:1143–1154

Norris KS, Møhl B (1983) Can odontocetes debilitate prey with sound? Am Nat 122:85–104

Oliveira C, Wahlberg M, Johnson M, Miller PJ, Madsen PT (2013) The function of male sperm whale slow clicks in a high latitude habitat: communication, echolocation, or prey debilitation? J Acoust Soc Am 133:3135–3144

Oliveira C, Wahlberg M, Silva MA et al (2016) Sperm whale codas may encode individuality as well as clan identity. J Acoust Soc Am 139:2860–2869

Pavan G, Hayward TV, Borsani JF, Priano M, Manghi M, Fossati C, Gordon J (2000) Time patterns of sperm whale codas recorded in the Mediterranean Sea 1985–1996. J Acoust Soc Am 107:3487–3495

Rendell L, Whitehead H (2003) Vocal clans in sperm whales (Physeter macrocephalus). Proc R Soc B 270:225–231

Rendell L, Mesnick SL, Dalebout ML, Burtenshaw J, Whitehead H (2012) Can genetic differences explain vocal dialect variation in sperm whales, Physeter macrocephalus? Behav Genet 42:332–343

Rødland ES, Bjørge A (2015) Residency and abundance of sperm whales (Physeter macrocephalus) in the Bleik Canyon. Norway Mar Biol Res 11:974–982

Schakner ZA, Lunsford C, Straley J, Eguchi T, Mesnick SL (2014) Using models of social transmission to examine the spread of longline depredation behavior among sperm whales in the Gulf of Alaska. PLoS One 9:e109079

Schulz T, Whitehead H, Gero S, Rendell L (2008) Overlapping and matching of codas in vocal interactions between sperm whales: insights into communication function. An Behav 76:1977–1988

Schulz TM, Whitehead H, Gero S, Rendell L (2011) Individual vocal production in a sperm whale (Physeter macrocephalus) social unit. Mar Mamm Sci 27:149–166

Teloni V, Mark JP, Patrick MJO, Peter MT (2008) Shallow food for deep divers: dynamic foraging behavior of male sperm whales in a high latitude habitat. J Exp Mar Biol Ecol 354:119–131

Tønnesen P, Gero S, Ladegaard M, Johnson M, Madsen PT (2018) First year sperm whale calves echolocate and perform long, deep dives. Behav Ecol Sociobiol 72:165

Waters S, Whitehead H (1990) Aerial behaviour in sperm whales, *Physeter macrocephalus*. Can J Zool 68:2076–2082

Watkins WA, Schevill WE (1977) Sperm whale codas. J Acoust Soc Am 62:1485–1490

Watwood SL, Miller PJ, Johnson M, Madsen PT, Tyack PL (2006) Deep-diving foraging behaviour of sperm whales (*Physeter macrocephalus*). J An Ecol 75:814–825

Weilgart L, Whitehead H (1993) Coda communication by sperm whales (*Physeter macrocephalus*) off the Galapagos Islands. Can J Zool 71:744–752

Weir CR, Frantzis A, Alexiadou P, Goold JC (2007) The burst-pulse nature of 'squeal' sounds emitted by sperm whales (*Physeter macrocephalus*). J Mar Biol Assoc UK 87:39–46

Werth AJ (2004) Functional morphology of the sperm whale (*Physeter macrocephalus*) tongue, with reference to suction feeding. Aq Mamm 30:405–418

Whitehead H (2002) Estimates of the current global population size and historical trajectory for sperm whales. Mar Ecol Prog Ser 242:295–304

Whitehead H (2003) Sperm whale societies: social evolution in the ocean. University of Chicago Press, Chicago

Whitehead H (2010) Conserving and managing animals that learn socially and share cultures. Learn Behav 38:329–336

Whitehead H (2016) Consensus movements by groups of sperm whales. Mar Mamm Scie 32:1402–1415

Whitehead H, Weilgart LS (1991) Patterns of visually observable behaviour and vocalizations in groups of female sperm whales. Behaviour 118:275–296

Whitehead H, Rendell L (2004) Movements, habitat use and feeding success of cultural clans of South Pacific sperm whales. J Anim Ecol 73:190–196

Whitehead H, Rendell L (2014) The cultural lives of whales and dolphins. University of Chicago Press, Chicago

Whitehead H, Dillon M, Dufault S, Weilgart L, Wright J (1998) Non-geographically based population structure of South Pacific sperm whales: dialects, fluke-markings and genetics. J Anim Ecol 67:253–262

Whitehead H, Coakes A, Jaquet N, Lusseau S (2008) Movements of sperm whales in the tropical Pacific. Mar Ecol Prog Ser 361:291–300

Whitehead H, Antunes R, Gero S, Wong SN, Engelhaupt D, Rendell L (2012) Multilevel societies of female sperm whales (*Physeter macrocephalus*) in the Atlantic and Pacific: why are they so different? Int J Primatol 33:1142–1164

Whiten A (2017) A second inheritance system: the extension of biology through culture. Interface Focus 7:20160142

Chapter 13
Pilot Whales: Delphinid Matriarchies in Deep Seas

Jim Boran and Sara Heimlich

Abstract Pilot whales are distributed worldwide with two species, the cold-water long-finned pilot whale, *Globicephala melas*, and the warm-water short-finned pilot whale, *G. macrorhynchus*. Long-finned pilot whales have an anti-tropical distribution in separated northern and southern hemisphere groupings, with the North Pacific group extinct, while short-finned pilot whales occur in all tropical oceans. These two species exhibit different life history patterns, with long-finned pilot whales maturing faster and reproducing up to the end of life and dying sooner, while short-finned pilot whales have long maturation (esp. for males) and a post-reproductive phase where females live up to 30 years beyond the birth of their last calf. Photo-identification studies of short-finned pilot whales in Macaronesia (the Canary Islands, Azores, and Madeira) show long-term relationships maintained over hundreds of kilometers and exhibit a wide variety of site fidelity patterns. Mixing between core residents and visiting transients in high productivity areas suggest more fluid interactions than those observed for resident ecotype killer whales or orca (*Orcinus orca*) that maintain isolated groups from other, co-occurring ecotypes. Four features of orca matrilineal society, (1) matrilineal associations between females, (2) male natal philopatry, (3) extra-group mating or natal exogamy, and (4) post-reproductive lifespan, are compared with short-finned pilot whales, and we conclude that pilot whale societies have numerous matrilineal features, albeit less specialized than those exhibited by orca. Divergent prey preferences most likely account for these differences.

Keywords Pilot whale · Photo-ID · Matrilineal · Orca · Macaronesia

---

J. Boran (✉)
University of Manchester, Manchester, UK
e-mail: jim.boran@manchester.ac.uk

S. Heimlich
Hatfield Marine Science Center, Oregon State University, Newport, OR, USA

© Springer Nature Switzerland AG 2019
B. Würsig (ed.), *Ethology and Behavioral Ecology of Odontocetes*, Ethology and Behavioral Ecology of Marine Mammals,
https://doi.org/10.1007/978-3-030-16663-2_13

13.1 Introduction

Pilot whales are distributed around much of the world, with two species subdividing the oceans into tropical and temperate waters (Fig. 13.1; Olson 2018). The genus *Globicephala* was first identified by Lesson (1828) and evolved within the family Delphinidae (LeDuc et al. 1999), along with the other dolphins and the killer whale or orca (*Orcinus orca*). Pilot whales are most closely related to the subfamily Globicephalinae that includes Risso's dolphins, false killer whales, pygmy killer whales, and the melon-headed whale. *Globicephala melas*, the long-finned pilot whale (Traill 1809), and *Globicephala macrorhynchus*, the short-finned pilot whale (Gray 1846), are mostly geographically isolated, but there are a number of areas where the two species overlap (Nores and Peréz 1988; van Bree et al. 1978).

Long-finned pilot whales are a cold-water species with an anti-tropical distribution between two groupings in the northern and southern hemispheres (Davies 1963). Long-finned populations of the southern hemisphere have a circumpolar distribution around Antarctica (highest latitude sightings: 63° S along the ice edge) and the southern tips of South America, Africa, and Australasia (van Waerebeek et al. 2010). In the northern hemisphere, long-finned pilot whales are only found in the North Atlantic. Long-finned pilot whales in different hemispheres share two of three haplotypes, but overall haplotype frequencies were still significantly different

Fig. 13.1 Nine short-finned pilot whales, *Globicephala macrorhynchus*, socialize in the waters off Tenerife, Canary Islands. Social relationships are strengthened through close body contact. This behavior increased in occurrence during summer months off Tenerife, perhaps when less familiar animals associated (Servidio 2014). © Teo Lucas

(Oremus et al. 2009), suggesting that any gene flow between these major groupings is restricted. There was a historical population in the North Pacific that appears to have gone extinct (Kasuya 1975). Long-finned pilot whales have been the subject of drive-fishery whaling in the North Atlantic in the Faroe Islands, where at times hundreds of boats herd pilot whale groups into bays for slaughter, using loud percussive-pipe-banging noises (Donovan et al. 1993; Gibson-Lonsdale 1990).

Short-finned pilot whales occur in tropical and warm temperate waters of the central Atlantic, Pacific, and Indian Oceans. They typically range between a maximum of 40° N and 40° S latitude, favoring water temperatures above 15 °C (Fullard et al. 2000). Short-finned pilot whales have also been the subject of drive-fishery whaling in the Pacific Ocean off Japan (Kasuya and Marsh 1984; Marsh and Kasuya 1984). Mixing between the Atlantic and Pacific populations is apparently limited by continental land masses, but with only 0.25% genetic divergence between Pacific and Atlantic short-finned pilot whales (Siemann 1994), suggesting that some mixing occurs, probably between the Indian and Atlantic Oceans at the Cape of Good Hope, South Africa (Findlay et al. 1992; van Bree et al. 1978).

Comparison of life history parameters found long-finned pilot whale males matured 3–7 years earlier than short-finned pilot whale males (11–16 years vs. 14–23 years; Kasuya et al. 1988). Long-finned pilot whales also have shorter lives (females, 42 years; males, 33 years) than short-finned pilot whales (females, 63 years; males, 46 years; Kasuya et al. 1988). In spite of their shorter lives, male long-finned pilot whales grow larger than short-finned males (5.4 vs. 4.5 m) and exhibit a larger size sexual dimorphism (1.43 vs. 1.27 male/female size ratios: Möller et al. 2012). Female long-finned pilot whales did not exhibit a strong post-reproductive period, with 96% reproducing until death (Kasuya et al. 1988; Martin and Rothery 1993; Pavelka et al. 2018) compared to 25% of short-finned pilot whale females who may live up to 30 years (mean = 14 years) beyond the birth of their last calf (Marsh and Kasuya 1984, 1986).

These life history features will likely affect the roles of post-reproductive females and adult males, which may in turn affect a range of other features of society such as social learning (Boran and Heimlich 1999) and culture (Whitehead and Rendell 2014). These features may depend on the importance of leadership roles for either post-reproductive females or adult males, which could be different between the two species.

The ecology of pilot whales is defined by their preference for specific temperatures (Fullard et al. 2000) and deep water environments around the 1000 m depth contour with bathymetric features along continental slopes and shelf edges and around oceanic archipelagos (Baird et al. 2002; Heide-Jørgensen et al. 2002; Mate et al. 2005; Wells et al. 2013) where they can locate their preferred prey: squid (Desportes and Mouritsen 1988; Gannon et al. 1997; Santos et al. 2014). Aguilar Soto et al. (2008) recorded acoustic behavior and dive profiles of 23 whales off Tenerife, Canary Islands, and documented hunting strategies with downward vertical sprints up to 6 m/s at the bottom of daytime deep dives (maximum depth: 1018 m), followed by sonar buzzing for deeper prey and eventually leveling out for prey capture adapted to large, fast, high-calorie squid prey. These amazing sprints to depths have earned them the nickname "cheetahs of the sea," and we urge a reading of the Aguilar Soto et al. (2008) article.

Further understanding of pilot whales has come from detailed, long-term studies of individuals in areas of predictable occurrence. These studies have been carried out on short-finned pilot whales around the Hawaiian Islands (Baird 2016; Mahaffy et al. 2015; van Cise et al. 2017) and long-finned pilot whales off Nova Scotia, Canada (Augusto et al. 2017; Augusto 2017; Ottensmeyer and Whitehead 2003), and the Straits of Gibraltar (de Stephanis et al. 2008; Verborgh et al. 2009, 2016).

13.2 Short-Finned Pilot Whales

With this brief introduction to the two species of pilot whales, we now explore some older and more recently-gathered details on the more tropical species, the short-finned pilot whale. We hope that the reader will appreciate the iterative advances of science, from almost 30 years ago to the present, and our increasing sophistication in gathering, analyzing, and interpreting long-term data sets. We argue that multi-decadal investigations are necessary for a more appropriate understanding of long-lived marine mammals (and other biological entities).

In 1989, we started a photo-ID study in the Canary Islands off the northwest coast of Africa (Heimlich-Boran 1993; Fig. 13.2). Our goal was to collect behavioral data

Fig. 13.2 Short-finned pilot whales, *Globicephala macrocephalus*, in their core area off the southwest coast of Tenerife, Canary Islands. Mount Teide, the tallest point of the Canaries, is 3718 m above sea level and drops to the 1000 m depth contour within a few kilometers from shore. © Boran/Heimlich

13 Pilot Whales: Delphinid Matriarchies in Deep Seas

to help understand biological data collected from whaling and strandings, as well as to explore the similarities between pilot whale and killer whale social organization as related large delphinids. We had been studying killer whales, also called orca (*Orcinus orca*), in the inshore waters of Washington state and British Columbia since 1976 (Balcomb et al. 1982) using photo-ID and behavioral observations of known individuals to explore behavioral ecology and matrilineal social structure (Bigg et al. 1990; Ford et al. 2000; Heimlich-Boran 1986a, b, 1988a, b). Applying a similar methodology to study pilot whales seemed a logical approach for investigating the hypothesis that pilot whales may have a similar social organization to orca. Of course, the pilot whales' behavioral ecology may also be significantly different due to specialized habitat and prey preferences. For one, orcas are not the amazing divers that pilot whales are.

Since the end of our research on pilot whales, a number of projects have been conducted in the eastern North Atlantic where short-finned pilot whales congregate around islands with nearshore deep waters such as the Canary Islands (Hartny-Mills 2015; Servidio 2014), Madeira (Alves 2013; Alves et al. 2013, 2015), and the Azores (Mendonça 2012; Silva et al. 2003, 2014).

This chapter presents a case study of the development of short-finned pilot whale research in the eastern North Atlantic and reviews the latest photographic identification research on a large population of pilot whales, to understand details of behavioral adaptations to their oceanographic habitat, and opportunities and constraints that may have led to a matriarchal social system. We explore four features of the matrilineal society of orca and examine their applicability to our understanding of pilot whales: (1) matrilineal associations between females, (2) natal philopatry of adult males, (3) natal group exogamy or extra-group mating outside the group, and (4) female post-reproductive lifespan.

13.2.1 Canary Islands

The Canary Islands are located in the northeast Atlantic Ocean off the coast of Morocco, northwest Africa, and spread over 500 km between 27° 37′ and 29° 23′ N latitude and 13° 30′ and 18° 16′ W longitude. The waters here are cooler than might be expected for such latitudes, ranging from 18 to 28 °C, due to the cold, south-flowing Canary Current (Barton et al. 1998). There have been several major pilot whale photo-ID studies in these islands for nearly 30 years, since 1989. As methods have developed, some elements of studies are incomparable, but together they form a picture of the social structure of pilot whales in this region.

Our research project started off the island of Tenerife in September 1989. Surveys were conducted on 200 days for 1131 h over a 22 month period between October 1989 and July 1991 (Heimlich-Boran 1993). There were 495 whales identified. Eight years after we completed our study, Antonella Servidio (2014) began her PhD research project with support from the Sociedad para el Estudio de los Cetáceos en el Archipiélago Canarias (SECAC) across the wider Canary Islands. For 13 years,

Fig. 13.3 Short-finned pilot whale, *Globicephala macrorhynchus*, adult matriarch, Splitfin, AKA Indio, and Gma029, first seen in 1989 (left) and last seen in 2017 (right), 28 years later (Hartny-Mills 2015; Vidal Matrin, SECAC pers. comm.). Splitfin was classified as "other" in 1989 because it had no adult male characteristics and no calves. No male characteristics have been observed and a lack of calves suggests she could be a post-reproductive matriarch. Photo: © Boran/Heimlich (1989); © SECAC (2017)

from 1999 to 2012, she conducted surveys covering over 110,000 km, surveying for 70,618 km on effort over 5436 h across all Canary Islands. Servidio analyzed 1310 well-marked individuals from 1999 to 2012. Six years after Servidio started her study, Lauren Hartny-Mills (2015) began a 4-year photo-ID study off Tenerife for her PhD research. Observations were conducted between July 2005 and July 2008 and identified 382 whales. This work was carried out using Tenerife whale-watching boats with experienced observers, as platforms of opportunity and employing "citizen science" to get the general public to submit photos of animals for identification. Photographs were collected on 2133 whale-watching excursions over 555 days.

We report here for the first time on 41 matches between individuals in our 1989–1991 catalog and the two subsequent studies. There were 23 individuals reidentified during Hartny-Mills' study (2005–2008) and 25 individuals reidentified during Servidio's study (1999–2012), with 7 of those seen in all 3 studies. Full sighting histories have not been compiled, but this gives a minimum "residency" of 14 years for Tenerife pilot whales. Along with opportunistic sightings in September 2017 of one distinctive individual with a dorsal fin injury (named "Splitfin"/"Indio"/"GMA029" in different studies: Fig. 13.3), there is now history of the same individual pilot whales occurring off the SW coast of Tenerife for 28 years (V. Martín pers. comm.).

The 41 matched whales were 6 males, 16 mothers, 16 "others," 1 juvenile, and 2 distinctively marked calves, now fully grown (Heimlich-Boran 1993). The whales came from 25 different pods (as defined by Heimlich-Boran 1993), with matches ranging from one to six individuals per pod, including two pairs of constant companions with an association coefficient of 1.0. This shows that the two individuals of each pair were always seen together in the 1990s and have maintained these relationships over at least 15 years. Thirty-five of the 41 matched whales were classified as "resident" in the 1990 study, but 6 whales were from 6 "visitor" pods

13 Pilot Whales: Delphinid Matriarchies in Deep Seas

who were only seen once in the 1990s. But even including these visitors seen only once, matched whales were seen on an above average number of sighting days (8.8 ± 7.1 vs. 5.0 ± 5.7 average days per individual) during the 1990 study. This suggests that the longer-term site fidelity patterns of "resident" and "visitor" are much more fluid than we had suspected. At least some "visitors" in 1990 returned to the study area at intervals greater than the 21-month duration of that study (Heimlich-Boran 1993).

Some resightings had locations that give an understanding of long-term habitat use. In 1999, eight of the animals were resighted off Gran Canaria, an island 90 km to the east of Tenerife. In 2004, 14 of the whales, including five of the 1999 animals, were resighted off southwest Tenerife in the whales' core area. Two of the 2004 whales were seen again in 2006 off La Gomera, 38 km across the channel from Tenerife. The same 14 whales were sighted again in 2009 in their Tenerife core area. The final resighting was in 2012 of two whales seen off northern Tenerife (Servidio 2014).

Our original analysis (Heimlich-Boran 1993) looked at simple sighting histories, but statistical techniques are now available to better analyze patterns of site fidelity. The 495 animals from the 1990s catalog were seen a varying number of times: 186 animals (38%) were seen on only 1 day, 1 highly distinctive individual was photographed on 28 days, and the remaining 308 animals had intermediate degrees of occurrence (average: 5.3 ± 6.0 days).

To better understand resident associations, we classified 79 of the 186 animals seen only once as residents because they were observed travelling in association with resident whales. Our interest at the time was maximizing contact with regularly occurring whales. The other animals seen only once, but only in discrete, independent groups, were classified as "visitors" in 15 "pods" of the animals in association on that 1 day. This is clearly different from the "pods" identified by cluster analysis.

Modern methods have considered these sighting histories separately to reveal a potentially wider variety of residency and site fidelity patterns (Whitehead 2008). Mahaffy (2012) rightly considered that Heimlich-Boran (1993) "ignored temporal components by including animals seen only once within the 'resident' category." This was a valid critique and has made some of our data noncomparable with other studies that have defined three categories of residency: (1) animals seen in a majority of study years or seasons, e.g., five times or more in at least 3 years and three seasons, (2) those animals seen only once over the duration of the study, and (3) an intermediate category between these two extremes (Alves 2013; Hartny-Mills 2015; Mahaffy et al. 2015; Servidio 2014).

Servidio (2014) defined four patterns of residency: (1) 50 "core residents" seen at least twice over a 4-year period and during all four seasons of the years, (2) 255 "residents" seen over 3–4 years and during two seasons, (3) 780 "transients" seen one or two times in 1 year or in one season, and (4) 156 "occasional visitors" of all other animals seen more than transients and less than residents. Off the islands of Fuerteventura and Lanzarote, 263 transients and 10 occasional visitors were seen. Core residents were primarily seen off southwest Tenerife (35 of 50), 12 covered both Tenerife and La Gomera, and another crossed between Tenerife and Gran

Canaria. Only two animals appeared to occur across all three islands. Nearly all individuals (134 of 137) resident or core resident off La Gomera and Gran Canaria were also seen off Tenerife. Thus, the core area off Tenerife appears to be the primary base for pilot whales exploring the surrounding Canary Islands.

The distribution of pilot whale sightings pooled into 1 km^2 grid cells showed a similar core area off southwest Tenerife along the 1000 m depth contour (Hartny-Mills 2015) to that area the whales were using in the early 1990s (Heimlich-Boran 1993). Whales were sighted in ocean depths between 800 and 1200 m with a mean depth of 1035 m. Whales were observed closest to shore off Tenerife when compared to off the other islands (Hartny-Mills 2015), indicating an important affiliation to this island.

Over one-half of all pilot whale sightings (58%) were off southwest Tenerife, with an encounter rate of 1.95 sightings/km of effort. The remaining encounters were dispersed across the Canaries with a significantly lower sighting rate of 1.14 sightings/km. There was one additional cluster of sightings of southwest La Gomera, where over 10% of encounters occurred (Servidio 2014). This was an area repeatedly surveyed during 1989–1991, but pilot whales were never observed in this area (Heimlich-Boran 1993).

Servidio (2014) examined seasonal distribution of encounters off Tenerife and found significantly larger groups in summer and autumn than during winter and spring. No other islands showed significant differences in seasonal occurrence. One of the few significant variations in behavior was an increase in "socializing" during the summer months off Tenerife, as might be expected if animals less familiar with each other are associating together (Servidio 2014), and much mating may be occurring in these larger groups.

Servidio (2014) used lagged identification rates in her analysis of pilot whale associations to investigate movements between islands. Analysis for data on interchange between the islands utilized interisland matches of a subset of 1241 individuals and she hypothesized that an Emigration + Re-immigration model (Whitehead 2007) best fit the data and predicted a core population of 254 that stayed in the Canary Islands for an average of 531 days and then left the area to return after 4087 days (11.2 years). Analysis of just the Tenerife sightings supported a similar model and estimated there were 279 core animals (out of a population of 717) that stayed off Tenerife for an average of 2049 days (5.6 years).

Servidio (2014) also calculated detailed social networks (Newman and Girvan 2014) for the core residents and residents observed off Tenerife and found that all core residents and nearly all residents were grouped in a central cluster of 549 animals which contained all age/sex classes (Servidio 2014). There were also 328 animals in 30 "satellite" clusters, 21 of which were composed exclusively of transients. However, nine clusters contained other classes: three clusters contained residents and occasional animals with transients and another six were of occasional animals and transients. Six of these mixed clusters had clear links to other islands: some travelled between Tenerife and La Gomera and others were predominately from Lanzarote to Fuerteventura. Additionally, there were 166 transients connected to some of the core clusters, although always with peripheral associations to the

13 Pilot Whales: Delphinid Matriarchies in Deep Seas

residents and core residents. These links were the result of associations between individual core residents and transients, showing that these classes of individuals are not as completely isolated as Heimlich-Boran (1993) proposed between residents and visitors. In fact, there were a number of examples of open associations where individuals had strong pair associations with an individual in their own cluster while also showing a strong association with an individual in a completely different cluster. This is considerably more fluid than has been observed in orca (Ford et al. 2000; Chap. 11).

In addition to network analysis, cluster analysis of association data defined 19 groups. Fourteen of these clusters had both adult males and females, but three clusters were lacking females and only included adult males and "indeterminates" (either females without a calf or adolescent males) and could have been all-male groups. Eight clusters included animals seen off both Tenerife and La Gomera, while another eight occurred predominately off Gran Canaria. Some of these multi-island individuals were seen multiple times up to 12 years, so they regularly returned to the region. Specific individuals had high levels of association that were significantly different from random and defined a "strongly differentiated" society (Servidio 2014).

The resulting picture is of pilot whales using the Canary Islands in complex ways, some maintaining an area as core residents off southwest Tenerife and others centering on other islands and associating with Tenerife core residents in specific pair associations. However, the links with adjacent islands show that the entire archipelago is potential territory for these animals, even though they were highly likely to be resighted off Tenerife. Alternatively, although core residents were seen multiple times, one model predicted that they may only stay in the area for 500 days and then leave and not return for 4000 days (11 years). However, when analyzing lagged association rates specifically for Tenerife residents, the estimate of residency duration was 20,059 days, or a remarkable 54.6 years (Servidio 2014)!

Servidio (2014) also reported on 11 matches between pilot whales in the Canaries and Madeira, 300 km apart, and three animals travelled back and forth three times during 2004. One pod was even sighted in both archipelagos over a 20-day period. The longest gap between resightings was 7 years (Servidio et al. 2007). These links between oceanic islands across hundreds of kilometers have highlighted the importance of international comparison of photo-identification catalogs to fully understand the distribution and behavior of short-finned pilot whales.

13.2.2 Madeira

Short-finned pilot whales off the Madeiran Islands archipelago (37.8° N, 17° W) were studied between 2003 and 2011 (Alves 2013; Alves et al. 2013, 2015). This study combined genetic biopsy sampling with photo-identified individuals, giving the potential of contributing to a better understanding of pilot whale social relationships.

A total of 347 well-marked individuals were identified. Resighting rates were low: 261 whales (71%) were seen only once over the 9 years of the study and were called "transients." Of the 103 animals seen more than once, 85 were seen in multiple years (range: 2–9 years; 3.9 ± 2.2 years). Longer-term persistence around Madeira was extended by opportunistic photographs of one individual initially sighted in 1997 and resighted 14 years later off Madeira (Alves et al. 2013).

Two additional classes were defined: (1) residents seen five times or more in at least 3 years of the study and over at least three seasons of the year and (2) visitors seen more than once but less frequently than the restrictions used to define residents, perhaps with higher rates of emigration and immigration than the residents. These three categories (including transients) of site fidelity are comparable to those used by Hartny-Mills (2015) and Servidio (2014), but different from our two visitor/resident categories (Heimlich-Boran 1993).

Social network analysis for 344 whales identified 124 individuals in a "core" social network composed of all 39 residents and primarily residents and visitors (Alves 2013). An additional 220 animals were in 44 "satellite" clusters, 85% of which were composed solely of transients. But, visitors occurred in seven of the satellite clusters, placing them in both core and satellite clusters.

Hierarchical cluster analysis identified eight "pods" based on association indices above 0.5 (i.e., whales were seen more than half of the time together). There were five resident and three transient pods of closely associating mixed age-sex groups with an average of 15 individuals each (Alves et al. 2013). Alves defined three "clans" (following de Stephanis et al. 2008) or pods that regularly associated with one another. This is different from orca acoustic clans defined by shared discrete calls (Ford and Fisher 1983) but characterizes the higher-level associations of pods that frequently associate with one another. One clan was composed of three resident pods, another of one resident and two visitor pods, and a third clan of one resident and one visitor pod (Alves et al. 2013). These mixed membership pods emphasize the high degree of mixing between residents and visitors.

Genetic relatedness gathered with skin biopsies of 32 whales examined the extent to which association patterns were linked to relatedness. Average relatedness decreased with increasing group size, indicating that smaller groups were increasingly genetically similar, and supported the natal philopatry hypothesis for both sexes.

When comparing residents, transients, and visitors, there were no significant genetic differences, in either nuclear or mitochondrial DNA, showing that there is gene flow between different communities. Interbreeding could be occurring seasonally. Although whales were seen off Madeira in every month, i.e., year-round presence, there was seasonal variation in group size. Average group size across the year was 18 whales, but groups during May through October were larger. Mixed groups, where animals of different residency types occurred together, were observed significantly more during July through December. This variation in associations implies that residents seasonally increase limited associations with the less frequently occurring "visitors." As associations between less related individuals, mixed sightings could be considered to provide opportunities for extra-group matings.

13.2.3 Macaronesia

The final identification study is a recent examination of photographically identified short-finned pilot whales across the larger oceanic region of Macaronesia (Fig. 13.4): the Atlantic island archipelagos of Madeira, the Azores, and the Canary Islands (Alves et al. 2019). The study covered 16 years between 1999 and 2015 and combined photo-ID catalogs from Madeira, the Canaries, and the Azores (Alves et al. 2019).

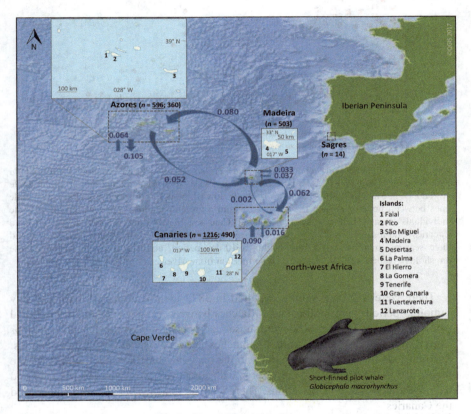

Fig. 13.4 Macaronesia is comprised of four island archipelagoes across the northeast Atlantic: the Azores, Madeira, the Canaries, and the Cape Verde Islands. Photographic and sighting data of short-finned pilot whales, *Globicephala macrorhynchus*, were obtained. *n* (underlined) indicates the number of naturally well-marked catalogued individuals used to estimate residency and transition probabilities, while the non-underlined indicates the number of additional well-marked whales used to help analyzing movement patterns. The arrows width and values illustrate heterogeneous transition probabilities between the Azores, Madeira, Canaries, and an outside area. These probabilities consider the proportion of the population moving between areas within a period of 10 days, using a parameterized Markov model. The 10-day period should allow movements between these areas, based on their distance and the mean travelling speed of 100 km per day. Illustration by Fishpics/ImagDOP. Reprinted with permission from Alves et al. (2019)

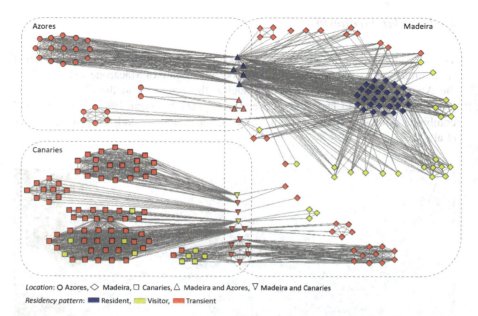

Fig. 13.5 Social network diagram for the 21 short-finned pilot whales, *Globicephala macrorhynchus*, captured in different areas of Macaronesia [9 in Madeira and the Azores (triangle up) and 12 in Madeira and the Canaries (triangle down)]. Nodes correspond to individuals, while lines between nodes represent the presence within a group. Symbol shapes and colors indicate individuals' area of capture (Azores, Madeira, Canaries, Madeira and Azores, or Madeira and Canaries) and residency pattern in each area (resident if captured ≥5 times in at least 3 years and three seasons, transient if captured within a 10-day period, and visitor if fell between these thresholds), respectively. Reprinted with permission from Alves et al. (2019)

There were 2120 individuals captured in 3872 identifications from 240 10-day sampling periods across all islands. Matches between catalogs found 21 individual whales sighted in Madeira who were subsequently resighted in adjacent archipelagos over 27 encounters. There were 12 matches between Madeira and the Canaries: 8 from Madeira to the Canaries and 4 in the opposite direction. Nine matches were found between Madeira and the Azores with four from the Azores to Madeira and five that travelled in both directions. There were no matches between the Azores and the Canaries.

Social network analysis (Newman and Girvan 2014) was conducted on 209 individuals that included the 21 matched whales, and all whales that associated in some manner with these matched whales reveal a complex pattern of mixing across regions (Fig. 13.5). All whales were classified into three categories of residency used by Alves et al. (2013) and Servidio (2014): residents seen at least five times in 3 years and in three seasons, visitors seen multiple times but less than residents, and transients seen only once. The strongest links were between Madeira and the Azores: five resident animals from Madeira travelled 1000 km back and forth to the Azores over 49 and 77 days, associating with a cluster of 15 Azorean transient whales.

Another four Madeiran transients associated independently with two small clusters of Azorean transients. Matches between Madeira and the Canary Islands occurred for two visitors and 10 transients that associated with four clusters of Canarian whales composed of 10 visitors and 67 transients (Alves et al. 2019).

Seasonal variation in whale encounters in some regions showed higher sighting rates during October through December, especially in the Azores and Madeira. Sightings in the Canaries were more evenly spread throughout the year (Alves et al. 2019). Understanding these seasonal patterns is essential to determine reasons behind interisland movements. The conclusion was that seasonal increases in sighting rate were primarily linked to an influx of new animals, as opposed to existing small groups joining. This emphasizes the emigration and immigration rates observed (Alves et al. 2019) and suggests there is a large oceanic population that may come to the islands seasonally and are recorded as transients. Alves et al. (2019) describe this as "spatial structuring" of whales across available habitats in nonrandom ways that imply that specific choices are being made because of differential benefits.

The Azores are located in one of the boundary areas of overlap between northern long-finned pilot whales and southern short-finned pilot whales (Prieto and Fernandes 2007). The observations of short-finned pilot whales in this study shows that the Azores has fewer resident animals than the Canaries or Madeira and may be primarily visited by a large, pelagic population. Matched whales from Madeira have travelled from 1000 km away, suggesting that these island archipelagos are desirable habitats. The submarine canyons off the islands form the core areas for resident animals, but there may not be enough resources to support large numbers of transient animals.

13.3 Comparative Social Structures: Are *Globicephala* Matrilineal?

Pilot whales are listed as one of the few mammals with a matrilineal social structure (Whitehead 1998), along with sperm whales (Richard et al. 1996), elephants (Lee et al. 2016; Wittemyer et al. 2009), some primates (Chapais et al. 1991), orca (Bigg et al. 1990; Heimlich-Boran 1986b, 1988b), and some humans (Behar et al. 2008). As part of our final hypothesis we want to carefully assess how our current understanding of pilot whale society compares to knowledge about orca society. One of the orcas' primary adaptations is the matrilineal society. We propose that there is more to orca society than matrilinearity, as not all of their specialized adaptations are shared by other matrilineal societies.

We explore four main features of the matrilineal society of orca and examine their applicability to our understanding of pilot whales: (1) matrilineal associations between females, (2) natal philopatry of adult males, (3) natal group exogamy or extra-group mating outside the group, and (4) female post-reproductive lifespan.

13.3.1 Matrilineal Associations with Female Relatives

To live in a matrilineal society is to live with female relatives. Ancestry follows the female line because female offspring remain with mothers and grandmothers. Female young exhibit "natal philopatry," by remaining with the group into which they were born. This occurs in many female mammals (Clutton-Brock and Lukas 2012; Greenwood 1980). It has been predicted that one sex should emigrate from their natal group to avoid mating with close relatives, usually males, to avoid deleterious effects of inbreeding (Clutton-Brock 1989; Pusey and Wolf 1996; Ralls et al. 1980).

Photo-ID studies of pilot whales and orca have shown persistent relationships between adult females, lasting for decades, and thereby provide support for a matrilineal society (Heimlich-Boran 1986b, 1988b). For at least one orca society, female natal philopatry has been confirmed from a complete absence of observed emigrations from clearly defined pods (Ford et al. 2000). Amos et al. (1993) showed that females within two pods killed in the Faroe Islands long-finned pilot whale hunt were close relatives sharing a number of alleles, although the validity of pods sampled in the hectic driving hunts has been questioned as artificial (Connor 2000).

Matrilineal associations determined by descent should be distinguished from matriarchal relationships based on the leadership of older, post-reproductive females over related adults: not all matrilineal societies feature matriarchies (e.g., chimpanzees vs. bonobos Furuichi 2011). However, many matrilineal societies function through matriarchal relationships: older elephant matriarchs have greater discrimination abilities to distinguish the calls of familiar and unfamiliar females in playback experiments (McComb et al. 2001), and resident orca matriarchs appear to lead groups when prey is scarce using their ecological knowledge of previous feeding success to locate alternate prey (Brent et al. 2015). We do not know what decisions pilot whale leaders (whether female or male) might need because we lack knowledge of the need for cooperative behavior in the detection, collection, or capture of squid nor of the navigational challenges of cross-ocean travel.

13.3.2 Male Natal Philopatry

The benefits of natal philopatry are the inclusive fitness benefits of living with kin, where high degrees of relatedness occur within groups (Hamilton 1964; West Eberhard 1975). Benefits include improved access to resources that require coordinated capture, improved access to mates, improved care of young via alloparenting, and improved shared defense from predators. Female natal philopatry assures that females will benefit from the increased likelihood of interacting with female kin to access these benefits. What is less clear is why some adult males would also exhibit natal philopatry? In 40 years of the annual photographic census of resident orca pods of the San Juan Island area between northwestern Washington State, USA, and southwestern British Columbia, Canada, there has not been one example of a resident male leaving his natal pod and permanently emigrating to another pod (Balcomb 2018; Ford et al. 2000; see also Chap. 11). There is, however, male emigration from natal pods for the transient, mammal-eating orca, including of all-male groups (Baird and Whitehead 2000; Baird and Dill 1996). In matrilineal sperm whales, males emigrate from natal pods and as adults exhibit a "roving" strategy to locate and gain access to receptive females (Whitehead 1990). This is similar to male strategies in elephants (Nyakaana and Arctander 1999), so male natal philopatry does not appear to be a required element of all matrilineal societies.

The required proof of male natal philopatry is that adult males and females in the same pod are related kin, for example, mothers and sons or brothers and sisters. In agreement with long-term observations, genetic studies of orca multilevel societies show that the most closely related individuals occur in the smallest groups, with relatedness decreasing through levels from mother-offspring pairs (highest relatedness) to matrilines to sub-pods to pods (Pilot et al. 2010). This is also the case for both short-finned (van Cise et al. 2017) and long-finned pilot whales (Amos et al. 1993; Augusto et al. 2017; Verborgh et al. 2009).

The genetic relatedness between adult males and adult females in the same social unit has been documented for orca (Barrett-Lennard 2000) and long-finned pilot whales (Amos et al. 1993). Barrett-Lennard (2000) examined genetic relatedness within Alaskan resident orca pods and found no cases of unique mitochondrial haplotypes between males and females, implying that male emigration was rare. Adult males in long-finned pilot whale pods also shared alleles with adult females in their pod, implying a close genetic relationship (Amos et al. 1993).

The existence of all-male groups, defined by cluster analysis in many photo-ID studies, suggests that there may be benefits for male pilot whales to group together and take on elements of a roving male strategy (Whitehead 1990). This led some authors to conclude that males disperse from their natal pods to form all-male groups that cooperatively search for receptive, nonkin females (Andersen and Siegismund 1994; Desportes et al. 1992). Mahaffy (2012) suggested that stability of pilot whale society fell between orca and sperm whale models because, although associations within matrilines were based on male natal philopatry, associations between matrilines were more fluid (as pointed out by Oremus 2008), as is the case for sperm whales (Richard et al. 1996). These all-male groups are rare in resident orca, but occur in mammal-eating transient whales (Baird and Whitehead 2000; Baird and Dill 1996). We do not know enough of the long-term demographics of pilot whale pods nor the conditions that might lead to the formation of all-male groups, such as the death of a matriarch with many male offspring. All-male groups could also be temporary seasonal aggregations that increase group size to improve males' access to extra-group mating opportunities, with a longer-term return to matrilineal (i.e., kin) associations. There is still much to be learned about the role of males in pilot whale societies.

13.3.3 Extra-Group Mating

The third feature of matrilineal societies is extra-group mating (Sillero-Zubiri et al. 1996) or natal exogamy (Wright et al. 2016), a necessary feature of a society with natal philopatry of both sexes. Mating should be avoided within stable groups of kin to avoid deleterious effects of inbreeding. Extra-group matings have been reported for long-finned pilot whale pods in the Faroe Islands and in genetic studies of orca (Barrett-Lennard 2000; Ford et al. 2011; Pilot et al. 2010). Pilot et al. (2010) found that 83% of orca calves had fathers outside their pod. Also, the majority of siblings were half-siblings, implying that the mother was mating with multiple males. All evidence points to the existence of "...male-mediated gene flow occurring without male dispersal" (Pilot et al. 2010, p. 27).

Observations of mating tend to be rare for cetaceans, and we observed only one mating in over 500 h with short-finned pilot whales in the Canary Islands (Heimlich-Boran 1993). Confirmation of extra-group mating in pilot whales awaits further genetic studies of known populations.

13.3.4 Post-reproductive Lifespan

The fourth feature in some matrilineal whales is presence of a female post-reproductive lifespan (PRLS). This was documented for short-finned pilot whale females living up to 15–20 years beyond their last ovulation (Marsh and Kasuya 1984, 1986). However, similar studies with long-finned pilot whales have failed to identify a post-reproductive phase in females (Kasuya et al. 1988; Martin and Rothery 1993; Sergeant 1962). Foote (2008) compared life history data from orca (from photo-ID data) and short-finned and long-finned pilot whales (from whaling data) and calculated mortality rates for the three species. He found that long-finned pilot whales die earlier than orca and short-finned pilot whales, and proposed that this earlier mortality was a possible explanation for why long-finned pilot whales are not post-reproductive (Foote 2008).

For orca, reproductive senescence was inferred from long-term observations of adult females during which no female older than 48 years produced offspring (Bigg et al. 1990). Current evidence suggests that orca females may live over half their lives in a post-reproductive state (Foster et al. 2012; Pavelka et al. 2018). Reproductive senescence is rare in mammals, with the majority of mammals reproducing until death (Packer et al. 1998). G. C. Williams (1957, p. 407) suggested: "...Any individual, of whatever age, who is caring for dependent offspring is acting in a way that promotes the survival of his own genes and is properly considered a part of the breeding population. No one is post-reproductive until his youngest child is self-sufficient." These post-reproductive females are the matriarchs of the group, displaying leadership roles (Brent et al. 2015) beyond contributing to the survival of their last offspring (Clutton-Brock 1984) toward guiding the entire group through their information (McAuliffe and Whitehead 2005): memories of past feeding success and past long-term interactions with cooperative and competitive associates (McComb et al. 2001). Pavelka et al. (2018) proposed that "post-fertile" lifespans evolved independently in primates (principally humans), short-finned pilot whales, and orca, perhaps for different evolutionary reasons. Orca reproductive histories suggest that there could be direct selection to stop reproducing and helping kin instead (Pavelka et al. 2018).

13.4 Multilevel Societies

The societies in which animals are grouped into increasing levels of hierarchy from mother-offspring pairs to matrilines to pods to communities have been termed multilevel societies (Whitehead et al. 2012) and are known in other mammals (e.g., elephants: Wittemyer et al. 2009). Support for the existence of multilevel societies in pilot whales has come from the variable patterns of site fidelity for identified whales. Pilot whales worldwide have two extreme patterns of site fidelity and distribution: many establish themselves as core residents in a specific area, while others are transient and cover vast areas of deep ocean, exploring multiple sites over long intervals. The link between these two are the visitors that stay in core areas a bit longer and associate with core residents and transients. All appear to live in similarly stable groups of mixed age and sex, but their mixing suggests a large degree of fluidity.

Core residents should benefit from residency by being able to forage repeatedly in regular areas and to learn local variations of prey distribution and behavior. We know little about available prey biomass in these core areas. Squid vertical migrations, from 300 to 400 m to the surface (Barham 1966) with a diurnal pattern of rising to the surface during the night, are in turn based on planktonic squid prey with their own temperature and depth preferences. The predictability and population dynamics of squid will be an important element for future pilot whale behavioral ecology research (Doubleday et al. 2016).

If there are only enough resources to support core residents, then there should be limits to visitors or transients becoming resident. Competitive interactions could be occurring at depth. However, if there are sufficient resources, especially to support large, seasonal assemblages, then there could be negligible costs to core residents where "patch richness is always large enough that a single patch, when available, can support more than the needs of the primary animals" (Johnson et al. 2002, p. 565).

The resident orca's extreme dependence on salmon abundance results in increased mortality in low prey years, indicating limitation by prey resources (Ford et al. 2010). Pilot whale residency in any area is likely related to foraging success as well, but we do not know what, if any, local knowledge of prey distribution and seasonal movements may be required to efficiently exploit squid prey. Squid populations are increasing due to global warming (Doubleday et al. 2016), and these heterogeneous patterns of site fidelity could be indications of

Fig. 13.6 A young short-finned pilot whale, *Globicephala macrorhynchus*, explores the camera. Interactions with humans, whether through whale-watching or through competition with fisheries, will continue to be an important element of the pilot whales' ongoing survival: © Boran/Heimlich

multiple strategies for locating prey in robust oceanic ecosystems that can support both residents and visitors.

There is still much to learn about the specifics of pilot whale relationships between kin and nonkin in matrilineal societies, but these studies have shown that complex associations are maintained over long time periods in core areas, allowing the experienced development of locally adapted hunting strategies (Aguilar Soto et al. 2008). Further analysis is required for the applicability of these hypotheses to long-finned pilot whale societies. Larger issues affecting these species are primarily anthropogenic, such as ship strikes (Carrillo and Ritter 2010), fisheries bycatch (Werner et al. 2018), and climate change (Sousa et al. 2018) (Fig. 13.6). Far-reaching effects of climate change may be found in reports of long-finned/short-finned second generation hybrids in the North Atlantic (Miralles et al. 2013, 2016). Warm water incursions due to climate change allow tropical short-finned pilot whales to expand their range north (Sabatier et al. 2015) and expand into the southern range of long-finned pilot whales. As mating occurs mainly during warmer months, opportunities for hybridization increase (Miralles et al. 2016). This scenario could lead to the diminution of numbers of long-finned pilot whales from the North Atlantic, perhaps similar to the disappearance of this species from the North Pacific 800–1200 years ago (Kasuya 1975). These changes could also be interpreted as signs of habitat expansion for short-finned pilot whales and their unique lifestyle.

References

Aguilar Soto N, Johnson MP, Madsen PT, Díaz F, Domínguez I, Brito A, Tyack P (2008) Cheetahs of the deep sea: deep foraging sprints in short-finned pilot whales off Tenerife (Canary Islands). J Anim Ecol 77(5):936–947. https://doi.org/10.1111/j.1365-2656.2008.01393.x

Alves F (2013) Population structure, habitat use and conservation of short-finned pilot whales (*Globicephala macrorhynchus*) in the Archipelago of Madeira. PhD Thesis. Universidade da Madeira

Alves F, Quérouil S, Dinis A, Nicolau C, Ribeiro C, Freitas L et al (2013) Population structure of short-finned pilot whales in the oceanic archipelago of Madeira based on photo-identification and genetic analyses: implications for conservation. Aquat Conserv Mar Freshwat Ecosyst 23 (5):758–776. https://doi.org/10.1002/aqc.2332

Alves F, Dinis A, Nicolau C, Ribeiro C, Kaufmann M, Fortuna C, Freitas L (2015) Survival and abundance of short-finned pilot whales in the archipelago of Madeira, NE Atlantic. Mar Mamm Sci 31(1):106–121. https://doi.org/10.1111/mms.12137

Alves F, Alessandrini A, Servidio A, Mendonça AS, Hartman KL, Prieto R et al (2019) Complex biogeographical patterns support an ecological connectivity network of a large marine predator in the north-east Atlantic. Divers Distrib 25:269–284. https://doi.org/10.1111/ddi.12848

Amos B, Schltterer C, Tautz D (1993) Social structure of pilot whales revealed by analytical DNA profiling. Science 260(5108):670–672. http://www.jstor.org/stable/2881258

Andersen LW, Siegismund HR (1994) Genetic evidence for migration of males between schools of the long-finned pilot whale *Globicephala melas*. Mar Ecol Prog Ser 105:1–7

Augusto JF (2017) Social structure of the pilot whales (*Globicephala melas*) off Cape Breton, Nova Scotia, Canada. Dalhousie University

Augusto JF, Frasier TR, Whitehead H (2017) Social structure of long-finned pilot whales (*Globicephala melas*) off northern Cape Breton Island, Nova Scotia. Behaviour 154 (5):509–540. https://doi.org/10.1163/1568539X-00003432

Baird RW (2016) The lives of Hawai"i"'s dolphins and whales: natural history and conservation. University of Hawai'i Press, Honolulu

Baird RW, Dill LM (1996) Ecological and social determinants of group size in transient killer whales. Behav Ecol 7(4):408–416. https://doi.org/10.1093/beheco/7.4.408

Baird RW, Whitehead H (2000) Social organization of mammal-eating killer whales: group stability and dispersal patterns. Can J Zool 78:2096–2105

Baird RW, Borsani JF, Hanson MB, Tyack PL (2002) Diving and night-time behavior of long-finned pilot whales in the Ligurian sea. Mar Ecol Prog Ser 237:301–305. https://doi.org/10.3354/meps237301

Balcomb KC (2018) The Center for Whale Research. https://www.whaleresearch.com/. Accessed 20 Aug 2018

Balcomb KC, Boran JR, Heimlich SL (1982) Killer whales in greater puget sound. Rep Int Whal Comm 32:681–685

Barham EG (1966) Deep scattering layer migration and composition: observations from a diving saucer. Science 151(3716):1399–1403. https://doi.org/10.1126/science.151.3716.1399

Barrett-Lennard LG (2000) Population structure and mating patterns of killer whales (*Orcinus orca*) as revealed by DNA analysis. PhD Thesis. The University of British Columbia

Barton ED, Aristegui J, Tett P, Canton M, García-Braun J, Hernández-León S et al (1998) The transition zone of the Canary Current upwelling region. Prog Oceanogr 41(4):455–504. https://doi.org/10.1016/S0079-6611(98)00023-8

Behar DM, Villems R, Soodyall H, Blue-Smith J, Pereira L, Metspalu E et al (2008) The dawn of human matrilineal diversity. Am J Hum Genet 82(5):1130–1140. https://doi.org/10.1016/j.ajhg.2008.04.002

Bigg MA, Olesiuk PF, Ellis GM (1990) Social organization and genealogy of resident killer whales (*Orcinus orca*) in the coastal waters of British Columbia and Washington State. Rep Int Whal Comm 12:383–405

Boran J, Heimlich S (1999) Social learning in cetaceans: hunting, hearing and hierarchies. In: Box H, Gibson K (eds) Mammalian social learning. Cambridge University Press, Cambridge, pp 282–307. https://www.nhbs.com/mammalian-social-learning-book

Brent LJN, Franks DW, Foster EA, Balcomb KC, Cant MA, Croft DP (2015) Ecological knowledge, leadership, and the evolution of menopause in killer whales. Curr Biol 25(6):746–750. https://doi.org/10.1016/j.cub.2015.01.037

Carrillo M, Ritter F (2010) Increasing numbers of ship strikes in the Canary Islands: proposals for immediate action to reduce risk of vessel-whale collisions. J Cetacean Res Manag 11 (2):131–138

Chapais B, Girard M, Primi G (1991) Non-kin alliances, and the stability of matrilineal dominance relations in Japanese macaques. Anim Behav 41(3):481–491. https://doi.org/10.1016/S0003-3472(05)80851-6

Clutton-Brock TH (1984) Reproductive effort and terminal investment in iteroparous animals. Am Nat 123(2):212–229

Clutton-Brock TH (1989) Female transfer and inbreeding avoidance in social mammals. Nature 337:70–72

Clutton-Brock TH, Lukas D (2012) The evolution of social philopatry and dispersal in female mammals. Mol Ecol 21(3):472–492. https://doi.org/10.1111/j.1365-294X.2011.05232.x

Connor RC (2000) Group living in whales and dolphins. In: Mann J, Connor RC, Tyack PL, Whitehead H (eds) Cetacean societies: field studies of dolphins and whales. University of Chicago Press, Chicago, pp 199–218

Davies JL (1963) The antitropical factor in cetacean speciation. Evolution 17:107–116

de Stephanis R, Verborgh P, Pérez S, Esteban R, Minvielle-Sebastia L, Guinet C (2008) Long-term social structure of long-finned pilot whales (Globicephala melas) in the Strait of Gibraltar. Acta Ethol 11:81–94. https://doi.org/10.1007/s10211-008-0045-2

Desportes G, Mouritsen R (1988) Diet of the pilot whale, Globicephala melas, around the Faroe Islands. ICES Report C.M. 1988/N: 12. Copenhagen

Desportes G, Andersen LW, Aspholm PE, Bloch D, Mouritsen R (1992) A note about a male-only pilot whale school observed in Faroe Islands. Fróðskaparrit 40:31–37

Donovan GP, Lockyer CH, Martin AR (1993) Biology of northern hemisphere pilot whales. Rep Int Whal Comm 14:x+479

Doubleday ZA, Prowse TAA, Arkhipkin A, Pierce GJ, Semmens J, Steer M et al (2016) Global proliferation of cephalopods. Curr Biol 26(10):R406–R407. https://doi.org/10.1016/j.cub.2016.04.002

Findlay KP, Best PB, Ross GJB, Cockcroft VG (1992) The distribution of small odontocete cetaceans off the coasts of South Africa and Namibia. S Afr J Mar Sci 12(1):237–270. https://doi.org/10.2989/02577619209504706

Foote AD (2008) Mortality rate acceleration and post-reproductive lifespan in matrilineal whale species. Biol Lett 4(2):189–191. https://doi.org/10.1098/rsbl.2008.0006

Ford JKB, Fisher HD (1983) Group-specific dialects of killer whales (Orcinus orca) in British Columbia. In: Payne RS (ed) Communication and behavior of whales. Westview Press, Boulder CO, pp 129–161

Ford JKB, Ellis GM, Balcomb KC (2000) Killer whales. University of British Columbia Press, Vancouver

Ford JKB, Ellis GM, Olesiuk PF, Balcomb KC (2010) Linking killer whale survival and prey abundance: food limitation in the oceans' apex predator? Biol Lett 6(1):139–142. https://doi.org/10.1098/rsbl.2009.0468

Ford MJ, Hanson MB, Hempelmann JA, Ayres KL, Emmons CK, Schorr GS et al (2011) Inferred paternity and male reproductive success in a killer whale (Orcinus orca) population. J Hered 102(5):537–553. https://doi.org/10.1093/jhered/esr067

Foster EA, Franks DW, Mazzi S, Darden SK, Balcomb KC, Ford JKB, Croft DP (2012) Adaptive prolonged postreproductive life span in killer whales. Science 337(6100):1313. https://doi.org/10.1126/science.1224198

Fullard KJ, Early G, Heide-Jorgensen MP, Bloch D, Rosing-Asvid A, Amos W (2000) Population structure of long-finned pilot whales in the North Atlantic: a correlation with sea surface temperature? Mol Ecol 9(7):949–958. https://doi.org/10.1046/j.1365-294x.2000.00957.x

Furuichi T (2011) Female contributions to the peaceful nature of bonobo society. Evol Anthropol 20:131–142. https://doi.org/10.1093/oso/9780198728511.003.0002

Gannon DP, Read AJ, Craddock JE, Fristrup KM, Nicolas JR (1997) Feeding ecology of long-finned pilot whales *Globicephala melas* in the western North Atlantic. Mar Ecol Prog Ser 148 (1–3):1–10. https://doi.org/10.3354/meps148001

Gibson-Lonsdale JJ (1990) Pilot whaling in the Faroe Islands–its history and present significance. Mammal Rev 20(1):44–52. https://doi.org/10.1111/j.1365-2907.1990.tb00102.x

Gray JE (1846) On the cetaceous animals. In: Richardson EJ, Gray JE (eds) The zoology of the voyage of H.M.S. Erebus and Terror, vol 1. Mammalia, birds. Longmans, Brown, Green and Longmans, London, pp 13–53

Greenwood PJ (1980) Mating systems, philopatry and dispersal in birds and mammals. Anim Behav 28(1960):1140–1162. https://doi.org/10.1016/S0003-3472(80)80103-5

Hamilton WD (1964) The genetical evolution of social behavior. J Theor Biol 7:17–52

Hartny-Mills L (2015) Site fidelity, social structure and spatial distribution of short-finned pilot whales, *Globicephala macrorhynchus*, off the South West Coast of Tenerife. PhD Thesis, University of Portsmouth

Heide-Jørgensen MP, Bloch D, Stefansson E, Mikkelsen B, Ofstad LH, Dietz R (2002) Diving behaviour of long-finned pilot whales *Globicephala melas* around the Faroe Islands. Wildl Biol 8(4):307–313

Heimlich-Boran JR (1986a) Fishery correlations with the occurrence of killer whales in greater Puget Sound. In: Kirkevold B, Lockard J (eds) Behavioral biology of killer whales. A. R. Liss, New York, pp 113–131

Heimlich-Boran SL (1986b) Cohesive relationship among Puget Sound killer whales. In: Kirkevold B, Lockard JS (eds) Behavioral biology of killer whales. A. R. Liss, New York, pp 251–284

Heimlich-Boran JR (1988a) Behavioral ecology of killer whales (*Orcinus orca*), in the Pacific Northwest. Can J Zool 66(3):565–578. https://doi.org/10.1139/z88-084

Heimlich-Boran SL (1988b) Association patterns and social dynamics of killer whales (*Orcinus orca*) in Greater Puget Sound. MA Thesis. San Jose State University

Heimlich-Boran JR (1993) Social organisation of the short-finned pilot whale, *Globicephala macrorhynchus*, with special reference to the comparative social ecology of delphinids. PhD Thesis. University of Cambridge. https://doi.org/10.17863/CAM.17314

Johnson DDP, Kays R, Blackwell PG, MacDonald DW (2002) Does the resource dispersion hypothesis explain group living? Trends Ecol Evol 17(12):563–570

Kasuya T (1975) Past occurrence of *Globicephala melaena* in the western North Pacific. Sci Rep Whales Res Inst 27(27):95–110

Kasuya T, Marsh H (1984) Life history and reproductive biology of the short-finned Pilot whale, *Globicephala macrorhynchus*, off the Pacific coast of Japan. Rep Int Whal Comm 6:259–310. http://ci.nii.ac.jp/naid/10019045962/en/

Kasuya T, Sergeant DE, Tanaka K (1988) Re-examination of life history parameters of long-finned pilot whales in the Newfoundland waters Canada. Sci Rep Whales Res Inst Tokyo 39:103–120

LeDuc RG, Perrin WF, Dizon AE (1999) Phylogenetic relationships among the delphinid cetaceans based on full Cytochrome B sequences. Mar Mamm Sci 15(3):619–648. https://doi.org/10.1111/j.1748-7692.1999.tb00833.x

Lee PC, Fishlock V, Webber CE, Moss CJ (2016) The reproductive advantages of a long life: longevity and senescence in wild female African elephants. Behav Ecol Sociobiol 70:337–345. https://doi.org/10.1007/s00265-015-2051-5

Lesson RP (1828) Cétacés. Complément des oeuvres de Buffon ou Histoire Naturelle des animaux rares découverts par les naturalistes et les voyageurs depuis la mort de Buffon, vol 1. Badouin Freres, Paris

Mahaffy SD (2012) Site fidelity, associations and long-term bonds of short-finned pilot whales off the Island of Hawai'i. MSc Thesis, Portland State University

Mahaffy SD, Baird RW, McSweeney DJ, Webster DL, Schorr GS (2015) High site fidelity, strong associations, and long-term bonds: short-finned pilot whales off the island of Hawai'i. Mar Mamm Sci 31(4):1427–1451. https://doi.org/10.1111/mms.12234

Marsh H, Kasuya T (1984) Changes in the ovaries of the short-Finned pilot whale, *Globicephala macrorhynchus*, with age and reproductive activity. Rep Int Whal Comm 6:311–335

Marsh H, Kasuya T (1986) Evidence for reproductive senescence in female cetaceans. Rep Int Whal Comm 8:57–74

Martin AR, Rothery P (1993) Reproductive parameters of female long-finned pilot whales (*Globicephala melas*) around the Faroe Islands. Rep Int Whal Comm 14:263–304

Mate BR, Lagerquist BA, Winsor M, Geraci JR, Prescott JH (2005) Movements and dive habits of a satellite-monitored long-finned pilot whale (*Globicephala melas*) in the Northwest Atlantic. Mar Mamm Sci 21:136–144. https://doi.org/10.1111/j.1748-7692.2005.tb01213.x

McAuliffe K, Whitehead H (2005) Eusociality, menopause and information in matrilineal whales. Trends Ecol Evol 20(12):650. https://doi.org/10.1016/j.tree.2005.09.003

McComb K, Moss C, Durant SM, Baker L, Sayialel S (2001) Matriarchs as repositories of social knowledge in African elephants. Science 292(5516):491–494. https://doi.org/10.1126/science.1057895

Mendonça AS (2012) Estudo da distribuição e abundância da Baleia pi-loto-tropical (*Globicephala macrorhynchus*) no Arquipélago dos Açores. PhD Thesis, University of Algarve, Portugal

Miralles L, Lens S, Rodríguez-Folgar A, Carrillo M, Martín V, Mikkelsen B, Garcia-Vazquez E (2013) Interspecific introgression in cetaceans: DNA markers reveal post-F1 status of a pilot whale. PLoS One 8(8):e69511. https://doi.org/10.1371/journal.pone.0069511

Miralles L, Oremus M, Silva MA, Planes S, Garcia-vazquez E (2016) Interspecific hybridization in pilot whales and asymmetric genetic introgression in northern *Globicephala melas* under the scenario of global warming. PLoS One 11(8):1–15. https://doi.org/10.1371/journal.pone.0160080

Möller LM, Wiszniewski J, Parra G, Beheregaray L (2012) Sociogenetic structure, kin associations and bonding in delphinids. Mol Ecol 21(3):745–764. https://doi.org/10.1111/j.1365-294X.2011.05405.x

Newman MEJ, Girvan M (2014) Finding and evaluating community structure in networks. Phys Rev E 69(2):26113. https://doi.org/10.1103/PhysRevE.69.026113

Nores C, Peréz C (1988) Overlapping range between *Globicephala macrorhynchus* and *Globicephala melaena* in the northeastern Atlantic. Mammalia 52(1):51–55

Nyakaana S, Arctander P (1999) Population genetic structure of the African elephant in Uganda based on variation at mitochondrial and nuclear loci: evidence for male-biased gene flow. Mol Ecol 8(7):1105–1115. https://doi.org/10.1046/j.1365-294X.1999.00661.x

Olson PA (2018) Pilot whales: *Globicephala melas* and *G. macrorhynchus*. In: Würsig B, Thewissen JGM, Kovacs KM (eds) Encyclopedia of marine mammals, 3rd edn. Academic, New York, pp 701–705. https://doi.org/10.1016/B978-0-12-373553-9.00197-8

Oremus M (2008) Genetic and demographic investigation of population structure and social system in four delphinid species. PhD Thesis, University of Aukland

Oremus M, Gales R, Dalebout ML, Funahashi N, Endo T, Kage T et al (2009) Worldwide mitochondrial DNA diversity and phylogeography of pilot whales (*Globicephala spp.*). Biol J Linn Soc 98(4):729–744. https://doi.org/10.1111/j.1095-8312.2009.01325.x

Ottensmeyer CA, Whitehead H (2003) Behavioural evidence for social units in long-finned pilot whales. Can J Zool 81(8):1327–1338. https://doi.org/10.1139/z03-127

Packer C, Tatar M, Collins A (1998) Reproductive cessation in mammals. Nature 392 (1989):807–811

Pavelka MSM, Brent LJN, Croft DP, Fedigan LM (2018) Post-fertile lifespan in female primates and cetaceans. In: Kalbitzer U, Jack KM (eds) Primate life histories, sex roles, and adaptability. Springer, Basel, pp 37–55. https://doi.org/10.1007/978-3-319-98285-4_3

13 Pilot Whales: Delphinid Matriarchies in Deep Seas 303

Pilot M, Dahlheim ME, Hoelzel AR (2010) Social cohesion among kin, gene flow without dispersal and the evolution of population genetic structure in the killer whale (*Orcinus orca*). J Evol Biol 23(1):20–31. https://doi.org/10.1111/j.1420-9101.2009.01887.x

Prieto R, Fernandes M (2007) Revision of the occurrence of the long-finned pilot whale *Globicephala melas* (Traill, 1809), in the Azores. Arquipélago 24(1861):65–69. http://repositorio.uac.pt/handle/10400.3/225

Pusey A, Wolf M (1996) Inbreeding avoidance in animals. Trends Ecol Evol 11(5):201–206. https://doi.org/10.1016/0169-5347(96)10028-8

Ralls K, Brugger K, Glick A (1980) Deleterious effects of inbreeding in a herd of captive Dorcas gazelle *Gazella dorcas*. Int Zoo Yearb 20(1):137–146. https://doi.org/10.1111/j.1748-1090.1980.tb00957.x

Richard KR, Dillon MC, Whitehead H, Wright JM (1996) Patterns of kinship in groups of free-living sperm whales (*Physeter macrocephalus*) revealed by multiple genetic analyses. Proc Natl Acad Sci 93:8792–8795

Sabatier E, Pante E, Dussud C, Van Canneyt O, Simon-Bouhet B, Viricel AA (2015) Genetic monitoring of pilot whales, *Globicephala* spp. (*Cetacea: Delphinidae*), stranded on French coasts. Mammalia 79(1):111–114. https://doi.org/10.1515/mammalia-2013-0155

Santos MB, Monteiro SS, Vingada JV, Ferreira M, Lopez A, Martínez Cedeira JA et al (2014) Patterns and trends in the diet of long-finned pilot whales (*Globicephala melas*) in the northeast Atlantic. Mar Mamm Sci 30(1):1–19. https://doi.org/10.1111/mms.12015

Sergeant DE (1962) The biology of the pilot or pothead whale *Globicephala melaena* (Traill) in Newfoundland waters. Bull Fish Res Bd Can 132:1–84

Servidio A (2014) Distribution, social structure and habitat use of short-finned pilot whale, *Globicephala macrorhynchus*, in the Canary Islands. PhD Thesis, University of St Andrews

Servidio A, Alves F, Dinis A, Freitas L, Martín V (2007) First record of movement of short-finned pilot whales between two Atlantic oceanic Archipelagos. In: The 17th biennial conference on the biology of marine mammals. The Society for Marine Mammalogy, Cape Town

Siemann LA (1994) Mitochondrial DNA sequence variation in North Atlantic long-finned pilot whales, *Globicephala melas*. PhD Thesis, Massachusetts Institute of Technology

Sillero-Zubiri C, Gottelli D, MacDonald DW (1996) Male philopatry, extra-pack copulations and inbreeding avoidance in Ethiopian wolves (*Canis simensis*). Behav Ecol Sociobiol 38 (5):331–340. http://www.jstor.org/stable/4601211

Silva MA, Prieto R, Magalhães S, Cabecinhas R, Cruz A, Gonçalves JM, Santos RS (2003) Occurrence and distribution of cetaceans in the waters around the Azores (Portugal), Summer and Autumn 1999–2000. Aquat Mamm 29(1):77–83. https://doi.org/10.1578/016754203101024095

Silva MA, Prieto R, Cascão I, Seabra MI, Machete M, Baumgartner MF, Santos RS (2014) Spatial and temporal distribution of cetaceans in the mid-Atlantic waters around the Azores. Mar Biol Res 10(2):123–137. https://doi.org/10.1080/17451000.2013.793814

Sousa A, Alves F, Dinis A, Bentz J, Cruz MJ, Nunes JP (2018) How vulnerable are cetaceans to climate change? Developing and testing a new index. Ecol Indic 98:9–18. https://doi.org/10.1016/j.ecolind.2018.10.046

Traill TS (1809) Description of a new species of whale, *Delphinus melas*. In a letter from Thomas Stewart Traill, M. D. to Mr. Nicholson. J Nat Philos Chem Arts 1809:81–83

van Bree PJH, Best BB, Ross GJB (1978) Occurrence of the two species of pilot whales (genus *Globicephala*) on the coast of South Africa. Mammalia 42(1872):323–328

van Cise AM, Martien KK, Mahaffy SD, Baird RW, Webster DL, Fowler JH et al (2017) Familial social structure and socially-driven genetic differentiation in Hawaiian short-finned pilot whales. Mol Ecol 26:6730–6741. https://doi.org/10.1111/mec.14397

van Waerebeek K, Leaper R, Baker AN, Papastavrou V, Thiele D, Findlay K et al (2010) Odontocetes of the Southern Ocean Sanctuary. J Cetacean Res Manag 11(3):315–346

Verborgh P, De Stephanis R, Pérez S, Jaget Y, Barbraud C, Guinet C (2009) Survival rate, abundance, and residency of long-finned pilot whales in the Strait of Gibraltar. Mar Mamm Sci 25(3):523–536. https://doi.org/10.1111/j.1748-7692.2008.00280.x

Verborgh P, Gauffier P, Esteban R, Giménez J, Cañadas A, Salazar-Sierra JM, de Stephanis R (2016) Conservation status of long-finned pilot whales, *Globicephala melas*, in the Mediterranean Sea. In: Notarbartolo Di Sciara G, Podestà M, Curry BE (eds) Advances in marine biology, vol 75. Academic, Oxford, pp 173–203. https://doi.org/10.1016/bs.amb.2016.07.004

Wells RS, Fougeres EM, Cooper AG, Stevens RO, Brodsky M, Lingenfelser R et al (2013) Movements and dive patterns of short-finned pilot whales (*Globicephala macrorhynchus*) released from a mass stranding in the Florida Keys. Aquat Mamm 39(1):61–72. https://doi.org/10.1578/AM.39.1.2013.61

Werner TB, Northridge S, McClellan K, Young N (2018) Mitigating bycatch and depredation of marine mammals in longline fisheries. ICES J Mar Sci 72(5):1576–1586

West Eberhard MJ (1975) The evolution of social behavior by kin selection. Q Rev Biol 50(1):1–33

Whitehead H (1990) Rules for roving males. J Theor Biol 145:355–368

Whitehead H (1998) Cultural selection and genetic diversity in matrilineal whales. Science 282 (5394):1708–1711

Whitehead H (2007) Selection of models of lagged identification rates and lagged association rates using AIC and QAIC. Commun Stat Simul Comput 36(6):1233–1246. https://doi.org/10.1080/03610910701569531

Whitehead H (2008) Analyzing animal societies: quantitative methods for vertebrate social analysis. University of Chicago Press, Chicago

Whitehead H, Rendell L (2014) The cultural lives of whales and dolphins. University of Chicago Press, Chicago

Whitehead H, Antunes R, Gero S, Wong SNP, Engelhaupt D, Rendell L (2012) Multilevel societies of female sperm whales (*Physeter macrocephalus*) in the Atlantic and Pacific: why are they so different? Int J Primatol 33(5):1142–1164. https://doi.org/10.1007/s10764-012-9598-z

Williams GC (1957) Pleiotropy, natural selection, and the evolution of senescence. Evolution 11 (4):398–411. http://www.jstor.org/stable/2406060

Wittemyer G, Okello JBA, Rasmussen HB, Arctander P, Nyakaana S, Douglas-Hamilton I, Siegismund HR (2009) Where sociality and relatedness diverge: the genetic basis for hierarchical social organization in African elephants. Proc R Soc B Biol Sci 276(1672):3513–3521. https://doi.org/10.1098/rspb.2009.0941

Wright BM, Stredulinsky EH, Ellis GM, Ford JKB (2016) Kin-directed food sharing promotes lifetime natal philopatry of both sexes in a population of fish-eating killer whales, *Orcinus orca*. Anim Behav 115:81–95. https://doi.org/10.1016/j.anbehav.2016.02.025

Chapter 14
Behavior and Ecology of Not-So-Social Odontocetes: Cuvier's and Blainville's Beaked Whales

Robin W. Baird

Abstract While beaked whales are the poorest-known family of cetaceans overall, the behavior and ecology of two species of beaked whales, Cuvier's (*Ziphius cavirostris*) and Blainville's (*Mesoplodon densirostris*), have been studied extensively for more than 15 years in multiple areas around the world. This research was largely initiated as a result of the susceptibility of both species to react to high-intensity navy sonars, sometimes resulting in the death of individuals. In this chapter long-term studies of both species in Hawai'i are reviewed, informed by research on these species elsewhere. Both species have small populations that are resident to the island slopes, evidenced by a combination of long-term photo-identification and shorter-term satellite tag deployments. The two species coexist by partitioning their habitat in three dimensions, with Cuvier's beaked whales being found in deeper water, and diving deeper, than Blainville's beaked whales. Diving and acoustic behavior of the two species appears to be driven in part by predator avoidance. Both species echolocate only at depth, foraging deep in the water column during the day and at night, with less time spent near the surface during the day in between the deep foraging dives. Ascent rates are also slower than descent rates. All of these factors are likely ways of minimizing detection from near-surface visually or acoustically oriented predators such as large sharks and killer whales. There appears to be no strong selective pressure for grouping in these species. Both are often found alone and on average are found in very small groups (medians: Cuvier's = 2; Blainville's = 3). Groups that do form appear to function in part to avoid predators (for females with small calves) and allow for mating opportunities (for adult males seeking mates). Individuals of both species tend to have ephemeral social relations, although one pair of subadult Cuvier's have been documented together over an 11-year period. Blainville's beaked whale males exhibit female defense polygyny, while sperm competition may play a role in the mating system of Cuvier's beaked whales. Studies of these species in multiple areas spanning the tropics to temperate waters in two different oceans

R. W. Baird (✉)
Cascadia Research Collective, Olympia, WA, USA
e-mail: rwbaird@cascadiaresearch.org

© Springer Nature Switzerland AG 2019
B. Würsig (ed.), *Ethology and Behavioral Ecology of Odontocetes*, Ethology and Behavioral Ecology of Marine Mammals,
https://doi.org/10.1007/978-3-030-16663-2_14

are beginning to earn them an important place in our overall understanding of cetacean ethology and behavioral ecology.

Keywords Citizen science · Predator avoidance · Mating system · Diving behavior · Niche partitioning · Residency · Association patterns · Hawai'i

Studies of the ethology and behavioral ecology of beaked whales (members of the family Ziphiidae) have lagged behind many other groups of cetaceans. The family Ziphiidae is the second-most speciose (after delphinids) taxonomic family of cetaceans, yet the poorest-known overall. There are 22 recognized species of beaked whales from 6 genera, as of 2018, and of those, 6 species (all from the genus *Mesoplodon*, with 15 recognized species) are known only from beach-cast specimens or skeletal remains. Such levels of obscurity reflect a combination of deepwater (usually open-ocean) habits, generally low abundance (at least for most species; Bradford et al. 2017), and very long dives, often exceeding 1 h (e.g., Hooker and Baird 1999; Minamikawa et al. 2007; Baird et al. 2008a; Schorr et al. 2014). Recent genetic studies have resulted in the recognition of new species (e.g., Perrin's beaked whale, *M. perrini*, Dalebout et al. 2002) and resurrection of long-lost species (e.g., Deraniyagala's beaked whale, *M. hotaula*, Dalebout et al. 2014) and revealed that there are undescribed species waiting to be recognized (e.g., a new species of *Berardius*, Morin et al. 2017). Of the 22 presently recognized species, only 2 have been studied based on large numbers of specimens (in both cases as they were targeted in whaling operations): the northern bottlenose whale (*Hyperoodon ampullatus*) and Baird's beaked whale (*B. bairdii*) (Benjaminsen and Christensen 1979; Kasuya 2017). Only four species, including the two above, have been the subject of long-term, in-depth studies in the wild, although two of those four have only been studied extensively in single localities: northern bottlenose whales in the Gully, off eastern Canada, and Baird's beaked whales off the Commander Islands, Russia (Whitehead et al. 1997; Gowans et al. 2000; Hooker et al. 2002; Fedutin et al. 2015).

Given their long dives and open-ocean habits, sightings of beaked whales tend to be infrequent and brief, limiting opportunities to study behavior and ecology with traditional observational methods.[1] Yet in recent years, our knowledge of behavior and ecology of beaked whales, in general, and of Cuvier's (*Ziphius cavirostris*) and Blainville's (*M. densirostris*) beaked whales, in particular (Fig. 14.1), has burgeoned. The impetus for most studies has been the evidence that at least some beaked whales are susceptible to effects of high-intensity sonar, occasionally stranding and dying as a result (Frantzis 1998; Balcomb and Claridge 2001; Fernández et al. 2005; Cox et al. 2006; Tyack et al. 2011). A wide range of research techniques includes traditional sighting

---

[1] Some encounters with both Blainville's and Cuvier's beaked whales in Hawai'i primarily involve the animals floating motionless at the surface for a few minutes, before they disappear on a long dive, usually not seen again during the encounter.

Fig. 14.1 Top: While often considered cryptic, Cuvier's beaked whales, *Ziphius cavirostris*, in Hawai'i have been documented breaching in 11.6% of our encounters. We were able to identify the individual on the right as HIZc007, based on scarring patterns. Bottom: A juvenile Blainville's beaked whale, *Mesoplodon densirostris*, with a suction-cup attached time-depth recorder. Photos by (top) Annie B. Douglas and (bottom) Robin W. Baird

surveys, photo-identification, tagging to examine movements and behavior, analysis of acoustic behavior from animal-borne tags or towed or fixed hydrophone systems, controlled exposure experiments to document reactions to sonar and the sounds of potential predators, and genetic analyses of biopsy samples. We now have in-depth multi-year studies of Cuvier's beaked whales around oceanic islands (i.e., Hawai'i, the Canary Islands) and on continental slopes (i.e., off southern California, North Carolina, and in the Ligurian Sea). For Blainville's beaked whales, studies have been undertaken around a number of oceanic islands, in Hawai'i and the western (Bahamas) and eastern (Macaronesia) North Atlantic. Studies in multiple ocean basins and diverse habitats, from the tropics to temperate areas, can provide valuable opportunities for comparisons.

My work with Cuvier's and Blainville's beaked whales in Hawai'i began in April 2002 as part of a collaboration with Dan McSweeney, who had been working along the west coast of Hawai'i Island since the early 1980s, taking photos of beaked whales whenever encountered. The work in Hawai'i is a multi-species study[2] of odontocetes that includes efforts with a dozen species that range from nearshore to over 4000 m depth. Our studies are from small 6–9 m vessels on the leeward (west and southwest) sides of the islands, ranging from Ni'ihau in the northwest to Hawai'i Island in the southeast. Working conditions in typical deepwater beaked whale habitat (i.e., >500 m depth for Blainville's beaked whales and >800 m for Cuvier's beaked whales) are best off Hawai'i Island. In addition to our encounters (Cuvier's $n = 78$; Blainville's $n = 58$), we have photographs from other researchers (primarily Dan McSweeney) and from citizen science contributors (Cuvier's $n = 45$ encounters; Blainville's $n = 105$ encounters).

Our ability to understand beaked whale behavioral ecology in Hawai'i is enhanced by a broad suite of methods to study and compare two sympatric species, both of which travel in small groups. As well as photo-identification and some biopsy sampling, during our encounters we've obtained short-term information on diving and surfacing behavior from 13 suction-cup attached time-depth recorders (Baird et al. 2006, 2008a) and longer-term information on movements (and some dive behavior) from 27 LIMPET satellite tag deployments (Schorr et al. 2009; Baird 2016). In Hawai'i, it is helpful for photo-identification that both species are regularly bitten by cookie-cutter sharks (*Isistius* spp.), and the white oval scars that accumulate over time (McSweeney et al. 2007) can be used to identify individuals and assess age class of individuals. The scars can remain visible for up to 20 years (Baird 2016), and they allow individual identification from various angles, similar to the use of pigmentation patterns to identify blue whales (*Balaenoptera musculus*) or rough-toothed dolphins (*Steno bredanensis*), rather than just perpendicular photos of the dorsal fin. With good photos we are able to sex adults, because only adult males have erupted teeth (a single pair erupting from the lower jaw, at the tip in Cuvier's beaked whales

---

[2]The methods of this work have been published in detail so are not repeated fully here. Readers wishing more information on methods for surveys and during encounters should see Baird et al. (2013) and Baird (2016); on photo-identification and association analyses, see McSweeney et al. (2007); and on satellite tagging, see Schorr et al. (2009), Baird et al. (2011), and Baird (2016).

and approximately mid-jaw in Blainville's beaked whales), visible even when the mouth is closed. For adult and subadult Blainville's beaked whales, jaw morphology also differs dramatically between the sexes; as a male begins to mature, the arch of the lower jaw enlarges well before the teeth erupt, so we are also able to identify subadult males even before teeth erupt. Even if no photo of the head is available to assess whether teeth are present (or absent), the extensive linear scarring on adult males, and other pigmentation patterns, can be used to sex adult individuals (McSweeney et al. 2007; Baird 2016; Coomber et al. 2016).

This review uses published and unpublished results from our Hawai'i work, as well as insights from studies of beaked whales elsewhere, to help elucidate what factors influence the behavior and ecology of Cuvier's and Blainville's beaked whales. First, I set the stage by laying out the evidence for small resident populations of both species. Second, using satellite tag data, I show that the two species coexist on the island slopes by partitioning their habitat horizontally as well as vertically and also show how their range extends offshore and among the islands, outside our study area. Third, I use information on diving and vocal behavior to show that predation risk seems to drive not only where they spend their time in the water column but what they do while diving. Lastly, using information on group composition and association patterns, I draw insights into social organization—how grouping patterns reflect: (a) solitary foraging, (b) minimizing the risk of predation to calves, and (c) males seeking out, or sequestering, females for the opportunity to mate.

14.1 Evidence for Small Resident Populations

There are three species of beaked whales in Hawaiian waters (Cuvier's, Blainville's, and Longman's (*Indopacetus pacificus*) beaked whales), and another two (Ginkgo-toothed beaked whales, *M. ginkgodens*, and Hubbs' beaked whales, *M. carlhubbsi*) that are suspected to occur based on acoustics (Baumann-Pickering et al. 2014) but that have not been seen (Baird 2016). Beaked whale density is low in Hawai'i, even in suitable habitats. Off west Hawai'i Island, through early 2018, we have 4126 h of search effort in beaked whale habitats (i.e., depths >500 m) and have sighted Cuvier's only once every 52.9 h and Blainville's once every 71.1 h.[3] Overall, there are 3.5 beaked whale sightings per 100 h of effort, but the study area is home to small resident populations of Cuvier's and Blainville's beaked whales, similar to what has been found for both species around other islands (e.g., Claridge 2013; Reyes 2017).

Photo-identifications allowed for estimating abundance of Cuvier's and Blainville's beaked whales from a Petersen capture–recapture model, with pooled data from 2003 to 2004 versus 2005–2006, two pairs of years where the sample sizes of identifications were high. For Cuvier's, abundance of marked (distinctive)

---

[3]As well as one sighting of Longman's beaked whales and eight sightings of unidentified beaked whales that were likely either Cuvier's or Blainville's.

individuals was estimated at only 55 (CV = 0.26), with 98.5% of individuals considered marked, while for Blainville's abundance of marked individuals was estimated at 125 (CV = 0.30), with 89.0% of individuals considered marked (Baird et al. 2009). Discovery curves, representing the rate at which new individuals are identified relative to the accumulation of identifications (i.e., not excluding re-sightings), also suggest that the populations of both species off Hawai'i Island are relatively small (Fig. 14.2). For Blainville's, this abundance estimate includes both a resident, island-associated population and some members of a pelagic, open-ocean population (Baird et al. 2011).

Long-term photo-identification data indicate a high level of site fidelity for Cuvier's and Blainville's beaked whales off west Hawai'i Island. For Cuvier's beaked whales, of the 40 individuals identified with excellent quality photos (thus maximizing the chances of being able to recognize them after even very long periods), 28 (70%) have been seen on more than one occasion, and 25 of those (89.3%) have been seen in more than 1 year. Over one-half (13 of 25) of those have been seen over spans of greater than 10 years, including females ($n = 6$) and males ($n = 5$). The longest span that an individual has been documented is 24.47 years, for a female first identified as a probable adult when seen in 1990. The adult sex ratio for Cuvier's beaked whales in Hawai'i appears to be close to 1:1 (54% female/46% male). Of the 37 known-sex distinctive adult individuals from Hawai'i Island, 20 are female and 17 male. Social network analyses show that the majority of individuals (59.7%) documented off Hawai'i Island are linked by association in the main cluster of the social network, indicating they are all part of the same island-associated population (Fig. 14.3).[4] It is likely that many of the individuals in isolated clusters are also part of the island-associated population, but given the small group sizes (median = 2, range = 1–5), the likelihood of individuals being linked by association is relatively small for those individuals seen on only one occasion.

For Blainville's beaked whales photo-identified off the island of Hawai'i, the situation is similar: of 75 individuals with excellent quality photos, 44 (58.7%) have been seen more than once, and of those 36 (81.8%) have been seen in more than 1 year. Eleven of the 36 (30.5%) have been seen for over 10 years (maximum = 21.02 years). This includes eight adult females and three adult males. The adult sex ratio for distinctive Blainville's beaked whales also appears to be approximately 1:1 (52% female/48% male). Thus, the predominance of adult females among those seen over spans of 10 years or more suggests higher site fidelity for females. One female Blainville's that has been repeatedly photo-identified off Hawai'i Island (HIMd133, over a 10-year span) has been recently documented off O'ahu associated with a group there, suggesting that there may be links among communities off different islands. Social network analyses for Blainville's beaked whales off Hawai'i and O'ahu show that the majority of individuals (73.3%) are linked by association in

---

[4]This analysis was undertaken with individuals considered at least slightly distinctive and with fair- or better-quality photos; see McSweeney et al. (2007) for definitions. This result is particularly surprising given the high proportion of sightings of lone individuals and a median group size of two.

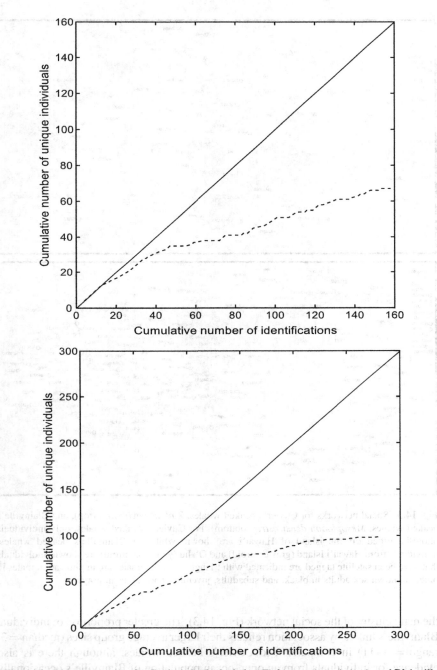

Fig. 14.2 Discovery curves for Cuvier's beaked whales, *Ziphius cavirostris* (top), and Blainville's beaked whales, *Mesoplodon densirostris* (bottom) off Hawai'i Island. Distinctive individuals with fair or better photo qualities are included, and a 1:1 line is shown. Time periods: Cuvier's photos 2002–2017; Blainville's 1986–2018, including individuals found in <3000 m only

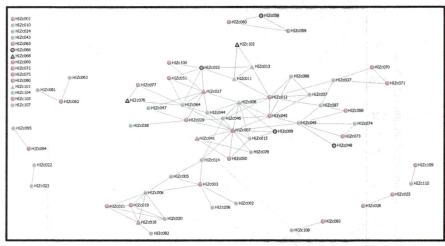

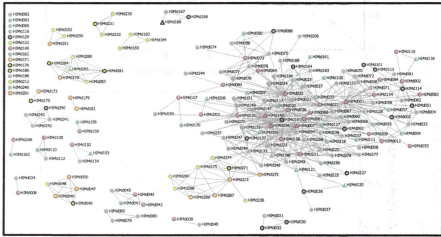

Fig. 14.3 Social networks for Cuvier's beaked whales, *Ziphius cavirostris* (top), and Blainville's beaked whales, *Mesoplodon densirostris* (bottom). For Cuvier's beaked whales, only individuals photo-identified off the island of Hawai'i are shown, while for Blainville's beaked whales, individuals from Hawai'i Island (gray centroid) and O'ahu (yellow centroid) are shown. Individuals that have been satellite tagged are indicated with triangles. Adult females are in pink, adult males in blue, unknown sex adults in black, and subadults, juveniles, and calves in gray

the main cluster of the social network (Fig. 14.3). The greater proportion of individual Blainville's linked by association reflects their larger average group sizes (median = 3, range = 1–11) in comparison to Cuvier's beaked whales, although there is also evidence for individuals from an open-ocean population of Blainville's occasionally using the study area (see below and Baird et al. 2011).

14.2 Evidence for Niche Partitioning and Residency from Satellite Tag Data

Location data from satellite tags provides evidence of population identity (e.g., insular or open-ocean), habitat use, and niche partitioning between the two species. Tag data are available from ten individual Cuvier's (seven adult females, two adult males, and one subadult) tagged off Hawai'i Island between 2006 and 2015, for periods ranging from 7.2 to 49.5 days. While there is a combined 237 days of movement data, two pairs of individuals were tagged in the same group and remained associated for part of the period of overlap, so we effectively have 209 days of location and movement data. For Blainville's beaked whales, locations from satellite tags are available from 15 individuals tagged off Hawai'i Island between 2006 and 2015, for periods ranging from 10.4 to 169.2 days. Of these, three individuals tagged in the same group remained associated (Schorr et al. 2009), and two others were thought to be from an open-ocean population (see below). This leaves us with 456 days of location and movement data from the insular population, from adult females (six) and males (four).

Satellite tag data provide a more complete picture than do sighting surveys of how the species spend their time. On average, Cuvier's beaked whales stayed relatively close to the area where they were tagged (grand median = 51.4 km). All individuals remained largely associated with Hawai'i Island and nearby islands to the north, using our study area off the west side of the island and also off the south, east, and north sides—the latter three areas we have never surveyed (Fig. 14.4). For Cuvier's, the greatest distance moved from where they were tagged was only 203 km, by an adult female. This individual moved to off the north side of Moloka'i (via the east side of Maui). Of the ten tagged individual Cuvier's beaked whales, eight were in the main cluster of the social network (Fig. 14.3), indicating they are part of the resident, island-associated population, and both of those that were not in the main cluster remained close to the island regardless and were thus also probably part of the insular population.

For Blainville's beaked whales, we used three different factors to assess whether individuals are part of the resident insular population or part of an open-ocean population: (1) sighting history or association with known-resident individuals, (2) depth where the group was first encountered, and (3) spatial use after tagging. Of the 15 individuals tagged off Hawai'i Island, 12 are part of the main cluster of the social network, indicating they are part of the resident insular population. These individuals were encountered at depths ranging from 636 to 1492 m (median = 1024 m) when tagged. One other tagged individual, not part of the main cluster, was encountered in 737 m depth, and during the 20 days of tag data, this whale remained in relatively shallow waters (median = 1114 m). In general, individuals remained relatively close to where they were tagged (grand median = 55.1 km). Within their range, areas of highest density tend to have relatively weak surface currents and a high density of mid-water micronekton (Abecassis et al. 2015). Of the known or suspected resident individuals, the greatest distances moved were by two adult males, with one moving to north of

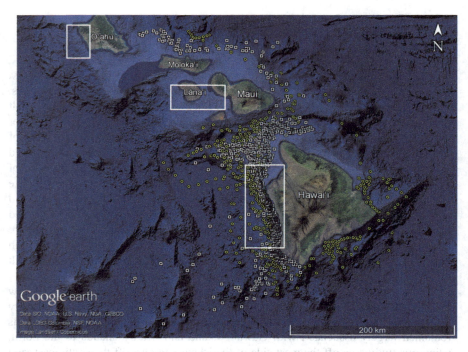

Fig. 14.4 Locations (produced using a switching state space model with 12-h time steps) from satellite tagged Cuvier's beaked whales, *Ziphius cavirostris* (yellow circles), and Blainville's beaked whales, *Mesoplodon densirostris* (white squares) including individuals known or thought to be from the island-associated populations and excluding individuals known to be acting in concert with others. The primary study areas are in white boxes

Maui, and the other as far as the east side of Oʻahu (Fig. 14.4), a distance of 291 km from where it was tagged.[5] Even individuals that are known to be part of the resident, island-associated population can move quite far offshore on occasion. One adult female (HIMd066), part of the resident social network and first documented off the island in 1997, moved as far as 192 km offshore, visiting several seamounts to the southwest of the island, before returning back to the island. By contrast, two other individuals that were not part of the main social cluster were encountered in deep water (3800 and 4000 m) and spent their time in very deep water (medians of 4702 and 4424 m, respectively), suggesting they are part of an open-ocean population. One individual in particular, tagged in 3800 m depth, moved about 1000 km west of the island in the first 20 days after tagging, remaining in deep water over the 39 days of tag data (Baird et al. 2011).

Although the two species broadly overlap along the slopes of the islands, spatial use data suggest they are able to coexist by partitioning the habitat between

---

[5]This individual, HIMd206 in our catalog, was our longest duration track (169 days). HIMd206 also moved offshore of the island for a short period, as far as 72.9 km from shore.

them.[6] On average, Cuvier's beaked whales use depths more twice that of the island-associated Blainville's beaked whales (grand median of 2333 and 1155 m, respectively), and this difference is statistically significant (Mann–Whitney U-test, $p = 0.0035$, $n = 10$ Cuvier's, $n = 11$ Blainville's). By contrast, Cuvier's beaked whales may be found closer to shore (grand median $= 15.0$ km) than Blainville's beaked whales (grand median $= 21.2$ km), although this difference was not significant (Mann–Whitney U-test, $p = 0.45$). This potentially somewhat contradictory finding, with Cuvier's being in deeper water but closer to shore, is perhaps explained by differences in area use. For example, Cuvier's beaked whales used nearshore deepwater areas off the east side of the island of Hawai'i and off the south side of Maui, areas that Blainville's beaked whales did not use. Similarly, Blainville's beaked whales extensively used a relatively shallow area far from shore off the northwestern part of Hawai'i Island, and this area was less frequently used by Cuvier's (Fig. 14.4).

14.3 Diving and Surfacing Behavior: Foraging at Depth and Predator Avoidance Near the Surface

Whalers knew that at least some species of beaked whales could remain underwater for extended periods, with a report of a harpooned northern bottlenose whale remaining submerged for 2 h (Irving 1939). Application of suction-cup attached data logging tags (Hooker and Baird 1999; Baird et al. 2006; Tyack et al. 2006), harpoon-attached data loggers (Minamikawa et al. 2007), and most recently depth-transmitting satellite tags (Schorr et al. 2014; Baird 2016; Joyce et al. 2017) confirmed such long dives and demonstrated that, as noted by Tyack et al. (2006), beaked whales are "extreme" divers, both in terms of durations and depths. In Hawai'i, we have dive data from Cuvier's and Blainville's beaked whales, with short-term high-resolution dive data available from suction-cup attached data loggers (Cuvier's $n = 2$, Blainville's $n = 7$; Baird et al. 2006, 2008a) and longer-term lower-resolution information (i.e., depths and durations of dives and lengths of "surface" periods) from depth-transmitting satellite tags (Cuvier's $n = 4$, Blainville's $n = 2$). Combined, this represents 766 h (~32 days) of dive data for Cuvier's beaked whales (332 dives >800 m) and 559 h (~23 days) of dive data for Blainville's beaked whales (205 dives > 800 m).

Cuvier's and Blainville's beaked whales typically perform at least three different types of dives: short and shallow dives during surfacing series that function primarily for gas exchange, very long and deep foraging dives, and intermediate dives that appear to serve some other purpose than either respiration or foraging. Between longer dives, in Hawai'i Cuvier's typically perform a series of 20–30 short dives (3–15 s in duration) that are only to 2–3 m in depth. For Blainville's beaked whales,

---

[6]Dive data from the two species also suggest niche partitioning; see the diving behavior section below.

these inter-ventilation dives are also only to 2–4 m deep, and the inter-breath intervals range from 5 to 42 s. It is during these relatively short surface periods that we have the opportunity to observe social behaviors, such as they are, and identify individuals as well as record information on the proximity of individuals at the surface.

Maximum dive depths and durations in Hawai'i were 2800 m and 117 min for Cuvier's beaked whales and 1599 m and 83.4 min for Blainville's beaked whales, and both species dive deeper than 800 m about ten times per 24 h. Based on studies of diving behavior in other sympatric odontocetes in Hawai'i, both species of beaked whales are regularly diving deeper than any other odontocetes (Baird 2016). Cuvier's beaked whales do not appear to prepare for such long dives or recover from them at the surface (Baird et al. 2006), while Blainville's beaked whales do both. From our observations with suction-cup attached time-depth recorders, Blainville's increase the number of breaths prior to long dives (we recorded 38 and 41 breaths; Baird et al. 2006) and also recover from long dives, with 15 or more breaths.

These deep dives are known to be foraging dives, based on studies with acoustic tags. Neither species echolocates near the surface; instead echolocation, including buzzes associated with prey captures, occurs at depths greater than about 200 m (Johnson et al. 2004; Tyack et al. 2006). Echolocation starts shallower on the descent of dives and ends deeper on the ascent, suggesting that the whales begin foraging during the descent and stop shortly after they begin to ascend toward the surface. Although they often forage in groups, results from such studies have shown that Blainville's beaked whales are individually catching multiple small prey (an average of about 25) on each foraging dive (Arranz et al. 2011; Madsen et al. 2013). There is limited information on what they are feeding on at such depths, but these deep foraging dives allow the whales to exploit a variety of deepwater cephalopods and fishes that rarely come close to the surface (West et al. 2017).[7] Such long and deep foraging dives occur at similar rates and to similar depths during day and night (Baird et al. 2008a), suggesting that their prey do not vertically migrate, at least on short time scales (but see below for lunar cycle-related movements).

The number of ≥ 800 m dives per hour is similar between species (grand mean $= 0.44$ dives per hour for Cuvier's and 0.41 dives per hour for Blainville's). There is significant evidence of niche partitioning in dive depths: the grand mean dive depths and durations for Cuvier's are 1284 m and 61.3 min, versus 1050 m and 49.9 m for Blainville's (Mann–Whitney U-tests, depth $p = 0.008$, duration $p = 0.01$). Given the water depths where they are typically found in Hawai'i (see above), Cuvier's are likely feeding primarily in mid-water, while Blainville's may feed at least occasionally on or near the bottom (see also Arranz et al. 2011). Lunar

---

[7]Beaked whales are suction feeders, retracting their tongue and using throat grooves to expand their gular region, creating suction (Heyning and Mead 1996). We documented one adult female Cuvier's completely missing her rostrum and two adult female Blainville's with rostrum deformities that might influence their ability to completely close the mouth (Dinis et al. 2017); but even with such deformities, the individuals appeared healthy and obviously able to feed.

cycles may influence the diving behavior of at least some species of beaked whales. One of the depth-transmitting satellite tags deployed on a Blainville's beaked whale in Hawai'i produced 11.8 days of dive and surfacing data over a 15.5-day span of time. From this, we found a significant positive relationship between deep-dive (i.e., dives > 500 m) depth and moon illuminated fraction, both overall (regression, $r^2 = 0.21$, $p = 0.0002$) and, more strongly, for nighttime dives when the moon was above the horizon ($r^2 = 0.366$, $p = 0.0017$), indicating that some vertical migration of prey occurs over longer time scales.

For species that feed at such depths, one obvious question is what happens when a female is accompanied by a small calf. In 2004 we deployed a suction-cup attached time-depth recorder/VHF radio tag on a female Blainville's that was accompanied by a calf estimated to be less than 2 months old. Using VHF radio tracking, we followed the group for over 5 h, and the calf was never seen at the surface while the female was down on any of her long dives, including dives of 48.0 and 47.5 min (to 1408 and 1380 m, respectively). Conditions were good during the period we followed the pair, suggesting we were unlikely to have missed surfacings by the calf (Baird et al. 2006). While it is hard to imagine such a young calf remaining beneath the surface for such periods, beaked whale calves are extremely precocious, even for cetaceans, with very high oxygen storage capabilities in the muscles (Velten 2012). Dunn et al. (2016) note that in 155 encounters with Blainville's beaked whale groups in the Bahamas with calves present, the calves were never seen at the surface alone. This is quite unlike the situation for sperm whales (*Physeter macrocephalus*), where calves of deep-diving mothers are often left with other subadults and adults in female-based matriarchies (Whitehead 1996).

While the very short and shallow dives are intervals between breaths when the whales are near the surface, and the long and deep dives function for foraging, what is the purpose of intermediate dives, typically to depths of 100–600 m for Cuvier's and 30–300 m for Blainville's? Unlike foraging dives, there is a diel pattern in the frequency of these intermediate dives, with an almost complete absence of them during the night (Baird et al. 2008a; see Fig. 14.5). This suggests that these intermediate dives are tied in some way to light levels; during the day, when they are not foraging and not breathing, they avoid near-surface waters. The silence of Blainville's beaked whales when near the surface, and during their long ascents toward the surface, has been suggested as a way of minimizing detection by near-surface predators (Aguilar de Soto et al. 2011). Predators such as killer whales (*Orcinus orca*) detect potential prey through passive listening and, like large sharks, also use vision. Beaked whales in Hawai'i are subject to predation pressure from both killer whales and large sharks (McSweeney et al. 2007; Baird 2016). Such a diel difference in occurrence of intermediate dives suggests that these dives represent avoidance of near-surface waters during the day to minimize detection by visually oriented predators (Baird et al. 2008a). Beaked whale descent rates are much faster than ascent rates (Baird et al. 2006; Tyack et al. 2006). It was originally thought that the slow ascents might function in reducing the likelihood of suffering decompression sickness (Hooker et al. 2009), but more recent models of blood and gas exchange (García-Párraga et al. 2018) suggest that slow ascents might not be physiologically needed.

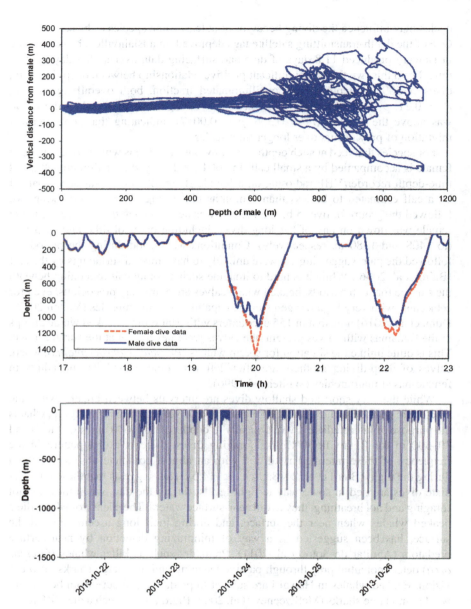

Fig. 14.5 Top: Dive data from a pair of Blainville's beaked whales, *Mesoplodon densirostris*, showing the depth of the adult male in relation to its vertical distance from the adult female. Middle: A 6-h time series of dive data from both individuals, showing intermediate dives (~50–200 m) and two deep foraging dives. Bottom: A ~5.5 day series of dive data from a juvenile Blainville's beaked whale (HIMd198), with nighttime periods indicated by shading. Note the relative lack of intermediate dives during nighttime periods

Slow and quiet ascents would, however, provide an opportunity for visual and potential acoustic detection of predators that might be closer to the surface while not advertising their presence to acoustically astute killer whales.

Based on observations at the surface, individuals tend to dive in synchrony, but less is known about what happens at depth. In Hawai'i we have dive data from two Blainville's beaked whales tagged in the same group, an adult male and female. In this case (Fig. 14.5), until the whales were about 500–600 m deep, they remained at similar depths (within ~30 m), but after that they diverged, the male foraged deeper than or shallower than the female. While we were not able to assess horizontal distances between the two, this is consistent with the whales remaining close to each other prior to the start of foraging (i.e., when they began echolocating), after which they were likely able to keep track of each other by listening to echolocation clicks. They were coordinating behavior, or, more likely, the male was following the female, given the mating system of this species (see below), rejoining on ascents. From tracking individuals through multiple long dives with the aid of VHF tags, individuals would always surface within 10–20 m of each other after the end of long dives. Work by Aguilar de Soto et al. (2018) also demonstrates that when foraging in a group, Cuvier's and Blainville's coordinate their dives and their vocal activity, in such a way that would minimize detection by potential predators such as killer whales while generally foraging on small individual prey several hundred meters apart.

14.4 Social Organization: Insights from Group Composition, Association Patterns, and Social Behavior

Knowing sex and approximate age of many individuals makes it possible to understand social organization. One limiting factor in interpreting association patterns is that we do not yet know precisely the age of sexual maturity for Cuvier's beaked whales. Few calves have been documented, and only one female has been known to have two calves (7 years apart). Therefore, it is unlikely that Cuvier's become pregnant when still nursing a calf, which helps in the interpretation of association patterns. Blainville's beaked whales are thought to give birth to their first calf when between 10 and 15 years of age (Claridge 2013).

14.4.1 Cuvier's Beaked Whales

Cuvier's beaked whales are not particularly social. Of the 18 species of odontocetes in Hawaiian waters, only pygmy sperm whales are found in smaller groups, on average (Baird 2016). In Hawai'i, the most frequently encountered groups are of lone individuals, representing 34.2% of all sightings, and the median group size is two (mean = 2.2, maximum = 5). Group sizes in the Canary Islands are similar (mode = 1, median = 2, max = 7, n = 357; C. Reyes and N. Aguilar de Soto, personal communication). Off Cape Hatteras, North Carolina, in an area with much higher density of Cuvier's beaked whales (McLellan et al. 2018), two is the most frequently encountered group size (Fig. 14.6), and lone individuals are uncommon (14.5% of sightings). While median group size off Cape Hatteras is only three, the maximum group size documented was eight (A.J. Read, unpublished), with a similar sample size (Hawai'i n = 79; North Carolina n = 83). Group sizes in the Ligurian Sea are somewhat in between (mode = 2, median = 2, max = 8, n = 100; Moulins et al. 2007; Tepsich et al. 2014; CIMA Foundation unpublished data). Such differences in group size likely reflect ecological differences among the areas. Although our understanding of deepwater ecology is limited, the area surrounding the Hawaiian Islands is generally oligotrophic, with only small increases in productivity associated with the islands, while the area off Cape Hatteras is highly productive. Whether such productivity extends into the bathypelagic ecosystem is unclear, but I suspect it does, allowing for the formation of slightly larger groups (Fig. 14.6).

Group size, as well as age and sex of the group members, can be used to help interpret the function or purpose of grouping or in this case lack of grouping. In Hawai'i, of 22 lone individuals for which we were able to determine age class, 8 were adult males, 8 adult females, and 6 subadults of unknown sex. Such a high presence of lone adults suggests that, at least in an area with relatively low abundance, there is no strong selective pressure for grouping, in terms of either potential benefits of group foraging (evidence from acoustic studies suggest the whales forage on small individual prey, even when in a group) or avoidance of predation (at least for lone adults). For pairs of individuals, of the 13 cases when we were able to confirm age class, there was only one single pair comprised of an adult male and adult female; other groups were an adult female with a juvenile or small calf (seven groups), pairs of subadults (two groups) or unknown adults with a subadult (two groups), and one adult female–adult female pair. Of 18 groups of 3 or more where all individuals were identified and sex of all adults was known, only 3 calves less than 1 year of age were documented, and all were in groups of 4 or 5 individuals that contained 2 adult females and 1 adult male (3 of 7 groups, ~43%). This suggests that females with small calves may be more likely to associate with other adults. One possible explanation is reduced predation risk—a female nursing a small calf that may not have completely developed its diving capabilities might be more likely to associate with other adults as a way of diluting the risk of predation to her calf or to benefit from the increased ability of the group to detect predators (Mann et al. 2000; Creel et al. 2014).

Five of the groups of three or more individuals had no adult males present; in each case there was a pair of adult females with one or two additional juvenile or subadult individuals. By contrast, there were seven groups with two adult males and one or two adult females, with no juveniles or subadults, presumably reflecting that females are

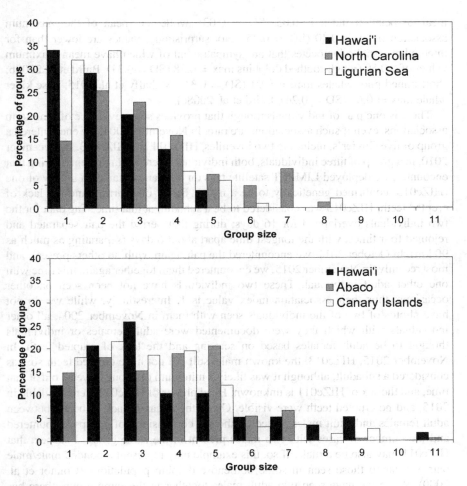

Fig. 14.6 Comparison of group sizes for Cuvier's beaked whales, *Ziphius cavirostris* (top), and Blainville's beaked whales, *Mesoplodon densirostris* (bottom) encountered during small-boat effort, each in three different study areas. North Carolina data provided by A. J. Read; Abaco, Bahamas, data provided by D.E. Claridge; Canary Islands data provided by C. Reyes and N. Aguilar de Soto; and Ligurian Sea data provided by M. Rosso. The x- and y-axis scales are the same for comparative purposes

more likely to come into estrus once their earlier calf has weaned and dispersed.[8] Association analyses provide a basis for understanding the details and nature of these groupings. Using a half-weight association index with all individuals in our larger dataset (i.e., including citizen science contributions) seen three or more times, the

---

[8]Much of this is speculative of course, since so little is known about the reproductive cycles of beaked whales.

mean association index is 0.05 (SD = 0.03), while the mean of the maximum association index is 0.40 (SD = 0.25). Not surprisingly, values are lower than for three other social odontocetes that are sympatric, all of which have mean maximum values of >0.50 (rough-toothed dolphins max = 0.58 (SD = 0.31), Baird et al. 2008b; short-finned pilot whales max = 0.91 (SD = 0.31), Mahaffy et al. 2015; false killer whale max = 0.64 (SD = 0.24), Baird et al. 2008c).

There is one pair of individuals though that provides some evidence of long-term associations, even if such associations are rare. In November 2004, we encountered a group of five Cuvier's, including two juveniles, HIZc011 and HIZc013. In December 2010, in a group of three individuals, both individuals were again present. During that encounter, we deployed LIMPET satellite tags on both and obtained a biopsy of one (HIZc013, confirmed genetically to be a male). Based on scarring and the lack of erupted teeth, HIZc013 was considered to be a subadult at that time. Tag data for the two individuals overlapped for 16 days; during that period the pair separated and rejoined four times, with the longest time apart about 6 days (separating as much as 90 km). In October 2011, we encountered the pair again, with no others present, and most recently, in November 2015, we encountered them together again, this time with one other adult individual. These two individuals have not been seen on other occasions; thus their association index value is 1. Interestingly, while we do not have photos of two of the individuals seen with them in November 2004, all other individuals with which they were documented were adult females or individuals thought to be adult females based on scarring and the lack of erupted teeth. In November 2015, HIZc013, the known male, still did not have erupted teeth so was considered a subadult, although it was likely a minimum of about 13 years old at that time, and the sex of HIZc011 is unknown. Head photos of HIZc011 were available in 2015, and no erupted teeth were visible. Given the apparent lack of bonds between adult females and adult males, based on the sex composition of groups encountered and the number of cases with two males present in the same group, I suspect that HIZc011 may also be a male. If so, this example may represent a bonded male/male pair, similar to those seen in some bottlenose dolphin populations (Connor et al. 2000). When we have seen two adult males together in the same group, there has always been at least one (and often two) adult female(s) present, and the males have always remained apart, typically on either side of the female(s). In the past I interpreted such separation as evidence of competition between the two males over access to the females, effectively some sort of prolonged posturing, while the individuals were sorting out who was dominant (or the female was trying to choose with whom to mate). The large number of linear scars in adult male Cuvier's beaked whales clearly indicates that direct male–male competition occurs, but there is also a suggestion that sperm competition may be important for this species (MacLeod 2010). It is possible that pairs of adult males at times cooperate to sequester a female from other competing males, and if both get to mate with the female, they thus have an increased likelihood of siring offspring.

Surprisingly, this type of time series is the best indicator of the age of sexual maturity for Cuvier's beaked whales—we are awaiting our next encounter of these two, with the hopes of documenting tooth eruption in HIZc013, to confirm the sex of HIZc011 and to see if they are still together. If they are still together, knowing the sex of HIZc011 should help confirm or refute the idea that males may occasionally form long-term bonds. Obtaining a genetic sample of HIZc011 to examine relatedness between the pair would also help elucidate the potential reasons for such apparently enduring bonds. We had one other case with two Cuvier's satellites tagged in the same group, an adult female and adult male tagged in the same group in May 2008. In this case, the pair remained together for the first 8 days before separating. Although they have been seen a combined 13 times (6 times for the adult male and 7 times for the adult female), this is the only encounter where the two were documented together, so associations over those 8 days likely represented a relatively rare association, rather than an enduring bond between them.

14.4.2 Blainville's Beaked Whales

Blainville's beaked whales are more social than Cuvier's, albeit only slightly; the mean group size in Hawai'i is 3.8, and both the median and modal group sizes are three (range $= 1-11$). Group sizes in other areas where this species has been studied are generally in the same range, although there is variability (Fig. 14.6). Lone individuals are much less common than for Cuvier's, representing just 12% of all sightings, and have included adult females (twice), an adult male (once), subadult male (once), and subadult of unknown sex (once). This lack of an obvious sex bias for lone individuals suggests they are not adult males moving between groups. For pairs of individuals, of the ten cases when we were able to confirm age class (and in most cases sex) of both individuals in the group, there were two pairs comprised of an adult male and an adult female; other groups were of an adult female with a juvenile or small calf (five groups), an adult female–adult female pair (one group), a subadult and a juvenile (one group), or an unknown adult with a subadult (one group). Of 34 groups of 2 or more individuals for which the sex of all adults in the group was known, there were only 2 groups with more than 1 adult male present: a group of 8 including 3 adult female/juvenile pairs and a group of 5 containing 3 adult females. In both cases, the adult males remained separated from each other during the encounters. Calves less than 1 year old were documented in ten different groups, and the median group size of these was six individuals (maximum $= 9$), double the overall median group size. This suggests that females with small calves preferentially associate with other females, possibly due to some advantage in terms of detecting predators or minimizing predation risk on their calf, as mentioned for Cuvier's above. Adult males were present in only three of the ten groups, and in all three cases, there were multiple adult females present (2, 3, 3), suggesting that adult males only associate with such groups if there are adult females present that might be in, or may be about to come into, estrus. In those same ten groups, adult females were present in all groups, and more than one adult female was present in five groups.

Using a half-weight association index with all individuals in our larger dataset (i.e., including citizen science contributions) seen three or more times, the mean association index for Blainville's beaked whales is 0.08 (SD $= 0.03$), while the mean of the maximum association index is 0.44 (SD = 0.22). Examining dyads illustrates the duration of bonds among different age/sex classes.[9] For example, between adult females and juveniles or calves ($n = 36$ dyads), at least some of which are likely mothers with their calves, 16 (44.4%) were seen together over periods greater than 2 months, and the maximum span of time together has been about 2.5 years, suggesting that calves disperse between 2 and 3 years of age.[10] For adult male–adult female dyads ($n = 23$), only three (13%) were longer than 2 months. Based on the number of intervening sightings of one or both individuals in the dyad without the other present, the three that were longer represented cases where the two individuals associated, disassociated for periods of from 1 to 11 years, and then reassociated, rather than reflecting a long-term association among them. Adult male dyads were only documented four times, and there were no repeated associations for them. Unlike the unusual example of the two subadult Cuvier's beaked whales appearing to remain associated for more than 10 years, there were no long-term associations among subadults or juveniles. Of the five dyads between subadults and juveniles meeting our criteria,[9] all were associated for less than 1 year. For Blainville's, the only age/sex classes that appeared to reassociate on a semi-regular basis are adult females with other adult females. Of the 39 adult female dyads that were seen together on 2 more occasions, almost half of cases where they reassociated (17, 43.6%) were greater than 2 months apart, and 12 of those were greater than 1 year. In most cases there are numerous intervening sightings of one or both individuals in the dyad without the other present, suggesting that these associations are not constant. Individual females reassociated with each other over spans ranging up to 17 years. These types of associations are consistent with the larger group sizes documented when calves are present, again suggesting that females seek out other females when they have small calves.

Like Cuvier's, Blainville's beaked whale males clearly fight repeatedly with other males, with extensive scarring focused on the front half of the body. Injuries occasionally include removal of tissue around the base of one or both teeth and tissue loss on the head. As previously suggested by Claridge (2006) and McSweeney et al. (2007), Blainville's beaked whales appear to exhibit female defense polygyny, where one male escorts one or more receptive females and denies access of other males to the group. Our sample size, adding 12 additional years to the study of McSweeney et al. (2007), supports this suggestion, based on the composition and stability of groups. The few examples we have of subadult males in Hawai'i suggest that they associate with adults of both sexes as well as younger animals. One subadult male, HIMd147, for which we have the longest sighting history, was initially found in

---

[9]There were 609 dyads documented, but these analyses, with the exception of adult male–adult male dyads, were restricted to cases where both members of the dyad have been seen at least three times (together or apart, $n = 167$ dyads).

[10]Note some of these were documented without the other individual present in intervening periods, but particularly for the citizen science contributions, we do not always have good identifications from all individuals of a group, so in some cases both members of a dyad may have been present and not documented.

larger groups composed of adult females occasionally with adult males present, but as a subadult has been seen either by itself or with one adult male. This individual lacks the linear scarring characteristic of adult males; however, other subadult males in Hawai'i exhibit some scarring consistent with occasional interactions with adult males, a likely indicator that they are close to reaching sexual maturity.

14.5 Conclusions

Our studies in Hawai'i and research elsewhere have provided evidence of small populations that show high site fidelity for Blainville's and Cuvier's beaked whales. Coexisting in the same bathypelagic ecosystem, the two species show clear signs of resource partitioning in three dimensions, both where they occur in relation to island slopes and in how deep they dive. Like most other animals, behavior is largely driven by competing demands to avoid predators, find food, and, at least for males, increase mating opportunities. Both species are somewhat asocial, and groups that form seem to do so for reasons other than cooperative foraging, in the case of females with small calves seemingly to reduce predation risk. Their vocal behavior, the depths they dive to when not foraging, and even their ascent rates from dives appear to be driven at least in part to minimize detection from predators. As research on these species has expanded in varied habitats worldwide, and with results coming in from multiple ongoing long-term studies, beaked whales are beginning to earn an important place in our overall understanding of cetacean ethology and behavioral ecology.

Let us not forget the conservation concerns that originally spurred on research on these species. Both Cuvier's and Blainville's, and at least some other species of beaked whales, show strong responses to anthropogenic sounds, at times resulting in the death of individuals (Balcomb and Claridge 2001; Aguilar Soto et al. 2006; Cox et al. 2006; Manzano-Roth et al. 2016; Cholewiak et al. 2017; Falcone et al. 2017). For those individuals repeatedly exposed to high-intensity anthropogenic sounds, cessation of foraging for periods of time has the potential to influence survival and reproduction (New et al. 2013). Even in areas where the species persist in spite of repeated exposure to sonar, such as around San Clemente Island off California (Falcone et al. 2009), it is possible such areas act as population sinks, with cryptic mortality (Faerber and Baird 2010), or reduced reproductive rates that could result in population declines (Moore and Barlow 2013). Given the low density of most beaked whale populations, and the evidence of population structure, detecting declines using abundance estimates from traditional line transect surveys will be difficult, if not impossible. For small populations, mark–recapture abundance estimates using photo-identification data will produce more precise estimates, potentially allowing for examining population trends. However, for assessing population-level effects of repeated disturbance, studies of age structure of populations (e.g., Claridge 2013) may be more useful. Combining such studies with satellite tagging and genetics to understand population structure and range may be necessary to help elucidate the details of beaked whale lives. Despite studies in multiple areas that

326

R. W. Baird

have been ongoing for 20 years or more, and tremendous advances in our understanding of 4 of the 22 recognized species of beaked whales, much remains to be learned, not only about the most well-known species of beaked whales but about the many others in this diverse family.

Acknowledgments Our work is a team effort, in the field and office. Dan McSweeney deserves special mention for systematically photographing beaked whales in Hawai'i since the mid-1980s and contributing them to this study, and I also thank Daniel Webster, Greg Schorr, and numerous staff and volunteers for their work in the field. Much of what we know about beaked whales in Hawai'i comes from citizen science and contributions of photographs by individuals who work or play on the water, and their efforts are greatly acknowledged as they play a major role in our understanding of these species. In the office, I particularly thank Sabre Mahaffy for her contributions both to understanding beaked whales in Hawai'i and to providing results from many of the analyses used in this chapter, as well as David Anderson for analyses of satellite tag data. I also thank a number of colleagues, Natacha Aguilar de Soto, Diane Claridge, Andy Read, Cris Reyes, Massimiliano Rosso, and Greg Schorr, for providing unpublished data and answering questions regarding their studies and Bernd Würsig and two anonymous reviewers for helpful comments on the manuscript.

References

Abecassis M, Polovina J, Baird RW, Copeland A, Drazen JC, Komokos R, Oleson E, Jia Y, Schorr GS, Webster DL, Andrews RD (2015) Characterizing a foraging hotspot for short-finned pilot whales and Blainville's beaked whales off the west side of the Island of Hawai'i with tagging and oceanographic data. PLoS One. https://doi.org/10.1371/journal.pone.0142628

Aguilar de Soto N, Madsen PT, Tyack P, Arranz P, Marrero J, Fais A, Revelli E, Johnson M (2011) No shallow talk: cryptic strategy in the vocal communication of Blainville's beaked whales. Mar Mamm Sci. https://doi.org/10.1111/j.1748-7692.2011.00495.x

Aguilar de Soto N, Visser F, Madsen P, Tyack P, Ruxton G, Arranz P, Alcazar J, Johnson M (2018) Beaked and killer whales show how collective prey behaviour foils acoustic predators. bioRxiv. https://doi.org/10.1101/303743

Aguilar Soto N, Johnson M, Madsen PT, Tyack PL, Bocconcelli A, Borsani JF (2006) Does intense ship noise disrupt foraging in deep-diving Cuvier's beaked whales (*Ziphius cavirostris*)? Mar Mamm Sci 22:690–699

Arranz P, Aguilar de Soto N, Madsen PT, Brito A, Bordes F, Johnson MP (2011) Following a foraging fish-finder: diel habitat use of Blainville's beaked whales revealed by echolocation. PLoS One. https://doi.org/10.1371/journal.pone.0028353

Baird RW (2016) The lives of Hawai'i's dolphins and whales: natural history and conservation. University of Hawai'i Press, Honolulu, 342 p

Baird RW, Webster DL, McSweeney DJ, Ligon AD, Schorr GS, Barlow J (2006) Diving behaviour of Cuvier's (*Ziphius cavirostris*) and Blainville's (*Mesoplodon densirostris*) beaked whales in Hawai'i. Can J Zool 84:1120–1128

Baird RW, Webster DL, Schorr GS, McSweeney DJ, Barlow J (2008a) Diel variation in beaked whale diving behavior. Mar Mamm Sci 24:630–642

Baird RW, Webster DL, Mahaffy SD, McSweeney DJ, Schorr GS, Ligon AD (2008b) Site fidelity and association patterns in a deep-water dolphin: rough-toothed dolphins (*Steno bredanensis*) in the Hawaiian Archipelago. Mar Mamm Sci 24:535–553

Baird RW, Gorgone AM, McSweeney DJ, Webster DL, Salden DR, Deakos MH, Ligon AD, Schorr GS, Barlow J, Mahaffy SD (2008c) False killer whales (*Pseudorca crassidens*) around the main Hawaiian Islands: long-term site fidelity, inter-island movements, and association patterns. Mar Mamm Sci 24:591–612

Baird RW, McSweeney DJ, Schorr GS, Mahaffy SD, Webster DL, Barlow J, Hanson MB, Turner JP, Andrews RD (2009) Studies of beaked whales in Hawai'i: population size, movements, trophic ecology, social organization and behaviour. Eur Cetacean Soc Spec Publ 51:23–25

Baird RW, Schorr GS, Webster DL, Mahaffy SD, McSweeney DJ, Hanson MB, Andrews RD (2011) Open-ocean movements of a satellite-tagged Blainville's beaked whale (*Mesoplodon densirostris*): evidence for an offshore population in Hawai'i? Aquat Mamm 37:506–511

Baird RW, Webster DL, Aschettino JM, Schorr GS, McSweeney DJ (2013) Odontocete cetaceans around the main Hawaiian Islands: habitat use and relative abundance from small-boat surveys. Aquat Mamm 39:253–269

Balcomb KC, Claridge DE (2001) A mass stranding of cetaceans caused by naval sonar in the Bahamas. Bahamas J Sci 8(2):2–12

Baumann-Pickering S, Roch MA, Brownell RL Jr, Simonis AE, McDonald MA, Solsona-Berga A, Oleson EM, Wiggins SM, Hildebrand JA (2014) Spatio-temporal patterns of beaked whale echolocation signals in the North Pacific. PLoS One 9:e86072

Benjaminsen T, Christensen I (1979) The natural history of the bottlenose whale, *Hyperoodon ampullatus* (Forster). In: Winn HE, Olla BL (eds) Behavior of marine animals. Plenum, New York

Bradford AL, Forney KA, Oleson EM, Barlow J (2017) Abundance estimates of cetaceans from a line-transect survey within the U.S. Hawaiian Islands Exclusive Economic Zone. Fish Bull 115:129–142

Cholewiak D, DeAngelis AI, Palka D, Corkeron PJ, Van Parijs SM (2017) Beaked whales demonstrate a marked acoustic response to the use of shipboard echosounders. R Soc Open Sci 4:170940

Claridge DE (2006) Fine-scale distribution and habitat selection of beaked whales. M.Sc. Thesis, University of Aberdeen, Scotland

Claridge DE (2013) Population ecology of Blainville's beaked whales (*Mesoplodon densirostris*). Ph.D. Thesis, University of St. Andrews, Scotland

Connor RC, Wells RW, Mann J, Read AJ (2000) The bottlenose dolphin: social relationships in a fission-fusion society. In: Cetacean societies: field studies of dolphins and whales. University of Chicago Press, Chicago, pp 91–126

Coomber F, Moulins A, Tepsich P, Rosso M (2016) Sexing free-ranging adult Cuvier's beaked whales (*Ziphius cavirostris*) using natural marking thresholds and pigmentation patterns. J Mammal 97:879–890

Cox TM, Ragen TJ, Read AJ, Vos E, Baird RW, Balcomb K, Barlow J, Caldwell J et al (2006) Understanding the impacts of anthropogenic sound on beaked whales. J Cetacean Res Manag 7:177–187

Creel S, Schuette P, Christianson D (2014) Effects of predation risk on group size, vigilance, and foraging behavior in an African ungulate community. Behav Ecol 25:773–784

Dalebout ML, Mead JG, Baker CS, Baker AN, van Helden AL (2002) A new species of beaked whale *Mesoplodon perrini* sp. n. (Cetacea: Ziphiidae) discovered through phylogenetic analyses of mitochondrial DNA sequences. Mar Mamm Sci 18:577–608

Dalebout ML, Baker CS, Steel D, Thompson K, Robertson KM, Chivers SJ, Perrin WF, Goonatilake M, Anderson RC, Mean JG, Potter CW, Thompson L, Jupiter D, Yamada TK (2014) Resurrection of *Mesoplodon hotaula* Deraniyagala 1963: a new species of beaked whale in the tropical Indo-Pacific. Mar Mamm Sci 30:1081–1108

Dinis A, Baird RW, Mahaffy SD, Martín V, Alves F (2017) Beaked whales with rostrum deformities: implications for survival and reproduction. Mar Mamm Sci. https://doi.org/10.1111/mms.12406

Dunn C, Claridge D, Durban J, Shaffer J, Moretti D, Tyack P, Rendell L (2016) Insights into Blainville's beaked whale (*Mesoplodon densirostris*) echolocation ontogeny from recordings of mother-calf pairs. Mar Mamm Sci. https://doi.org/10.1111/mms.12351

Faerber MM, Baird RW (2010) Does a lack of observed beaked whale strandings in military exercise areas mean no impacts have occurred? A comparison of stranding and detection probabilities in the Canary and main Hawaiian Islands. Mar Mamm Sci 26:602–613

Falcone EA, Schorr GS, Douglas AB, Calambokidis J, Henderson E, McKenna MF, Hildebrand J, Moretti D (2009) Sighting characteristics and photo-identification of Cuvier's beaked whales (*Ziphius cavirostris*) near San Clemente Island, California: a key area for beaked whales and the military? Mar Biol 156:2631–2640

Falcone EA, Schorr GS, Watwood SL, DeRuiter SL, Zerbini AN, Andrews RD, Morrissey RP, Moretti DJ (2017) Diving behaviour of Cuvier's beaked whales exposed to two types of military sonar. R Soc Open Sci 4:170629

Fedutin ID, Filatova OA, Mamaev EG, Burdin AM, Hoyt E (2015) Occurrence and social structure of Baird's beaked whales, *Berardius bairdii*, in the Commander Islands, Russia. Mar Mamm Sci 31:853–865

Fernández A, Edwards JF, Rodríguez F, Espinosa de los Monteros A, Herráez P, Castro P, Jaber JR, Martín V, Arbelo M (2005) "Gas and fat embolic syndrome" involving a mass stranding of beaked whales (Family Ziphiidae) exposed to anthropogenic sonar signals. Vet Pathol 42:446–457

Frantzis A (1998) Does acoustic testing strand whales? Nature 392:29

García-Párraga D, Moore M, Fahlman (2018) Pulmonary ventilation–perfusion mismatch: a novel hypothesis for how diving vertebrates may avoid the bends. Proc R Soc B 285(1877):20180482. https://doi.org/10.1098/rspb.2018.0482

Gowans S, Whitehead H, Arch J, Hooker SK (2000) Population size and residency patterns of northern bottlenose whales (*Hyperoodon ampullatus*) using the Gully. J Cetacean Res Manag 2:201–210

Heyning JE, Mead JG (1996) Suction feeding in beaked whales: morphological and observation evidence. Contrib Sci 464:1–12

Hooker SK, Baird RW (1999) Deep-diving behaviour of the northern bottlenose whale, *Hyperoodon ampullatus* (Cetacea: Ziiphidae). Proc R Soc Lond B 266:671–676

Hooker SK, Whitehead H, Gowans S, Baird RW (2002) Fluctuations in distribution and patterns of individual range use of northern bottlenose whales. Mar Ecol Prog Ser 225:287–297

Hooker SK, Baird RW, Fahlman A (2009) Could beaked whales get the bends? Effect of diving behaviour and physiology on modelled gas exchange for three species: *Ziphius cavirostris*, *Mesoplodon densirostris* and *Hyperoodon ampullatus*. Respir Physiol Neurobiol 167:235–246

Irving L (1939) Respiration in diving mammals. Am J Physiol 123:112–134

Johnson M, Madsen PT, Zimmer WMX, Aguilar de Soto N, Tyack PL (2004) Beaked whales echolocate on prey. Proc R Soc Lond B 271:S383–S386

Joyce TW, Durban JW, Claridge DE, Dunn CA, Fearnbach H, Parsons KM, Andrews RD, Balance LT (2017) Physiological, morphological, and ecological tradeoffs influence vertical habitat use of deep-diving toothed whales in the Bahamas. PLoS One 12:e0185113

Kasuya T (2017) Small cetaceans of Japan: exploitation and biology. CRC Press, Boca Raton

MacLeod CD (2010) The relationship between body mass and relative investment in testes mass in cetaceans: implications for inferring interspecific variations in the extent of sperm competition. Mar Mamm Sci 26:370–380

Madsen PT, Aguilar de Soto N, Arranz P, Johnson M (2013) Echolocation in Blainville's beaked whales (*Mesoplodon densirostris*). J Comp Physiol A 199:451–469

Mahaffy SD, Baird RW, McSweeney DJ, Webster DL, Schorr GS (2015) High site fidelity, strong associations, and long-term bonds: short-finned pilot whales off the island of Hawai'i. Mar Mamm Sci 31:1427–1451

Mann J, Connor RC, Barre LM, Heithaus MR (2000) Female reproductive success in bottlenose dolphins (*Tursiops* sp.): life history, habitat, provisioning, and group-size effects. Behav Ecol 11:210–219

Manzano-Roth R, Henderson EE, Martin SW, Martin C, Matsuyama BM (2016) Impacts of U.S. Navy training events on Blainville's beaked whale (*Mesoplodon densirostris*) foraging dives in Hawaiian waters. Aquat Mamm 42:507–518

McLellan WA, McAlarney RJ, Cummings EW, Read AJ, Paxton CGM, Bell JT, Pabst DA (2018) Distribution and abundance of beaked whales (Family Ziphiidae) off Cape Hatteras, North Carolina, USA. Mar Mamm Sci. https://doi.org/10.1111/mms.12500

McSweeney DJ, Baird RW, Mahaffy SD (2007) Site fidelity, associations and movements of Cuvier's (*Ziphius cavirostris*) and Blainville's (*Mesoplodon densirostris*) beaked whales off the island of Hawai'i. Mar Mamm Sci 23:666–687

Minamikawa S, Iwasaki T, Kishiro T (2007) Diving behaviour of a Baird's beaked whale, *Berardius bairdii*, in the slope water region of the western North Pacific: first dive records using a data logger. Fish Oceanogr 16:573–577

Moore JE, Barlow JP (2013) Declining abundance of beaked whales (Family Ziphiidae) in the California current large marine ecosystem. PLoS One 8:e52770

Morin PA, Baker CS, Brewer RS, Burdin AM, Dalebout ML, Dines JP, Fedutin ID, Filatova OA, Hoyt E, Jung J-L, Lauf M, Potter CW, Richard G, Ridgway M, Robertson KM, Wade PR (2017) Genetic structure of the beaked whale genus *Berardius* in the North Pacific, with genetic evidence for a new species. Mar Mamm Sci 31:96–111

Moulins A, Rosso M, Nani B, Würtz M (2007) Aspects of distribution of Cuvier's beaked whale (*Ziphius cavirostris*) in relation to topographic features in the Pelagos Sanctuary (north-western Mediterranean Sea). J Mar Biol Assoc UK 87:177–186

New LF, Moretti DJ, Hooker SK, Costa DP, Simmons SE (2013) Using energetic models to investigate the survival and reproduction of beaked whales (family Ziphiidae). PLoS One 8: e68725

Reyes C (2017) Abundance estimate, survival and site fidelity patterns of Blainville's (*Mesoplodon densirostris*) and Cuvier's (*Ziphius cavirostris*) beaked whales off El Hierro (Canary Islands). Master's Thesis, University of St. Andrews, Scotland

Schorr GS, Baird RW, Hanson MB, Webster DL, McSweeney DJ, Andrews RD (2009) Movements of satellite-tagged Blainville's beaked whales off the island of Hawai'i. Endanger Species Res 10:203–213

Schorr GS, Falcone EA, Moretti DJ, Andrews RD (2014) First long-term behavioral records from Cuvier's beaked whales (*Ziphius cavirostris*) reveal record-breaking dives. PLoS One 9:e92633

Tepsich P, Rosso M, Halpin PN, Moulins A (2014) Habitat preferences of two deep-diving cetacean species in the northern Ligurian Sea. Mar Ecol Prog Ser 508:247–260

Tyack PL, Johnson M, Aguilar Soto N, Sturlese A, Madsen PT (2006) Extreme diving of beaked whales. J Exp Biol 209:4238–4253

Tyack PL, Zimmer WMX, Moretti D, Southall BL, Claridge DE, Durban JW, Clark CW, D'Amico A, DiMarzio N, Jarvis S, McCarthy E, Morrissey R, Ward J, Boyd IL (2011) Beaked whales respond to simulated and actual navy sonar. PLoS One 6:e17009

Velten BP (2012) A comparative study of the locomotor muscle of extreme deep-diving cetaceans. M.Sc. Thesis, University of North Carolina Wilmington

West KL, Walker WA, Baird RW, Mead JG, Collins PW (2017) Diet of Cuvier's beaked whales *Ziphius cavirostris* from the North Pacific and a comparison with their diet world-wide. Mar Ecol Prog Ser 574:227–242. https://doi.org/10.3354/meps12214

Whitehead H (1996) Babysitting, dive synchrony, and indications of alloparental care in sperm whales. Behav Ecol Sociobiol 38:237–244

Whitehead H, Gowans S, Faucher A, McCarrey S (1997) Population analysis of northern bottlenose whales in the Gully, Nova Scotia. Mar Mamm Sci 13:173–185

Chapter 15
Common Bottlenose Dolphin Foraging: Behavioral Solutions that Incorporate Habitat Features and Social Associates

Randall S. Wells

Abstract Common bottlenose dolphins (*Tursiops truncatus*) live in a large variety of habitats, where they confront a wide range of ecological challenges to which they have developed diverse behavioral solutions. They inhabit shallow marsh creeks, estuaries, bays, open coasts, islands, shelves, and deep open ocean. Abiotic factors such as physiography, salinity, temperature, depth, tidal excursions, and currents influence ecological factors that in turn help shape behaviors of bottlenose dolphins, within their morphological and physiological constraints. Among the ecological factors of greatest importance for influencing bottlenose dolphin behavior is its prey, and foraging serves as the focus of this review. Bottlenose dolphins consume a wide variety of prey, primarily fish and squid, that typically are taken in one bite. Prey vary in size, energy content, behavior, schooling tendency, speed, maneuverability, seasonal availability, sensory abilities, sound production, defenses, location in the water column, and use of habitat features or structures. The availability of potential prey to the dolphins is dictated largely by the dolphins' biology and the development of appropriate skills for detecting, capturing, and handling prey. The interplay of characteristics of the fish, features of their environment, and capabilities of the dolphins themselves shape the dolphins' foraging behaviors and influence dolphin sociality.

Keywords Common bottlenose dolphin · Foraging behavior · Feeding behavior · Passive listening · Cooperative behavior · Observational learning · Human interactions

Foraging involves a suite of behaviors from searching for and detecting prey, through pursuit, capture, and handling (e.g., Nowacek 2002). Foraging behaviors

---

R. S. Wells (✉)
Chicago Zoological Society's Sarasota Dolphin Research Program, c/o Mote Marine Laboratory, Sarasota, FL, USA
e-mail: rwells@mote.org

© Springer Nature Switzerland AG 2019
B. Würsig (ed.), *Ethology and Behavioral Ecology of Odontocetes*, Ethology and Behavioral Ecology of Marine Mammals,
https://doi.org/10.1007/978-3-030-16663-2_15

are influenced by a number of factors such as innate physical abilities, sensory systems, ability to learn, familiarity with prey, familiarity with area and habitat, ability to work with conspecifics, and habitat features that can facilitate prey detection or capture. Bottlenose dolphins employ a variety of senses in the first stage of foraging, detecting prey. The relative utility of dolphin sensing capabilities varies with habitat. They can use vision in clear water, but must rely on acoustics when underwater visibility is poor, as is the case in many of their inshore habitats. This can take the form of passive listening, in which dolphins listen for sounds produced by the prey fish themselves. Alternatively, active echolocation can be used to find prey, balanced against the risk that at least some prey species are able to sense echolocation and initiate evasive maneuvers.

In shallow inshore waters, prey detection may benefit from individual dolphin residency to a well-established home range, familiarity with the available prey and conditions, and experience with the most effective approaches for finding prey individually. Bottlenose dolphins in many areas have demonstrated long-term residency to well-defined home ranges (Wells and Scott 2018). In Sarasota Bay, Florida, such residency has been documented across more than four decades, and involves at least five generations within a given lineage, including individuals up to 67 years of age (Wells 2014). In deep open waters, where prey schools tend to be mobile and patchy but rich, integrated sensory systems from bottlenose dolphins working in groups may extend prey finding abilities over a broader area (Norris and Dohl 1980). Groups tend to be more variable and reach larger sizes in offshore waters; group size may be constrained at least in part by the ability of the food patches to meet energetic needs of the group (Wells et al. 1980; Scott and Chivers 1990).

Once a prey item has been detected, pursuit can involve a direct chase in open waters, or in the case of prey using habitat features for cover, specialized behaviors may be employed for flushing prey, as described below. In a direct chase, a bottlenose dolphin tends to have an advantage in terms of speed, but its larger size constrains its maneuverability, so prey may have an advantage in terms of tight maneuvering.

Prey capture can involve a variety of specialized techniques, depending on prey type and habitat. In inshore waters, bottlenose dolphins often feed individually. Individual dolphins may use features of the habitat as barriers to restrict prey movements for capture. Where barriers are absent, such as in deep offshore waters, or even in some cases when barriers exist, bottlenose dolphins can work together to limit prey options for escape. Coordinated foraging by dolphins in groups is often used with schooling prey.

While most bottlenose dolphin capture of fish involves head-first grasping and ingestion of whole prey with minimal oral manipulation, specialized prey handling techniques have also been developed. Bottlenose dolphins at multiple sites in the Gulf of Mexico feed on marine catfish by severing the head from the remainder of the body (Ronje et al. 2017). Special care must be taken when feeding on these potentially deadly fish. Strong, sharp, venomous dorsal and pectoral spines attached to the head can cause serious injury or death, as evidenced by stranding reports describing tissue trauma or secondary infections from catfish spines (Ronje et al. 2017). The exact mechanism for separating heads from tails is not understood, but based on observations of catfish carcasses, it seems likely that fish are grasped from behind, and a combination of biting and dolphin head motion against the resistance of the water results in the separation. On occasion, tailless catfish have been observed at the surface near dolphins, still exhibiting swimming motions. Catfish do not appear to be a frequent item in the diet of most dolphins, based on stomach contents (Barros and Wells 1998; McCabe et al. 2010). However, this may be a somewhat biased assessment given the importance of otoliths for identifying stomach contents and the fact that catfish heads, with their ear bones, are not consumed. One stranded bottlenose dolphin was found with 72 headless catfish in its stomach. Catfish have been consumed when other prey fish have been abundant. The reason for selecting such a dangerous prey requiring special handling is unclear, but may be related to the ease of finding them given that the fish make sounds (Barros 1993) that make them easy to find, they are often found in schools, and gravid catfish can have very high energy content (Ronje et al. 2017). Occasional removal of heads before consumption of other large fish has also been reported for bottlenose dolphins in Patagonia and elsewhere (Caldwell and Caldwell 1972; Würsig 1986). In one unusual case in Costa Rica, a male and female bottlenose dolphin shared prey, passing a fish back and forth multiple times before consuming it (Fedorowicz et al. 2003).

Within the framework of general foraging patterns described above, bottlenose dolphins have developed a number of specialized techniques for prey detection, pursuit, and capture. Most techniques have been described for dolphins in shallow water situations, during daylight hours, due to the difficulties of observing dolphin behavior in the dark or in deep water. However, thanks to technological advances, behavioral data are becoming available from indirect observations. Foraging occurs throughout day and night in at least some shallow water habitats, such as Sarasota Bay, Florida, with dolphins eating small proportions of their total daily intake in brief bouts (Wells et al. 2013). Satellite-linked telemetry shows continued movements and dives through typical feeding areas during both day and night (Wells et al. 2013). Acoustic recordings from a hydrophone array in the dolphins' home range documented the nighttime occurrence of echolocation clicks consistent with those used during foraging, with additional evidence from forestomach temperature telemetry indicating ingestion of prey at night and during the day (Wells et al. 2013).

Technology has also provided insights into foraging behaviors of deepwater bottlenose dolphins. Off the Hawaiian Islands, bottlenose dolphins occur in well-defined home ranges mostly in waters less than 1000 m deep and feed during day and night, including nearshore and reef fishes in daytime (Baird 2016). Using satellite-linked telemetry, Baird documented presumed nighttime feeding dives to 752 m. Similarly, bottlenose dolphins off the island of Bermuda outfitted for up to 23 hrs with short-term digital archival acoustic tags (Fig. 15.1) were recorded making feeding buzzes both in surface waters and at depths of up to 500 m. The deepest foraging dives occurred at night, presumably to feed on organisms associated with the deep scattering layer that migrates vertically relative to light levels (F. Jensen, pers. comm.). Surface feeding was observed directly during the day. Over the next 1.5 months, satellite-linked time-depth-recording tags on the same dolphins showed them to make presumed foraging dives as deep as 1000 m, remaining submerged for up to 13.5 min (Wells et al. 2017).

Fig. 15.1 Common bottlenose dolphin, *Tursiops truncatus*, outfitted with DTAG on its back, cranial to the dorsal fin, and a satellite-linked time-depth-recording tag trailing from its dorsal fin, for studying behaviors including foraging in the deep waters off Bermuda. Photo courtesy of Dolphin Quest Bermuda

The most complete and detailed foraging behavior descriptions result from direct observations of bottlenose dolphins in shallow water. In some cases, the animals have developed complex techniques specific to certain sites, while other complex techniques are seen across much of the species range. The shallow water seafloor, with its associated physiography, structures, and vegetation, provides numerous opportunities for dolphins to make use of these environmental features to enhance foraging success.

Nowacek (2002) studied sequential foraging behavior of bottlenose dolphins in shallow waters of Sarasota Bay, Florida, from the novel perspective of an overhead video camera suspended 50 m above the water from an aerostat tethered to a small houseboat, from which hydrophones were also deployed, to link acoustic and visual records. The overhead system allowed the collection of detailed and continuous behavior records of distinctive well-known individuals, even when the animals were submerged. He examined patterns of transitions between events and states to define progressive stages of foraging. The dolphins of Sarasota Bay fed primarily as individuals. Specific feeding patterns varied from individual to individual, but general patterns emerged. Following detection of the possible presence of prey fish, dolphins went through stages of active search, terminal pursuit, and capture. Several iterations of the initial stages may occur before prey are actually captured—the choice of which behaviors are used in a foraging sequence can be influenced by a variety of factors including habitat, prey type, and individual preferences (Nowacek 2002).

Prey fish detection in shallow habitats may occur in several ways. In Sarasota Bay, vision underwater is limited by turbidity, so initial detection is likely acoustically

15 Common Bottlenose Dolphin Foraging: Behavioral Solutions that... 335

mediated, through passive listening or active echolocation. Many fish produce noise, and based on examination of dolphin stomach contents, Barros and Odell (1990) and Barros (1993) hypothesized that dolphins use sounds of soniferous fishes to facilitate hunting. Gannon et al. (2005) advanced this hypothesis through playbacks of recorded fish sounds to free-ranging dolphins in Sarasota Bay. Quantitative fish surveys demonstrated that bottlenose dolphins select soniferous fishes disproportionately to their availability in Sarasota Bay (McCabe et al. 2010). However, in areas with better underwater visibility, such as Turneffe Atoll, Belize, passive listening for soniferous fishes appears to be less important (Eierman and Connor 2014).

As part of the initial detection process, when a Sarasota Bay dolphin senses the possibility of a hidden prey, it may engage in rooting, drifting, looking back, or bottom disturbance behaviors (Nowacek 2002). Looking back is a form of inspection of a possible prey situation. During rooting, the dolphin is oriented almost vertically in the water column with its rostrum close to or digging into the bottom. Drifting is a variant of rooting, in which the dolphin remains above the sea floor. Rooting is likely a variant and predecessor to such behaviors as "crater feeding" in the Bahamas (Rossbach and Herzing 1997), in which a dolphin, after scanning with echolocation from side to side, may burrow nearly to pectoral fin depth in a sandy sea floor, presumably to obtain prey. Crater feeding was documented for 18% of identifiable dolphins, and half of these engaged in it repeatedly, suggesting its importance as a foraging strategy. It may also be similar to a behavior referred to as "drilling" observed in shallows in places such as Barataria Bay in the northern Gulf of Mexico, in which a dolphin is vertical in the water column, swishing its tail back and forth at the surface, presumably seeking prey in the seafloor (pers. obs.).

Bottom disturbance can be used to flush or concentrate prey. It involves a dolphin creating a small and local cloud of sediment by swimming near the bottom, often repeated in succession, and often accompanied by production of a bubble cloud (Nowacek 2002). In the Florida Keys, a variant of this behavior, mud plume feeding, is used by individual dolphins for capturing prey in shallow water (<1 m deep). The dolphin uses downward fluke thrusts near the seafloor to create a 5–10 m-long linear or curvilinear plume of mud (Lewis and Schroeder 2003). The dolphin immediately turns into the plume and lunges through the surface, usually on its right side. It is unclear what fish are targeted, but ballyhoo (*Hemiramphus brasiliensis*) have been observed jumping ahead of a lunging dolphin on nearly one-half of observations. It may be that fish are attracted to food stirred up by the plume, or the plume may concentrate fish as they seek to hide in the mud cloud.

The foraging strategy of kerplunking may be employed by some individuals to flush prey in shallow water (<2.5 m deep), by causing a fish to move rapidly when it was previously motionless in the cluttered environment of a seagrass bed. For this behavior, a dolphin raises its tail flukes out of water and then forcefully brings the flukes through the water's surface, sometimes multiple times in succession as the dolphin moves in a line or semicircle (Wells 2003). Kerplunking creates a geyser 1–2 m high, a subsurface bubble cloud and trail, and a loud "kerplunking" sound. Most kerplunking in Sarasota Bay occurs within one dolphin body length of seagrass beds. Kerplunking is not unique to Sarasota Bay—it has been described from other sites along the west coast of Florida and for Indo-Pacific bottlenose dolphins (*Tursiops aduncus*) in Western Australia (Connor et al. 2000). Connor et al. (2000) suggested that sound pressure or particle displacement from kerplunks may evoke a startle response in the fish, making them more detectible, and may be able to reveal the location of fish in a broader area than the dolphin's echolocation cone. In Sarasota Bay, there is indication for observational learning of the behavior by offspring and nonrelatives (Wells 2003).

Active searching for targeted prey immediately precedes prey capture and can involve scanning with echolocation to localize a prey fish (Nowacek 2002). Side-swimming, in which an animal is rotated 90° with respect to its longitudinal axis and swims with normal fluke motion, is often associated with this stage in Sarasota Bay.

Terminal pursuit involves a variety of behaviors culminating in prey capture. Sometimes, individual prey capture is as straightforward as the dolphin simply overtaking a fish during the chase (Bel'kovich et al. 1991; Nowacek 2002). Dolphins may approach prey in normal orientation, inverted, or on their sides, with orientation perhaps related to optimizing vision or echolocation as they close on the prey and/or the need for quick, tight maneuvers (Leatherwood 1975). Pinwheeling, in which a dolphin tucks its head and spins, rotating around the midpoint of its body, allows a dolphin to make an extremely tight turn in response to evasive maneuvers by fish. Fishwhacking, during which a side-swimming dolphin uses a forceful, fast (<1 s) dorsal or ventral thrust of its flukes to strike one or more fish, often propels fish into the air (Wells et al. 1987). Fishwhacking may be an energy-efficient means of catching fast-moving or schooling prey, by increasing surface area of dolphin's body to be able to disturb, disorient, or even make contact with or disable prey, thereby making them easier to capture (Nowacek 2002). Fishwhacking may be performed by individuals or simultaneously by multiple dolphins. The final step in terminal pursuit involves rapid acceleration or lunging to grasp prey between the dolphin's teeth.

In shallow water habitats, individuals are able to use physical features of the habitat as obstacles to restrict prey fish movements. Bel'kovich et al. (1991) described dolphins in the Black Sea driving fish toward shore and herding fish toward the wing of stationary fishing nets, using it as an obstacle against which fish could be driven and in some cases pushed toward shore. Torres and Read (2009) described herding of fish in Florida Bay up against mud banks, mangrove islands, or seagrass beds. In Sarasota Bay, Florida, some individual dolphins or lineages engage in driving fish up against, and chasing them along, structures such as seawalls, using the wall, the seafloor, and the water's surface to limit the prey's escape opportunities.

Many of the behaviors described above can be used by individuals alone or simultaneously by multiple animals in a group. Beyond engaging in simultaneous behaviors that may or may not benefit both parties, bottlenose dolphins also engage in coordinated group foraging behaviors, in some cases with clear differentiation of roles for specific individuals. Some of the more complex group foraging patterns also make use of specific features associated with shallow water habitats.

Cooperative herding by bottlenose dolphins of schools of fish such as mullet (family Mugilidae), menhaden (*Brevoortia* sp.), or catfish into a gradually tightening ball has been described by a number of observers (Morozov 1970; Caldwell and Caldwell 1972; Leatherwood 1975; Bel'kovich et al. 1991). Würsig (1986) noted that some of the same dolphins that engaged in individual foraging near shore also foraged cooperatively in more open waters, moving in a line abreast, and joined with other dolphins. Dolphins encircling the fish from above, around, and below take advantage of stragglers, or some individuals charge through the school, while others maintain the tight ball. In some cases, the water's surface and/or shoreline help to constrain fish movements. Leatherwood (1975) described a sweeping herding pattern, in which dolphins in a crescentic formation drove schools of small fish ahead of them, picking off stragglers or occasionally darting into the school.

Feeding success appears to increase when dolphin groups use habitat features to constrain fish movements. Rossbach (1999) described bottlenose dolphin groups in the Bahamas swimming rapidly in waters 3–4 m deep, actively herding fish together into a circle of about 8–10 m diameter. They drove the fish into the seagrass and then slowly captured individual small fish. In the Black Sea, bottlenose dolphin groups drive fish schools into shallows and force them to the surface or against the beach (Bel'kovich et al. 1991). There is no evidence of specialization of dolphin roles in this herding. Torres and Read (2009) described similar behaviors for dolphin groups in Florida Bay, where dolphins herded fish up against mud banks, mangrove islands, or seagrass beds.

More extreme variants of these strategies incorporating habitat features involve dolphins forcing the fish out of water and in some cases following them ashore for capture. Leatherwood (1975) reported groups of dolphins driving fish against or onto mud banks in the northern Gulf of Mexico, where they would slide out to capture fish. Hoese (1971) described dolphins in tidal creeks of salt marshes in Georgia working together to create pressure waves and push schools of primarily mullet or menhaden onto the banks. The dolphins then follow fish onto the banks, often on their right sides, to remove fish from the mud, and then slide back into the water. Hoese suggested that this coordinated behavior passed from generation to generation through social learning. Petricig (1995) described similar behavior for dolphins in South Carolina, breaking the behavior down into phases of location and approach, setup, charge, landing, and exit. Subsequent work at the same site by Duffy-Echevarria et al. (2008) found that strand feeding is typically performed by 3–4 dolphins, that not all adults in a group engage in the behavior, and that individuals do not have specific preferred stranding positions.

In the upper Florida Keys, bottlenose dolphins engage in cooperative foraging that incorporates bottom disturbance, known as mud ring feeding (Torres and Read 2009). One individual encircles a school of mullet with a ring of mud, and members of the dolphin group catch the fish as they leap out of the ring. Engleby and Powell (2019) provide a detailed description of this shallow water behavior, in depths less than 2 m, involving 3–4 dolphins on average. As the ringmaker begins to circle the school of fish with strong fluke beats creating mud boils, other members of the group concentrate the fish, and then position themselves where the ring will be completed, with heads out of the water and mouths open. The ringmaker closes the ring, spiraling on its side with a final fluke thrust toward the center of the ring, and ends up aligned

Fig. 15.2 Upon completion of the mud ring, a final fluke thrust by the ringmaker sends mullet (Mugilidae) leaping in a predictable direction, toward waiting dolphins. Photo by Brian Skerry, National Geographic

with the other waiting dolphins. The final fluke thrust apparently causes the fish to jump predictably toward the dolphins, as they attempt to avoid the sound and/or pressure of the fluke-generated boils, as sensed by their lateral line or swim bladder (Fig. 15.2). Observations of similar behavior in clearer waters 450 km to the north suggest the mud boils are by-products due to sediments and not crucial to the prey capture technique. More than 19% of dolphins identified by Engleby and Powell in the study area engage in mud ring feeding, some of them repeatedly, but it was rarely possible to identify ringmakers. Engleby and Powell support the suggestion of Torres and Read (2009) that social learning and cultural transmission are likely important for the development of this technique, which takes advantage of or requires specific habitat features, such as depths of less than 2 m.

Bel'kovich et al. (1991) described dolphins in the Black Sea driving fish schools against "walls" formed by other dolphins. Similarly, near Cedar Keys, Florida, groups of 3–6 dolphins engage in cooperative foraging in which one "driver" dolphin herds fish (primarily mullet) in circles toward a tight barrier of dolphins (Gazda et al. 2005). As the fish are driven into this barrier, and sometimes after a tail slap, they leap and are caught by lunging dolphins. In each of two groups observed repeatedly by Gazda et al. (2005), the driver was always the same, suggesting a clear division of labor with role specialization. Foraging success varied with group stability. Follow-up studies demonstrated that drivers have higher foraging success than barrier dolphins (Gazda 2016).

15 Common Bottlenose Dolphin Foraging: Behavioral Solutions that... 339

Bottlenose dolphins have also learned to use human activities as barriers to restrict fish movements. In addition to the use of fixed nets as walls to direct herded fish as described by Bel'kovich et al. (1991), in several parts of the world, dolphins predictably drive fish schools toward humans actively working in the water with nets. Busnel (1973) related several ancient accounts of such interactions and described in detail a more recent symbiotic fishing cooperative from the coast of Mauritania, where mixed groups of Atlantic humpback dolphins (*Sousa teuszii*) and presumed bottlenose dolphins apparently responded to mullet jumping or the similar sound of fishermen slapping the water as nets are being set in shallow water. The dolphins rush in around the nets and the standing fishermen, catching mullet and chasing mullet into the nets.

Along the southern coast of Brazil (Laguna, Santa Catarina), bottlenose dolphins interact predictably with cast-netting fishermen (Simões-Lopes et al. 1998). These dolphins drive schools of mullet toward a line of fishermen standing in the water and indicate with stereotyped head or tail slaps when and where the fishermen should cast their nets; because of turbidity, the fishermen are unable to see the fish themselves. The dolphins stop about 4 m away from the fishermen, the fishermen cast their nets to cover the space in between, and the dolphins open their mouths to catch the disoriented and isolated escaping fish. This cooperative behavior is performed by 45% of the local dolphins at Laguna, and those dolphins that engage in cooperative behavior tend to associate more closely with one another than with noncooperative dolphins (Daura-Jorge et al. 2012). It has been suggested that social learning is important for maintaining this specialized behavior. In both cases described above, these activities are believed to improve fishing success of dolphins and humans and are mutualistic.

Other dolphin foraging strategies that take advantage of human activities, especially fishing, to improve their foraging success do not provide benefits to the humans and in some cases result in damage to fishing gear and/or risks to dolphins. Interactions with fishing operations take many forms, including taking fish stirred up by nets, falling out of nets, or discarded by fishers as bycatch (Leatherwood 1975; Caldwell and Caldwell 1972; Read et al. 2003); with dolphins actively plucking fish from nets (depredation, Leatherwood 1975; Read et al. 2003), taking bait or catch from hook and line gear (depredation, Zollett and Read 2006; Powell and Wells 2011; Baird 2016), removing bait from crab traps or scavenging discarded bait (Noke and Odell 2002), and taking fish proffered by humans (provisioning; Cunningham-Smith et al. 2006; Powell and Wells 2011). Risks from fishing gear include becoming entangled in nets and drowning (Zollett and Read 2006), becoming entangled in crab trap float lines (Wells et al. 2008; Noke and Odell 2002), and becoming entangled in, becoming hooked by, or ingesting hook and line gear, leading to serious injury or death (Wells et al. 2008; Stolen et al. 2013).

Using an overhead video camera suspended from an aerostat off North Carolina (the same system used by Nowacek 2002), Read et al. (2003) observed bottlenose dolphins patrolling gillnets set for Spanish mackerel (*Scomberomorus maculatus*). Although dolphins are killed in such nets, during the research no entanglements were observed. However, dolphins removed mackerel and bluefish (*Pomatomus saltatrix*) from the nets, and holes were found in the net following depredation. Several of the dolphins engaged in begging from the boat, waiting for fish to be discarded.

Bottlenose dolphins are attracted to fishing trawlers in many parts of the world and take fish from the actively trawled net as well as during cleaning and discarding of bycatch (Greenman and McFee 2014). Off Savannah, Georgia, dolphins behind shrimp trawlers approach most closely when fishermen are manipulating and cleaning nets, leading to begging behavior on more than 89% of trawling days (Kovacs and Cox 2014). However, only a portion of the dolphins approach the boats during haulback and net cleaning. Similar findings were reported by Gonzalvo et al. (2008) relative to bottom trawling in the Mediterranean Sea.

Provisioning of dolphins, either through direct feeding or inadvertently through discarding of fish, can decrease dolphin awareness of threats in the environment, such as predators or fishing gear. Provisioning can lead to the ingestion of inappropriate items that can adversely impact their health, or attract them to situations where they are at increased risk from fishing gear, boat collisions, or retaliation/vandalism (Cunningham-Smith et al. 2006; Powell and Wells 2011). Christiansen et al. (2016) noted that dolphins in Sarasota Bay, Florida, conditioned to human interactions through direct or indirect food provisioning, were more likely to be injured by human interactions, when compared to unconditioned animals. Thus, conditioning could lead to a decrease in survival, which could ultimately affect population dynamics (Christiansen et al. 2016).

There is an innate component to bottlenose dolphin foraging behaviors, but there are also important learning and experience elements in developing foraging behaviors, especially with regard to the interplay of prey and habitat features. Some of the more complex behaviors described above, including those that involve human activities, have developed fairly recently in the evolutionary history of bottlenose dolphins. The impressive ability of bottlenose dolphins to learn is well known from decades of working with and observing bottlenose dolphins under human care. Under these circumstances, behavior is readily shaped through reinforcement with a food reward (Ramirez 1999). Performance of a desired behavior results in obtaining fish, which leads to repeating the behavior. In the wild, successful foraging behaviors inherently involve a food reward.

Bottlenose dolphin abilities for social learning through observation are well known for dolphins under human care (Pryor 1973). Individuals exhibit new (to them) behaviors after observing others perform the behavior (Norris 1974). In the wild, evidence for social learning comes from several sources, including observations of highly coordinated foraging patterns. Christiansen et al. (2016) noted that the association with already conditioned dolphins strongly affected the probability of dolphins becoming conditioned to human interactions, suggesting that conditioning is at least partly a learned behavior. Wells (2003) suggested that one of the reasons behind the 1.5–4.5 years of calf rearing extending beyond nutritional weaning for bottlenose dolphins in Sarasota Bay is the need for calves to learn survival skills from the mother and her close associates. It is not uncommon to observe Sarasota Bay mothers and their most recent calves engaging in bouts of foraging behaviors, during which perfectly performed behaviors by the mother, such as kerplunking, are followed by incomplete versions of the behavior by the younger animals. Such observations are among those leading Wells (2003) to suggest the occurrence of

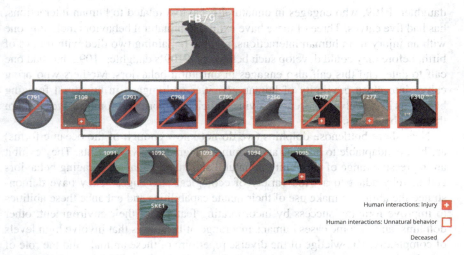

Fig. 15.3 Four-generation maternal lineage of long-term resident Sarasota Bay common bottlenose dolphins, *Tursiops truncatus*, related to 39-year-old female FB79, indicating those individuals that engage in unnatural behaviors related to human interactions, those with evidence of injuries from human interactions, and whether the individual is still alive. Figure prepared by René Byrskov and Katherine McHugh

cultural transmission of knowledge in bottlenose dolphins, as exhibited through foraging behaviors. Following the definition of Whiten et al. (1999, p. 682) of cultural behavior as "being transmitted repeatedly through social or observational learning to become a population-level characteristic," patterns of occurrence of foraging behaviors within and across a variety of bottlenose dolphin research sites around the world support vertical and horizontal transmission of behaviors (Wells 2003). Whitehead et al. (2004) provide strong additional evidence for the idea that culture is an important determinant of behavior in cetaceans.

While social learning can be an important factor in acquiring natural foraging behaviors, it appears that it can also lead to the development of unnatural foraging behaviors that can place bottlenose dolphins at risk. In Sarasota Bay, Florida, research initiated in 1970 has provided detailed records for members of a long-term resident community of about 170 bottlenose dolphins (Wells 2014). Some maternal lineages are more inclined than others to engage in unnatural behaviors involving interactions with humans, including behaviors such as patrolling near fishing boats, lines, or piers, scavenging discarded bait or catch, depredating bait or catch from active fishing gear, begging, accepting food from humans, and interacting with fixed fishing gear such as crab pots, among others. Some of these behaviors lead to injury or death (Powell and Wells 2011; Christiansen et al. 2016). In an example of apparent transmission of risky behaviors vertically through four generations, a 39-year-old female (FB79) that frequently engages in unnatural behaviors has been observed with nine calves over the course of her life (Fig. 15.3). Seven of these engaged in unnatural behaviors, and at least three of these exhibited injuries or died from human interactions. One daughter, F109, who engages in unnatural behaviors related to human interactions, has had five calves. Three of these have exhibited unnatural behaviors including one with an injury from human interactions, and the remaining two died within days of birth, before they could develop such behaviors. F109's daughter, 1092, has had one calf to date, and this calf also engages in unnatural behaviors. Mothers who bring calves into close proximity of human activities and engage in unnatural foraging behaviors in the calves' presence provide sufficient opportunity for the calves to learn the risky behaviors.

Near-shore bottlenose dolphins (we do not know enough of the oceanic forms) are highly adaptable to humans and to human-degraded environments. They exhibit an impressive range of variability in natural and human-adapted foraging behaviors and thereby adapt to a wide variety of ecological challenges. They have demonstrated the ability to make use of their innate capabilities and enhance these abilities to improve foraging success by incorporating features of their environment, other dolphins, and in some cases humans in a range of behaviors that involve high levels of complexity. Knowledge of the diverse repertoire of these animals and the role of learning in developing foraging behaviors not only leads to a fuller appreciation of their amazing abilities but can also aid in designing conservation strategies for protection of the animals and their habitats.

References

Baird RW (2016) The lives of Hawai'i's dolphins and whales: natural history and conservation. University of Hawai'i Press, Honolulu, HI

Barros NB (1993) Feeding ecology and foraging strategies of bottlenose dolphins on the central east coast of Florida. PhD dissertation, University of Miami, Coral Gables, FL, 328p

Barros NB, Odell DK (1990) Food habits of bottlenose dolphins in the southeastern United States. In: Leatherwood S, Reeves RR (eds) The bottlenose dolphin. Academic Press, San Diego, CA, pp 309–328

Barros NB, Wells RS (1998) Prey and feeding patterns of resident bottlenose dolphins (*Tursiops truncatus*) in Sarasota Bay, Florida. J Mammal 79(3):1045–1059

Bel'kovich VM, Ivanova EE, Yefremenkova OV, Kozarovitsky LB, Kharitonov SP (1991) Searching and hunting behavior in the bottlenose dolphin (*Tursiops truncatus*) in the Black Sea. In: Pryor K, Norris KS (eds) Dolphin societies: discoveries and puzzles. University of California Press, Berkeley, CA, pp 38–67

Busnel R-G (1973) Symbiotic relationship between man and dolphins. Trans N Y Acad Sci 35: 112–131

Caldwell DK, Caldwell MC (1972) The world of the bottlenose dolphin. J. B. Lippincott, Philadelphia

Christiansen F, McHugh KA, Bejder L, Siegal EM, Lusseau D, Berens McCabe E, Lovewell G, Wells RS (2016) Food provisioning increases the risk of injury and mortality in a long-lived marine top predator. R Soc Open Sci 3:160560. https://doi.org/10.1098/rsos.160560

Connor RC, Heithaus MR, Berggren P, Miksis JL (2000) "Kerplunking": surface fluke-splashes during shallow-water bottom foraging by bottlenose dolphins. Mar Mamm Sci 16:646–653

Cunningham-Smith P, Colbert DE, Wells RS, Speakman T (2006) Evaluation of human interactions with a provisioned wild bottlenose dolphin (*Tursiops truncatus*) near Sarasota Bay, Florida, and efforts to curtail the interactions. Aquat Mamm 32:346–356

15 Common Bottlenose Dolphin Foraging: Behavioral Solutions that...

Daura-Jorge FG, Cantor M, Ingram SN, Lusseau D, Simões-Lopes PC (2012) The structure of a bottlenose dolphin society is coupled to a unique foraging cooperation with artisanal fishermen. Biol Lett 8:702–705

Duffy-Echevarria EE, Connor RC, St. Aubin DJ (2008) Observations of strand-feeding behavior by bottlenose dolphins (*Tursiops truncatus*) in Bull Creek, South Carolina. Mar Mamm Sci 24: 202–206

Eierman LE, Connor RC (2014) Foraging behavior, prey distribution, and microhabitat use by bottlenose dolphins *Tursiops truncatus* in a tropical atoll. Mar Ecol Prog Ser 503:279–288

Engleby LK, Powell JR (2019) Detailed observations and mechanisms of mud ring feeding by common bottlenose dolphins (*Tursiops truncatus*) in Florida Bay, Florida, U.S.A. Mar Mamm Sci. https://doi.org/10.1111/mms.12583

Fedorowicz SM, Beard DA, Connor RC (2003) Food sharing in wild bottlenose dolphins. Aquat Mamm 29:355–359

Gannon DP, Barros NB, Nowacek DP, Read AJ, Waples DM, Wells RS (2005) Prey detection by bottlenose dolphins (*Tursiops truncatus*): an experimental test of the passive-listening hypothesis. Anim Behav 69:709–720

Gazda SK (2016) Driver-barrier feeding behavior in bottlenose dolphins (*Tursiops truncatus*): new insights from a longitudinal study. Mar Mamm Sci 32:1152–1160

Gazda SK, Connor RC, Edgar RK, Cox F (2005) A division of labour with role specialization in group–hunting bottlenose dolphins (*Tursiops truncatus*) off Cedar Key, Florida. Proc R Soc B Biol Sci 272:135–140

Gonzalvo J, Valls M, Cardona L, Aguilar A (2008) Factors determining the interaction between common bottlenose dolphins and bottom trawlers off the Balearic Archipelago (western Mediterranean Sea). J Exp Mar Biol Ecol 367:47–52

Greenman JT, McFee WE (2014) A characterisation of common bottlenose dolphin (*Tursiops truncatus*) interactions with the commercial shrimp trawl fishery of South Carolina, USA. J Cetacean Res Manag 14:69–79

Hoese HD (1971) Dolphin feeding out of water in a salt marsh. J Mammal 52:222–223

Kovacs C, Cox T (2014) Quantification of interactions between common bottlenose dolphins (*Tursiops truncatus*) and a commercial shrimp trawler near Savannah, Georgia. Aquat Mamm 40:81–94

Leatherwood S (1975) Some observations of feeding behavior of bottle-nosed dolphins (*Tursiops truncatus*) in the northern Gulf of Mexico and (*Tursiops* cf. *T. gilli*) off southern California, Baja California, and Nayarit, Mexico. Mar Fish Rev 37:10–16

Lewis JS, Schroeder WW (2003) Mud plume feeding, a unique foraging behavior of the bottlenose dolphin in the Florida Keys. Gulf Mex Sci 21:92–97

McCabe EJB, Gannon DP, Barros NB, Wells RS (2010) Prey selection by resident common bottlenose dolphins (*Tursiops truncatus*) in Sarasota Bay, Florida. Mar Biol 157:931–942

Morozov DA (1970) Dolphin hunting. Rybnoe Khoziaistvo 46:16–17

Noke WD, Odell DK (2002) Interactions between the Indian River Lagoon blue crab fishery and the bottlenose dolphin, *Tursiops truncatus*. Mar Mamm Sci 18:819–832

Norris KS (1974) The porpoise watcher. Norton, New York

Norris KS, Dohl TP (1980) The structure and function of cetacean schools. In: Herman LM (ed) Cetacean behavior: mechanisms and functions. Wiley, New York, pp 211–261

Nowacek DP (2002) Sequential foraging behaviour of bottlenose dolphins, *Tursiops truncatus*, in Sarasota Bay, FL. Behaviour 139:1125–1145

Petricig RO (1995) Bottlenose dolphins (*Tursiops truncatus*) in Bull Creek, South Carolina. Ph. D. dissertation, University of Rhode Island, p 298

Powell JR, Wells RS (2011) Recreational fishing depredation and associated behaviors involving common bottlenose dolphins (*Tursiops truncatus*) in Sarasota Bay, Florida. Mar Mamm Sci 27:111–129

Pryor KW (1973) Behavior and learning in porpoises and whales. Naturwissenschaften 60:412–420

Ramirez K (1999) Animal training. Shedd Aquarium, Chicago, IL

Read AJ, Waples DM, Urian KW, Swanner D (2003) Fine-scale behavior of bottlenose dolphins around gill nets. Proc R Soc Lond B (Suppl) 270:S90–S92

Ronje EI, Barry KP, Sinclair C, Grace MA, Barros N, Allen J, Balmer B, Panike A, Toms C, Mullin KD, Wells RS (2017) A common bottlenose dolphin (*Tursiops truncatus*) prey handling technique for marine catfish (Ariidae) in the northern Gulf of Mexico. PLoS One 12:e0181179

Rossbach KA (1999) Cooperative feeding among bottlenose dolphins (*Tursiops truncatus*) near Grand Bahama Island, Bahamas. Aquat Mamm 25:163–167

Rossbach KA, Herzing DL (1997) Underwater observations of benthic-feeding bottlenose dolphins (*Tursiops truncatus*) near Grand Bahama Island, Bahamas. Mar Mamm Sci 13:498–504

Scott MD, Chivers SJ (1990) Distribution and herd structure of bottlenose dolphins in the eastern tropical Pacific Ocean. In: Leatherwood S, Reeves RR (eds) The bottlenose dolphin. Academic Press, San Diego, CA, pp 387–402

Simões-Lopes PC, Fabián ME, Menegheti JO (1998) Dolphin interactions with the mullet artisanal fishing on southern Brazil: a qualitative and quantitative approach. Rev Bras Zool 15:709–726

Stolen M, Noke Durden W, Mazza T, Barros N, St. Leger J (2013) Effects of fishing gear on bottlenose dolphins (*Tursiops truncatus*) in the Indian River Lagoon system, Florida. Mar Mamm Sci 29:356–364

Torres LG, Read AJ (2009) Where to catch a fish? The influence of foraging tactics on the ecology of bottlenose dolphins (*Tursiops truncatus*) in Florida Bay, Florida. Mar Mamm Sci 25:797–815

Wells RS (2003) Dolphin social complexity: lessons from long-term study and life history. In: de Waal FBM, Tyack PL (eds) Animal social complexity: intelligence, culture, and individualized societies. Harvard University Press, Cambridge, MA, pp 32–56

Wells RS (2014) Social structure and life history of common bottlenose dolphins near Sarasota Bay, Florida: insights from four decades and five generations. In: Yamagiwa J, Karczmarski L (eds) Primates and cetaceans: field research and conservation of complex mammalian societies, Primatology monographs. Springer, Tokyo, pp 149–172

Wells RS, Scott MD (2018) Bottlenose dolphin: common bottlenose dolphin (*Tursiops truncatus*). In: Würsig B, Thewissen JGM, Kovacs K (eds) Encyclopedia of marine mammals, 3rd edn. Academic Press/Elsevier, San Diego, CA, pp 118–125

Wells RS, Irvine AB, Scott MD (1980) The social ecology of inshore odontocetes. In: Herman LM (ed) Cetacean behavior: mechanisms and functions. Wiley, New York, pp 263–317

Wells RS, Scott MD, Irvine AB (1987) The social structure of free-ranging bottlenose dolphins. In: Genoways H (ed) Current mammalogy, vol 1. Plenum, New York, pp 247–305

Wells RS, Allen JB, Hofmann S, Bassos-Hull K, Fauquier DA, Barros NB, DeLynn RE, Sutton G, Socha V, Scott MD (2008) Consequences of injuries on survival and reproduction of common bottlenose dolphins (*Tursiops truncatus*) along the west coast of Florida. Mar Mamm Sci 24: 774–794

Wells RS, McHugh KA, Douglas DC, Shippee S, Berens McCabe EJ, Barros NB, Phillips GT (2013) Evaluation of potential protective factors against metabolic syndrome in bottlenose dolphins: feeding and activity patterns of dolphins in Sarasota Bay, Florida. Front Endocrinol 4(139):1–16. https://doi.org/10.3389/fendo.2013.00139

Wells R, Fahlman A, Moore M, Jensen F, Sweeney J, Stone R, Barleycorn A, Trainor R, Allen J, McHugh K, Brenneman S, Allen A, Klatsky L, Douglas D, Tyson R (2017) Bottlenose dolphins in the Sargasso Sea – ranging, diving, and deep foraging. 22nd Biennial Conference on the Biology of Marine Mammals, 22–27 Oct 2017, Halifax, Nova Scotia, Canada

Whitehead H, Rendell L, Osborne RW, Würsig B (2004) Culture and conservation of non-humans with reference to whales and dolphins: review and new directions. Biol Conserv 120:427–437

Whiten A, Goodall J, McGrew WC, Nishida T, Reynolds V, Sugiyama Y, Tutin CEG, Wrangham RW, Boesch C (1999) Cultures in chimpanzees. Nature 399:682–685

Würsig B (1986) Delphinid foraging strategies. In: Schusterman RJ, Thomas JA, Wood FG (eds) Dolphin cognition and behavior: a comparative approach. Lawrence Erlbaum Associates, Hillsdale, NJ, pp 347–359

Zollett EA, Read AJ (2006) Depredation of catch by bottlenose dolphins (*Tursiops truncatus*) in the Florida king mackerel (*Scomberomorus cavalla*) troll fishery. Fish Bull 104:343–349

Chapter 16
The Indo-Pacific Bottlenose Dolphin (*Tursiops aduncus*)

Richard C. Connor, Mai Sakai, Tadamichi Morisaka, and Simon J. Allen

Abstract The behavioral ecology of *Tursiops aduncus* (Indo-Pacific bottlenose dolphin) is usually reviewed alongside the much more widely studied *T. truncatus* (common bottlenose dolphin). However, the smaller, typically shallow water *T. aduncus* has been closely scrutinized in Australian and Japanese waters. As a result, there now exists a robust body of information spanning all three of Hinde's levels of social analysis—interactions, relationships, and social structure—that may be unmatched in any other cetacean. Research on *T. aduncus* has contributed significantly to the social complexity hypothesis of large brain evolution and our understanding of delphinid mating systems, communication, and individual differences in foraging tactics within populations. Here, we focus on behavioral research at two primary sites, Shark Bay in Australia and Mikura Island in Japan, with additional observations of importance from other locales in each region.

Keywords Bottlenose dolphin · Indo-Pacific · Behavioral ecology · Mating strategy · Delphinids · Social complexity · Alliances

Indo-Pacific bottlenose dolphins (*Tursiops aduncus*) typically occur near shore in shallow tropical and temperate waters of the Indian and western Pacific Oceans. They are smaller than *T. truncatus*, at less than 2.7 m in length, and are generally

---

R. C. Connor (✉)
Biology Department, UMASS-Dartmouth, Dartmouth, MA, USA
e-mail: rconnor@umassd.edu

M. Sakai
Department of Fisheries, Faculty of Agriculture, Kindai University, Osaka, Japan

T. Morisaka
Cetacean Research Center, Graduate School of Bioresources, Mie University, Tsu, Japan

S. J. Allen
School of Biological Sciences, University of Bristol, Bristol, UK

© Springer Nature Switzerland AG 2019
B. Würsig (ed.), *Ethology and Behavioral Ecology of Odontocetes*, Ethology and Behavioral Ecology of Marine Mammals,
https://doi.org/10.1007/978-3-030-16663-2_16

smaller in the tropics (Fig. 16.1a). Where abundant, *T. aduncus* are prominent in coastal waterways, often surfing waves along beaches (Fig. 16.1b) and foraging around estuaries, seagrass beds, and rocky reefs. Most populations consist primarily of resident animals (e.g., Möller et al. 2002; Shirakihara et al. 2002; Tsai and Mann 2013), but some individuals have been documented across sites separated by several hundred kilometers (e.g., Brown et al. 2016; Tsuji et al. 2018). *T. aduncus* eat small fish and cephalopods primarily, although some may target larger fish and occasionally sharks, rays, and crustaceans. After spending several years with their mothers, *T. aduncus* achieve sexual maturity at 10–15 years of age and may live 40–50 years (Karniski et al. 2018).

The earliest studies of *T. aduncus*, indeed, some of the earliest studies of any odontocete, are those of Tayler and Saayman in wild and captive settings in South Africa (Tayler and Saayman 1972; Saayman and Tayler 1973). Observations on wild populations were limited to grouping patterns, but Tayler and Saayman's (1973) descriptions of a captive dolphin infant retrieving milk from its mother and then spitting it out to imitate a human viewer exhaling cigarette smoke, and using

Fig. 16.1 (a) Adult Indo-Pacific bottlenose dolphin, *Tursiops aduncus*, from Shark Bay, Western Australia. Note the robust body, medium-length rostrum/beak, large fins relative to body size and ventral speckling characteristic of *T. aduncus* bottlenose dolphins in warmer climes. (b) *T. aduncus* surfing the beaches of southern New South Wales. (c) Alliances of alliances, Shark Bay: a second-order alliance comprising two trios (first-order alliances) of adult males following two female consorts (top left of frame); a first-order alliance and female consort with calf of weaning age (center); and a second-order alliance of three trios and two consorts (bottom right). (d) A first-order alliance following their female consort who is between bouts of foraging with the aid of a sponge tool (photos: S. Allen)

16 The Indo-Pacific Bottlenose Dolphin (*Tursiops aduncus*) 347

objects and blowing bubbles to imitate scuba divers cleaning their tank, are among the most widely cited examples of dolphin imitation.

The most detailed behavioral work on wild *T. aduncus* has been conducted in Australia and Japan. We review the long-term research in Shark Bay, Western Australia, and then survey notable observations from other Australian locations and Japan, where key studies based on subsurface observations have been conducted.

16.1 Australia

16.1.1 Shark Bay, Western Australia

The >35-year Shark Bay *T. aduncus* study is the longest running and most in-depth for the species, spanning Hinde's (1976) three levels of social analysis: interactions, relationships, and structure. Using a range of approaches and tools, studies of social and other behaviors have been integrated increasingly with demographic, life history, genetic, and ecological analyses. The original study off the east side of Peron Peninsula, which bisects Shark Bay, was joined in 2007 by a comparative study in the western gulf, and the two sites have proven highly complementary (e.g., Krützen et al. 2014).

16.1.1.1 Social Structure and Mating System

The Shark Bay dolphins have a social and mating system within a dynamic fission-fusion grouping pattern in which both sexes exhibit natal philopatry (Connor et al. 2000a; Tsai and Mann 2013). Females may first conceive at 10 or 11 years of age and give birth 1 year later (Karniski et al. 2018). Weaning age ranges from 2.6 to 8.6 years (mean = 4.0) and increases with maternal age (Karniski et al. 2018). In the transition to independence, young juvenile dolphins increase their same sex associations (Krzyszczyk et al. 2017), anticipating differences in associations among adults (Smolker et al. 1992). A female's reproductive success is enhanced if she has successful female relatives or if she associates with successful females; interestingly, this second factor is actually enhanced for female associates that are less closely related (Frère et al. 2010). Females typically become attractive to males when their infants are about 2.5 years of age (Connor et al. 1996). Thereafter, for varying periods of time over months, possibly extending for a time after they conceive, females become involved in the males' complex system of nested alliances (Connor et al. 1996; Connor and Krützen 2015; Galezo et al. 2018).

Males cooperate in pairs and, more often, trios (= first-order alliances) to consort individual females for periods of minutes to weeks (Connor et al. 1992a, b, 1996, 2011; Fig. 16.1c, d). Many consortships are established and maintained by aggressive herding (Connor et al. 1992a). While males are coercing females to remain with them, we have no evidence that males are physically forcing copulation; indeed, females may role away from mounting males. On the other hand, in a manner similar to chimpanzees, males may use intimidation to coerce females into copulating (Connor and Vollmer 2009).

During the year they conceive, females typically occur in consortships with a number of alliances over months. The costs from male aggression, and possibly reduced foraging efficiency of females being herded away from their core ranges, led Connor et al. (1996) to predict infanticide in bottlenose dolphins (*Tursiops* spp.). Subsequently, evidence for infanticide was reported in a number of locations (e.g., Patterson et al. 1998; Dunn et al. 2002; Robinson 2014). Females can reduce infanticide risk by mating with many males to confuse paternity. Escaping from her male consorts after mating with them, having multiple attractive periods and increasing range size during cycling may help females mate with more males (Connor et al. 1996; Wallen et al. 2016). Male associations with females decline post-conception (Wallen et al. 2017), but at a rate that is consistent with females continuing to be attractive and confusing paternity (contra Wallen et al. 2017), particularly if such behavior is more likely among a subset of females that have failed to mate with many males prior to conception.

Most males belong to groups of 4–14 (= second-order alliances), where they find their first-order alliance partners (herding by males from different second-order alliances occurs infrequently, Connor et al. 2011). Males in second-order alliances cooperate in contests for females against other groups. One first-order alliance may even recruit another within the same second-order alliance to attack another alliance to steal their female (Connor et al. 1992b). These second-order alliances also serve a defensive function, especially during the mating season, when most males in a second-order alliance may be consorting females. Connor and Vollmer (2009) suggested that the importance of such defensive associations has increased the need for aggressive herding in Shark Bay compared to other locations without second-order alliances. If males simply followed their female consorts rather than herding them, they would often be led away from second-order alliance partners as their females traveled to different favored foraging areas, rendering the males more vulnerable to attack from other alliances.

The stability of first-order alliances within second-order alliances varies and is not strongly related to second-order alliance size (Connor et al. 2011). Further, the rate that males consort females correlates with the stability of their first-order alliances (Connor et al. 2001; Connor and Krützen 2015), a possible indication of dominance relationships, which otherwise are only known from one study of captive *T. truncatus* (Samuels and Gifford 1997). The most stable pairs and trios can persist for decades, as can second-order alliances (with the expected attrition as males die)—considered the "core" unit of male social structure (Connor and Krützen 2015; Fig. 16.1c).

Relatedness is a significant factor in subadult male associations but not among adults (Gerber et al. 2019). As males mature and consolidate their second-order alliance membership, they associate with a larger number of age-mates. The relatedness signal that shows up in subadults may be a by-product of the bisexual philopatry, as immature male associates may often be the offspring of related females (Gerber et al. 2019).

A third level of alliance is indicated by preferential association between second-order alliances and, sometimes, second-order alliances and "lone trios" (typically the

16 The Indo-Pacific Bottlenose Dolphin (*Tursiops aduncus*) 349

aged remnants of previously larger second-order alliances), as well as cooperation in conflicts with other groups over females. The functional redundancy of third-order alliances may reflect a kind of "insurance," given the disparity in second-order alliance sizes and the fact that second-order alliance partners are not always present to provide aid in conflicts (Connor et al. 2011). Third-order alliance fights can involve over 30 male dolphins.

16.1.1.2 Social Interactions

Shark Bay dolphins conduct interactions and relationships with a wide range of motor and vocal behaviors. Affiliative interactions often include gentle contact behaviors, such as use of the pectoral fin to "pet" another dolphin and rubbing their body against another dolphin's fins or body. Such behaviors are analogous to primate grooming.

One kind of gentle contact behavior, "contact swimming," is rare in males but occurs often between females being harassed by males (Connor et al. 2006a). In contact swimming, one female places her pectoral fin against the side of another, swimming close to, but slightly behind her. Such episodes are often brief but can last for up to 20 min. Contact swimming can occur when one, both, or neither of the females are being herded by males in the group. Bouts of contact swimming may include partner and role switching between more than two females. Interestingly, contact swimming occurs between familiar females and those rarely observed together (and are only together because both are being herded by males in the same second-order alliance). Given the similarity in appearance to calves swimming in infant position, Connor et al. (2006a) suggested that contact swimming may reflect support from the lead female to the trailing female, who is responsible for maintaining contact.

The benefits of primate grooming include parasite removal and stress reduction (e.g., Aureli et al. 1999; Zamma 2002). Dolphin affiliative contact behaviors are unlikely to involve parasite removal, although removal of dead skin may be beneficial (dolphins shed skin cells much faster than humans, Hicks et al. 1985). However, as illustrated by the typical context of contact swimming, stress reduction seems an obvious plausible benefit, and role switching during and across bouts of petting or contact swimming might represent a kind of reciprocity.

Sexual behavior is used as a social tool in dolphins and is found in many non-conceptive contexts and, to varying degrees, between all age and sex classes (Connor et al. 2000a; Mann 2006). The two most common and obvious sexual behaviors are mounting and "goosing", where one dolphin probes the genital area of another with its rostrum. The same behavior, e.g., goosing, may be performed gently, accompanied by petting and rubbing, or roughly, accompanied by threats and direct aggression, indicating that sex is used in both affiliative and agonistic interactions. Connor and Vollmer (2009) showed that males tended to mount other males but goose females more, suggesting that the origins of goosing might lie in stimulating female receptivity.

Consortship initiation may include long chases and direct aggression in the form of biting and hitting with peduncles and flukes. Females sometimes attempt to escape from males by "bolting" (accelerating rapidly away from the males). Males use a vocalization, "pops," to keep a consorted female close (Connor and Smolker 1996; Vollmer et al. 2015). Pops are low-frequency, low repetition rate click trains (Connor and Smolker 1996) that are reinforced by other vocal and physical threats (burst-pulsed vocalizations, head-jerks, and charging at the female), as well as direct aggression. Cycling females have significantly more new tooth rake marks than non-cycling females (Scott et al. 2005).

Males in the Shark Bay nested alliance system maintain individually distinctive "signature" whistles for communication (King et al. 2018). Synchronous movement also plays an important role in male alliance behavior. Males often surface synchronously side-by-side with first-order alliance partners. In trios, the pair that is found together most often synchronizes more when all three are present (Connor et al. 2006b). Sometimes two males from different first-order alliances within a second-order alliance surface side-by-side synchronously, especially when they are excited around a consorted female, suggesting a reduced tension function. The side-by-side synchronous surfacing by male dolphins in alliance contexts is similar to human synchrony and raises the question of why such a potent visual signal is so rarely observed in other primates that form coalitions (see Connor et al. 2006b; Connor 2007).

Males also perform synchronous displays around females, often in consorting context. This suggests that, while males are coercing a female's proximity and possibly intimidating her into copulating, they may still need to impress a female to achieve a reasonable chance of fertilizing her egg (females may have internal mechanisms to choose among males, e.g., Firman et al. 2017; see also Chap. 4). Some synchronous displays are observed repeatedly, while at the other extreme, some have been observed only once. Simple side-by-side synchrony might be produced by mutual entrainment (Wilson and Cook 2016), but it is difficult to explain some of the more elaborate displays this way. Connor and Krützen (2015) suggested that some male displays might be creative, with one dolphin closely following the movements of another. Herman (2002) showed that captive *Tursiops* are capable of creating synchronous displays. Links to captioned videos of the behaviors and vocalizations described here may be found at www.dapinc.org.

16.1.1.3 Ecology

Shark Bay dolphins feed on a wide range of schooling and solitary fish, and occasionally cephalopods and crustaceans, throughout the water column, sometimes catching jumping and skittering fish in the air and probing for partially buried prey in the substrate (Connor et al. 2000a, b; Fig. 16.2a, b). The incidence of shark bite scars is unusually high (74%, Heithaus 2001) compared to other dolphin populations (eclipsed only by the recently reported 89% of snubfin dolphins in Cygnet Bay, Western Australia, Smith et al. 2018), indicating significant predation risk. The

Fig. 16.2 (a) Shark Bay *T. aduncus* captures a fleeing garfish (*Hyporhamphus* sp.) amidst its final leap. (b) This *T. aduncus* (bottom left of frame subsurface) would be feeding on the tiger prawn (*Penaeus* sp.) it had flushed from the substrate and pursued to the surface if not for the untimely theft from the crested tern (*Thalasseus* sp.) above. (c) An adult *T. aduncus* intentionally beaches itself in pursuit of a mullet (*Mugil* sp.), while her calf looks on, Peron Peninsula. Beaching, where individuals chase fish through the shallows and even onto a beach to capture them (Sargeant et al. 2005), is a specialized foraging technique developed by calves of beaching mothers (photos: S. Allen). (d) Another specialized foraging technique is shelling, where dolphins lift large trumpet (*Syrinx aruanus*) or baler (*Melo amphora*) shells out of the water to dislodge prey that are hiding inside (Allen et al. 2011; photo: A. Pierini, Dolphin Innovation Project)

dynamic fission-fusion grouping pattern in Shark Bay (2.4 events/hour in one study, Connor and Krützen 2015) reflects a continually shifting balance between the benefits of grouping (socializing, reducing predation risk, and feeding on schooling fish) and reducing competition for food, a balance that tips easily either way given the low cost of locomotion in dolphins, relative to many terrestrial mammals (Connor 2000).

Some behavioral variation among dolphins in Shark Bay, both across a larger spatial scale and within a given area, has an ecological component. Adult females within a given area differ in the number of their same sex associates and activity budgets, and their home ranges vary by an order of magnitude (Watson-Capps 2005), variation likely linked to individual differences in learned foraging techniques (Connor et al. 2000a, b; Mann and Sargeant 2003). One specialized foraging technique that is especially time consuming is "sponging," where a dolphin carries a marine sponge on its rostrum to probe for prey in the substrate (Krützen et al. 2005, 2014; Mann et al. 2008, 2012; Smolker et al. 1997; Fig. 16.1d). Sponging females conduct their limited social lives preferentially with other "sponger" females (Mann et al. 2012; Kopps et al. 2014). At the other end of the spectrum, it seems likely that highly sociable females with large home ranges and a large number of associates might feed to a greater degree on schooling fishes (Connor et al. 2000a), but this has not been tested.

The homophily observed in female spongers may also play an important role in the development of alliances. As mentioned, age similarity trumps relatedness in alliance affiliation as males mature and, as in the case of female spongers, male spongers also associate preferentially (Bizzozzero et al. 2019).

Second-order alliances that have extensively overlapping ranges often differ in foraging habitat use; for example, one forages more over shallow seagrass beds, while another favors an adjacent channel (O'Brien et al. in review). While essentially random demographic variation may impact the size of second-order alliances (e.g., number of male age-mates maturing together in an area), foraging differences, to the extent that they impact the costs of grouping, might also play a role. This possibility is of broad interest because it would require an extension of the standard behavioral ecology paradigm, where resources and shelter impact female distribution, which, in turn, determines male strategies. In Shark Bay, some of the variation we see in male alliance size may be influenced directly by variation in resource use. However, given the importance of being in an alliance for male reproductive success, we would expect to see males shift foraging style or habitat for the chance to join an alliance (O'Brien et al. in review). Males clearly strive to join alliances in Shark Bay, and solitary males have lower reproductive success (Krützen et al. 2004; Connor and Krützen 2015). The greater proportion of female calves born to sponging mothers that become spongers themselves indicates that the conflict between optimizing their social lives and foraging preferences is reduced relative to males, but it is not entirely absent. Complete freedom in foraging behavior may be limited not only by predation risk but by males herding females away from favored foraging areas where their specialized learned techniques are useless. We have observed sponging females foraging without sponges on shallow seagrass banks while males were herding them there.

Along with complex social lives, the array of specialized foraging techniques and links to different habitats has been another exciting area of discovery in Shark Bay. In addition to the well-studied sponge carrying, foraging techniques (see Figs. 16.1d and 16.2a–d, for example) include swimming into shallows to be fed by people (Connor and Smolker 1985; Foroughirad and Mann 2013) and kerplunking, where a percussive, bubble-forming tail slap may prompt a startle reaction by prey that was otherwise concealed (Connor et al. 2000a, b). Several of these techniques are linked to specific habitats. For instance, while spongers are more likely to forage in deeper channels (Sargeant et al. 2007; Tyne et al. 2012), kerplunking occurs over shallow seagrass beds (Connor et al. 2000b).

Sponging and beaching behaviors (Figs. 16.1d and 16.2c), both typically transmitted from the mother to offspring, are observed more often in females, and none of the techniques appear to have a male bias (Kopps et al. 2014; Mann et al. 2008; Sargeant et al. 2005). The demonstration, via fatty acid signature analysis, that sponging dolphins have a different diet than non-sponging individuals foraging in the same

16 The Indo-Pacific Bottlenose Dolphin (*Tursiops aduncus*) 353

habitat (Krützen et al. 2014) remains the only empirical demonstration that one of the specialized foraging techniques in Shark Bay is a resource specialization.

Variation is also found on a broader spatial scale. From south to north along the eastern shore of Peron Peninsula, males form more trios (relative to pairs) to consort females, consort females at a higher rate, have a greater coverage of new tooth rake marks on their dorsal fins, and make greater seasonal movements (Connor et al. 2017; Hamilton et al. 2019). Shallow seagrass-covered banks subdivided by deeper channels characterize the southern part of the study area, while the north is mostly open embayment plain. Presently, the weight of evidence favors the "rate of interaction" model over food or predator distribution (discussed in Connor et al. 2017), in which males form larger alliances when they encounter each other more often in competition for estrus females (Whitehead and Connor 2005; Connor and Whitehead 2005). More than one environmental factor may influence the rate that males encounter each other. For example, the southern area may be marginal habitat with a lower density of dolphins that encounter each other less often, while the open northern waters may conduct sound further, effectively increasing encounter rate (see Connor et al. 2017).

16.1.1.4 Cognition

Humans and some species of dolphins have the largest relative brain sizes among mammals and are the only mammals with multilevel male alliances within social groups. This convergence led Connor and co-authors (Connor et al. 1992b; Connor 2007; Connor and Mann 2006; Randic et al. 2012) to explore the cognitive implications of a multilevel alliance system in an unbounded social network where individuals show strong preferences and avoidances within second-order alliances, recruit alliance members for attacks on others, rush to a conflict to aid second- or third-order alliance members; and where allies are often separated from each other and may have only partial information about the relationships and strength of rival groups. We identify three major insights that have emerged from considerations of the Shark Bay dolphin society:

1. Negotiating multilevel alliances should be more cognitively demanding than a single level because decisions at one level may significantly impact other levels. If, for example, a trio evicts a member to form a pair, so each member enjoys a greater probability of fathering a consorted female's calf, the drop in their second-order alliance size from six to five males may weaken their ability to defend their females from thefts by other groups. Biological anthropologists, focusing on the cognitive challenges of relationships within one level of nonhuman primate alliances or resulting from conflict between human groups (warfare), largely overlooked the challenges of a nested alliance structure (see Connor et al. 1992b; Connor 2007).
2. Societies of terrestrial mammals are "bounded"; individuals are members or they are not (think of a baboon troop, a wolf pack, or a chimpanzee community).

Primatologists especially, evaluating the cognitive challenges of social living, focused on the number of group members (as equal to the number of social relationships) and whether individuals had knowledge of "third-party" relationships (it would be obviously unwise to attack a subordinate individual when her larger, dominant friend is close by). In the dolphins' unbounded society, range overlap varies tremendously and, thus, so will individuals' knowledge of third-party relationships. Connor (2007) turned the primatologist logic on its head and emphasized third-party knowledge uncertainty. Your own allies may be strengthening their bonds at your expense while they are out of sight and sound, and the small, weak alliance that you easily defeated at the edge of your range 2 years ago may have doubled in size.

3. In the context of adaptations to the environment, Randic et al. (2012) pointed out a previously unrecognized convergence in the "big three" in mammalian brain evolution: dolphins, humans, and elephants all have adaptations to reduce the cost of locomotion. Elephants have the lowest cost of locomotion of any terrestrial mammal, humans have clear adaptations that reduced locomotion costs relative to our nearest relatives, and dolphins have much lower costs of locomotion than terrestrial mammals (see Connor 2000 and discussion in Randic et al. 2012). In all three mammals, low costs of locomotion allowed larger ranges, resulting in contact with a larger number of individuals (for a given population density) and, hence, greater cognitive challenges in the social domain.

16.1.2 Other Australian Study Sites

Outside of Shark Bay, the most detailed description of social structure and mating behavior comes from a relatively small population (ca. 120–140 individuals) in Port Stephens, a ca. 160 km^2 embayment on the mid-north coast of New South Wales. A larger "eastern community" of at least 90 dolphins inhabits a marine sand and seagrass habitat and is socially segregated from a smaller western community of about 30 individuals that forage in an estuarine mud and mangrove habitat (Möller et al. 2002, 2006; Wiszniewski et al. 2009). Learned foraging tactics in the different habitats may have driven this division (Wiszniewski et al. 2009). At a finer scale, relatedness plays a role in associations within several clusters of females, as well as their male associates (the latter likely owing to bisexual philopatry, Wiszniewski et al. 2011).

Möller et al. (2001) describe consortships between single females and pairs, trios, and one quadruplet of strongly associating but mostly unrelated males (Fig. 16.3a). Some consortships in Port Stephens are coerced, as indicated by observations of captures and escape attempts by the female (Möller et al. 2001). Males in larger groups (3–4) enjoyed greater reproductive success (Wiszniewski et al. 2012). The males of Port Stephens may also have second-order alliances, but this is presently uncertain.

An unusual case of object manipulation/play coincided with an influx of large blubber jellies (*Catostylus mosaicus*) to western Port Stephens. Individuals were

Fig. 16.3 (a) All four members of the allied quadruplet of adult male *T. aduncus* in Port Stephens, New South Wales. From left to right of frame are George, Paul, John, and Ringo Starr. (b) A pair of allied male *T. aduncus* pushing jelly blubbers (*Catostylus mosaicus*) to the surface on their rostra/melons in western Port Stephens. Note the individual on the right of the frame, Kenny (Norris), already has a jelly rolling down his caudal peduncle. (c) An adult female *T. aduncus*, newborn beside her, incapacitates an octopus with a specific prey-handling technique off Bunbury in southwestern Australia (photos: S. Allen). (d) A tail-walking dolphin in the Port River Estuary of Adelaide, South Australia (photo: M. Bossley, WDC)

observed pushing the jellies to the surface on their rostra/melons and then rolling the jellies down their backs and over their peduncles, sometimes raising their peduncles/tails above the surface as the jelly slid off (Allen pers. obs., Fig. 16.3b). We speculate that the dolphins engaged in this behavior for the sensation of the jelly and/or the "tingle" of the nematocysts (stinging cells) on their skin.

Foraging driven social segregation within a *T. aduncus* population can change rapidly. Despite considerable home-range overlap, dolphins foraging in association with prawn trawlers in Moreton Bay, Queensland, were socially segregated from those that did not (Chilvers and Corkeron 2001), but when trawler presence diminished, the social partition disappeared (Ansmann et al. 2012). Prawn trawler association fits the "producer-scrounger" model (Barnard and Sibly 1981), but there is an historical record of a cooperative fishing association between *T. aduncus* and aboriginal Australians (Hall 1985), similar to that which still occurs between fishers and *T. truncatus* in southern Brazil (e.g., Daura-Jorge et al. 2012).

Finn et al. (2009) detailed the stepwise process used by *T. aduncus* in Spencer Gulf, South Australia, for removing the ink and calcareous cuttlebone from giant cuttlefish (*Sepia apama*) before consumption. In southwestern Australia, Sprogis et al. (2017) outlined two methods, "shaking" and "tossing," by which *T. aduncus* incapacitate captured octopus (Fig. 16.3c). Dolphins who fail to disable octopus prey have died of suffocation (Stephens et al. 2017).

A remarkable case of the rise and fall of an arbitrary cultural fad was reported recently from the Port River estuary in South Australia (Bossley et al. 2018). A young dolphin that became trapped in a harbor was held for 1 month in a dolphinarium for rehabilitation, where it received no training but was able to observe the show dolphins perform their routines, which included tail walking. The dolphin was first observed tail walking 7 years after its release, but another dolphin performed the vast majority of tail walking, which was observed in 11 individuals, including 6 adult females and 5 juveniles. The occurrence of tail walking peaked 23–24 years after the original dolphin's stint in captivity and 15–16 years after tail walking was first observed in the wild. Thereafter the behavior declined. The authors argue that the temporary spread of tail walking in this population was most likely a case of social learning, given that tail walking is virtually unheard of in wild dolphins and is energetically costly and of no obvious adaptive value (Bossley et al. 2018; Fig. 16.3d).

16.2 Japan

Photo-identification studies on Indo-Pacific bottlenose dolphins have been conducted in seven distinct locations in Japan (e.g., Shirakihara et al. 2002; Kogi et al. 2004; Morisaka et al. 2013; Funasaka et al. 2016). Genetic differentiation suggests strong site fidelity and limited gene flow among populations, including possible multiple founding events from large, genetically diverse southern populations (Hayano 2013; Chen et al. 2017). Significant differences in the characteristics of whistles were found among Amakusa-Shimoshima, Mikura, and Ogasawara Islands (Morisaka et al. 2005a). In keeping with the behavioral ecology theme, we focus on underwater observations around Mikura Island and boat-based studies around the Amakusa-Shimoshima Islands.

16.2.1 Mikura Island

Mikura is a small (ca. 20 km^2), volcanic island located about 220 km south of Tokyo and thus represents a markedly different habitat from those of the Australian populations. Underwater video-identification studies have been conducted since 1994 on an estimated 160 *T. aduncus* within 300 m of the island's coastline, at water depths 2–45 m, during spring to early autumn (Kogi et al. 2004). Ad lib observations have been supplemented recently by fecal sampling to examine genetic information (Sakai et al. 2016; Kita et al. 2017).

16.2.1.1 Ecology and Social Structure

Mikura *T. aduncus* sometimes forage during the day for cephalopods (particularly octopus), flying fish (Exocoetidae) and other teleosts, but the timing of defecation suggests that nocturnal feeding predominates (Suzuki 2005). Morisaka et al. (2015) recorded continuous nocturnal vocalizations in shallower water, where dolphins are common during the day, which included burst-pulsed sounds with continuously decreasing inter-pulse intervals indicative of foraging activity.

Female Mikura dolphins first give birth at ~10 years of age (mean = 10.3, range = 7–13 years), calves are typically weaned after ~3.5 years (mean = 3.4, range = 2–5 years). The population is quite closed, with no immigration of unidentified adults for more than 20 years (Kogi et al. 2004; Kogi 2013). Tsuji et al. (2018) report 41 (25 male and 16 female) emigration events from Mikura to adjacent islands as far as 390 km distant. No sex difference was found in emigrants, but adults tend to emigrate more than younger animals. Population density around Mikura Island is high (6.5 individuals/km^2), which may be driving this emigration, and would thereby appear to be a source for neighboring populations (cf. Manlik et al. 2018).

Underwater observations dictated a different group definition than typical of boat-based studies (e.g., Smolker et al. 1992; Allen et al. 2012). A group was defined as all video-recorded dolphins from the time a videographer entered the water to when he/she returned to the boat. Nagata (2006) documented relatively low association indices between female and male pairs, although some male pairs associated for at least 9 years. During underwater interactions, several dolphins typically swam line abreast; with subgroups of several such lines—pairs of more closely positioned animals were identifiable within lines (Miyazaki 2009). Ueda (2013) analyzed female relationships based on associations in lines and found that females without calves tended to associate, as did those with calves, similar to findings that shared reproductive state enhanced female associations in Port Stephens *T. aduncus* (Möller and Harcourt 2008).

16.2.1.2 Social Interactions

Underwater observations have revealed detailed characteristics of many social interactions at this study site. Affiliative petting or "flipper rubbing," for example, involves one dolphin (the "rubber") rubbing its pectoral fin over various parts of a partner's (the "rubbee") body (Sakai et al. 2006b; Fig. 16.4a). The rubbee tends to be more active than the rubber, and individuals often switch roles during bouts. Flipper rubbing occurs most often between individuals of the same sex and age class and is also common in mother–calf dyads. The left pectoral fin is used more frequently and for longer periods, and configurations during flipper rubbing events revealed that this asymmetry owed not to the laterality of the rubber but by a preference for use of the left eye by both dolphins during bouts (Sakai et al. 2006a), possibly reflecting a left-eye advantage in conspecific recognition. Significant left-side bias was also observed

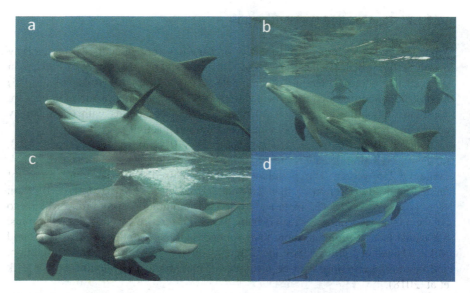

Fig. 16.4 (a) Flipper-to-body rubbing behavior of *T. aduncus* around Mikura Island. (b) Synchronous breathing between Mikura *T. aduncus*; three dolphins yonder breathe synchronously, while two synchronous dolphins are in the foreground. (c) A *T. aduncus* calf in echelon position next to its mother near Mikura Island. (d) A *T. aduncus* calf in infant position below its mother (photos: M. Sakai & T. Morisaka)

in flipper rubbing events initiated by the rubbee, determining its position during rubbing and thus enhancing the behavioral asymmetry by choosing the left side of the rubber, perhaps ensuring longer affiliative contact.

Dyadic synchronous breathing was common among individuals of the same age and sex class, as well as in mother–calf and escort–calf pairs (Sakai et al. 2010; Fig. 16.4b). The interindividual distance during synchronous breathing was less for mother–calf pairs than for other pairs and for female compared to male pairs. The size of a dolphin's "personal space" (Hall 1966; the region surrounding each person or that area which a person considers their domain or territory) may vary with age and sex. Personal space for female–female pairs is smaller than for male–male pairs in adults and subadults. Interestingly, the time differences between breaths for synchronous surfacing (0.63 ± 0.48 s) were intermediate between those reported for Shark Bay males as "social synch" (0.14 ± 0.11 s) and "nonsocial synch" (0.87 ± 0.56 s). However, the subjects in Connor et al. (2006b) were exclusively alliance members, while the Sakai et al. (2010) study included all male pairs. Allied males may engage in more precise synchrony than those not as tightly bonded.

Contact swimming (as per Shark Bay) was recorded more often between females than males (Chaturaphatranon 2011). Contact swimming appears similar to the echelon position (Fig. 16.4c) of mother–calf pairs, in which the calf gains a hydrodynamic benefit. Indeed, the fluke stroke frequency of the actor (0.45/s) was significantly lower than that of the receiver (1.11/s).

As might be expected, neonates tended to swim closer to their mothers than 1–3-year-old calves, and they utilize the echelon position (Fig. 16.4c) before gradually transitioning to the infant position (Fig. 16.4d; Masaki 2003). Primiparous mothers engaged in rubbing and synchronous breathing with their calves less than parous mothers. The distance between parous mothers and their calves always increased during development, but not in primiparous mother–calf pairs, suggesting a role for maternal experience (Kimura 2009). "Escorting," or "babysitting"/alloparental care, where dolphins other than the mother swam with dependent calves, was observed frequently. Young nulliparous females babysat more than other sex-age classes. Calves used echelon position significantly more than infant position during babysitting, and a preliminary genetic study found no role for kin selection in babysitting (Haraguchi 2005).

Sakai et al. (2016) documented the adoption of a calf following the death of its mother, by a subadult female. On 3 of 18 observation days post-adoption, the calf was observed swimming in the suckling position, and milk was seen leaking from the female's mammary slit. A 5-year dataset revealed no significant social or kin relationships between the biological mother and allomother (Sakai et al. 2016). Ridgway et al. (1995) reported that orphans in captivity immediately tried to nurse from allomothers, and dry adult dolphin females were brought into lactation by repeated nursing attempts. Thus, nursing attempts by the calf may have been critical in the adoption process. Several kinds of epimeletic or helping behaviors have been reported in dolphins (reviewed in Sakai et al. 2016). These behaviors suggest that dolphins have the capacity for empathic perspective taking (de Waal 2008), but the precise nature of these cognitive abilities remains unknown. The cognitive characteristics that evoke adoption behaviors in dolphins need additional study to determine the extent to which they reflect social cognition or more proximate responses.

Groups of between 2 and 14 Mikura Island *T. aduncus* have been recorded engaging in sociosexual behavior (Fig. 16.5a). Such bouts included mounting, in which one dolphin attempts to insert its penis into the genital slit of another dolphin, rubbing the genital area of another dolphin, and contact between the rostrum or melon and genital area and rostrum-to-genital propulsion (recall "goosing" in Shark Bay above for these latter two). Sociosexual behavior was observed among males, sometimes among subadult females, and between mothers and their male calves (Shimomaki 2001; Jiroumaru 2008).

Play behaviors including object carrying were also observed, using seaweed, plastic bags, and, on occasion, octopus (Morisaka, Sakai pers. obs.; Fig. 16.5b). Object carrying as a component of play has been documented across many delphinids in both wild and captive settings (Greene et al. 2011). A rare spontaneous ejaculation event was also filmed in this population (Morisaka et al. 2013; Fig. 16.5c). Dudzinski et al. (2003) documented two cases of apparently inquisitive behavior toward a dead conspecific. Repeated investigative behaviors such as echolocation and scanning toward the genital and chest area occurred first, followed by male erections. The carrying of dead calves has also been observed frequently, primarily by adult females, followed by other, mostly male, dolphins (Sakai, Morisaka pers. obs.).

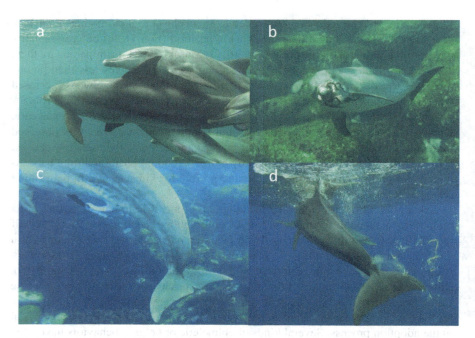

Fig. 16.5 (**a**) Sociosexual behavior of male *T. aduncus* near Mikura Island. (**b**) A female Mikura *T. aduncus* holds (or is held by?) an octopus for play and as food. (**c**) Spontaneous ejaculation by a 16-year-old male Mikura *T. aduncus*. (**d**) Defecation toward a swimmer by *T. aduncus*, Mikura Island (photos: M. Sakai & T. Morisaka)

Suzuki (2005) observed defecation by Mikura dolphins immediately in front of other dolphins and human swimmers (Fig. 16.5d). All defecating dolphins changed direction just after defecation; in most events, the dolphin first engaged in "inquiring" behaviors (echolocation and/or circle swim around the human swimmer) followed by defecation just in front of the swimmer, implying that defecation by Mikura dolphins may function as a mild threat behavior.

16.2.2 Amakusa-Shimoshima Island

At Amakusa-Shimoshima Island, western Kyushu, boat-based photo-identification studies have been conducted since 1994 (Shirakihara et al. 2002). A resident population of *T. aduncus* occurs along the northern coast of the island, where there are estuarine features. Similar to recently reported large group sizes of South African *T. aduncus* (Bouveroux et al. 2018), group size commonly exceeded 100 individuals (Shirakihara et al. 2002). In contrast to Bouveroux et al. (2018), who invoked predation risk from large white sharks (*Carcharodon carcharias*) as a driver of larger aggregations, Shirakihara et al. (2002) suggested that large groups are likely

16 The Indo-Pacific Bottlenose Dolphin (*Tursiops aduncus*)   361

formed in response to rich food sources. Even the usually solitary finless porpoises (*Neophocaena phocaenoides*) occurring in this area sometimes form groups of more than 100 individuals (Yoshida et al. 1997).

Ambient ocean noise levels around Amakusa-Shimoshima Island are much higher than those around Mikura Island and Ogasawara Island, because of rich biotic sounds produced by crustaceans, especially snapping shrimp (Alpheidae; Morisaka et al. 2005b). The resident *T. aduncus* tend to produce lower, less frequency-modulated, high amplitude whistles, probably to overcome these noisy conditions (Morisaka et al. 2005b). The range over which dolphins can recognize each other's whistles around this island was estimated to be comparable to the diameter of the majority of groups. Thus, the Amukusa-Shimoshima Island dolphins should be able to hear the whistles produced by most group members even in large but cohesive groups (Morisaka 2009; Morisaka et al. unpub. data).

In the year 2000, most individuals moved 60 km to the southern part of the island, and then all but 20 returned the following year. Nishita et al. (2015) found that male–male pairs in the new, southern community had higher associations before splitting, indicating long-term relationships, but that was not the case for male–female and female–female pairs. Nishita et al. (2017b) documented ten male pairs that associated across multiple years and which sometimes flanked single females, suggesting consortship behavior, while Nishita et al. (2017a) found associations among females were based on shared reproductive status, similar to Ueda (2013) for Mikura Island. Females that lost infants within 1 year of birth changed their associations accordingly. These features, also of other populations, are maintained even when ecological factors favor much larger group sizes.

16.3 Discussion

The Indo-Pacific bottlenose dolphin *Tursiops aduncus* is usually compared with its congener, *T. truncatus* (e.g., Connor et al. 2000a). The two longest running *Tursiops* studies in the world are in Sarasota Bay, Florida (see Wells, this volume), and Shark Bay, Australia. The most distinctive difference between the two species and populations is the markedly more complex male alliance formation in Shark Bay. The Sarasota males form first-order alliances of two males only, whereas the Shark Bay males form three alliance levels, with trios predominating at the first-order level (Owen et al. 2002; Connor et al. 2011; Connor and Krützen 2015).

The leading model to explain differences in alliance formation between populations is the rate of interaction model (Connor and Whitehead 2005; Whitehead and Connor 2005). This model posits that selection for alliance formation is greater when individuals encounter each other more often in competition for resources (estrus females, in this case). The encounter rate should increase with population density. Given that population density decreases with body size (Cotgreave 1993; White et al. 2007) and that *T. aduncus* are smaller than *T. truncatus* (Hale et al. 2000), we could expect that, if other factors were equal, *T. aduncus* populations exhibit higher densities and more complex alliance formation than *T. truncatus*. However, other factors may impact encounter rates, including detection distance, as well as day and home range (Connor and Whitehead 2005), and these factors will likely vary among habitats. Further, population densities of both *Tursiops* species vary between locations, and possibly even within study sites (Connor et al. 2017), impacting selection for alliance formation. The single report of second-order alliances in *T. truncatus* is instructive; Ermak et al. (2017) report second-order alliances (based on associations between male pairs) in an unusually high-density population in the St. Johns River, Florida. They further suggest that their discovery was evidence against the habitat structure hypothesis (Connor et al. 2000a), where males would form alliances more in open habitats versus subdivided habitats, in which males could hide with female consorts. This conclusion depends on the movements of dolphins in the habitat; if dolphins seeking prey must travel along a narrow river, their encounter rates with others might be unusually high.

The differences in alliance structure may also impact the dolphins' communication systems. The complex alliances in Shark Bay, where males often switch first-order alliance partners and interact with a larger number of second- and third-order allies, may favor the retention of individual identity signals compared to Sarasota, where an alliance "badge" may be more practical in the strongly bonded stable pairs (King et al. 2018).

Learned foraging behaviors that vary within and between habitats can lead to social segregation (e.g., Moreton Bay, Chilvers and Corkeron 2001; Port Stephens, Wiszniewski et al. 2009; Shark Bay, Mann et al. 2012). Specialized foraging tactics like those described in Shark Bay are also reported in populations of *T. truncatus* (Wells, this volume). A case of social segregation based on association with human fisheries was recently reported in *T. truncatus* (Kovacs et al. 2017). More broadly, the phenomenon may be widespread in delphinids and could, for example, account for the culturally and often genetically distinct ecotypes in killer whales (Riesch et al. 2012) and even the remarkable speciation of delphinids (Connor and Krützen 2015).

As a result of recent advances in technology, including hydrophone arrays for localizing sound production to individuals and drones for observing behavior, the future of *T. aduncus* research is bright. One of the broader questions to address is whether *T. aduncus* in other populations also lead apparently more complex social lives than found in *T. truncatus*. If so, such differences may be represented in measures of the size of the brain or brain components and in cognitive abilities. As noted by Connor (2007), inshore delphinids are ecological outliers in the Delphinidae; and some offshore species living in large, fast-moving groups may be even more complex socially, but we are fortunate to have a species like *T. aduncus* that combines astonishing social complexity with ease of observation by humans.

References

Allen SJ, Bejder L, Krützen M (2011) Why do Indo-Pacific bottlenose dolphins (*Tursiops* sp.) carry conch shells (*Turbinella* sp.) in Shark Bay, Western Australia? Mar Mamm Sci 27:449–454

16 The Indo-Pacific Bottlenose Dolphin (*Tursiops aduncus*)

Allen SJ, Cagnazzi DD, Hodgson AJ, Loneragan NR, Bejder L (2012) Tropical inshore dolphins of north-western Australia: unknown quantities along a changing coastline. Pac Conserv Biol 18:56–63

Ansmann IC, Parra GJ, Chilvers BL, Lanyon JM (2012) Dolphins restructure social system after reduction of commercial fisheries. Anim Behav 84:575–581

Aureli F, Preston SD, de Waal FB (1999) Heart rate responses to social interactions in free-moving rhesus macaques (*Macaca mulatta*): a pilot study. J Comp Psychol 113:59–65

Barnard CJ, Sibly RM (1981) Producers and scroungers: a general model and its application to captive flocks of house sparrows. Anim Behav 29:543–550

Bizzozzero MR, Allen SJ, Gerber L, Wild S, King SL, Connor RC, Friedman WR, Wittwer S, Krützen M (2019) Tool use and social homophily among male bottlenose dolphins. Proc R Soc Lond B 286:20190898

Bossley M, Steiner A, Brakes P, Shrimpton J, Foster C, Rendell L (2018) Tail walking in a bottlenose dolphin community: the rise and fall of an arbitrary cultural 'fad'. Biol Lett 14:20180314. https://doi.org/10.1098/rsbl.2018.0314

Bouveroux TN, Caputo M, Froneman PW, Plön S (2018) Largest reported groups for the Indo-Pacific bottlenose dolphin (*Tursiops aduncus*) found in Algoa Bay, South Africa: trends and potential drivers. Mar Mamm Sci 34:645–665

Brown AM, Bejder L, Pollock KH, Allen SJ (2016) Site-specific assessments of the abundance of three inshore dolphin species to inform conservation and management. Front Mar Sci 3:4

Chaturaphatranon K (2011) Contact swimming behavior in wild Indo-Pacific bottlenose dolphins (*Tursiops aduncus*). M.S. thesis, Tokyo Institute of Technology, Tokyo, Japan

Chen I, Nishida S, Yang W, Isobe T, Tajima Y, Hoelzel AR (2017) Genetic diversity of bottlenose dolphin (*Tursiops* sp.) populations in the western North Pacific and the conservation implications. Mar Biol 164:202

Chilvers BL, Corkeron PJ (2001) Trawling and bottlenose dolphins' social structure. Proc R Soc B 268:1901–1905

Connor RC (2000) Group living in whales and dolphins. In: Mann J, Connor R, Tyack P, Whitehead H (eds) Cetacean societies: field studies of whales and dolphins. University of Chicago Press, Chicago, IL, pp 199–218

Connor RC (2007) Complex alliance relationships in bottlenose dolphins and a consideration of selective environments for extreme brain size evolution in mammals. Philos Trans R Soc B 362:587–602

Connor RC, Krützen M (2015) Male dolphin alliances in Shark Bay: changing perspectives in a thirty-year study. Anim Behav 103:223–235

Connor RC, Mann J (2006) Social cognition in the wild: Machiavellian dolphins? In: Hurley S, Nudd M (eds) Rational animals. Oxford University Press, Oxford, pp 329–367

Connor RC, Smolker RA (1985) Habituated dolphins (*Tursiops* sp.) in Western Australia. J Mammal 36:304–305

Connor RC, Smolker RA (1996) "Pop" goes the dolphin: a vocalization male bottlenose dolphins produce during consortships. Behaviour 133:643–662

Connor RC, Vollmer N (2009) Sexual coercion in dolphin consortships: a comparison with chimpanzees. In: Muller MN, Wrangham RW (eds) Sexual coercion in primates: an evolutionary perspective on male aggression against females. Harvard University Press, Cambridge, MA, pp 218–243

Connor RC, Whitehead H (2005) Alliances II: Rates of encounter during resource utilization: a general model of intrasexual alliance formation in fission-fusion societies. Anim Behav 69:127–132

Connor RC, Smolker RA, Richards AF (1992a) Two levels of alliance formation among male bottlenose dolphins (*Tursiops* sp.). Proc Natl Acad Sci USA 89:987–990

Connor RC, Smolker RA, Richards AF (1992b) Dolphin alliances and coalitions. In: Harcourt AH, de Waal FBM (eds) Coalitions and alliances in animals and humans. Oxford University Press, Oxford, pp 415–444

Connor RC, Richards AF, Smolker RA, Mann J (1996) Patterns of female attractiveness in Indian Ocean bottlenose dolphins. Behaviour 133:37–69

Connor RC, Wells R, Mann J, Read A (2000a) The bottlenose dolphin: social relationships in a fission-fusion society. In: Mann J, Connor R, Tyack P, Whitehead H (eds) Cetacean societies: field studies of whales and dolphins. University of Chicago Press, Chicago, IL, pp 91–126

Connor RC, Heithaus MR, Berggren P, Miksis JL (2000b) 'Kerplunking': surface fluke-splashes during shallow water bottom foraging by bottlenose dolphins. Mar Mamm Sci 16:646–653

Connor RC, Heithaus MR, Barre LM (2001) Complex social structure, alliance stability and mating access in a bottlenose dolphin 'super-alliance'. Proc R Soc Lond B 268:263–267

Connor RC, Mann J, Watson-Capps J (2006a) A sex-specific affiliative contact behavior in Indian Ocean bottlenose dolphins, *Tursiops* sp. Ethology 112:631–638

Connor RC, Smolker RA, Bejder L (2006b) Synchrony, social behavior and alliance affiliations in Indian Ocean bottlenose dolphins (*Tursiops aduncus*). Anim Behav 72:1371–1378

Connor RC, Watson-Capps JJ, Sherwin WB, Krützen M (2011) New levels of complexity in the male alliance networks of Indian Ocean bottlenose dolphins (*Tursiops* sp.). Biol Lett 7:623–626

Connor RC, Cioffi WR, Randic S, Allen SJ, Watson-Capps JJ, Krützen M (2017) Male alliance behavior and mating access varies with habitat in a dolphin social network. Sci Rep 7:46354

Cotgreave P (1993) The relationship between body-size and population abundance in animals. Trends Ecol Evol 8:244–248

Daura-Jorge FG, Cantor M, Ingram SN, Lusseau D, Simões-Lopes PC (2012) The structure of a bottlenose dolphin society is coupled to a unique foraging cooperation with artisanal fishermen. Biol Lett 8:702–705

de Waal FBM (2008) Putting the altruism back into altruism: the evolution of empathy. Annu Rev Psychol 59:279–300

Dudzinski KM, Sakai M, Masaki K, Kogi K, Hishii T, Kurimoto M (2003) Behavioural observations of bottlenose dolphins towards two dead conspecifics. Aquat Mamm 29:108–116

Dunn DG, Barco SG, Pabst DA, McLellan WA (2002) Evidence for infanticide in bottlenose dolphins of the Western North Atlantic. J Wildl Dis 38:505–510

Ermak J, Brightwell K, Gibson Q (2017) Multi-level dolphin alliances in northeastern Florida offer comparative insight into pressures shaping alliance formation. J Mammal 98:1096–1104

Finn J, Tregenza T, Norman M (2009) Preparing the perfect cuttlefish meal: complex prey handling by dolphins. PLoS One 4:e4217

Firman RC, Gasparini C, Manier MK, Pizzari T (2017) Postmating female control: 20 years of cryptic female choice. Trends Ecol Evol 32:368–382

Foroughirad V, Mann J (2013) Human fish provisioning has long-term impacts on the behaviour and survival of bottlenose dolphins. Biol Conserv 160:242–249

Frère CH, Krützen M, Mann J, Connor RC, Bejder L, Sherwin WB (2010) Social and genetic interactions drive fitness variation in a free-ranging dolphin population. Proc Natl Acad Sci USA 107:19949–19954

Funasaka N, Okabe H, Oki K, Tokutake K, Kawazu I, Yoshioka M (2016) The occurrence and individual identification study of Indo-Pacific bottlenose dolphins *Tursiops aduncus* in the waters around Amami Oshima Island, southern Japan: a preliminary report. Mamm Study 41:163–169

Galezo A, Krzyszczyk E, Mann J (2018) Sexual segregation in Indo-Pacific bottlenose dolphins is driven by female avoidance of males. Behav Ecol 29(2):377–386

Gerber L, Connor R, King S, Allen S, Wittwer S, Bizzozzero M, Friedman W, Kalberer S, Sherwin W, Wild S, Willems E, Krutzen M (2019) Multi-level cooperation in wild male bottlenose dolphins is predicted by longterm friendships. Behav Ecol

Greene WE, Melillo-Sweeting K, Dudzinski KM (2011) Comparing object play in captive and wild dolphins. J Comp Psychol 24:292–306

Hale PT, Barreto AS, Ross GJB (2000) Comparative morphology and distribution of the aduncus and truncatus forms of bottlenose dolphin *Tursiops* in the Indian and Western Pacific Oceans. Aquat Mamm 26:101–110

16 The Indo-Pacific Bottlenose Dolphin (*Tursiops aduncus*)

Hall ET (1966) The hidden dimension. Doubleday, Garden City, NY, 201p

Hall HJ (1985) Fishing with dolphins?: affirming a traditional aboriginal fishing story in Moreton Bay, SE. Queensland. In: Focus on Stradbroke: new information on North Stradbroke Island and Surrounding areas, 1974–1984. Boolarong Publications, Brisbane, pp 16–22

Hamilton RA, Borcuch T, Allen SJ, Cioffi WR, Bucci V, Krützen M, Connor RC (2019) Aggression varies with consortship rate and habitat in a dolphin social network. Behav Ecol and Sociobiol. in review

Haraguchi R (2005) Behavior and kinship of allomaternal care in Indo-Pacific bottlenose dolphins (*Tursiops aduncus*). B.S. thesis, Tokyo University of Agriculture and Technology (In Japanese)

Hayano A (2013) Genetic composition of genus *Tursiops* in Japanese waters inferred from mitochondrial DNA analysis. Kaiyo Monthly 45:341–347 (In Japanese)

Heithaus MR (2001) Shark attacks on bottlenose dolphins (*Tursiops aduncus*) in Shark Bay, Western Australia: attack rate, bite scar frequencies, and attack seasonality. Mar Mamm Sci 17:526–539

Herman LH (2002) Vocal, social, and self-imitation by bottlenosed dolphins. In: Dautenhahn K, Nehaniv CL (eds) Imitation in animals and artifacts. MIT Press, Cambridge, MA, pp 63–108

Hicks BD, Aubin DJS, Geraci JR, Brown WR (1985) Epidermal growth in the bottlenose dolphin, *Tursiops truncatus*. J Invest Dermatol 85:60–63

Hinde RA (1976) Interactions, relationships and social structure. Man 11:1–17

Jiroumaru M (2008) Sociosexual behaviors of Indo-Pacific bottlenose dolphins around Mikura Island. B.S. thesis, Teikyo University of Science (In Japanese)

Karniski C, Krzyszczyk E, Mann J (2018) Senescence impacts reproduction and maternal investment in bottlenose dolphins. Proc R Soc B 285:1123

Kimura Y (2009) Difference of calf-caring behaviors with experience of giving birth. M.S. thesis, Mie University (In Japanese)

King SL, Friedman WR, Allen SJ, Gerber L, Jensen FH, Wittwer S, Connor RC, Krützen M (2018) Bottlenose dolphins retain individual labels in multilevel alliances. Curr Biol 28:1993–1999

Kita YF, Inoue-Murayama M, Kogi K, Morisaka T, Sakai M, Shiina T (2017) Kinship analysis of Indo-Pacific bottlenose dolphin (*Tursiops aduncus*) in Mikura Island. DNA Polymorph 25:52–57 (In Japanese)

Kogi K (2013) Indo-Pacific bottlenose dolphins around Mikurashima Island. Kaiyo Monthly 45:215–225 (In Japanese)

Kogi K, Hishii T, Imamura A, Iwatani T, Dudzinski KM (2004) Demographic parameters of Indo-Pacific bottlenose dolphins (*Tursiops aduncus*) around Mikura Island, Japan. Mar Mamm Sci 20:510–526

Kopps AM, Krützen M, Allen SJ, Bacher K, Sherwin WB (2014) Characterizing the socially transmitted foraging tactic "sponging" by bottlenose dolphins (*Tursiops* sp.) in the Western Gulf of Shark Bay, Western Australia. Mar Mamm Sci 30:847–863

Kovacs CJ, Perrtree RM, Cox TM (2017) Social differentiation in common bottlenose dolphins (*Tursiops truncatus*) that engage in human-related foraging behaviors. PLoS One 12:e0170151

Krützen M, Barre LM, Connor RC, Mann J, Sherwin WB (2004) O father: where art thou? – Paternity assessment in an open fission-fusion society of wild bottlenose dolphins (*Tursiops* sp.) in Shark Bay, Western Australia. Mol Ecol 13:1975–1990

Krützen M, Mann J, Heithaus MR, Connor RC, Bejder L, Sherwin WB (2005) Cultural transmission of a foraging strategy involving tool use in bottlenose dolphins. Proc Natl Acad Sci USA 102:8939–8943

Krützen M, Kreicker S, MacLeod CD, Learmonth J, Kopps AM, Walsham P, Allen SJ (2014) Cultural transmission of tool use by Indo-Pacific bottlenose dolphins (*Tursiops* sp.) provides access to a novel foraging niche. Proc R Soc Lond B 281:20140374

Krzyszczyk E, Patterson EM, Stanton MA, Mann J (2017) The transition to independence: sex differences in social and behavioural development of wild bottlenose dolphins. Anim Behav 129:43–59

Manlik O, Chabanne D, Daniel C, Bejder L, Allen SJ, Sherwin WB (2018) Demography and genetics suggest reversal of dolphin source-sink dynamics, with implications for conservation. Mar Mamm Sci. https://doi.org/10.1111/mms.12555

Mann J (2006) Establishing trust: sociosexual behaviour and the development of male-male bonds among Indian Ocean bottlenose dolphin calves. In: Vasey P, Sommer V (eds) Homosexual behaviour in animals: an evolutionary perspective, Chapter 4. Cambridge University Press, Cambridge, pp 107–130

Mann J, Sargeant BL (2003) Like mother, like calf: the ontogeny of foraging traditions in wild India Ocean bottlenose dolphins (*Tursiops* sp.). In: Fragaszy D, Perry S (eds) The biology of traditions: models and evidence. Cambridge University Press, Cambridge, pp 236–266

Mann J, Sargeant BL, Watson-Capps JJ, Gibson QA, Heithaus MR, Connor RC et al (2008) Why do dolphins carry sponges? PLoS One 3:e3868

Mann J, Stanton M, Patterson EM, Bienenstock EJ, Singh LO (2012) Social networks reveal cultural behavior in tool-using dolphins. Nat Commun 3:980

Masaki K (2003) Developmental changes of mother-calf distances and positions in bottlenose dolphins. M.S. thesis, Mie University (In Japanese)

Miyazaki Y (2009) Group swimming behaviors in the wild Indo-Pacific bottlenose dolphins. M.S. thesis, Tokyo Institute of Technology (In Japanese)

Möller LM, Harcourt RG (2008) Shared reproductive state enhances female associations in dolphins. Res Lett Ecol 2008:498390. https://doi.org/10.1155/2008/498390

Möller LM, Beheregaray LB, Harcourt RG, Krützen M (2001) Alliance membership and kinship in wild male bottlenose dolphins (*Tursiops aduncus*) of southeastern Australia. Proc R Soc Lond B 268:1941–1947

Möller LM, Allen SJ, Harcourt RG (2002) Group characteristics, site fidelity and seasonal abundance of bottlenosed dolphins (*Tursiops aduncus*) in Jervis Bay and Port Stephens, South-Eastern Australia. Aust Mammal 24:11–22

Möller LM, Beheregaray LB, Allen SJ, Harcourt RG (2006) Association patterns and kinship in female Indo-Pacific bottlenose dolphins (*Tursiops aduncus*) of southeastern Australia. Behav Ecol Sociobiol 61:109–117

Morisaka T (2009) Acoustic communication by dolphins and its constraints. Mamm Sci 49:121–127 (In Japanese)

Morisaka T, Shinohara M, Nakahara F, Akamatsu T (2005a) Geographic variations in the whistles among three Indo-Pacific bottlenose dolphin *Tursiops aduncus* populations in Japan. Fish Sci 71:568–576

Morisaka T, Shinohara M, Nakahara F, Akamatsu T (2005b) Effects of ambient noise on the whistles of Indo-Pacific bottlenose dolphin population. J Mammal 86:541–546

Morisaka T, Sakai M, Kogi K, Nakasuji A, Sakakibara K, Kasanuki Y, Yoshioka M (2013) Spontaneous ejaculation in a wild Indo-Pacific bottlenose dolphin (*Tursiops aduncus*). PLoS One 8:e72879

Morisaka T, Sakai M, Kogi K (2015) Detection of the nighttime distribution of Indo-Pacific bottlenose dolphins (*Tursiops aduncus*) around Mikura Island with stationed acoustic buoys. Bull Inst Oceanic Res Dev Tokai Univ 36:1–7

Nagata K (2006) Individual relationships of Indo-Pacific bottlenose dolphins around Mikura Island. M.S. thesis, Mie University (In Japanese)

Nishita M, Shirakihara M, Amano M (2015) A community split among dolphins: the effect of social relationships on the membership of new communities. Sci Rep 5:17266

Nishita M, Shirakihara M, Amano M (2017a) Patterns of association among female Indo-Pacific bottlenose dolphins (*Tursiops aduncus*) in a population forming large groups. Behaviour 154:1013–1028

Nishita M, Shirakihara M, Iwasa N, Amano M (2017b) Alliance formation of Indo-Pacific bottlenose dolphins (*Tursiops aduncus*) off Amakusa, Western Kyushu, Japan. Mamm Study 42:125–130

16 The Indo-Pacific Bottlenose Dolphin (*Tursiops aduncus*)

O'Brien O, Allen SJ, Krützen MK, Connor RC (in review) Alliance specific habitat use by male dolphins in Shark Bay, Western Australia.

Owen EC, Wells RS, Hofmann S (2002) Ranging and association patterns of paired and unpaired adult male Atlantic bottlenose dolphins, *Tursiops truncatus*, in Sarasota, Florida, provide no evidence for alternative male strategies. Can J Zool 80:2072–2089

Patterson I, Reid R, Wilson B, Grellier K, Ross H, Thompson P (1998) Evidence for infanticide in bottlenose dolphins: an explanation for violent interactions with harbour porpoises? Proc R Soc Lond B 265:1167–1170

Randic S, Connor RC, Sherwin WB, Krützen M (2012) A novel mammalian social structure in Indo-Pacific bottlenose dolphins (*Tursiops* sp.): complex male alliances in an open social network. Proc R Soc Lond B 279:3083–3090

Ridgway S, Kamolnick T, Reddy M, Curry C, Tarpley RJ (1995) Orphan-induced lactation in *Tursiops* and analysis of collected milk. Mar Mamm Sci 11:172–182

Riesch R, Barrett-Lennard LG, Ellis GM, Ford JK, Deecke VB (2012) Cultural traditions and the evolution of reproductive isolation: ecological speciation in killer whales? Biol J Linn Soc 106:1–17

Robinson KP (2014) Agonistic intraspecific behavior in free-ranging bottlenose dolphins: calf-directed aggression and infanticidal tendencies by adult males. Mar Mamm Sci 30:381–388

Saayman GS, Tayler CK (1973) Social organization of inshore dolphins (*Tursiops aduncus* and *Sousa* sp.) in the Indian Ocean. J Mammal 54:993–996

Sakai M, Hishii T, Takeda S, Kohshima S (2006a) Laterality of flipper rubbing behaviour in wild bottlenose dolphins (*Tursiops aduncus*): caused by asymmetry of eye use? Behav Brain Res 170:204–210

Sakai M, Hishii T, Takeda S, Kohshima S (2006b) Flipper rubbing behaviors in wild bottlenose dolphins (*Tursiops aduncus*). Mar Mamm Sci 22:966–978

Sakai M, Morisaka T, Kogi K, Hishii T, Kohshima S (2010) Fine-scale analysis of synchronous breathing in wild Indo-Pacific bottlenose dolphins (*Tursiops aduncus*). Behav Process 83:48–53

Sakai M, Kita YF, Kogi K, Shinohara M, Morisaka T, Shiina T, Inoue-Murayama M (2016) A wild Indo-Pacific bottlenose dolphin adopts a socially and genetically distant neonate. Sci Rep 6:23902

Samuels A, Gifford T (1997) A quantitative assessment of dominance relations among bottlenose dolphins. Mar Mamm Sci 13:70–99

Sargeant BL, Mann J, Berggren P, Krützen M (2005) Specialisation and development of beach hunting, a rare foraging behavior, by wild bottlenose dolphins (*Tursiops sp*). Can J Zool 83:1400–1410

Sargeant BL, Wirsing AJ, Heithaus MR, Mann J (2007) Can environmental heterogeneity explain individual foraging variation in wild bottlenose dolphins (*Tursiops* sp)? Behav Ecol Sociobiol 61:679–688

Scott EM, Mann J, Watson JJ, Sargeant BL, Connor RC (2005) Aggression in bottlenose dolphins: evidence for sexual coercion, male-male competition and female tolerance though analysis of tooth rake scars and behavior. Behaviour 142:21–44

Shimomaki M (2001) Sociosexual behavior of bottlenose dolphins (*Tursiops aduncus*) resident around Mikura Island in the Izu Islands. M.S. thesis, The University of Tokyo

Shirakihara M, Shirakihara K, Tomonaga J, Takatsuki M (2002) A resident population of Indo-Pacific bottlenose dolphins (*Tursiops aduncus*) in Amakusa, western Kyushu, Japan. Mar Mamm Sci 18:30–41

Smith F, Allen SJ, Bejder L, Brown AM (2018) Shark bite injuries on three inshore dolphin species in tropical Northwestern Australia. Mar Mamm Sci 34:87–99

Smolker RA, Richards AF, Connor RC, Pepper J (1992) Association patterns among bottlenose dolphins in Shark Bay, Western Australia. Behaviour 123:38–69

Smolker RA, Richards AF, Connor RC, Mann J (1997) Sponge carrying by dolphins (Delphinidae, *Tursiops* sp.): a foraging specialization involving tool use? Ethology 103:454–465

Sprogis KR, Raudino HC, Hocking D, Bejder L (2017) Complex prey handling of octopus by bottlenose dolphins (*Tursiops aduncus*). Mar Mamm Sci 33:934–945

Stephens N, Duignan P, Symons J, Holyoake C, Bejder L, Warren K (2017) Death by octopus (*Macroctopus maorum*): laryngeal luxation and asphyxiation in an Indo-Pacific bottlenose dolphin (*Tursiops aduncus*). Mar Mamm Sci 33:1204–1213

Suzuki A (2005) Study on defecation behavior and its social function in a dolphin, *Tursiops aduncus*. M.S. thesis, Toho University (In Japanese)

Tayler CK, Saayman GS (1972) The social organization and behavior of dolphins (*Tursiops aduncus*) and baboons (*Papio ursinus*): Some comparisons and assessments. Ann Cape Providence Mus Nat Hist 9:11–49

Tayler CK, Saayman GS (1973) Imitative behaviour by Indian Ocean bottlenose dolphins (*Tursiops aduncus*) in captivity. Behaviour 44:286–298

Tsai Y-JJ, Mann J (2013) Dispersal, philopatry, and the role of fission-fusion dynamics in bottlenose dolphins. Mar Mamm Sci 29:261–279

Tsuji K, Kogi K, Sakai M, Morisaka T (2018) Emigration of Indo-Pacific bottlenose dolphins (*Tursiops aduncus*) from Mikura Island, Japan. Aquat Mamm 43:585–593

Tyne JA, Loneragan NR, Kopps AM, Allen SJ, Krützen M, Bejder L (2012) Ecological characteristics contribute to sponge distribution and tool use in bottlenose dolphins Tursiops sp. Mar Ecol Prog Ser 444:143–153

Ueda N (2013) Individual relationships among female Indo-Pacific bottlenose dolphins around Mikura Island. Mikurensis 2:13–28 (In Japanese)

Vollmer NL, Hayek LAC, Heithaus MR, Connor RC (2015) Further evidence of a context-specific agonistic signal in bottlenose dolphins: the influence of consortships and group size on the pop vocalization. Behavior 152:1979–2000

Wallen MM, Patterson E, Krzyszczyk E, Mann J (2016) Ecological costs to females in a system with allied sexual coercion. Anim Behav 115:227–236

Wallen MM, Krzyszczyk E, Mann J (2017) Mating in a bisexually philopatric society: bottlenose dolphin females associate with adult males but not adult sons during estrous. Behav Ecol Sociobiol 71(10):153

Watson-Capps JJ (2005) Female mating behavior in the context of sexual coercion and female ranging behavior of bottlenose dolphins (*Tursiops* sp.) in Shark Bay, Western Australia (Doctoral thesis). Washington, DC: Georgetown University

White EP, Ernest SKM, Kerkhoff AJ, Enquist BJ (2007) Relationships between body size and abundance in ecology. Trends Ecol Evol 22:323–330

Whitehead H, Connor RC (2005) Alliances I: how large should alliances be? Anim Behav 69:117–126

Wilson M, Cook PF (2016) Rhythmic entrainment: why humans want to, fireflies can't help it, pet birds try, and sea lions have to be bribed. Psychon Bull Rev 23:1647–1659

Wiszniewski J, Allen SJ, Möller LM (2009) Social cohesion in a hierarchically structured embayment population of Indo-Pacific bottlenose dolphins. Anim Behav 77:1449–1457

Wiszniewski J, Beheregaray L, Corrigan S, Möller LM (2011) Male reproductive success increases with alliance size in a population of Indo-Pacific bottlenose dolphins (*Tursiops aduncus*). J Anim Ecol 81:423–431

Wiszniewski J, Brown C, Möller LM (2012) Complex patterns of male alliance formation in a dolphin social network. J Mammal 93:239–250

Yoshida H, Shirakihara K, Kishino H, Shirakihara M (1997) A population size estimate of the finless porpoise, *Neophocaena phocaenoides*, from aerial sighting surveys in Ariake Sound and Tachibana Bay, Japan. Res Popul Ecol 39:239–247

Zamma K (2002) Grooming site preferences determined by lice infection among Japanese macaques in Arashiyama. Primates 43:41–49

Chapter 17
Spinner Dolphins of Islands and Atolls

Marc O. Lammers

Abstract Spinner dolphins, *Stenella longirostris*, occur globally in tropical and subtropical waters and form island-associated populations in many parts of the world. These populations are closely tied to island resources, relying on enhanced aggregations of mesopelagic prey and nearshore habitats to conduct a highly stereotyped daily behavioral cycle. To exploit these resources, spinner dolphins have adapted their social structure, foraging ecology, reproductive patterns, predator avoidance behavior, and communication in unique ways. Spinner dolphins are a gregarious species with individuals relying on the dynamics of the pod for nearly every aspect of their lives. Because of a daily tendency to visit the same coastal waters and engage in acrobatic displays, many populations have seen a steady rise in popular and commercial dolphin-watching by humans. This has led to management concerns about the potential impacts that chronic interactions may have on the dolphins' ability to conduct normal daily behaviors.

Keywords Island-associated · Diel cycle · Social structure · Foraging · Resting · Predator avoidance · Acoustic signaling · Acrobatic behavior · Conservation

17.1 Introduction

The spinner dolphin (*Stenella longirostris*) is one of the most common small odontocete species in tropical and subtropical waters. Several subspecies are recognized, including the globally distributed Gray's spinner (*S. l. longirostris*), the eastern tropical Pacific's (ETP) endemic eastern spinner (*S. l. orientalis*) and Central American spinner (*S. l. centroamericana*), and the dwarf spinner of central Southeast Asia (*S. l. roseiventris*) (Perrin 2009). An additional form of spinner dolphin in offshore waters of the ETP, the

M. O. Lammers (✉)
National Oceanic and Atmospheric Administration, Hawaiian Islands Humpback Whale
National Marine Sanctuary, Kihei, HI, USA
e-mail: marc.lammers@noaa.gov

© Springer Nature Switzerland AG 2019
B. Würsig (ed.), *Ethology and Behavioral Ecology of Odontocetes*, Ethology and
Behavioral Ecology of Marine Mammals,
https://doi.org/10.1007/978-3-030-16663-2_17

"whitebelly," is possibly a fifth subspecies. I will abbreviate the term "spinner dolphin" to "spinners" from here on to reflect common practice when discussing the species.

Spinners are generally pelagic, but coastal populations exist in the eastern Pacific, Indian Ocean, Southeast Asia, and elsewhere (Perrin et al. 1989, 1999). Pelagic populations, such as in the ETP, often occur in waters with a shallow thermocline (Reilly 1990). The thermocline concentrates pelagic organisms in and above it, upon which the dolphins feed. In certain coastal populations, such as those in Southeast Asia, dwarf spinners exploit mainly benthic and reef fishes and invertebrates (Perrin et al. 1999).

Across their global distribution, spinners exhibit an affinity for island archipelagos. In many parts of the world, island-associated populations occur that are distinct from their pelagic conspecifics, sometimes with sufficient genetic differentiation to be managed as unique stocks (Carretta et al. 2011). These island-associated populations often display high site fidelity to specific islands within an archipelago (Oremus et al. 2007; Andrews et al. 2010; Viricel et al. 2016) and even to individual coastlines (Marten and Psarakos 1999). In general, the island coastlines preferred by spinners are characterized by shallow water habitat in close proximity to steep island slopes, where the "island-mass effect" results in localized areas of biological productivity (Gilmartin and Revelante 1974; Gove et al. 2016). It is this enhanced productivity and protective nearshore habitat that tie these populations to particular islands, thereby forgoing a more pelagic existence (Andrews et al. 2010; Viricel et al. 2016).

Island-associated spinner populations have been studied in several parts of the world, including archipelagos in the south Atlantic (De Lima Silva and Da Silva 2009), south Pacific (Gannier and Petiau 2006; Oremus et al. 2007; Cribb et al. 2012), Indian Ocean (Anderson 2005; Condet and Dulau-Drouot 2016; Viricel et al. 2016), the Red Sea (Notarbartolo di Sciara et al. 2008), and other locations. However, a disproportionate amount of knowledge about behavior and ecology of island spinners has come from studies of the populations of the Hawaiian archipelago in the central north Pacific. Beginning in the late 1970s and continuing through modern times, research groups have focused efforts on studying spinners across this more than 1500-mile-long (2400 km) chain of islands and atoll. Hawaiian spinners are genetically distinct from other populations in the Pacific Ocean (Andrews et al. 2010) and occur off Hawai'i Island, Maui, Kaho'olawe, Lāna'i, Moloka'i, Oahu, Kaua'i and Ni'ihau in the Main Hawaiian Islands, as well as at French Frigate Shoals, Pearl and Hermes Reef, Midway Atoll, and Kure Atoll in the Northwestern Hawaiian Islands, spanning a latitudinal range of more than 10 degrees. Genetic differentiation is present among many of these island clusters, including between the islands of Hawai'i and Maui that are separated by only 46 km, thereby underscoring the island fidelity that characterizes some of these populations (Andrews et al. 2010).

17.2 The Daily Cycle

Island-associated spinners are characterized by a behavioral cycle first described by Norris and Dohl (1980) and later detailed by Norris et al. (1994a) for the population along the Kona coast of the island of Hawai'i. Dolphins off the Kona coast feed in

17 Spinner Dolphins of Islands and Atolls

deep offshore waters at night and spend much of daytime close to shore, often in bays that tend to be sheltered by the prevailing winds. After they enter the nearshore area in the morning, they descend into rest and tend to rest for much of the day, but with intermittent bouts of social, sexual, and aerial activity. Resting groups can be as few as 20 dolphins but can also be as large as 100 or more. Groups at rest are in tight formation, tend to dive synchronously, and are relatively quiet, with only occasional sounds produced. While resting, spinner dolphins appear to rely mainly on vision. They tend to prefer sandy substrate, not bottom with coral outcroppings or rocks, presumably so that they cannot be surprised by tiger sharks (*Galeocerdo cuvier*) coming from hiding places. In the afternoon, the dolphin group becomes more active, with increasing social, sexual, aerial, and vocal activity, before moving out of the nearshore shallows and traveling to deeper offshore waters to feed. As they do so, they are often joined by other groups who spent their time in other nearshore areas, and the foraging group becomes large again. This is therefore a diel fission/fusion society, to be described later. The diel cycle is quite consistent, except that a particular nearshore area or bay is not be visited each and every day (although usually over 50% of days), and there are seasonal shifts depending on sunrise and sunset, i.e., day length (Würsig et al. 1994a). This basic cycle has since then been documented in numerous other island-associated spinner populations (Notarbartolo di Sciara et al. 2008; De Lima Silva and Da Silva 2009; Cribb et al. 2012).

Wells and Norris (1994a) suggested that availability of prey and resting habitat are the primary limiting factors to influence occurrence of spinners along the Kona coast. Other studies in Hawai'i (Karczmarski et al. 2005; Tyne et al. 2015) but also in French Polynesia (Poole 1995; Gannier and Petiau 2006) and Reunion Island (Condet and Dulau-Drouot 2016) have similarly highlighted the relationship between shallow water habitats with light-colored bottom substrate in the proximity to deepwater prey and the occurrence of spinners. Thorne et al. (2012) used predictive modeling to establish that vicinity to deepwater foraging areas, depth, proportion of bays with shallow depths, and substrate rugosity were important factors influencing spinner occurrence in Hawai'i, with distance to the 100 m depth contour a strong predictor of daytime resting habitat.

It would be misleading to characterize spinner occurrence as limited to coastlines resembling the one found off Kona, which is unique in its availability of numerous protected coves in close proximity to steep bathymetry and vast stretches of calm, clear leeward waters. Rather, spinners are more adaptable and often exploit habitats not considered ideal by the criteria laid out above. For example, off the island of Oahu, spinners often rest outside protected coves while transiting or milling along coastlines with little or no lee (Lammers 2004). Similarly, spinners in the Maui Nui region (Maui, Lāna'i, Kaho'olawe, and Moloka'i) rest in the channel between Maui and Lāna'i, tens of kilometers from deep offshore waters (McElligott 2018). Thus, although the general cycle of nighttime foraging and daytime socializing and resting is consistent among spinner populations, its implementation is adapted to the resources available in the habitat.

17.3 Social Structure

Spinners have high sociality, and individuals do not occur alone. Rather, they are herd animals dependent on the dynamic of the group for nearly all life functions. A spinner pod ranges in size from a minimum of three to several hundred individuals, depending on location and behavioral context. Such an obligate social lifestyle has led not only to social structure in spinner groups but also to complex social dynamics.

A pod of several dozen or more spinners is composed of multiple sub-pods. The behavior of individuals is largely regulated at the level of the sub-pod, rather than the pod as a whole. Individual dolphins reflect the behavioral state of the sub-pod they belong to, which may or may not be the same as that of other sub-pods. Thus, as an example, while a pod of ~100 animals might occupy the same area or be traveling in the same direction, it is likely that there are between three and four sub-pods of 20–40 dolphins, with individuals from sub-pods intermixing. Depending on which stage in the daily behavioral cycle the pod is in, all the sub-pods may exhibit the same behavioral state (e.g., socializing), or some sub-pods may be more or less active, coordinated, and/or synchronized than others (behavioral descriptions can be found in Norris et al. 1994a).

A likely reason for the behavioral state regulated at the level of the sub-pod is the need for coordination and synchrony during rest, which may be achieved optimally in groups of certain size. Because resting spinners rely primarily on vision for vigilance, synchronized behavior helps maximize efficiency of information transfer among pod-mates via a sensory integration system (or SIS) (Norris and Schilt 1988). An SIS is created when individuals swim in coordinated fashion as parts of a supra-individual signaling system that allows the perception, amplification, and transfer of environmental information from collective sensory windows (Norris and Johnson 1994). The SIS can provide an early warning of predators and allows for information to travel rapidly across a sub-pod, facilitating fast communal response.

Sub-pods appear to be organized around coalitions of males, with females and younger animals joining coalitions (Östman 1994). Some degree of social interaction among individuals in a sub-pod is nearly continuous, with various forms of caressing, rubbing, genital-to-genital contact, and/or chasing being the norm rather than the exception (Johnson and Norris 1994). Tactile interactions likely play important roles in establishing, reaffirming, and/or strengthening social relationships among individuals. The majority of affiliative and sexual behaviors generally occur between males and females, while overt aggressive behaviors are most common between females, suggesting that females may actively compete for access to males (Östman 1994). Associations among individuals range in longevity from momentary to lifelong, with the strongest levels of affiliation among male spinner dolphins and the lowest among females (Östman 1994; Marten and Psarakos 1999).

Associations among and within sub-pods are marked by characteristic delphinid fission-fusion dynamics (Gowans et al. 2007). Sub-pods typically aggregate into larger pods for one to several hours in the morning before fragmenting and becoming more distributed during mid-day rest. The larger pod is then typically reassembled in the afternoon prior to departing toward offshore evening feeding areas, although

17 Spinner Dolphins of Islands and Atolls 373

often with a different complement of sub-pods than in the morning (Lammers 2004). Similarly, during periods of pod assembly and presumably at night, the makeup of sub-pods is reshuffled through the intermixing of individuals. Thus, long-term affiliations among individuals, but also fluid group composition, are generally hallmarks of spinner pod dynamics.

Exceptions to the fission-fusion pattern occur in smaller, isolated populations of spinners. Karczmarski et al. (2005) reported that the population at Midway Atoll in the Northwestern Hawaiian Islands has an absence of fission-fusion dynamics for the mostly closed population occupying the atoll's lagoon during the day. Rather, spinners there live in a stable society with strong geographic fidelity, no inter-individual variation in the structure of groups, and stable associations from day to day. This difference from aforementioned spinner dolphin social patterns is thought to result from the relatively close proximity of prey resources to daytime resting areas coupled with large distances to other atolls, leading to stability rather than variability in social dynamics of the group. It is too far for dolphins to go to easily exchange with individuals of another group.

17.4 Reproduction

Reproductive strategies have been determined for the pelagic eastern and whitebelly spinner dolphins in the ETP based on comparisons of sexual dimorphism and male testes weight (Perrin and Mesnick 2003). There is great variation in sexual strategies by population. Perrin and Mesnick (2003) concluded that the eastern spinner, which exhibits a higher degree of sexual dimorphism and lower testes weight, likely has a polygynous mating system characterized by intense pre-mating competition among males. On the other hand, the more offshore whitebelly spinner dolphin is characterized by reduced sexual dimorphism and greater testes weight, indicating an emphasis on sperm competition among males and therefore a polygynandrous, or promiscuous, mating system with both males and females mating with multiple partners. This difference in mating strategies may explain the canted dorsal fin and exaggerated anal keel among mature eastern spinner males where a strong polygyny presides but comparatively absent among whitebelly spinners where multi-mate sexuality appears to occur. Perrin and Mesnick (2003) further speculated that the different mating strategies likely are tied to differences in resource availability, with eastern spinners less limited by prey abundance and therefore able to devote more energy toward competition for mates than whitebelly spinners, which occupy less productive, patchy offshore environments and therefore must devote more energy on foraging.

Island-associated spinners, such as the ones off Hawaii, are thought also to have a polygynandrous mating system, based on the prevalence of sexual behaviors among males and females and testes weights that are similar to those of whitebelly spinners (Wells and Norris 1994b; Perrin and Mesnick 2003). Like whitebelly spinners, Hawaiian spinners also exhibit a "seasonally diffuse" reproductive pattern with a slight bimodal tendency. Calving in both forms takes place year-round, but a modest peak occurs in late spring and summer and another in midwinter (Barlow 1984;

374 M. O. Lammers

Wells and Norris 1994b). Wells and Norris (1994b) measured hormonal levels in both males and females in a small captive colony of spinners and concluded that the temporal spreading of sexual readiness among females likely explains the diffuse bimodal appearance of newborn calves in the population.

17.5 Foraging

Spinner dolphins are nocturnal feeders who exploit the nightly vertical migration of oceanic micronekton. In pelagic waters this community is termed the deep scattering layer or mesopelagic community, while around islands it is typically referred to as the mesopelagic boundary community (MBC) because it lies at the interface between coastal and oceanic ecosystems (Reid et al. 1991). The composition of MBC is diverse and varies regionally but is made up of numerous species of deepwater fish, shrimp, and squid, as well as the larval stages of pelagic and coastal organisms. The MBC community off Hawai'i resides along island slopes at depths of 400–700 m during daytime hours and migrates to within 10 m of the surface at night (Reid et al. 1991; Benoit-Bird and Au 2004). In addition, the MBC also migrates horizontally toward shore to exploit nearshore densities of zooplankton (Benoit-Bird et al. 2001, 2008; Benoit-Bird and Au 2006), which brings elements of the MBC, particularly myctophids, into waters as shallow as 23 m and as close as 500 m from shore during the middle of the night (Benoit-Bird and Au 2006).

Spinners are specialized in feeding on members of the MBC measuring approximately 2–10 cm (Benoit-Bird 2004). Fish in the family Myctophidae typically make up more than 50% of stomach contents, and other mesopelagic fish, squid, and shrimp families compose the rest (Perrin et al. 1973; Würsig et al. 1994b). Benoit-Bird (2004) measured caloric content of spinner prey and concluded that spinners need to consume a minimum of 1.25 larger prey items per minute to meet energy needs. As a result, spinner feeding success is determined by their ability to find and exploit high densities of prey.

The spinner feeding cycle begins once the pod leaves the protection of shallow, nearshore waters used for daytime resting. The pod typically leaves resting habitat in a burst of energetic swimming and acrobatic leaping before settling into a more sustained travel speed (Würsig et al. 1994a). At this stage, larger pods often spread out widely and/or break apart, emphasizing the importance of the sub-pod and bonds among individuals in regulating group dynamics. Following a large pod of animals as they transit to their offshore feeding ground typically results in the pod gradually separating and the observer tracking only a handful of animals by the time extended foraging dives over deep waters occur. These dives generally begin within an hour of sunset, when the MBC is not yet near the surface, supporting the suggestion that spinner dolphins probably dive relatively deep (150–250 m) to meet the rising layer of prey (Würsig et al. 1994b).

How spinner dolphins find their prey remains a mystery. Au and Benoit-Bird (2008) measured the target strength of individual MBC animals using a simulated dolphin echolocation click and concluded that spinners could likely detect their prey at a range of 30–57 m, depending on the density of prey animals. It is unclear if this range is sufficient for spinner dolphins to detect concentrations of prey at depth in open waters. Afternoon tracking of pods on the island of Oahu revealed that spinners traveled to preferred locations along the coast to begin foraging, suggesting that they anticipated where prey first became available at shallower depths (Lammers 2004). These locations often coincided with areas being exploited by other odontocete species, such as spotted dolphins (*Stenella attenuata*) and common bottlenose dolphins (*Tursiops truncatus*), indicating that these were perhaps early evening hot spots of MBC prey occurrence.

Efforts to study how spinners exploit prey at night have largely relied on active acoustic methods of observation (i.e., sonar). By using vessel-based, downward-oriented sonar, Benoit-Bird and Au (2003) showed that spinners follow the diel horizontal migration of their prey toward shore, rather than feed offshore the entire night. This effectively brings spinners back into shallow, nearshore waters during the middle of the night and then offshore again during pre-dawn hours as the MBC retreats to deeper waters. The same study also showed that spinners track the vertical migration of their prey and exploit areas in the MBC that have the highest prey density. Moreover, Benoit-Bird and Au (2009) used data from a multi-beam echosounder to assert that foraging dolphin groups engage in cooperative and coordinated prey herding to help maximize prey density and therefore feeding success (Benoit-Bird and Au 2009).

17.6 Predation

Norris and Dohl (1980) reported that spinner dolphins in Hawai'i are likely attacked by large sharks, as evidenced by scarring patterns on various parts of the body. Norris (1994) further suggested that marine mammals, including false killer whales (*Pseudorca crassidens*), pygmy killer whales (*Feresa attenuata*), short-finned pilot whales (*Globicephala macrorhynchus*), and killer whales (*Orcinus orca*), might also target spinners occasionally. However, it is presently unclear whether and how they do so. Perhaps the most common predator of spinners is the cookie-cutter shark (*Isistius brasiliensis*), a vertically migrating, small, bioluminescent squaloid shark thought to mimic squid, thereby attracting feeding dolphins to it at night (Jones 1971). It has teeth only on the lower jaw that are used to scoop out disks of blubber and flesh approximately 5 cm in diameter. Because they are generally not deadly, both fresh and healed wounds from this shark are common among spinners.

Anti-predator strategies have played a large role in shaping spinner behavioral patterns. Norris and Schilt (1988) suggested that the first line of defense is simply schooling behavior, which offers relative safety through the dilution of risks to individuals, provides early warning of an impending threat through shared vigilance, and creates a confusion effect, whereby a predator's ability to track individual prey is confounded. The benefits of aggregation relative to predation explain not only the habit by spinners to form large pods (Norris and Dohl 1980) but also the tendency to take part in mixed-species associations in certain parts of the world (Scott and Cattanach 1998; Kiszka et al. 2011). Kiszka et al. (2011) reported that over 25% of spinner pods off the island of Mayotte in the southwestern Indian Ocean included pantropical spotted dolphins (*Stenella attenuata*), presumably to increase the size of the pod. They attributed this association to an anti-predator strategy for transiting between resting areas.

Besides aggregating into groups, spinners actively mitigate predation by carefully selecting the environment they occupy. Island-associated spinners typically seek out shallow, coastal habitats during the resting phase of their daily cycle. Because resting is characterized by reduced rates of echolocation, it is possible that spinners defend themselves against a surprise attack by reducing the dimensional properties of the environment they must monitor visually (Würsig et al. 1994a). Thus, by occupying shallow waters, they effectively eliminate the depth dimension and therefore the likelihood of an attack from below. Further, by occupying bays and/or the nearby shoreline, they reduce the threat of an attack coming from the flank facing the shore. Scarring from shark attacks is more common among island-associated populations of spinners than pelagic ones (Norris 1994). However, rather than offering this as evidence of lower rates of predation in open waters, Norris (1994) suggested that higher attack survivorship among island-associated spinner dolphins is a more likely explanation, supporting the proposed benefit of seeking nearshore resting habitats.

17.7 Communication

Spinners face the challenge of managing a life dependent on social connections in the vast open sea. Communication is the key to mediating social processes, so it is not surprising that spinners and other odontocetes with gregarious lives have evolved a complex system of signaling. Because the sea is a largely dark and opaque environment, acoustic communication plays a central role among delphinids (Lammers and Oswald 2015). This is because water is comparatively transparent to sound, which allows acoustic signals to transmit over much greater ranges than light-dependent signals and cues. As a result, highly social delphinids such as spinners have evolved a complex form of acoustic signaling using frequency-modulated whistles and short pulses or clicks (Fig. 17.1). Spinner whistles typically range in duration from tens to hundreds of milliseconds and are composed of a fundamental frequency contour with multiple harmonics (Lammers et al. 2003). Some whistles are simple frequency sweeps, while others are more complex with inflections, breaks, and steps in the contour. How whistles are used in communication among dolphins is a debated topic in the literature, but there is general agreement that they play a role in maintaining contact among dispersed pod-mates and in the identification of individuals (Lammers and Oswald 2015). Spinner dolphin clicks are only tens of microseconds long and are produced in "trains" with inter-click intervals between a few and hundreds of milliseconds (Lammers et al. 2004). They have broadband energy mostly or exclusively at frequencies above the human hearing range (i.e., >15 kHz). Click trains are differentiated into those

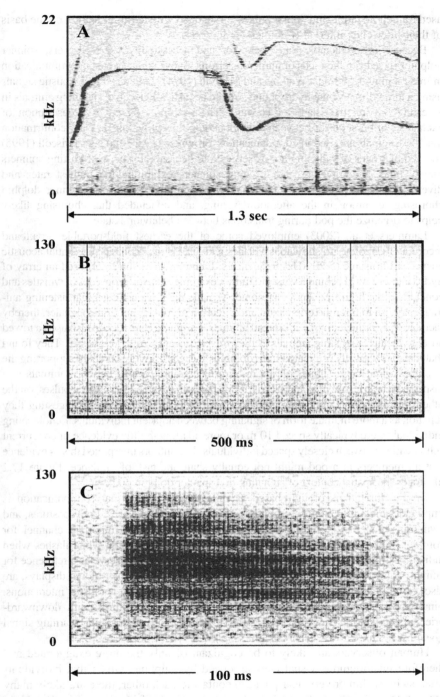

Fig. 17.1 Spectrogram examples of a spinner dolphin, *Stenella longirostris*, whistle with a harmonic (**a**), an echolocation click train (**b**), and a burst-pulse click train (**c**). From Lammers et al. 2006

used for echolocation and "burst pulses" associated with communication on the basis of their inter-click interval (Lammers et al. 2004).

Because of predictable daily behavior and accessibility to researchers, spinner dolphins have been the focus of numerous efforts to investigate acoustic communication in free-ranging delphinids. Watkins and Schevill (1974) studied spinner acoustic signals using a three-dimensional hydrophone array to acoustically localize signaling animals in a resting bay on the island of Hawai'i. Their efforts resulted in a description of exchanges of burst pulses between individual dolphins, providing an early confirmation that these signals are used in communication. Brownlee (1983) and later Driscoll (1995) showed that rates of signaling are closely tied to behavioral state, with resting spinners mostly quiet and traveling and foraging spinners exhibiting the highest rates and diversity of signaling. Driscoll (1995) further examined periods of spinner dolphin chorusing, common in the afternoon/evening, and concluded that chorusing likely helps to organize the pod during transitions between behavioral states.

Lammers et al. (2003) employed some of the earliest field-portable broadband recording technology to show that whistles and burst pulses contain substantial acoustic energy at ultrasonic (>20 kHz) frequencies. Lammers and Au (2003) used an array of hydrophones towed behind a vessel to further examine acoustic properties of whistles and concluded that high-frequency harmonics provide navigational cues that listening animals may use to infer the orientation and direction of movement of the signaler, thereby facilitating coordination of widespread animals. Lammers et al. (2006) also used a towed array to localize signaling animals as they traveled near the research vessel. They found that whistles typically originated from dolphins spaced relatively far apart, supporting the hypothesis that these signals play a role in maintaining contact between animals in a dispersed group (Brownlee and Norris 1994; Janik and Slater 1998). Burst pulses, on the other hand, usually came from animals spaced closer to one another, suggesting they function as a more intimate form of signaling between adjacent individuals. Echolocating individuals were typically spaced 10 m or more apart, with little evidence of concurrent echolocation between closely spaced individuals. The authors interpreted this as evidence that all members in a pod might not equally share the task of vigilance. Figure 17.2 illustrates the spatial context of signaling in a spinner dolphin pod.

Despite limitations of light-based signaling in the sea, visual communication is important among delphinids. Among spinners, coloration patterns, body postures, and swimming patterns strongly point to visual signaling as an important channel for communication (Würsig et al. 1990). Spinners underwater exhibit white flashes when individuals momentarily turn to expose bellies, which probably serve as a reference for animals at the edge of visual range. Distinct postures, such as open-mouth displays, are also common among socializing spinners, especially during agonistic interactions. Finally, certain swimming patterns, such as exaggerated S-turns with downward-oriented pectoral fins, have been suggested to be analogous to a similar warning signal given by certain territorial sharks (Johnson and Norris 1994).

Human observers are likely to be cognizant of only the more exaggerated and therefore unambiguous visual cues produced by dolphins. Given that individuals often swim within or very near physical contact to each other, there are likely many more subtle signals exchanged, both visual and physical, that a non-delphinid

Fig. 17.2 A hypothetical representation of the spatial occurrence of acoustic signals in a spinner dolphin, *Stenella longirostris*, pod. The white sinusoids represent whistle production, the black bars represent the production of burst pulses, and the yellow cones are the occurrence of echolocation click trains (Photo courtesy of Andre Seale). From Lammers et al. (2006)

observer would not register. At times, interactions among spinner dolphins are subtle, with a fin lightly rubbing against a pod-mate or an eye partially or fully closed. Whether such cues have meaning to nearby animals is speculative, but the potential is present for a vast array of meaningful visual and physical cues to help mediate social relationships.

17.8 Aerial Acrobatics

A review of spinner dolphin behavior would be lacking without a discussion of their aerial acrobatics. Simply put, spinner dolphins are master acrobats with few rivals in the cetacean world, save perhaps dusky dolphins (*Lagenorhynchus obscurus*; see Würsig and Würsig 2010). Most odontocete species engage in some form of aerial behaviors, and some may leap higher or further than spinner dolphins, but none match them in their ability to rotate, twist, and somersault, often in extended bouts of aerial displays (Fig. 17.3).

Fig. 17.3 A collated sequence of a spinner dolphin, *Stenella longirostris*, performing a spinning leap. This sequence was taken from Twitter, with permission by the Wyland Foundation. In this compilation of a video sequence, the leap begins on the right side, and the dolphin turns on its axis two times, until it lands to the left. Leaps can be in this vertical stance, but also horizontally to the water, and can incorporate over three turns of the body axis while in-air

Spinner dolphins engage in a wide range of above-water behaviors, spanning from subtle "nose-outs," whereby the rostrum is extended just above the water's surface, to high-energy "tail-over-head" leaps, in which the dolphin throws its tail forward in a somersault (Norris and Dohl 1980). Aerial behaviors can occur at any time but are most common during periods of socializing and traveling. They are least common during periods of rest, although individuals in a resting group may leap or spin in response to an approaching vessel, as if to alert the vessel of the pod's presence. The variety, intensity, and frequency of aerial activities, combined with other behavioral metrics, can thus be used as indicators of a spinner dolphin pod's behavioral state (Lammers 2004).

Invariably, the question arises: Why do they do it? Probably the most common answer is to rid themselves of parasites, including remoras (Hester et al. 1963; Norris and Dohl 1980; Norris et al. 1994b). However, this answer alone is rather unsatisfying, as most odontocetes face the problem of parasites yet do not display the acrobatic aptitude of spinner dolphins. Thus, additional explanations are necessary that account for the spinner dolphin's distinct behavioral ecology. Norris et al. (1994b) proposed that the bubble plume produced underwater by reentering spinning dolphins would offer a strong target for echolocating pod-mates and therefore serve as a beacon of sorts to help define the envelope of the pod or perhaps act as a target to more distant subgroups. Similarly, the reentry sound produced by a leaping dolphin could serve as an

17 Spinner Dolphins of Islands and Atolls

omnidirectional cue to nearby animals. Still, these explanations probably cannot reconcile the large variety of aerial behaviors in the spinner dolphin repertoire. Perhaps an additional clue lies in the fact that behavioral state can be inferred from the pod's aerial displays. Thus, it may be that spinners use aerial displays to regulate the pod's state in their perpetual effort to achieve and maintain group cohesion. These acrobatic leaps are perhaps designed as a signal of group social cohesion for entering or leaving a bay, or for signaling an after-feeding "party mood" as has been suggested for dusky dolphins (Würsig and Würsig 1980).

17.9 Anthropogenic Impacts

In many places where spinner dolphins occur, humans have taken great interest in them. This is not surprising, considering spinner dolphins' propensity for visiting the same nearshore water almost daily and while there performing acrobatic feats. In some parts of the world, a steady increase in popular and commercial interest in viewing and interacting with spinners has led to management concerns (Delfour 2007; Notarbartolo di Sciara et al. 2008; Cribb et al. 2012; Heenehan et al. 2015, 2017; Tyne et al. 2015, 2017). In Hawai'i, where long-term population estimates are available, there is evidence that the population has declined over several decades as a result of long-term, sustained anthropogenic pressure (Tyne et al. 2014).

The primary conservation concern is the disruption that chronic and sustained human attention can have on spinner dolphin abilities to enter and maintain rest during their daily cycle. In locations such as the Kona coast of Hawai'i, spinner dolphins in resting bays are nearly always surrounded by swimmers or vessels (Courbis and Timmel 2009; Tyne et al. 2018), which is of concern because once displaced from their preferred resting bay, they are unlikely to rest elsewhere (Tyne et al. 2015). Reduced periods of rest likely impact nighttime foraging success, affect vigilance and/or raise stress hormones, which can in turn affect rates of survival and reproduction (Tyne et al. 2015; Forney et al. 2017). To the casual observer, such impacts are generally not apparent when viewing or interacting with spinner dolphin pods. This is because spinners often approach and interact with vessels or swimmers, leaving the impression that the pod welcomes the attention, which at certain times may be true, at least for a part of the group. However, these responses can be misleading, particularly when interactions with humans become frequent and chronic and are unregulated.

The potential impacts to spinners resulting from anthropogenic activities can be relayed through an (somewhat "nightmarish") analogy. Imagine living in a quiet neighborhood occasionally visited by an ice-cream truck in the afternoon hours. The children hear the distinctive jingle and excitedly rush toward the truck to indulge in a harmless ice-cream treat. As time passes, however, additional ice-cream trucks begin to visit the neighborhood but now extend their presence into the evening hours, disrupting family bedtime routines by riling up the children who now want ice cream before bed. As additional time passes, even more ice-cream trucks begin showing up in the neighborhood at all hours of the night waking up the kids and disrupting the family's sleep. Powerless to evict the ice-cream trucks, the family must decide whether to stay or find a new neighborhood. Similarly, spinners placed under increasing human pressure in their preferred resting bays eventually must decide whether to abandon the neighborhood for perhaps a less ideal habitat but one with no or fewer human disturbances (i.e., ice-cream trucks).

In many parts of the world, spinner conservation is ultimately a matter of managing interactions with people. Given adequate space, particularly during periods of rest, spinners are likely to be resilient to certain levels of interactions. As with many wildlife-viewing experiences, adapting practices to the behavioral and ecological needs of the animals is the key to making these interactions sustainable. In the case of spinners, this means ensuring they are given the opportunity to get a good day's rest, and so be ready for their all-important nighttime foraging.

References

Anderson RC (2005) Observations of cetaceans in the Maldives, 1990–2002. J Cetacean Res Manag 7:119–135

Andrews KR, Karczmarski L, Au WWL, Rickards SH, Vanderlip CA, Bowen BW, Grau EG, Toonen RJ (2010) Rolling stones and stable homes: social structure, habitat diversity and population genetics of the Hawaiian spinner dolphin (*Stenella longirostris*). Mol Ecol 19:732–748

Au WWL, Benoit-Bird KJ (2008) Broadband backscatter from individual Hawaiian mesopelagic boundary community animals with implications for spinner dolphin foraging. J Acoust Soc Am 123:2884–2894

Barlow J (1984) Reproductive seasonality in pelagic dolphins (*Stenella* spp.): implications for measuring rates. In: Perrin WF, Brownell Jr RL, DeMaster DP (eds) Reproduction in whales, dolphins and porpoises. Rep Int Whaling Comm Spec Iss 6:191–198

Benoit-Bird KJ (2004) Prey caloric value and predator energy needs: foraging predictions for wild spinner dolphins. Mar Biol 145:435–444

Benoit-Bird KJ, Au WWL (2003) Prey dynamics affect foraging by a pelagic predator (*Stenella longirostris*) over a range of spatial and temporal scales. Behav Ecol Sociobiol 53:364–373

Benoit-Bird KJ, Au WWL (2004) Diel migration dynamics of an island-associated sound-scattering layer. Deep-Sea Res Part I 51:707–719

Benoit-Bird KJ, Au WWL (2006) Extreme diel horizontal migrations by a tropical nearshore resident micronekton community. Mar Ecol Prog Ser 319:1–14

Benoit-Bird KJ, Au WWL (2009) Cooperative prey herding by the pelagic dolphin, *Stenella longirostris*. J Acoust Soc Am 125:125–137

Benoit-Bird KJ, Au WWL, Brainard RE, Lammers MO (2001) Diel horizontal migration of the Hawaiian mesopelagic boundary community observed acoustically. Mar Ecol Prog Ser 217:1–14

Benoit-Bird KJ, Zirbel MJ, McManus MA (2008) Diel variation of zooplankton distributions in Hawaiian waters favors horizontal diel migration by midwater micronekton. Mar Ecol Prog Ser 367:109–123

Brownlee SM (1983) Correlations between sounds and behavior in wild Hawaiian spinner dolphins (*Stenella longirostris*). Masters Thesis, University of California Santa Cruz, 26 p

Brownlee SM, Norris KS (1994) The acoustic domain. In: Norris KS, Würsig B, Wells RS, Würsig M (eds) The Hawaiian spinner dolphin. University of California Press, Berkeley, CA, pp 161–185

Carretta JV, Forney KA, Oleson E, Martien K, Muto MM, et al (2011) U.S. Pacific marine mammal stock assessments: 2010. U.S. Department of Commerce, National Oceanic and Atmospheric Administration, National Marine Fisheries Service, Southwest Fisheries Science Center. 352 p. http://www.nmfs.noaa.gov/pr/pdfs/sars/po2010.pdf

Condet M, Dulau-Drouot V (2016) Habitat selection of two island-associated dolphin species from the south-west Indian Ocean. Cont Shelf Res 125:18–27

Courbis S, Timmel G (2009) Effects of vessels and swimmers on behavior of Hawaiian spinner dolphins (*Stenella longirostris*) in Kealake'akua, Honaunau, and Kauhako bays, Hawai'i. Mar Mamm Sci 25:430–440

Cribb N, Miller C, Seuront L (2012) Site fidelity and behaviour of spinner dolphins (*Stenella longirostris*) in Moon Reef, Fiji Islands: implications for conservation. J Mar Biol Assoc UK 92:1793–1798

De Lima Silva FJ, Da Silva JM (2009) Circadian and seasonal rhythms in the behavior of spinner dolphins (*Stenella longirostris*). Mar Mamm Sci 25:176–186

Delfour F (2007) Hawaiian spinner dolphins and the growing dolphin watching activity in Oahu. J Mar Biol Assoc UK 87:109–112

Driscoll AD (1995) The whistles of Hawai'ian spinner dolphins (*Stenella longirostris*). Masters Thesis, University of California Santa Cruz, 84 p

Forney KA, Southall BL, Slooten E, Dawson S, Read AJ, Baird RW, Brownell RL (2017) Nowhere to go: noise impact assessments for marine mammal populations with high site fidelity. Endanger Species Res 32:391–413

Gannier A, Petiau E (2006) Environmental variables affecting the residence of spinner dolphins (*Stenella longirostris*) in the Bay of Tahiti (French Polynesia). Aquat Mamm 32:202–211

Gilmartin M, Revelante N (1974) The 'island mass' effect on the phytoplankton and primary production of the Hawai'ian Islands. J Exp Mar Biol Ecol 16:18–204

Gove JM, McManus MA, Neuheimer AB, Polovina JJ, Drazen JC, Smith CR, Merrifield MA, Friedlander AM, Ehses JS, Young CW, Dillon AK, Williams GJ (2016) Near-island biological hotspots in barren ocean basins. Nat Commun 7:10581. https://doi.org/10.1038/ncomms10581

Gowans S, Würsig B, Karczmarski L (2007) Delphinid social strategies: an ecological approach. Adv Mar Biol 53:195–294

Heenehan HL, Basurto X, Bejder L, Tyne J, Higham JES, Johnston DW (2015) Using Ostrom's common-pool resource theory to build toward an integrated ecosystem-based sustainable cetacean tourism system in Hawai'i. J Sustain Tour 23:536–556

Heenehan HL, Van Parijs SM, Bejder L, Tyne J, Johnston DW (2017) Differential effects of human activity on Hawaiian spinner dolphins in their resting bays. Glob Ecol Conserv 10:60–69

Hester FJ, Hunter JR, Whitney RR (1963) Jumping and spinning behavior in the spinner porpoise. J Mammal 44:586–588

Janik VM, Slater PJB (1998) Context specific use suggests that bottlenose dolphin signature whistles are cohesion calls. Anim Behav 56:829–838

Johnson CM, Norris KS (1994) Social behavior. In: Norris KS, Würsig B, Wells RS, Würsig M (eds) The Hawaiian spinner dolphin. University of California Press, Berkeley, CA, pp 243–286

Jones EC (1971) *Isistius brasiliensis*, a squaloid shark, the probable cause of crater wounds on fishes and cetaceans. Fish Bull 69:791–798

Karczmarski L, Würsig B, Gailey G, Larson KW, Vanderlip C (2005) Spinner dolphins in a remote Hawaiian atoll: social grouping and population structure. Behav Ecol 16:675–685

Kiszka J, Perrin WF, Pusineri C, Ridoux V (2011) What drives island-associated tropical dolphins to form mixed-species associations in the southwest Indian Ocean? J Mammal 92(5):1105–1111

Lammers MO (2004) Occurrence and behavior of Hawaiian spinner dolphins (*Stenella longirostris*) along Oahu's leeward and south shores. Aqua Mamm 30:237–250

Lammers MO, Au WWL (2003) Directionality in the whistles of Hawaiian spinner dolphins (*Stenella longirostris*): a signal feature to cue direction of movement? Mar Mamm Sci 19:249–264

Lammers MO, Oswald JN (2015) Analyzing the acoustic communication of dolphins. In: Herzing DL, Johnson CM (eds) Dolphin communication & cognition. MIT Press, Cambridge, MA

Lammers MO, Au WWL, Herzing DL (2003) The broadband social acoustic signaling behavior of spinner and spotted dolphins. J Acoust Soc Am 114:1629–1639

Lammers MO, Au WWL, Aubauer R (2004) A comparative analysis of echolocation and burst-pulse click trains in *Stenella longirostris*. In: Thomas J, Moss C, Vater M (eds) Echolocation in bats and dolphins. University of Chicago Press, Chicago, pp 414–419

Lammers MO, Schotten M, Au WWL (2006) The spatial context of whistle and click production in pods of Hawaiian spinner dolphins (*Stenella longirostris*). J Acoust Soc Am 119:1244–1250

Marten K, Psarakos S (1999) Long-term site fidelity and possible long-term associations of wild spinner dolphins (*Stenella longirostris*) seen off Oahu, Hawaii. Mar Mamm Sci 15:1329–1336

McElligott M (2018) Behavioral and Habitat-use patterns of spinner dolphins (*Stenella longirostris*) in the Maui Nui region using acoustic data. Unpublished Masters Thesis. University of Hawaii at Manoa, 74 p

Norris KS (1994) Predators, parasites, and multispecies aggregations. In: Norris KS, Würsig B, Wells RS, Würsig M (eds) The Hawaiian spinner dolphin. University of California Press, Berkeley, CA, pp 287–300

Norris KS, Dohl TP (1980) Behavior of the Hawaiian spinner dolphin, *Stenella longirostris*. Fish Bull 77:821–849

Norris KS, Johnson M (1994) Schools and schooling. In: Norris KS, Würsig B, Wells RS, Würsig M (eds) The Hawaiian spinner dolphin. University of California Press, Berkeley, CA, pp 232–242

Norris KS, Schilt CR (1988) Cooperative societies in three-dimensional space: on the origins of aggregation, flocks and schools, with special reference to dolphins and fish. Ethol Sociobiol 9:149–179

Norris KS, Würsig B, Wells RS, Würsig M (1994a) The Hawaiian spinner dolphin. University of California Press, Berkeley, CA

Norris KS, Würsig B, Wells RS (1994b) Aerial behavior. In: Norris KS, Würsig B, Wells RS, Würsig M (eds) The Hawaiian spinner dolphin. University of California Press, Berkeley, CA, pp 103–121

Notarbartolo di Sciara G, Hanafy MH, Fouda MM, Afifi A, Costa M (2008) Spinner dolphin (*Stenella longirostris*) resting habitat in Samadai Reef (Egypt, Red Sea) protected through tourism management. J Mar Biol Assoc UK 89:211–216

Oremus M, Poole MM, Steel D, Baker CS (2007) Isolation and interchange among insular spinner dolphin communities in the South Pacific revealed by individual identification and genetic diversity. Mar Ecol Prog Ser 336:275–289

Östman JSO (1994) Social organization and social behavior of Hawai'ian spinner dolphins (*Stenella longirostris*). Unpublished PhD dissertation. University of California, Santa Cruz

Perrin WF (2009) Spinner dolphin *Stenella longirostris*. In: Perrin W, Würsig B, Thewissen J (eds) Encyclopedia of marine mammals. Academic Press, Amsterdam, pp 1100–1103

Perrin WF, Mesnick SL (2003) Sexual ecology of the spinner dolphin, *Stenella longirostris*: geographic variation in mating system. Mar Mamm Sci 19:462–483

Perrin WF, Warner RR, Fiscus CH, Holts DB (1973) Stomach contents of porpoise, *Stenella* spp., and yellowfin tuna, *Thunnus albacares*, in mixed-species aggregations. Fish Bull 71:1077–1092

Perrin WF, Miyazaki N, Kasuya T (1989) A dwarf form of the spinner dolphin (*Stenella longirostris*) from Thailand. Mar Mamm Sci 5:213–227

Perrin WF, Dolar ML, Robineau D (1999) Spinner dolphins (*Stenella longirostris*) of the western Pacific and Southeast Asia: pelagic and shallow-water forms. Mar Mamm Sci 15:1029–1053

Poole MM (1995) Aspects of the behavioral ecology of spinner dolphins (*Stenella longirostris*) in the nearshore waters of Mo'orea, French Polynesia. Unpublished PhD dissertation. University of California, Santa Cruz, 177 p

Reid SB, Hirota J, Young RE, Hallacher LE (1991) Mesopelagic-boundary community in Hawaii: micronekton at the interface between neritic and oceanic ecosystems. Mar Biol 109:427–440

Reilly S (1990) Seasonal changes in distribution and habitat differences among dolphins in the eastern tropical Pacific. Mar Ecol Prog Ser 66:1–11

Scott MD, Cattanach KL (1998) Diel patterns in aggregations of pelagic dolphins and tunas in the eastern Pacific. Mar Mamm Sci 14:401–428

Thorne LH, Johnston DW, Urban DL, Tyne J, Bejder L, Baird RW, Yin S, Rickards SH, Deakos MH, Mobley JR, Pack AA, Hill MC, Fahlman A (2012) Predictive modeling of spinner dolphin (*Stenella longirostris*) resting habitat in the main Hawaiian Islands. PLoS ONE 7(8):e43167

Tyne JA, Pollock KH, Johnston DW, Bejder L (2014) Abundance and survival rates of the Hawai'i island associated spinner dolphin stock. PLoS One 9:1–10

Tyne JA, Johnston DW, Rankin R, Loneragan NR, Bejder L (2015) The importance of spinner dolphin (*Stenella longirostris*) resting habitat: implications for management. J Appl Ecol 52:621–630

Tyne JA, Johnston DW, Christiansen F, Bejder L (2017) Temporally and spatially partitioned behaviours of spinner dolphins: implications for resilience to human disturbance. R Soc Open Sci 4:160626

Tyne JA, Christiansen F, Heenehan HL, Johnston DW, Bejder L (2018) Chronic exposure of Hawaii island spinner dolphins (*Stenella longirostris*) to human activities. R Soc Open Sci 5:171506

Viricel A, Simon-Bouhet B, Ceyrac L, Dulau-Drouot V, Berggren P, Amir OA, Jiddawi NS, Mongin P, Kiszka JJ (2016) Habitat availability and geographic isolation as potential drivers of population structure in an oceanic dolphin in the Southwest Indian Ocean. Mar Biol 163(10):219

Watkins WA, Schevill WE (1974) Listening to Hawaiian spinner porpoises, *Stenella* cf. *Longirostris*, with a three-dimensional hydrophone array. J Mammal 55(2):319–328

Wells RS, Norris KS (1994a) The island habitat. In: Norris KS, Würsig B, Wells RS, Würsig M (eds) The Hawaiian spinner dolphin. University of California Press, Berkeley, CA, pp 31–53

Wells RS, Norris KS (1994b) Patterns of reproduction. In: Norris KS, Würsig B, Wells RS, Würsig M (eds) The Hawaiian spinner dolphin. University of California Press, Berkeley, CA, pp 186–200

Würsig B, Würsig M (1980) Behavior and ecology of the dusky dolphin, *Lagenorhynchus obscurus*, in the South Atlantic. Fish Bull 77:871–890

Würsig B, Würsig M (2010) The dusky dolphin: master acrobat off different shores. Elsevier Academic Press, San Diego, CA

Würsig B, Kieckhefer TR, Jefferson TA (1990) Visual displays for communication in cetaceans. In: Thomas J, Kastelein R (eds) Sensory abilities of cetaceans. Plenum, New York, NY, pp 545–559

Würsig B, Wells RS, Norris KS, Würsig M (1994a) A spinner dolphin's day. In: Norris KS, Würsig B, Wells RS, Würsig M (eds) The Hawaiian spinner dolphin. University of California Press, Berkeley, CA, pp 65–102

Würsig B, Wells RS, Norris KS (1994b) Food and feeding. In: Norris KS, Würsig B, Wells RS, Würsig M (eds) The Hawaiian spinner dolphin. University of California Press, Berkeley, CA, pp 216–231

Chapter 18
Dusky Dolphins of Continental Shelves and Deep Canyons

Heidi C. Pearson

Abstract Dusky dolphins (*Lagenorhynchus obscurus*) exhibit highly flexible foraging and social strategies. Studies in three distinct environments offer a natural experiment for understanding influences shaping dusky dolphin societies. In shallow bays off Patagonia, Argentina, dusky dolphins form small traveling groups during the day in search of small, schooling fish, but fission-fusion of large groups enhances predator detection/avoidance and mating opportunities. Predation risk is also minimized by resting in small groups near shore at night. In the deep open waters off Kaikoura, New Zealand, large mixed age and sex groups and satellite mating and nursery groups occur. Loosely coordinated subgroups forage nocturnally on the deep scattering layer. Large group formation is again an anti-predation strategy. In the shallow wintertime habitat of Admiralty Bay, New Zealand, coordinated bait-ball foraging occurs but in smaller groups than off Patagonia. Outside of the breeding season and in the absence of predation risk, Admiralty Bay grouping patterns are driven by opportunities to secure prey and social partners. Compared to many other delphinids, dusky dolphins are more gregarious yet more loosely bonded. The social brain hypothesis helps to explain the evolution of large relative brain size and complex sociality in dusky dolphins. Bycatch, habitat loss, climate change, and whale-watching are current threats to the species. Application of new technology and research on female behavior, culture, and lesser-studied populations will help to fill knowledge gaps and advance conservation strategies.

Keywords *Lagenorhynchus obscurus* · Argentina · New Zealand · Foraging · Behavioral flexibility · Grouping patterns · Fission-fusion · Social structure · Intelligence

---

Electronic Supplementary Material The online version of this chapter (https://doi.org/10.1007/978-3-030-16663-2_18) contains supplementary material, which is available to authorized users.

H. C. Pearson (✉)
Department of Natural Sciences, University of Alaska Southeast, Juneau, AK, USA
e-mail: hcpearson@alaska.edu

© Springer Nature Switzerland AG 2019
B. Würsig (ed.), *Ethology and Behavioral Ecology of Odontocetes*, Ethology and Behavioral Ecology of Marine Mammals,
https://doi.org/10.1007/978-3-030-16663-2_18

18.1 Introducing the Dusky Dolphin

The dusky dolphin (*Lagenorhynchus obscurus*) is a small-bodied (69–85 kg, 1.7–2.1 m long) gregarious species restricted to temperate waters of the southern hemisphere off New Zealand, southern Australia, South Africa, Namibia, Argentina, Chile, Peru, and some sub-Antarctic oceanic islands (summarized in Cipriano and Webber 2010). Dusky dolphins are "semi-pelagic," inhabiting shallow nearshore zones of the continental shelf in addition to deep offshore zones of the continental slope reaching ca. 1500 m deep (Würsig et al. 2007).

Similar to other delphinids, dusky dolphins have a slow life history. Longevity is at least 35 years (Reeves et al. 2002; Orbach et al. 2018), with males and females reaching sexual maturity at 4–5 and 7–8 years off Peru (Van Waerebeek and Würsig 2018) and New Zealand (Cipriano 1992), respectively. Given an 11–13-month gestation period followed by 12-month lactation and at least 4-month resting periods (Cipriano 1992; van Waerebeek and Read 1994), it follows that an individual female will typically calve every 2–3 years. There is much for a dusky dolphin to learn about its ecological and social world, and this slow life history provides the necessary time to do so.

Perhaps the most outwardly notable behavioral attributes of the dusky dolphin are its high degree of sociability (Fig. 18.1) and propensity to perform a wide variety of leaps (Fig. 18.2). Dusky dolphins also exhibit a remarkable degree of flexibility in grouping patterns, likely in response to changing and variable socio-ecological conditions. They exhibit a high degree of fission-fusion dynamics as individuals typically join and split from groups of various sizes and composition numerous times throughout the day, all the while maintaining preferential and long-lasting bonds with certain individuals (Pearson et al. 2017b). This behavioral and social flexibility

Fig. 18.1 A large mixed age and sex dusky dolphin, *Lagenorhynchus obscurus*, group off Kaikoura, New Zealand, during the austral summer. Several mother-calf pairs are shown in this portion of the group as indicated by the calves' small body size and rostra breaking the surface of the water as they swim next to their mothers

Fig. 18.2 The primary leap types of dusky dolphins, *Lagenorhynchus obscurus*, are (clockwise from top left) as follows: "clean" headfirst reentry (parabolic arching leap with minimal splashing), noisy (reentry to water accompanied by a large splash and loud slapping noise akin to a "belly, side, or back flop"), acrobatic (somersaults, flips, and spins), and coordinated (the same leap performed synchronously by ≥ 2 dolphins; Pearson 2017)

is tightly coupled with the evolution of a large and complex brain. With an encephalization quotient (Jerison 1973) of 4.7, the dusky dolphin has evolved one of the largest relative brain sizes of any mammal (Marino et al. 2004).

In this chapter, I discuss how ecology has influenced the behavioral and social flexibility that is inherent to dusky dolphins. Using case studies of research conducted at three study sites, I provide evidence for the ways in which dusky dolphins use complex patterns of thinking to navigate their social world. I then examine dusky dolphin behavior in a comparative context with other delphinids. Finally, I describe conservation challenges and strategies for dusky dolphins and present ideas for advancing the state of knowledge for this species through technological advancement and the filling of data gaps.

18.2 Three Dusky Dolphin Study Sites

Prey availability, access to mates, and predation risk are three primary influences on animal behavior. Dusky dolphins offer a natural experiment for advancing knowledge of how ecology shapes behavior, as they have been studied in three distinct habitats (Fig. 18.3). Variation in grouping patterns and activity budgets is likely related to the unique combination of prey availability, mating opportunities, and

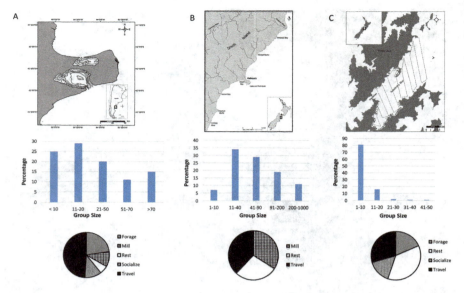

Fig. 18.3 Dusky dolphin, *Lagenorhynchus obscurus*, daytime grouping patterns and activity budgets vary according to site. (**a**) In Golfo Nuevo, Argentina, during January–June, most groups contain <20 individuals, and traveling is the predominant behavior (Degrati et al. 2008, map courtesy G. Garaffo and from Würsig and Pearson 2014). (**b**) Off Kaikoura, New Zealand, most groups year-round contain >40 individuals (Cipriano 1992), and traveling is the predominant summertime activity (Markowitz 2004, map from Childerhouse and Baxter 2010). (**c**) In Admiralty Bay, New Zealand, during austral winter, most groups contain <10 individuals, and resting is the predominant behavior. Survey transect lines (dotted lines) and mussel farms (black polygons) shown in map (Pearson 2009)

predation risk at each site. Below, I present the current state of knowledge of dusky dolphin populations in Argentina and New Zealand.[1]

18.2.1 Península Valdés, Argentina

Behavioral studies of dusky dolphins have been conducted off Península Valdés, Patagonia, since the 1970s, first in Golfo San José (42°20S, 64°20′W) from 1973 to 1976 (Würsig and Würsig 1980) and more recently in Golfo Nuevo (42°44′S, 64°29′W; e.g., Garaffo et al. 2007; Dans et al. 2008; Degrati et al. 2008, 2012,

---

[1]To enhance flow and readability of this chapter, I primarily use present tense. However, in some cases, the present-day situation is different from that previously described in the citations provided. For example, the coordinated bait-ball foraging described by Würsig and Würsig (1980) in Golfo San José, Argentina, during 1973–1976 is now a rare occurrence (Würsig 2010); this behavior may also be declining in Admiralty Bay, New Zealand (Piwetz 2018).

18 Dusky Dolphins of Continental Shelves and Deep Canyons 391

2013; Vaughn-Hirshorn et al. 2013). These semi-enclosed bays are large and relatively shallow (Golfo San José = 705 m^2, 30–60 m typical depth; Golfo Nuevo = 2500 m^2, 40–100 m typical depth), and there is evidence that individuals move outside of and between the two bays (Würsig and Bastida 1986). While the majority of groups off Península Valdés contain ≤20 individuals, foraging groups may contain 300 individuals (Würsig and Würsig 1980; Degrati et al. 2008); in Golfo San José, these large foraging groups may constitute the entire population in the bay at that time (Würsig and Würsig 1980).

The primary prey of dusky dolphins in Golfo San José and Golfo Nuevo are Argentine anchovies (*Engraulis anchoita*), which often occur in widely distributed prey balls (Würsig and Würsig 1980; Koen Alonso et al. 1988; Degrati et al. 2008; Romero et el. 2012; Vaughn-Hirshorn et al. 2013). Stable isotope analysis indicates that Patagonian squid (*Loligo gahi*) and butterfish (*Stromateus brasiliensis*) are also eaten (Loizaga De Castro et al. 2016). A patchy distribution of prey is evidenced by the finding in Golfo Nuevo that the primary daytime activity of dusky dolphins is traveling, likely in search of food (Degrati et al. 2008). Foraging strategies differ between winter and summer, with increased diving activity (likely for squid) occurring during winter and surface feeding (for anchovy) occurring during summer (Würsig and Würsig 1980; Degrati et al. 2012).

When searching for prey during the summer, dusky dolphins in Golfo San José form smaller groups separated by ≤1 km to increase the search field while still remaining in acoustic contact (Würsig and Würsig 1980). Once located, individuals typically start to feed through synchronous, coordinated swimming around and below the prey ball (Box 18.1). This movement increases cohesion of and decreases oxygen availability to the prey ball (thereby making the fish easier to capture) while simultaneously forcing it upward toward the surface barrier (Norris and Dohl 1980; Würsig and Würsig 1980; Würsig 1986; Vaughn-Hirshorn et al. 2013). "Clean" headfirst reentry leaps (Fig. 18.2) are also common during bait-ball feeding as this behavior enables an individual to quickly take a breath while gaining momentum to reach greater foraging depths (Würsig and Würsig 1980; Pearson 2017). Although foraging efficiency on large prey balls may be increased via formation of large groups (Würsig and Würsig 1980; Vaughn et al. 2010a), scramble competition can ensue if increased dolphin group size leads to rapid depletion of prey balls (Gowans et al. 2008). Although not specifically tested, a positive correlation would be expected between prey ball size and foraging group size.

Box 18.1 Coordination vs. Cooperation

There has been uncertainty if dusky dolphin bait-ball feeding behavior is "coordinated" (Vaughn et al. 2007, 2010b; Vaughn-Hirshorn et al. 2013) or "cooperative" (Würsig and Würsig 1980; Würsig 1986). Cooperation occurs when the collective action of ≥2 individuals leads to positive fitness benefits that outweigh individual costs of participating in the group (Dugatkin et al. 1992). As we currently lack data to quantify the individual fitness outcomes

(continued)

Box 18.1 (continued)

resulting from bait-ball feeding, I use the term "coordinated." However, based on the reasoning below, I speculate that it is indeed cooperative.

Evidence of cooperation arises from studies of bait-ball feeding at both Península Valdés sites and Admiralty Bay, New Zealand (Fig. 18.4). During bait-ball feeding, individuals incur costs by exhibiting "temporary restraint" as they perform herding passes without prey capture attempts; i.e., they "take turns" feeding on the prey ball instead of attacking the prey ball at once (Würsig and Würsig 1980; Würsig et al. 1989; Vaughn-Hirshorn et al. 2013). It can be surmised that individuals in these feeding groups have some understanding of the benefits of this behavior and that per capita energy intake (and ultimately, fitness) can be increased by working together as opposed to feeding individually. Similar evidence for cooperative foraging involving temporary restraint comes from spinner dolphins (*Stenella longirostris*) feeding on the deep scattering layer off Oahu, HI. Here, group members exhibited distinct foraging formations and synchronous behaviors to herd prey after which pairs of dolphins "took turns" moving from the group periphery to the center to capture prey (Benoit-Bird and Au 2009).

Fig. 18.4 A dusky dolphin, *Lagenorhynchus obscurus*, group herds a large pilchard school toward the surface in Admiralty Bay, New Zealand. Shown are four individuals exhibiting coordinated behavior through synchronous swimming in a clockwise direction around and underneath the bait ball. Image is a digital still taken from video; courtesy R. Vaughn-Hirshorn (Vaughn et al. 2007)

At the conclusion of coordinated foraging bouts, a high degree of sociosexual behavior involving copulations and aerial displays such as head-over-tail leaps, backslaps, headslaps, tailslaps, and spyhops occurs (Würsig and Würsig 1980). During this time, individuals likely form and reaffirm social bonds. This may be a particularly important social opportunity for individuals that may have been separated for prolonged periods. According to the social facilitation hypothesis, bond formation may ultimately function to enhance foraging efficiency, as individuals with stronger bonds may be more effective hunting partners (Norris and Dohl 1980; Würsig and Würsig 1980). This has been quantified for common bottlenose dolphins (*Tursiops truncatus*) in Cedar Key, FL, where prey capture rates were higher in a hunting group with more stable group membership (Gazda et al. 2005). The large dusky dolphin group disintegrates at the conclusion of this social bout, eventually forming small, nearshore resting groups during the night (Würsig and Würsig 1980; Würsig et al. 1989). Due to the fission-fusion nature of the society, group size and composition is most likely different pre- vs. post-feeding (Würsig and Pearson 2014).

Dusky dolphins have a multi-mate or polygynandrous mating system that includes a high degree of sperm competition, as indicated by large relative testis size comprising up to 5.4% of body weight (Cipriano 1992; van Waerebeek and Read 1994; Markowitz et al. 2010a). While mating behavior has not been specifically examined off Península Valdés, copulation was observed most frequently in large groups post-feeding in Golfo San José (Würsig and Würsig 1980). Calving is seasonal with the majority of newborn calves observed during November–February (Würsig and Würsig 1980; Degrati et al. 2008), though calving may span spring through autumn (Dans et al. 1997).

The primary natural threats to dusky dolphins off Península Valdés are killer whales (*Orcinus orca*) and sharks (e.g., broadnose sevengill shark, *Notorynchus cepedianus*; Würsig and Würsig 1980; Crespi et al. 2003). Although these predators may be rarely observed (Vaughn et al. 2010b), predators can still influence prey behavior through the "ecology of fear" (Brown et al. 1999; Srinivasan et al. 2018; also Chap. 7). This is evidenced by the finding that more vulnerable groups (i.e., mother-calf pairs, small groups, resting groups) occur in shallow, nearshore waters (Würsig and Würsig 1980; Garaffo et al. 2007). For mothers, male harassment is an additional threat (Degrati et al. 2008; Weir et al. 2010). Mother-calf pairs may thus form smaller groups of <20 individuals as they segregate from larger groups to decrease harassment risk. The increased vulnerability inherent to small mother-calf groups can then be offset by movement to shallow waters (Degrati et al. 2008).

Seasonal differences in foraging strategies drive association patterns in Golfo Nuevo. During the summer, prey is more available at the surface, which attracts larger dusky dolphin groups. This large group formation is likely a result of mutual attraction to the prey resource rather than attraction to conspecifics. Consequently, bonds are weak overall and the social structure is fluid. During the winter, prey occurs in deeper waters and is less available. Dusky dolphins form smaller groups, and the foraging strategy switches from surface feeding to diving. In contrast to summer, bonds are stronger and longer-lasting, perhaps because individuals are attracted to each other rather than to a common, easily accessible resource (Degrati et al. 2018).

In summary, the behavior of dusky dolphins off Península Valdés is characterized by a high degree of fission-fusion dynamics as group size and composition changes throughout the day. Large group formation enhances daytime coordinated foraging on patchily distributed prey and provides mating opportunities. Small group formation occurs to increase detection of prey and, for mothers, to decrease risk of male harassment. Movement to shallow waters decreases predation risk, particularly for resting groups and mother-calf pairs. Bond formation is driven by seasonal differences in foraging strategies, and bonds are stronger during the cold season when prey is less available.

18.2.2 Kaikoura, New Zealand

Dusky dolphins have been studied off Kaikoura, New Zealand (42°30S, 173°36′W), since 1984 (Würsig and Würsig 2010). This body of work has revealed that this population exhibits foraging and social strategies quite different from Península Valdés. The open habitat along the coastline is dominated by expansive waters of the 1200-m-deep Kaikoura Canyon starting <500 m from shore (Lewis and Barnes 1999). An associated upwelling system creates primary productivity that supports a rich biodiversity of marine life including ca. 1000–2000 dusky dolphins present at any given time (Markowitz 2004). Within this open population (Orbach et al. 2018), grouping patterns are characterized by large mixed age- and sex-class (hereafter referred to as "mixed") groups accompanied by smaller satellite groups composed of mating adults, non-mating adults, and mother-calf pairs (Markowitz 2004; Würsig et al. 2007).

The 24-h behavioral cycle of dusky dolphins off Kaikoura is driven by nocturnal foraging strategies. During the late afternoon, nearshore resting groups become more active and begin to move offshore. As groups move offshore, group fission occurs as individuals spread out in search of lanternfish (family Myctophidae) and squid (*Nototodarus* sp. and *Todaroides* sp.) in the deep scattering layer (DSL; Cipriano 1992; Markowitz 2004). While direct observation of DSL feeding has not occurred, acoustic studies have shown that feeding occurs once the DSL has risen to <130 m; this temporal window typically spans 7–9 h and 12–13 h in the summer vs. winter, respectively. Once this depth threshold is crossed, dusky dolphins appear to feed in loosely coordinated groups of 2–5 individuals (Benoit-Bird et al. 2004, 2009; Dahood and Benoit-Bird 2010).

After nighttime feeding, dusky dolphins move toward the shore during early morning in small traveling groups. Once nearshore, group fusion occurs, resulting in formation of large mixed groups (Video S1) in addition to smaller satellite groups. Group fusion is accompanied by an increased level of social activity and leaps, likely as previously separated individuals "reunite" after foraging in a manner akin to Península Valdés. Typically, one large mixed group of 200–300 individuals forms, although groups may reach ≥1000 individuals (Markowitz 2004). Within these large groups, individuals move together and remain within the group "envelope" (Norris and Schilt 1988). Activity then wanes toward a midday resting period. Toward

18 Dusky Dolphins of Continental Shelves and Deep Canyons

midafternoon, the activity level begins to increase once again, as evidenced by an increased degree of leaping (esp. noisy leaping, Fig. 18.2), which may serve to coordinate group movement in preparation for offshore, nighttime feeding (Markowitz 2004). Mean daytime depth of dusky dolphins in large mixed groups is 5.6 ± 5.33 m, indicative of nonfeeding behavior (Pearson et al. 2017a).

Similar to Península Valdés, reproduction is seasonal off Kaikoura. Observations of copulations and newborn calves peak during austral summer (Cipriano 1992; Markowitz et al. 2010a). While mating occurs in large mixed groups, some adults form smaller mating groups separate from the main group. Off Kaikoura, the male dusky dolphin mating strategy is one of scramble competition or "mating of the quickest," where males use speed and agility (but rarely aggression) to outmaneuver competitors and gain access to estrus females performing evasive maneuvers (e.g., belly-ups, rapid changes in direction, clean reentry leaps) that likely function in mate choice (Markowitz et al. 2010a; Orbach et al. 2014). Mean mating group size is six individuals, typically consisting of a single female. As mating group size increases, males incur increased energetic expenditure, decreased ability to monopolize females, and decreased copulation rates (Orbach et al. 2014, 2015).

Predation risk is relatively high off Kaikoura. The deep waters of the Kaikoura Canyon attract a myriad of predators such as great white sharks (*Carcharodon carcharias*), broadnose sevengill sharks, shortfin mako (*Isurus oxyrinchus*), Pacific sleeper sharks (*Somniosus pacificus*), blue sharks (*Prionace glauca*), and killer whales (Srinivasan and Markowitz 2010). Large group formation and protection via the dilution effect (Dehn 1990) is one anti-predation strategy for dusky dolphins off Kaikoura (Würsig et al. 2007).

For mother-calf pairs, the anti-predation benefits received from large group formation must be balanced with increased costs of male harassment. Mothers may thus alter their spatial positioning to minimize risk of male harassment. Strategies include aggregation or "clumping" of mother-calf pairs with other mother-calf pairs in the group periphery and close mother-calf proximity (i.e., closer mother-calf distance as compared to mean group nearest neighbor distance and closer mother-calf distance when the calf's nearest neighbor was a non-mother; Fanucci-Kiss 2015). These results are consistent with the infant safety hypothesis, which predicts that mothers will "keep their calves close" and reduce proximity to their offspring as a strategy to reduce male harassment (Otali and Gilchrist 2006). Mother-calf pairs also form nursery groups. Benefits of nursery group formation include "quiet" time to rest and save energy, especially important given the higher energetic demands of lactation, and offspring socialization (Weir et al. 2010). Nursery groups typically occur in shallow (≤ 20 m) waters where killer whales and blue sharks are less abundant. These waters also provide a refuge from male harassment, as nursery groups are encountered more frequently in shallow waters than mating, adult, and large mixed groups (Weir et al. 2008).

Based on a 30-year photo-identification dataset, the mean association index for dusky dolphins off Kaikoura is 0.04 ± 0.03 SD, indicating that, on average, any two individuals are observed together 4% of the time. However, the mean maximum association index is 0.32 ± 0.03 SD, indicating that, on average, an individual is

396 H. C. Pearson

sighted with its closest associate or "best buddy" 32% of the time. These association indices indicate a rather weak level of bonding throughout the society as a whole. However, some degree of social affinity is present as individuals are observed to form temporally stable associations for 8 years and exhibit preferences for occurring in large vs. small groups (Orbach et al. 2018).

In summary, the behavior of dusky dolphins off Kaikoura is driven by offshore, nocturnal foraging and large group formation to reduce predation risk in this open habitat. Mating occurs in large mixed groups and satellite mating groups. Mothers may form smaller nursery groups nearshore to avoid male harassment and decrease predation risk. A low level of social bonding occurs overall, and while group fission and fusion is present, it occurs to a lesser degree than off Península Valdés, Argentina.

18.2.3 Admiralty Bay, New Zealand

Dusky dolphins have been studied in Admiralty Bay (40°56'S, 173°53'E) since the late 1990s (Markowitz et al. 2004). This relatively small (117 km^2), shallow (typical depth = <50 m, maximum depth = 105 m), semi-enclosed bay is located at the tip of the South Island in the Marlborough Sounds (McFadden 2003). Compared to Península Valdés and Kaikoura, Admiralty Bay is unique because it is a seasonal habitat for dusky dolphins. During the austral spring through autumn (May through October), dusky dolphins occur in Admiralty Bay to exploit rich prey resources (Vaughn et al. 2007; Pearson 2009). In this non-mating habitat, calves are rarely observed, and genetic evidence indicates a predominance of males (Shelton et al. 2010). Grouping patterns within this open population (Pearson et al. 2017b) are characterized by relatively small (mean = 7 ± 6.0 SD individuals) groups that increase in size (maximum = 50 individuals) during foraging, though not to the same degree as in Golfo San José, Argentina (Pearson 2009). The relatively small size of Admiralty Bay may physically limit group sizes from reaching the sizes observed in Península Valdés and Kaikoura.

Dusky dolphins in Admiralty Bay primarily prey on New Zealand pilchard (*Sardinops neopilchardus*), a small, schooling fish (Vaughn et al. 2007). Similar to Golfo San José, dusky dolphins in Admiralty Bay exhibit coordinated bait-ball foraging during the daytime that often involves multi-species associations with seabirds (e.g., Australasian gannets, *Morus serrator*; shearwaters, *Puffinus* spp.; Vaughn et al. 2007, 2008, 2010a). Dolphins feed by preying on existing prey balls, actively herding fish into stationary prey balls, or feeding at depth. Mean feeding group size is 8.3 ± 5.0 SD dolphins. Mean duration of bait-ball feeding bouts is 7 ± 8.4 SD min, and larger groups (maximum = 30 individuals) appear better able to herd fish into stationary prey balls (Vaughn et al. 2007, 2010a).

Differences in prey patch size and patchiness may explain the observed differences in feeding behavior between Península Valdés and Admiralty Bay. Off Península Valdés, prey balls are larger. Since larger groups are better able to herd and contain larger prey balls, it follows that feeding group size would be larger off Península Valdés. Increased feeding bout duration in Península Valdés may also be attributed to larger prey ball size (Vaughn et al. 2010b). Alternatively, since prey are more widely distributed and unpredictable off Península Valdés, it may be more beneficial for individuals to expend additional handling time (thus lengthening feeding bout duration) to contain more mobile prey balls, despite potentially increased costs of scramble competition (Vaughn et al. 2010a). In contrast, observations from fine-scale video analysis in Admiralty Bay indicate a low degree of scramble competition as the number of herding passes per individual (costs) did not increase with increased group size and per capita prey capture events (benefits) did not decrease with increased group size (Vaughn-Hirshorn et al. 2013).

Predation risk is relatively low in Admiralty Bay. While killer whales occur in the region, they rarely occur during spring through fall when dusky dolphins are present. There is also an absence of shark predators. In response to this low predation risk, dusky dolphins do not form larger groups during resting and do not rest more often in nearshore habitats (Pearson 2009).

In the absence of predation risk, grouping patterns are driven by foraging and social strategies (Würsig and Pearson 2014). Foraging has a positive influence on group size, while socializing has a negative influence on group size (Pearson 2009). Group size is not static, however, as individuals join and split from groups every 5 ± 7.6 SD min (Pearson 2008). These dynamic grouping patterns enable individuals to associate with a wide variety of conspecifics, as evidenced by the relatively low mean association index of 0.3 ± 0.01 SD. However, this is juxtaposed against stronger bond formation throughout the society as the mean maximum association index is 0.4 ± 0.02 SD, associations remain nonrandom for 6 years, and preferential associations are formed during all behavioral states (i.e., foraging, resting, socializing, traveling). In particular, association during foraging may enhance coordinated foraging strategies through social facilitation (as in Golfo San José), and association during socializing may be used to form and renew bonds (Pearson et al. 2017b).

Within this ever-changing group environment, leaping provides important signaling functions. For example, noisy leaping appears to function as a signal for group members to transition to a more active behavioral state (e.g., from resting to traveling). Clean leaping likely functions as a signal for groups to split, perhaps to reduce scramble competition during foraging. Finally, coordinated leaping appears to have both signaling and social facilitation functions, as it signals the end of a feeding bout and promotes group fusion which may result in bond formation or renewal (Würsig and Würsig 1980; Pearson 2017).

Photo-identification studies in Admiralty Bay and Kaikoura show that some of the same individuals travel between the two habitats (Markowitz 2004; Weir 2008). This annual movement pattern may be a cultural tradition whereby the same individuals travel from Kaikoura to Admiralty Bay to exploit prey resources (Whitehead et al. 2004). While further study is required to deem this behavior a cultural tradition, if the data affirm this, then it is likely to be a form of horizontal cultural transmission (Cavalli-Sforza et al. 1982) as the majority of individuals that were sampled for genetic study in Admiralty Bay were male and unrelated (Shelton et al. 2010).

Further, some of the same males have been observed together in mating groups off Kaikoura during the summer and in foraging groups in Admiralty Bay the following winter. This indicates that individuals retain their social bonds as they move between the two habitats and that the same males may coordinate behaviors in mating groups and foraging groups (Markowitz et al. 2010a).

Dusky dolphins traveling between Kaikoura and Admiralty Bay switch lifestyles. The ability for a single individual to alter its foraging and social strategies on a biannual basis indicates a high degree of behavioral flexibility. Yet, within these ever-changing ecological conditions, dusky dolphins retain the same overall level of association between the two habitats, reinforcing the importance of social bonding to dusky dolphin society (Whitehead et al. 2004; Pearson et al. 2017b; Orbach et al. 2018).

In summary, dusky dolphins form relatively small groups in Admiralty Bay to take advantage of seasonally available prey resources. In the absence of mating opportunities and predation, groups are influenced by prey availability and social strategies. Differences in coordinated foraging strategies between Golfo San José and Admiralty Bay are likely due to differences in prey patchiness and patch size. Fission-fusion dynamics are intermediate between Golfo San José and Kaikoura. While some individuals form strong and lasting bonds, social patterns are characterized by relatively weak bonds overall.

18.3 Inside the Dusky Dolphin Mind

Brain tissue is energetically expensive, and thus there must be selective pressure for a species' investment in large brain size and associated complex cognition (Dunbar 1998). Various hypotheses have been proposed for the evolution of large brains, including ecological intelligence (Clutton-Brock and Harvey 1980), Machiavellian intelligence (Humphrey 1976; Byrne and Whiten 1989), brain size/environmental change (Sol et al. 2005), social learning/culture (van Schaik 2006), and social brain (Jolly 1966; Dunbar 1998).

For dusky dolphins (and other cetaceans, Fox et al. 2017), the social brain hypothesis offers a compelling explanation for the evolution of large brain size (summarized in Pearson and Shelton 2010). According to this hypothesis, large brain size evolved in a positive feedback loop with complex sociality. Large brain size provides the cognitive power necessary to solve ecological problems in a social context, thereby creating increased social complexity, which selects for increased brain size. While the need to solve complex ecological problems (e.g., finding patchy food resources) may have been the initial selective force for large brain size, conducting this problem-solving within an ever-changing social environment likely provided the evolutionary pressure that "ratcheted up" the need for increasing brain size and complex cognition (Dunbar 2003; Pearson and Shelton 2010; Dunbar and Shultz 2017; Würsig 2018).

Components of the dusky dolphin "toolbox" of complex cognition include coordinated foraging, complex association patterns, and behavioral flexibility. Coordinated bait-ball feeding shows evidence of forethought and problem-solving. In

18 Dusky Dolphins of Continental Shelves and Deep Canyons

Golfo San José, if a prey ball was too large for a given group to contain, individuals performed noisy leaps (presumably) to signal other groups to join (Würsig and Würsig 1980; Würsig 1986). This chain of behaviors requires an individual to assess a situation, identify a problem, develop a solution, and then act on it, all of which indicate complex thought processes. Dusky dolphins also form long-term bonds within a highly dynamic fission-fusion framework. This requires a high degree of "social bookkeeping" as an individual must keep track of associates (e.g., who is friend vs. foe) within the ever-changing social framework. Finally, dusky dolphins traveling between Kaikoura and Admiralty Bay exhibit a strong degree of behavioral flexibility as a change in foraging strategy is accompanied by a change in social strategy. While we have no scientific way of understanding all workings of a dusky dolphin's (or any animal's) mind, behavioral evidence can be used to infer the various complex thought processes that might occur within an individual over the course of a day or year (Box 18.2).

Box 18.2 The Life of a Dusky Dolphin

Presented below are two hypothetical examples of a day and year, respectively, in the life of a New Zealand dusky dolphin. While rooted in data and the findings presented in this chapter, some is conjecture as there is still much to be discovered about the life of an *individual* dusky dolphin. These examples illustrate the behavioral flexibility inherent to dusky dolphin society as an individual navigates its complex three-dimensional environment in search of prey and social partners. Throughout a given day or year, an individual is faced with many decisions, such as where and how to feed and with whom to form bonds, choices which likely impact fitness.

A Day in the Life of a Dusky Dolphin

It is just after summertime dawn and a male is traveling west toward the Kaikoura Peninsula. He is satiated after feeding on lanternfish and squid during the night. His feeding strategy changed as the depth of the deep scattering layer (DSL) changed, feeding alone when the DSL was deep and with 3–4 other individuals as the DSL rose toward the surface and shallower dives allowed more time for group coordination. While his foraging companions were different from the night before, they were not "strangers" as he remembered them from previous interactions occurring within his broad social network. As his traveling group of 10 or so individuals approaches the coastline, he meets other small groups also returning from the nighttime feast. Many of these reunions are accompanied by acrobatic leaps. Eventually, these smaller groups merge into one large group of approximately 300 individuals including single adults and juveniles of both sexes and mother-calf pairs. Within this large group, the male interacts with a variety of conspecifics (male and female), some whom he encountered during the previous day and some whom he has not encountered for several weeks. These interactions include synchronous leaping, flipper rubbing, and copulation, which all serve

(continued)

Box 18.2 (continued)

to renew social bonds. His companions now are different from those with whom he interacted in his morning traveling group. Toward late morning, the activity level of the group lessens and the male begins to rest. While never totally asleep, this period of reduced activity enables him to recover from the previous night's foraging activities. He feels safe from predators knowing he is surrounded by hundreds of companions and that some group members remain vigilantly "on watch." Toward midafternoon, the activity level of the group begins to heighten once again. The male swims near the front of the group and occasionally performs noisy leaps to signal the group's direction of travel, gradually moving offshore to prepare for the upcoming nighttime feast once again.

A Year in the Life of a Dusky Dolphin

It is midday in Admiralty Bay during austral winter, and the male dusky dolphin is working with a dozen conspecifics to herd a large pilchard prey ball. After remaining in Kaikoura during the peak mating season (Nov–Feb) and beyond, he traveled north to Admiralty Bay in search of its rich prey resources. Here, the male has a very different lifestyle than Kaikoura, coordinating feeding behavior during the daytime on schooling prey and forming groups that are typically an order of magnitude smaller than off Kaikoura. Despite the small group size, the male still encounters a wide variety of social partners as he joins and splits from groups while foraging, resting, traveling, and socializing throughout the day. Yet, throughout the wintertime foraging season, there are certain individuals with whom he preferentially associates, doing so year after year as they reunite in Admiralty Bay each winter. Further, some (but not all) of these preferred partners are also preferred partners in Kaikoura. This continuity amidst an ever-changing social world provides a sense of stability for the male and allows him to be a more effective hunter as he can predict the behavior of his foraging partners to more efficiently corral prey. However, toward autumn when the seasonal prey resources begin to dwindle, the male travels back to Kaikoura in time for the start of the peak mating season once again.

18.4 Comparison with Other Delphinids

Our understanding of dusky dolphin behavioral and social strategies may be deepened through interspecific comparison with other delphinids for which similar data are available (Table 18.1). A qualitative analysis reveals two important differences between the societies of dusky, spinner (*Stenella longirostris*), bottlenose (*Tursiops* sp.), and Atlantic spotted (*Stenella frontalis*) dolphins.

First, dusky dolphins, as a whole, are more gregarious than many other species. This may be attributed to two main factors: habitat and predation risk. The deep,

Table 18.1 Notable aspects of dusky dolphin behavior and ecology in comparison to other delphinids

	Dusky dolphins (*Lagenorhynchus obscurus*)			Spinner dolphins (*Stenella longirostris*), Midway Atoll, Hawaii[d]	Bottlenose dolphins (*Tursiops* sp.), Doubtful Sound, New Zealand[e]	Atlantic spotted dolphins (*Stenella frontalis*), Little Bahama Bank, Bahamas[f]
	Península Valdés, Argentina[a]	Kaikoura, New Zealand[b]	Admiralty Bay, New Zealand[c]			
Group size: mean, maximum	10–12 (typical), 300	150 (winter), 65 (non-winter); >1000	7, 50	211, 260	17, 65	11, 56
Association index: mean (SD), mean maximum (SD)	0.07 (0.05, warm season), 0.11 (0.08, cold season); 0.51 (0.31, warm season), 0.60 (0.27, cold season)	0.04 (0.03), 0.32 (0.15)	0.03 (0.01), 0.40 (0.20)	0.37 (0.09), 0.61 (0.14)	0.47 (0.04), 0.63 (0.08)	0.09 (0.09) to 0.12 (0.13), N/A
Duration of bonds	170 days (cold season only)	≥ 6 years	≥ 6 years	≥ 2 years	6 months (all), 14 years (male–male)	≥ 10 years
Preferred associations? (Y/N)	Y (cold season only)	N	Y	Y	Y	Y
Habitat type	Shallow, semi-enclosed bay	Coastal waters and deep canyon	Shallow, semi-enclosed bay	Shallow lagoon surrounded by deep open waters	Fjord (deep, coastal)	Shallow sandbank surrounded by deep open waters
Predation risk (high/low)	High	High	Low	High	Low	High
Resident population? (Y/N)	N	N	N	Y	Y	Y

(continued)

Table 18.1 (continued)

	Dusky dolphins (*Lagenorhynchus obscurus*)			Spinner dolphins (*Stenella longirostris*), Midway Atoll, Hawaii[d]	Bottlenose dolphins (*Tursiops* sp.), Doubtful Sound, New Zealand[e]	Atlantic spotted dolphins (*Stenella frontalis*), Little Bahama Bank, Bahamas[f]
	Península Valdés, Argentina[a]	Kaikoura, New Zealand[b]	Admiralty Bay, New Zealand[c]			
Cooperative/ coordinated foraging? (Y/N)	Y	Y	Y	Y	N	N/A
Prey type	Small, schooling fish	Non-schooling fish and squid in the deep scattering layer (DSL)	Small, schooling fish	Non-schooling fish and squid in the DSL	Solitary, demersal fish	Schooling and bottom-dwelling fish (daytime), non-schooling fish and squid in the DSL (nighttime)

[a]References: Würsig and Würsig (1980), Würsig and Bastida (1986), Würsig et al. (1997), Degrati et al. (2008, 2018), Vaughn et al. (2010b), Vaughn-Hirshorn et al. (2013), Würsig and Pearson (2014)

[b]References: Cipriano (1992), Benoit-Bird et al. (2004), Markowitz (2004), Srinivasan and Markowitz (2010), Orbach et al. (unpubl. data)

[c]References: Pearson (2009), Vaughn et al. (2007), Pearson et al. (2017a, b)

[d]References: Karczmarski et al. (2005)

[e]References: Williams et al. (1993), Lusseau et al. (2003), Lusseau and Wing (2006), Pirotta et al. (2014)

[f]References: Herzing (1996), Elliser and Herzing (2014a, b), Melillo-Sweeting et al. (2014)

Unless otherwise noted, data are for all age-sex classes combined at each site open predator-rich habitat off Kaikoura physically "allows" large group formation to decrease predation risk. In contrast, species living in more enclosed habitats with lower predation risk (e.g., bottlenose dolphins in Doubtful Sound, Lusseau et al. 2003; Pirotta et al. 2014) form smaller groups. One might expect large groups to also form among Atlantic spotted dolphins in the Bahamas, another open habitat with high predation risk. However, there dolphins typically occur on a shallow sandbank and avoid deep waters where predators occur (Elliser and Herzing 2014a, b), which may preclude selective pressure for large group formation. Midway Atoll has a unique combination of high predation risk and small, enclosed habitat. Medium-sized groups of spinner dolphins occur there, perhaps due to high geographic insularity and limited availability of resting habitat (Karczmarski et al. 2005).

Second, dusky dolphin societies are more loosely bonded than those of spinner, bottlenose, and Atlantic spotted dolphins. This may be related to the large and open societies formed by dusky dolphins vs. the smaller resident societies formed by the other delphinids. In an open society, individuals have more opportunities to interact with a greater number of conspecifics. As a result, an individual may interact with a larger number of conspecifics but at a lower level. Initially, this conclusion is surprising given the high degree of coordinated foraging observed in dusky dolphins, as one might expect that this would drive tight bond formation. However, due to the high degree of fission-fusion dynamics in dusky dolphin society, it may be important for an individual to maintain a large social network to ensure that a "familiar" companion is present when needed. The after-feeding "parties" by dusky dolphins described for Argentina after coordinated bait-ball feeding may be integral to social integrity needed for dolphins that only closely interact now and then (Würsig and Würsig 1980; Würsig 1979).

18.5 Conservation

Anthropogenic threats influence dusky dolphin behavior, and thus knowledge of these threats is imperative to their ethology. Primary threats to dusky dolphins include bycatch, habitat loss, climate change, and tourism (Van Waerebeek and Würsig 2018). Similar to 42% ($n = 16$) of all delphinid species, the conservation status for the dusky dolphin is currently listed as "data deficient" on the International Union for the Conservation of Nature Red List (IUCN 2018). However, conservation action need not wait until a species is declining or threatened. Indeed, "managing for robustness" (p. 600, Würsig et al. 2002) and protecting healthy populations using proactive approaches is a critical component of conservation.

Based on available data, bycatch poses the greatest threat to dusky dolphins off South America. Since the 1960s, dusky dolphins off Peru have been afflicted by high (e.g., 7000/year captured during the early 1990s; Van Waerebeek and Reyes 1994; Van Waerebeek and Würsig 2018), largely unknown, and potentially unsustainable levels of bycatch in gillnet and longline shark fisheries, in addition to direct harpooning for bait and human consumption. Despite Peruvian legislation enacted in the mid-1990s to reduce cetacean catch in these fisheries, a high level of take continues (Mangel et al. 2010, 2013; Markowitz et al. 2010b; Reeves et al. 2013). However, the use of modern acoustic deterrent devices ("pingers") is one promising strategy for reducing bycatch of dusky dolphins and other small cetaceans in Peruvian fisheries (Mangel et al. 2013). Off Patagonia, although occurring in lower absolute numbers than off Peru, dusky dolphins are threatened by high rates of bycatch from mid-water trawling fisheries that approach threshold levels for maximum annual removal (Dans et al. 2003).

Habitat loss poses a threat to dusky dolphins, particularly in New Zealand. Since its inception in the 1970s, the mussel farm industry has flourished and is a top industry in New Zealand (Childerhouse and Baxter 2010). The Marlborough Sounds, including Admiralty Bay, is prime habitat for green-lipped mussel (*Perna canaliculus*) farms, which also overlaps with prime dusky dolphin foraging habitat. As a result, human-wildlife conflict has escalated since the region was first documented to be an important wintertime foraging habitat for dusky dolphins in the late 1990s (Markowitz et al. 2004). Mussel farms cause habitat loss by excluding dusky dolphins and interfering with foraging behavior. After correcting for area, dusky dolphins spend significantly more time outside vs. inside mussel farms (Markowitz et al. 2004, 2010b; Pearson et al. 2012). Similarly, Chilean dolphins (*Cephalorhynchus eutropia*) in Yaldad Bay (Ribeirio et al. 2007) and bottlenose dolphins (*Tursiops* sp.) in Shark Bay, Australia (Watson-Capps and Mann 2005), avoid using areas within shellfish farms. For dusky dolphins, the three-dimensional obstruction created by the network of mussel farm lines extending vertically through the water column also hinders coordinated bait-ball foraging (Pearson et al. 2012). Acoustic deterrence from mussel farm tenders is another potential yet unquantified disturbance. While efforts to halt permits for growth of the mussel farm industry in Admiralty Bay have been successful so far (Childerhouse and Baxter 2010; B. Würsig pers. comm.), there is still cause for concern in light of the New Zealand government's goal of doubling the mariculture industry (i.e., mussel, oyster, and salmon farming) from NZ$500 million (current) to NZ$1 billion (target) per annum by the year 2025 (Aquaculture New Zealand 2018). However, there appear to be species-specific reactions as other New Zealand dolphin species (e.g., common bottlenose dolphins, *T. truncatus*) do not appear to avoid or be displaced by mussel farms (pers. obsv.).

Climate change poses a growing and relatively unknown threat to dusky dolphins (Würsig 2010). A relative decline in dusky dolphin abundance in Admiralty Bay appears to be linked to changing sea surface temperatures. A 6-year time series has shown that relative dusky dolphin abundance decreased as sea surface temperatures decreased. This could be attributed to a shift in their primary prey species in this region (i.e., pilchard) due to changing temperatures. It is also possible that this apparent population decline is related to bottom-up effects from mussel farming (e.g., nutrient depletion) impacting the prey base (Srinivasan et al. 2012). Further research is needed to understand how climate change is impacting dusky dolphins throughout their range and if/how it is acting synergistically with other threats.

Globally, whale-watching is a growing industry that attracts 13 million passengers annually at a valuation of US$2.1 billion (O'Connor 2009). In New Zealand and Patagonia, there are active whale-watching industries based on viewing (New Zealand and Patagonia) and swimming with (New Zealand only) dusky dolphins (Dans et al. 2008; Buurman 2010). While often labeled as "eco-friendly," such activities have the potential to impact dusky dolphins in the short and long term. Off Kaikoura, dusky dolphins decrease resting and increase milling and traveling behaviors in the presence of vessels (Lundquist et al. 2012). In Golfo Nuevo, dusky dolphins feed and socialize less and travel more in the presence of vessels (Dans et al. 2008). These studies indicate that short-term negative energetic effects are occurring that may lead to long-term consequences such as reduced fitness. Considering these results, a precautionary approach should be taken, for example, by enacting/maintaining limitations on the number of permits issued, proximity and number of vessels per dusky dolphin group, and mandatory rest periods (Dans et al. 2008; Childerhouse and Baxter 2010; Lundquist et al. 2013). New Zealand is at the forefront of such mandated activities, and scientists/conservationists are grateful for this, as surely may be the dolphins (Würsig et al. 2007; Department of Conservation 2009; Childerhouse and Baxter 2010).

18.6 Future Research Directions

New technology holds promise for advancing the state of knowledge of dusky dolphin behavior. It is challenging to collect data at the level of the individual, given the large and fluid groups formed by dusky dolphins. However, advancements in bio-logger technology are facilitating collection of fine-scale data on individual behavior. For example, the recent development of a noninvasive animal-borne camera suitable for small (<2 m) cetaceans has provided new glimpses of how dusky dolphins interact and care for their young (Video S2, Pearson et al. 2017a). Continued use of noninvasive bio-loggers will also deepen understanding of foraging and mating strategies. The use of unmanned aerial systems holds great promise for obtaining fine-scale data on group size and spatial positioning and fundamental data challenging to obtain via other methods yet critical to understanding the mechanisms of grouping behavior (as used for dusky dolphins, Weir et al. 2018).

In societies that lack paternal care, female distribution is driven by food availability and predation risk, while male distribution is driven by the distribution of females (Wrangham 1979; Krebs and Davies 1993). Thus, in order to fully understand selective pressures that have shaped dusky dolphin society, future research should focus on unraveling factors driving female distribution. Some headway has been made to understand distribution of mother-calf pairs in large groups (Fanucci-Kiss 2015). However, a rich area of future research is to determine if/how females move between large mixed groups, mating groups, and nursery groups off Kaikoura. Importantly, understanding factors driving female grouping patterns will also help to understand the evolution of large brains (Lindefors 2005).

Additional information is needed on individuals (e.g., how many, how often, age-sex class) that travel between Kaikoura and Admiralty Bay and how this is linked to foraging and social strategies. These data will help to determine if this is a cultural tradition (Whitehead et al. 2004) as speculated, but not confirmed. Photo-identification and focal follow data may be used to answer this and the aforementioned question regarding female distribution.

While much knowledge is gained by long-term studies in New Zealand and Patagonia, more information on lesser-known dusky dolphin populations (e.g., off Peru and South Africa) is warranted. Much has been learned about life history, reproduction, and feeding biology from stranded or bycaught animals (Best and Meÿer 2010), but dedicated behavioral studies are needed. Distinct ecological attributes of Península Valdés, Kaikoura, and Admiralty Bay have shaped dusky dolphin society in each region. Exciting discoveries will be made through behavioral studies of dusky dolphins in other parts of their range.

Acknowledgments I thank Bernd and Mel Würsig for introducing me to dusky dolphins and for their generous support and hospitality during graduate studies and field research. I also thank Alaska NASA EPSCoR, Earthwatch Institute, Encounter Foundation, Herchel Smith-Harvard Undergraduate Science Research Program, Marlborough District Council, National Geographic Society/Waitt Fund Grant, New Zealand Department of Conservation, Texas A&M University, Texas A&M University at Galveston (TAMUG), Tom Slick Graduate Fellowship, University of Alaska Southeast (UAS), and University of Sydney for funding. I am also grateful for the multitude of research assistants and students at TAMUG and UAS who assisted with data collection and analysis over the years.

References

Aquaculture New Zealand (2018) https://www.aquaculture.org.nz/industry/. Accessed 25 June 2018
Benoit-Bird KJ, Au WWL (2009) Cooperative prey herding by the pelagic dolphin, *Stenella longirostris*. J Acoust Soc Am 125:125–137
Benoit-Bird KJ, Würsig B, McFadden CJ (2004) Dusky dolphin (*Lagenorhynchus obscurus*) foraging in two different habitats: active acoustic detection of dolphins and their prey. Mar Mamm Sci 20:215–231
Benoit-Bird KJ, Dahood AD, Würsig B (2009) Using active acoustics to compare lunar effects on predator–prey behavior in two marine mammal species. Mar Ecol Prog Ser 395:119–135
Best PB, Meÿer MA (2010) Neglected but not forgotten-Southern Africa's dusky dolphins. In: Würsig B, Würsig M (eds) The dusky dolphin: master acrobat off different shores. Academic Press, San Diego, pp 291–312
Brown JS, Laundre JW, Gurung M (1999) The ecology of fear: optimal foraging, game theory, and trophic interactions. J Mammal 80:385–399
Buurman D (2010) Dolphin swimming and watching: one tourism operator's perspective. In: Würsig B, Würsig M (eds) The dusky dolphin: master acrobat off different shores. Academic Press, San Diego, pp 277–290
Byrne RW, Whiten A (eds) (1989) Machiavellian intelligence. Social expertise and the evolution of intellect in monkeys, apes, and humans. Clarendon, Oxford
Cavalli-Sforza LL, Feldman MW, Chen KH, Dornbusch SM (1982) Theory and observation in cultural transmission. Science 218:19–27

Childerhouse S, Baxter A (2010) Human interactions with dusky dolphins: a management perspective. In: Würsig B, Würsig M (eds) The dusky dolphin: master acrobat off different shores. Academic Press, San Diego, pp 245–276

Cipriano FW (1992) Behavior and occurrence patterns, feeding ecology, and life history of dusky dolphins (*Lagenorhynchus obscurus*) off Kaikoura, New Zealand. Dissertation, University of Arizona

Cipriano F, Webber M (2010) Dusky dolphin life history and demography In: Würsig B, Würsig M (eds) The dusky dolphin: master acrobat off different shores. Academic Press, San Diego, p. 21–48

Clutton-Brock TH, Harvey PH (1980) Primates, brains and ecology. J Zool 190:309–323

Crespi Abril AC, García NA, Crespo EA, Coscarella MA (2003) Consumption of marine mammals by broadnose sevengill shark *Notorynchus cepedianus* in the northern and central Patagonian shelf. Lat Am J Aquat Mamm 2:101–107

Dahood AD, Benoit-Bird KJ (2010) Dusky dolphins foraging at night. In: Würsig B, Würsig M (eds) The dusky dolphin: master acrobat off different shores. Academic Press, San Diego, pp 99–114

Dans SL, Crespo EA, Pedraza SN, Alonso MK (1997) Notes on the reproductive biology of female dusky dolphins (*Lagenorhynchus obscurus*) off the Patagonia coast. Mar Mamm Sci 13:303–307

Dans SL, Alonso MK, Pedraza SN, Crespo EA (2003) Incidental catch of dolphins in trawling fisheries off Patagonia, Argentina: can populations persist? Ecol Appl 13:754–762

Dans SL, Crespo EA, Pedraza SN, Degrati M, Garaffo GV (2008) Dusky dolphin and tourist interaction: effect on diurnal feeding behavior. Mar Ecol Prog Ser 369:287–296

Degrati M, Dans SL, Pedraza SN, Crespo EA, Garaffo GV (2008) Diurnal behavior of dusky dolphins, *Lagenorhynchus obscurus*, in Golfo Nuevo, Argentina. J Mammal 89:1241–1247

Degrati M, Dans SL, Garaffo GV, Crespo EA (2012) Diving for food: a switch of foraging strategies of dusky dolphins in Argentina. J Ethol 30:361–367

Degrati M, Dans SL, Garaffo GV, Cabreira AG, Castro Machado F, Crespo EA (2013) Sequential foraging of dusky dolphins with an inspection of their prey distribution. Mar Mamm Sci 29:691–704

Degrati M, Coscarella MA, Crespo EA, Dans SL (2018) Dusky dolphin group dynamics and association patterns in Península Valdés, Argentina. Mar Mamm Sci. doi: https://doi.org/10.1111/mms.12536

Dehn MM (1990) Vigilance for predators: detection and dilution effects. Behav Ecol Sociobiol 26:337–342

Department of Conservation (2009) New measures for management of commercial dolphin watching off Kaikoura. Available from https://www.doc.govt.nz/news/media-releases/2009/new-measures-for-management-of-commercial-dolphin-watching-off-kaikoura/. Accessed 8 Oct 2018

Dugatkin LA, Mesterton-Gibbons M, Houston AI (1992) Beyond the prisoner's dilemma: toward models to discriminate among mechanisms of cooperation in nature. Trends Ecol Evol 7:202–205

Dunbar RIM (1998) The social brain hypothesis. Evol Anthropol 6:178–190

Dunbar RM (2003) The social brain: mind, language, and society in evolutionary perspective. Annu Rev Anthropol 32:163–181

Dunbar RIM, Shultz S (2017) Why are there so many explanations for primate brain evolution? Philos Trans R Soc Lond Ser B Biol Sci 372:20160244

Elliser DR, Herzing DL (2014a) Long-term social structure of a resident community of Atlantic spotted dolphins, *Stenella frontalis*, in the Bahamas 1991–2002. Mar Mamm Sci 30:308–328

Elliser DR, Herzing DL (2014b) Social structure of Atlantic spotted dolphins, *Stenella frontalis*, following environmental disturbance and demographic changes. Mar Mamm Sci 30:329–347

Fanucci-Kiss A (2015) Mother knows best: ecological factors shaping maternal sociality in dusky dolphins (*Lagenorhynchus obscurus*). BA Thesis, Harvard University

Fox KCR, Muthukrishna M, Shultz S (2017) The social and cultural roots of whale and dolphin brains. Nat Ecol Evol 1:1699–1705

Garaffo GV, Dans SL, Pedraza SN, Crespo EA, Degrati M (2007) Habitat use by dusky dolphin in Patagonia: how predictable is their location? Mar Biol 152:165–177

Gazda SK, Connor RC, Edgar RK, Cox F (2005) A division of labour with role specialization in group-hunting bottlenose dolphins (*Tursiops truncatus*) off Cedar Key, Florida. Proc R Soc Lond B Biol Sci 272:135–140

Gowans S, Würsig B, Karczmarski L (2008) The social structure and strategies of delphinids: predictions based on an ecological framework. Adv Mar Biol 53:195–294

Herzing DL (1996) Vocalizations and associated underwater behavior of free-ranging Atlantic spotted dolphins, *Stenella frontalis* and bottlenose dolphins, *Tursiops truncatus*. Aquat Mamm 22:61–79

Humphrey NK (1976) The social function of intellect. In: Batespn PPG, Hinde RA (eds) Growing points in ethology. Cambridge University Press, Cambridge, pp 303–317

IUCN (2018) The IUCN Red List of threatened species. Version 2017–3. www.iucnredlist.org Accessed 01 July 2018

Jerison HJ (1973) Evolution of the brain and intelligence. Academic Press, New York

Jolly A (1966) Lemur social behavior and primate intelligence. Science 153:501–506

Karczmarski L, Würsig B, Gailey G, Larson KW, Vanderlip C (2005) Spinner dolphins in a remote Hawaiian atoll: social grouping and population structure. Behav Ecol 16:675–685

Koen-Alonso M, Crespo AD, Garcia NA, Pedraza SN, Coscarella MA (1988) Diet of dusky dolphins, *Lagenorhynchus obscurus*, in waters off Patagonia, Argentina. Fish Bull 96:366–374

Krebs JR, Davies NB (eds) (1993) An introduction to behavioral ecology. Blackwell Scientific, Oxford

Lewis KB, Barnes PM (1999) Kaikoura Canyon, New Zealand: active conduit from near-shore sediment zones to trench-axis channel. Mar Geol 162:39–69

Lindefors P (2005) Neocortex evolution in primates: the 'social brain' is for females. Biol Lett 1:407–410

Loizaga De Castro R, Saporiti F, Vales DG, García NA, Cardona L, Crespo EA (2016) Feeding ecology of dusky dolphins *Lagenorhynchus obscurus*: evidence from stable isotopes. J Mammal 97:310–320

Lundquist D, Gemmell NJ, Würsig B (2012) Behavioral responses of dusky dolphin groups (*Lagenorhynchus obscurus*) to tour vessels off Kaikoura, New Zealand. PLoS One 7:e41969

Lundquist D, Gemmell NJ, Würsig B, Markowitz T (2013) Dusky dolphins movement patterns: short-term effects of tourism. NZ J Mar Freshw Res 47:430–449

Lusseau SM, Wing SR (2006) Importance of local production versus pelagic subsidies in the diet of an isolated population of bottlenose dolphins *Tursiops* sp. Mar Ecol Prog Ser 321:283–293

Lusseau D, Schneider K, Boisseau OJ, Haase P, Slooten E, Dawson SM (2003) The bottlenose dolphin community of doubtful sound features a large proportion of long-lasting associations. Behav Ecol Sociobiol 54:396–405

Mangel JC, Alfaro-Shigueto J, Van Waerebeek K, Cáceres C, Bearhop S, Witt MJ, Godley BJ (2010) Small cetacean captures in Peruvian artisanal fisheries: high despite protective legislation. Biol Conserv 143:136–143

Mangel JC, Alfaro-Shigueto J, Witt MJ, Hodgson DJ, Godley BJ (2013) Using pingers to reduce bycatch of small cetaceans in Peru's small-scale driftnet fishery. Oryx 47:596–606

Marino L, McShea DW, Uhen MD (2004) Origin and evolution of large brains in toothed whales. Anat Rec A Discov Mol Cell Evol Biol 281A:1247–1255

Markowitz TM (2004) Social organization of the New Zealand dusky dolphin. Dissertation, Texas A&M University

Markowitz TM, Harlin AD, Würsig B, McFadden CJ (2004) Dusky dolphin foraging habitat: overlap with aquaculture in New Zealand. Aquat Conserv 14:133–149

Markowitz TM, Markowitz WJ, Morton LM (2010a) Mating habits of New Zealand dusky dolphins. In: Würsig B, Würsig M (eds) The dusky dolphin: master acrobat off different shores. Academic Press, San Diego, pp 151–176

Markowitz TM, Dans SL, Crespo EA, Lundquist DL, Duprey NMT (2010b) Human interactions with dusky dolphins: harvest, fisheries, habitat alteration, and tourism. In: Würsig B, Würsig M (eds) The dusky dolphin: master acrobat off different shores. Academic Press, San Diego, pp 211–244

McFadden CJ (2003) Behavioral flexibility of feeding dusky dolphins (*Lagenorhynchus obscurus*) in Admiralty Bay, New Zealand. MS thesis, Texas A&M University

Melillo-Sweeting K, Turnbull SD, Guttridge TL (2014) Evidence of shark attacks on Atlantic spotted dolphins (*Stenella frontalis*) off Bimini, The Bahamas. Mar Mamm Sci 30:1158–1164

Norris KS, Dohl TP (1980) The structure and functions of cetacean schools. In: Hermann LM (ed) Cetacean behavior: mechanisms and functions. Wiley Interscience, New York, pp 211–261

Norris KS, Schilt CR (1988) Cooperative societies in three-dimensional space: on the origins of aggregations, flocks, and schools, with special reference to dolphins and fish. Ethol Sociobiol 9:149–179

O'Connor S (2009) Whale watching worldwide. Tourism numbers, expenditures and expanding economic benefits. A special report from IFAW – the international fund for animal welfare. Yarmouth, MA, prepared by Economists at Large

Orbach DM, Packard JM, Würsig B (2014) Mating group size in dusky dolphins (*Lagenorhynchus obscurus*): costs and benefits of scramble competition. Ethology 120:804–815

Orbach DM, Rosenthal GG, Würsig B (2015) Copulation rate declines with mating group size in dusky dolphins (*Lagenorhynchus obscurus*). Can J Zool 93:503–507

Orbach DN, Pearson HC, Beier-Engelhaupt A, Deutsch S, Srinivasan M, Weir JS, Yin S, Würsig B (2018) Long-term assessment of spatio-temporal association patterns of dusky dolphins (*Lagenorhynchus obscurus*) off Kaikoura, New Zealand. Aquat Mamm 44(6):608–619

Otali E, Gilchrist JS (2006) Why chimpanzee (*Pan troglodytes schweinfurthii*) mothers are less gregarious than nonmothers and males: the infant safety hypothesis. Behav Ecol Sociobiol 59:561–570

Pearson HC (2008) Fission-fusion sociality in dusky dolphins (Lagenorhynchus obscurus), with comparisons to other dolphins and great apes, Dissertation. Texas A&M University, College Station, TX

Pearson HC (2009) Influences on dusky dolphin fission-fusion dynamics in Admiralty Bay, New Zealand. Behav Ecol Sociobiol 63:1437–1446

Pearson HC (2017) Unraveling the function of dolphin leaps using the dusky dolphin (*Lagenorhynchus obscurus*) as a model species. Behaviour 154:563–581

Pearson HC, Shelton DE (2010) A large-brained social animal. In: Würsig B, Würsig M (eds) The dusky dolphin: master acrobat off different shores. Academic Press, San Diego, pp 333–354

Pearson HC, Vaughn-Hirschorn RL, Srinivasan M, Würsig B (2012) Avoidance of mussel farms by dusky dolphins (*Lagenorhynchus obscurus*) in New Zealand. NZ J Mar Freshw Res 46:567–574

Pearson HC, Jones P, Srinivasan M, Lundquist D, Pearson CJ, Machovsky-Capuska GE (2017a) Testing and deployment of C-VISS (Cetacean-borne Video camera and Integrated Sensor System) on wild dolphins. Mar Biol 164:42

Pearson HC, Markowitz TM, Weir JS, Würsig B (2017b) Dusky dolphin (*Lagenorhynchus obscurus*) social structure characterized by social fluidity and preferred companions. Mar Mamm Sci 33:251–276

Pirotta E, New L, Harwood J, Lusseau D (2014) Activities, motivations and disturbance: an agent-based model of bottlenose dolphin behavioral dynamics and interactions with tourism in Doubtful Sound, New Zealand. Ecol Model 282:44–58

Piwetz S (2018) Effects of human activities on coastal dolphin behavior. Dissertation, Texas A&M University

Reeves RR, Stewart BS, Clapham PJ, Powell JA (eds) (2002) Guide to marine mammals of the world. AA Knopf, New York

Reeves RR, McClellan K, Werner TB (2013) Marine mammal bycatch in gillnet and other entangling net fisheries, 1990 to 2011. Endanger Species Res 20:71–97

Ribeirio S, Viddi FA, Cordeiro JL, Freitas TRO (2007) Fine-scale habitat selection of Chilean dolphins (*Cephalorhynchus eutropia*): interactions with aquaculture activities in southern Chiloé Island, Chile. J Mar Biol Assoc UK 87:119–128

Romero MA, Dans SL, García N, Svendsen GM, González R, Crespo EA (2012) Feeding habits of two sympatric dolphin species off North Patagonia, Argentina. Mar Mamm Sci 28:364–377

Shelton DE, Harlin-Cognato AD, Honeycutt RL, Markowitz TM (2010) In: Würsig B, Würsig M (eds) The dusky dolphin: master acrobat off different shores. Academic Press, San Diego, pp 195–210

Sol D, Duncan RP, Blackburn TM, Cassey P, Lefebvre L (2005) Big brains, enhanced cognition, and response of birds to novel environments. Proc Natl Acad Sci USA 102:5460–5465

Srinivasan M, Markowitz TM (2010) Predator threats and dusky dolphin survival strategies. In: Würsig B, Würsig M (eds) The dusky dolphin: master acrobat off different shores. Academic Press, San Diego, pp 133–150

Srinivasan M, Pearson HC, Vaughn-Hirshorn RL, Würsig B, Murtugudde R (2012) Using climate downscaling to hypothesize impacts on apex predator marine ecosystem dynamics. N Z J Mar Freshw Res 46:575–584

Srinivasan M, Swannack TM, Grant WE, Rajan J, Würsig B (2018) To feed or not to feed? Bioenergetic impacts of fear-driven behaviors in lactating dolphins. Ecol Evol 8:1384–1398

van Schaik C (2006) Why are some animals so smart? Sci Am 294(4):64–71

Van Waerebeek K, Read AJ (1994) Reproduction of dusky dolphins, *Lagenorhynchus obscurus*, from coastal Peru. J Mammal 75:1054–1062

Van Waerebeek K, Reyes JC (1994) Post-ban small cetacean takes off Peru: a review. Rep Int Whal Comm 15:503–520

Van Waerebeek K, Würsig B (2018) Dusky dolphin, *Lagenorhynchus obscurus*. In: Würsig B, Thewissen JGM, Kovacs KM (eds) Encyclopedia of marine mammals, 3rd edn. Academic Press, San Diego, pp 277–281

Vaughn RL, Shelton DE, Timm LL, Watson LA, Würsig B (2007) Dusky dolphin (*Lagenorhynchus obscurus*) feeding tactics and multi-species associations. NZ J Mar Freshw Res 41:391–400

Vaughn RL, Würsig B, Shelton DS, Timm LL, Watson LA (2008) Dusky dolphins influence prey accessibility for seabirds in Admiralty Bay, New Zealand. J Mammal 89:1051–1058

Vaughn R, Würsig B, Packard J (2010a) Dolphin prey herding: prey ball mobility relative to dolphin group and prey ball sizes, multispecies associates, and feeding duration. Mar Mamm Sci 26:213–225

Vaughn RL, Degrati M, McFadden CJ (2010b) In: Würsig B, Würsig M (eds) The dusky dolphin: master acrobat off different shores. Academic Press, San Diego, pp 115–132

Vaughn-Hirshorn RL, Muzi E, Richardson JL, Fox GJ, Hansen LN, Salley AM, Dudzinski KM, Würsig B (2013) Dolphin underwater bait-balling behaviors in relation to group and prey ball sizes. Behav Process 98:1–8

Watson-Capps JJ, Mann J (2005) Effects of aquaculture on bottlenose dolphin (*Tursiops* sp.) ranging in Shark Bay, Western Australia. Biol Conserv 124:519–526

Weir JS (2008) Dusky dolphin nursery groups off Kaikoura, New Zealand. MS thesis, Texas A&M University

Weir JS, Duprey NMT, Würsig B (2008) Dusky dolphin (*Lagenorhynchus obscurus*) subgroup distribution: are shallow waters a refuge for nursery groups? Can J Zool 86:1225–1234

Weir J, Deutsch S, Pearson HC (2010) Dusky dolphin calf rearing. In: Würsig B, Würsig M (eds) The dusky dolphin: master acrobat off different shores. Academic Press, San Diego, pp 177–194

Weir JS, Fiori L, Orbach DN, Piwetz S, Protheroe C, Würsig B (2018) Dusky dolphin (*Lagenorhynchus obscurus*) mother-calf pairs: an aerial perspective. Aquat Mamm 44:603–607

Whitehead H, Rendell L, Osborne RW, Würsig B (2004) Culture and conservation of non-humans with reference to whales and dolphins: review and new directions. Biol Conserv 120:427–437

Williams JA, Dawson SM, Slooten E (1993) Abundance and distribution of bottlenose dolphins (*Tursiops truncatus*) in Doubtful Sound, New Zealand. Can J Zool 71:2080–2088

18 Dusky Dolphins of Continental Shelves and Deep Canyons

Wrangham RW (1979) On the evolution of ape social systems. Soc Sci Inf 18:335–368

Würsig B (1979) Dolphins. Sci Am 240:136–148

Würsig B (1986) Delphinid foraging strategies. In: Schusterman RJ, Thomas JA, Wood FG (eds) Dolphin cognition and behavior: a comparative approach. Lawrence Erlbaum Associates, Hillsdale, pp 347–359

Würsig B (2010) Social creatures in a changing sea: concluding remarks. In: Würsig B, Würsig M (eds) The dusky dolphin: master acrobat off different shores. Academic Press, San Diego, pp 355–358

Würsig B (2018) Intelligence. In: Würsig B, Thewissen JGM, Kovacs KM (eds) Encyclopedia of marine mammals, 3rd edn. Academic Press, San Diego, pp 512–517

Würsig B, Bastida R (1986) Long-range movement and individual associations of two dusky dolphins (*Lagenorhynchus obscurus*) off Argentina. J Mammal 67:773–774

Würsig B, Pearson HC (2014) Dusky dolphins: flexibility in foraging and social strategies. In: Yamagiwa J, Karczmarski L (eds) Primates and cetaceans: field research and conservation of complex mammalian societies. Springer, New York, pp 25–42

Würsig B, Würsig M (1980) Behavior and ecology of the dusky dolphin, *Lagenorhynchus obscurus*, in the South Atlantic. Fish Bull 77:871–890

Würsig B, Würsig M (eds) (2010) The dusky dolphin: master acrobat off different shores. Academic Press, San Diego

Würsig B, Würsig M, Cipriano F (1989) Dusky dolphins in different worlds. Oceanus 32:71–75

Würsig B, Cipriano F, Slooten E, Constantine R, Barr K, Yin S (1997) Dusky dolphins (*Lagenorhynchus obscurus*) off New Zealand: status of present knowledge. Rep Int Whal Commn 47:715–722

Würsig B, Reeves RR, Ortega-Ortiz JG (2002) Global climate change and marine mammals. In: Evans PGH, Raga JA (eds) Marine mammals. Springer, Boston, pp 589–608

Würsig B, Duprey N, Weir J (2007) Dusky dolphins (*Lagenorhynchus obscurus*) in New Zealand waters: present knowledge and research goals. DOC Res Dev Ser 270:1–28

Chapter 19
Cetacean Sociality in Rivers, Lagoons, and Estuaries

Dipani Sutaria, Nachiket Kelkar, Claryana Araújo-Wang, and Marcos Santos

Abstract Cetaceans of rivers, lagoons, and estuaries are isolated or partially isolated from marine species and have evolved in habitats with typically low predation pressure, but particularly high fluctuating physical and biological environments. In this chapter we explore if life in these generally murky, restricted habitats has affected the evolution of grouping and sociality in seven cetacean taxa: *Platanista*, *Pontoporia*, *Inia*, *Lipotes*, *Neophocaena*, *Orcaella*, and *Sotalia*. We suggest that there is a gradual increase in social complexity and communication as we move from the obligate or true river dolphins to the more facultative estuarine/brackish-water species.

Keywords Dolphins · Rivers · Lagoons · Estuaries · Sociality · Behavior · South Asia · South America · Southeast Asia · *Platanista* · *Pontoporia* · *Inia* · *Lipotes* · *Neophocaena* · *Orcaella* · *Sotalia*

D. Sutaria (✉)
College of Science and Engineering, James Cook University, Townsville, QLD, Australia
e-mail: Dipani.Sutaria@jcu.edu.au

N. Kelkar
Ashoka Trust for Research in Ecology and the Environment (ATREE), Bangalore, Karnataka, India

C. Araújo-Wang
Botos do Cerrado – Pesquisas Ambientais, Goiânia, Brazil

CetAsia Research Group, Thornhill, ON, Canada

M. Santos
Departamento de Oceanografia Biológica, Laboratório de Biologia da Conservação de Mamíferos Aquáticos, Instituto Oceanográfico, Universidade de São Paulo, São Paulo, SP, Brazil

© Springer Nature Switzerland AG 2019
B. Würsig (ed.), *Ethology and Behavioral Ecology of Odontocetes*, Ethology and Behavioral Ecology of Marine Mammals,
https://doi.org/10.1007/978-3-030-16663-2_19

414   D. Sutaria et al.

19.1 Introduction

Sociality is a complex outcome of interactions between evolutionary traits and more proximate biological (life history) and environmental factors. Cetaceans of rivers or lagoons are isolated or partially isolated from sympatric marine species. River and lagoon cetacean species and populations have evolved in environments character-ized by spatial limitation, fluctuating prey availability, often a lack of predation pressure, and seasonally dynamic physical and biological environments. In rivers, the linear nature of habitats, patchiness of food resources at confluences and in deep pools, and highly variable water depth are three key factors that could influence movement, grouping, and social behavior. To understand if these differences influ-ence sociality, we review available information from seven cetacean taxa (*Platanista*, *Pontoporia*, *Inia*, *Lipotes*, *Neophocaena*, *Orcaella*, and *Sotalia*) across parts of South America, South Asia, and Southeast Asia.

Fox et al. (2017) offered that encephalization (a brain to body weight ratio) in dolphins is a correlate of sociality and grouping behavior. The diversity of commu-nication signals might also promote sociality under specific conditions, such as the importance of communication while under predation risk (Morisaka and Connor 2007). We review small-bodied cetaceans (adult body size usually less than 2.5 m), reaching sexual maturity between 7 and 10 years and calving about once every 3 years (Taylor et al. 2007). Natural changes in water area, seasonal flows, natural flood pulsing, and prey distribution across dry and wet seasons are common features of their habitats. They generally appear to have few natural predators (Hoy 1923; Pilleri 1972; Stacey and Arnold 1999), although there may be unrecognized preda-tion by freshwater sharks, caiman, etc. on dolphins that venture into salt water, i.e., Irrawaddy dolphins, Guiana dolphins (Khan et al. 2011; Santos et al. 2009), and franciscana in the coastal waters of South America (Di Beneditto 2004) that are more susceptible to predation pressure. We explore grouping, sociality, and sociosexual behaviors across these seven taxa in ascending order of their currently accepted ages of divergence, i.e., from most basal to most recent (based on Yan et al. 2005; Costeur et al. 2018).

19.1.1 Platanista

Platanista is one of the most basal odontocetes and includes the Ganges River dolphin or Susu *P. g. gangetica* and the Indus River dolphin or Bhulan *P. g. minor* distributed across the Indus-Ganga-Brahmaputra (IGB) rivers and their trib-utaries. The species as a whole is listed as "Endangered" on the IUCN Red List (Braulik and Smith 2017), but despite numerous dams and barrages fragmenting river flows across their ranges, populations have persisted at high densities in some reaches (e.g., 1600–1800 Indus dolphins in 700 km of the Indus; Braulik et al. 2015). Ganges dolphin populations are estimated at 3000–3500 in about 5000 river kms of

19 Cetacean Sociality in Rivers, Lagoons, and Estuaries 415

the Ganga Brahmaputra basin (Sinha and Kannan 2014). The taxonomy of *Platanista* is in flux, and we expect that the two subspecies may be separated into two species in future.

Given the murky, sediment-rich rivers they evolved in, *Platanista* are "almost blind" due to evolutionary regression of the optic nerve and eye lens (Herald 1969; Waller 1983). There are no confirmed records of predation on *Platanista*. It has also not been possible to observe *Platanista* underwater in the wild, the primary reason for their poorly known social behavior. Observations on captive Indus and Ganges dolphins kept in aquaria in the USA, Switzerland, and Japan in the late 1960s and 1970s (Herald 1969; Pilleri et al. 1971a, b; Pilleri 1974) showed that there was frequent body rubbing, stroking and caressing, chasing, biting (near dorsal hump, pectoral fins, and tail flukes), and rostrum-body contact, especially in young and subadult animals (Pilleri et al. 1971a, 1980; Gihr et al. 1976). Males, whose ranges were relatively larger than young females, also displayed more affiliative behaviors. Sexually excited behaviors (chasing, male biting female's rostrum) occurred at the start of April (in Switzerland: Pilleri et al. 1971a). Pilleri et al. (1980) documented basic play behaviors in captive animals.

In their natural habitat, *Platanista* are generally solitary (Reeves and Brownell 1989), except for mother-calf associations, mating pairs, and nonsocial feeding aggregations (up to 30–40 animals) in deep pools with high prey availability (Smith and Reeves 2012; Nowak 2003; Moreno 2004; Lal Mohan and Kelkar 2015). Sexual dimorphism is limited to body size and shape, adult females being slightly longer and with more curved rostra than adult males. Most descriptions of mating wild *Platanista* are from the peak dry season and pre-monsoon (March–June; Pilleri 1972; Sinha and Kannan 2014; Lal Mohan and Kelkar 2015), extending to August in the Brahmaputra River (Kasuya and Haque 1972; Haque et al. 1977). Pilleri (1972) suggested that a female might mate with up to three males, but it was not clarified how males were identified. In river areas with high dolphin densities, mating aggregations form—and *Platanista* are then more acrobatic (breaching high, tail slapping) than usual. Kelkar and others observed a mating pair spin out of water and twist screw-like in midair, jointly somersaulting, while surrounded by 5–6 dolphins. Lal Mohan and Kelkar (2015) suggested a "lek-like" mating system in *Platanista*, but perhaps only at high dolphin densities (Haque et al. 1977; Lal Mohan and Kelkar 2015).

Gestation of *Platanista* is about 8–10 months (Kasuya 1972), with calving peaks in February–March in the Ganges River (Sinha and Kannan 2014; Lal Mohan and Kelkar 2015), and May–December in the Indus (Pilleri 1970, 1972) and Brahmaputra Rivers (Haque et al. 1977). Apparently, a female can mate even when her suckling calf has not yet been weaned (Pilleri 1972). The mating season and calving periods are likely tuned to hydrological dynamics and prey availability peaks (rising floods in June–July and receding floods in October) in the Indus-Ganges-Brahmaputra basins. These rivers have peak flow (flooding) from June to September, despite the variable contributions of glacial melt and monsoon precipitation (Zhang et al. 2013a) in a lean period from November to April. Observations from the Ganges (Dey, S.; pers. comm.) suggest that adults might monopolize rich foraging spots such as river confluences and supplant subadults or juveniles. Mother-calf pairs were seen in shallow river pools in the Gandak River (Choudhary et al. 2012), and Lal Mohan and Kelkar (2015) suggested that they were there to capture concentrated prey and perhaps to avoid male harassment. Strong site fidelity in *Platanista* for stable channels (e.g., near towns/embankments) is noted at multiple sites over many years (Lal Mohan and Kelkar 2015; Smith and Reeves 2012).

Acoustic frequencies and sound source levels of Ganges River dolphins from the Ganges in Bhagalpur (Bihar, India; Kelkar et al. 2018) differed significantly from the Bangladesh Sundarbans (Jensen et al. 2013) and from captive Indus dolphins (Pilleri et al. 1977). These differences could result from varying ambient underwater noise levels in artificial, nontidal, or tidal environments or the overlapping ranges of *Platanista* and Irrawaddy dolphins, *Orcaella*, in Bangladesh, unlike elsewhere, as frequencies may have diverged to avoid interference (Morisaka, T.; pers. comm.).

As for all dolphins, *Platanista* clicks have dual function for communication and echolocation during navigation/foraging (Zbinden et al. 1978), but variation in modulation of click trains has not been studied. *Platanista* emit "burst-pulses" or rapidly repeated click trains (Pilleri et al. 1971b, 1977) that could serve communication functions. Burst-pulses were recorded in May 2014 (mating season) in the Ganges at Bhagalpur (Morisaka, T., pers. comm.). Though burst-pulses are at times used by other dolphin species to indicate aggression (Overstrom 1983), their function in *Platanista* is not understood. Peculiar click trains were recorded near mother-calf pairs, but whether they were contact calls needs further study. Jaw-snapping sounds (Andersen and Pilleri 1970) were perhaps related to social signaling.

19.1.2 Pontoporia blainvillei

Toninha or franciscana (also termed La Plata River dolphin), *Pontoporia blainvillei*, dwell in coastal shallow waters from southeastern Brazil to Uruguay close to riverine discharges and riverine systems in Argentina (Cremer and Simões-Lopes 2005; Santos et al. 2009; Bordino et al. 1999). Occurring in usually murky water up to 30 m deep (Pinedo et al. 1989; Praderi et al. 1989), their distribution strongly overlaps with important human fishery areas in Brazil, Uruguay, and Argentina, which exposes them to a high risk of incidental capture (Ott et al. 2002).

Franciscana are sexually dimorphic, with females larger than males (Kasuya and Brownell 1979). They exhibit apparent cooperative foraging of up to 30 dolphins (Bordino et al. 1999) with fish and cephalopod prey associated with the bottom of the water column (Danilewicz et al. 2002). Franciscana tend to associate in small groups of two to three individuals (Bordino et al. 1999; Wells et al. 2013), but records of larger groups exist (Secchi et al. 2001; Di Beneditto and Ramos 2001; Santos et al. 2009). Mating behavior occurs in spring and summer. Franciscana are prey of several shark species, based on stomach content analyses (e.g., Brownell 1975; Praderi 1985; Di Beneditto 2004), as well as of killer whales (*Orcinus orca*, Ott and Danilewicz 1998; Santos and Ferreira Neto 2005), and grouping may be related to predation risk in coastal populations.

Although their social organization remains poorly known, satellite-tracking data of two separate pairs of female-male adults showed that they moved in close proximity for several weeks along the coast of Argentina, possibly supporting the hypothesis of a single-male mating system (Wells et al. 2013), which is rare among marine mammals (Connor et al. 2000; Wells et al. 2013). A relatively small testis size suggests that sperm competition does not occur in this species (Danilewicz et al. 2002). There are no detailed observations on social behavior of franciscana.

19.1.3 Inia geoffrensis

The boto (or Amazon River dolphin), *Inia geoffrensis*, is distributed across six countries (Brazil, Bolivia, Colombia, Ecuador, Peru, Venezuela) and three major river basins (Amazon, Orinoco, Araguaia-Tocantins) (Best and Da Silva 1993). Solitary animals and pairs comprise the majority of sightings throughout their range, although larger group sizes (up to 40 individuals) foraging or socializing may occur, in river confluences, lakes, and river margins (e.g., Vidal et al. 1997; McGuire and Winemiller 1998; Aliaga-Rossel 2002; Martin et al. 2004; Araújo and da Silva 2014; Gómez-Salazár et al. 2012). Groups are predominantly small (up to five individuals) and can vary with water levels across dry-wet seasons (Aliaga-Rossel 2002; da Silva et al. 2010; Gómez-Salazár et al. 2012). In the Central Amazon (Brazil), males and females differ in habitat use, males tend to occupy the main rivers and channels, and females, with their calves, are more commonly found in lakes or low-current habitats such as flooded forest (*várzea*) (Martin and da Silva 2004). The high occurrence of mother-calf pairs and immature botos in floodplain habitats is likely due to prey availability, lower current, and protection from potential aggression by adult males (McGuire and Winemiller 1998; Martin and da Silva 2004; Mintzer et al. 2016). This spatial sexual segregation contributes to differences in home range sizes for both sexes, with males dispersing more than females. In concordance with sexual segregation in habitat use, a high degree of female philopatry has been described by genetic studies of mtDNA of animals from the Amazon, with males being more dispersive (Hollatz et al. 2011). However, in at least one area of the Amazon (Mamirauá Reserve), some animals show a high degree of residency, remaining in one area year-round (Martin and da Silva 2004).

Botos are sexually dimorphic, with males pinker in colour, 16% larger and 55% heavier than females, suggesting a polygynous mating system where males compete for females, so aggressiveness among males and occassionally towards females is expected (Martin et al. 2008). This is supported by the large amounts of scarring found on males due to male-male interactions. During aggressive behavior, botos may produce low-frequency sounds as a possible indication of competitive advantage compared with high-frequency sounds emitted during play and other behavioral contexts (Nunes 2015). Males may also compete for females by displaying

Fig. 19.1 A group of botos, *Inia geoffrensis*, socializing in the Tocantins River, Brazil. Photo by Claryana Araujo-Wang

various objects in a sociosexual context—this was first observed in the Amazon (Martin et al. 2008) and then later reported for botos in the Tocantins River (Araújo and Wang 2012; Santos et al. 2012). During object-carrying behavior, botos raise their heads out of the water while holding or throwing various objects (e.g., branches, stones, leaves, sticks) with their mouths. Botos may also drag the objects as they swim with their heads raised above the surface of the water. Because object-carrying is usually performed by adult males with other females and young males present, it is possible that such behavior is similar to lek-like mating and important for mating success. The mode of transmission of this behavior among individuals is unclear, but due to the presence of younger males in groups of object-carrying males and the social interactions among them, horizontal transmission (i.e., young males learning from interactions with older males) is possible (Fig. 19.1 of socializing botos). Santos et al. (2014) observed a group of six dolphins in the lower Tocantis River displaying sexual behaviours by rubbing against each other and one individual frequently breaching and exposing its penis to the group.

Botos are not known to form long-term social groups. Except during socializing and courtship situations, females do not commonly interact with males and are usually associated only with their offspring (Araújo-Wang, unpublished data). Play-like behaviours such as body rubbing, tail-slap, rolling, and spy-hoping, tossing inanimate objects such as leaves and playing with fish (e.g., small puffer fish) that are not consumed have been observed (e.g., Araújo-Wang 2017; Santos et al. 2014). Females often become pregnant while still lactating, and older calves may remain associated with their mother (although to a lower degree) until they become independent (Martin and da Silva 2018). In the central Brazilian Amazon, a female performed apparent epimeletic (nurturant) behavior toward a dead calf (Martin and da Silva 2018), but whether there were other females in the group is unknown as it was not documented.

Frequent 'aggressive' interactions between males and young calves occur, but young calves do not always appear harmed, and their mothers may not react negatively toward the aggressive males in such instances (Araújo-Wang, unpublished data). Such apparent aggression toward calves could indicate play, bonding or some lower-level social interactions. However, an instance of potential infanticide behaviour was described by Bowler et al. (2018) in the Napo River (Peru), during which an adult male (of a group of six individuals including a mother-neonate) seemed to target a neonate. An adult female was observed to try and keep the neonate safe, while the male showed forceful and apparently aggressive swimming around the neonate.

An unusual behavior of aerial urination by males has been described for botos in the Tocantins River (Araújo and Wang 2012), in which sometimes another male actively positions his beak close to or into the stream of the urine splashing on the surface of the water, probably receiving chemical cues from the urinating male. Such behaviours have not been observed in females, and females have also not been recorded in groups of males involved in such behavior (Araújo-Wang 2017).

Competition for resources in feeding aggregations (for provisioned and unprovisioned botos) is seen mainly between males, where males show aggressive behaviour and even supplant behaviour in order to monopolize food resources (Santos et al. 2014; Alves et al. 2013; Araújo-Wang, unpublished data). Such all-male groups appear to be structured by social dominance hierarchy (Alves et al. 2013). Larger males in a provisioned feeding aggregation at four human-provisioned locations in Amazons State, central Amazon (Brazil) competed, bit, pushed and rammed into others to supplant them. In unprovisioned botos of the Tocantins River, large males have frequently been observed "stealing" fish from smaller male botos (Araújo-Wang, unpublished data). In this situation, the behaviour of "stealing" fish is not necessarily accompanied by the highly aggressive behaviours such as biting, as seen in provisioned dolphin groups. Usually the larger male "waits" in the periphery for other males to catch fish and then swims quickly towards them and steals the fish in mouth. These types of dominant, aggressive and supplanting behaviours have not been reported in wild female groups, or in cases of provisioned dolphins where females are present (Santos et al. 2014). With high degrees of sexual dimorphism, aggression by males for food and the possibility of infanticide to monopolise mating opportunities, and diverse socio-sexual object-carrying displays, botos seem to be the most socio-sexual of the seven species reviewed here.

19.1.4 Lipotes vexillifer

The baiji (forming the monotypic family Lipotidae) once occurred over a great length of the Yangtze River and its large floodplain oxbow lakes such as Dongting (Tung Ting) and Boyang, from the mouth of the river at Shanghai almost up to the far-inland Three Gorges area (Hoy 1923; Yang et al. 2000). But, it has not been reported since 2006 (Turvey et al. 2007) and is considered the first cetacean species to become extinct from anthropogenic impacts (Turvey et al. 2007; Zhou 2009). Due to this sad fact, we rely substantially only on published captive direct and photo/videographed observations of *Lipotes* maintained at the Institute of Hydrobiology in Wuhan, China, 1980–2003. Of six captive *Lipotes*, one male, "Qi Qi," was the last survivor and lived for over 22 years in captivity, and a female, "Zhen Zhen," lived for >2 years (Curry et al. 2013); the rest died within 1 year of being brought in from nature. Wild observations are extremely few but are also summarized here (Zhou et al. 1979, 1998; Liu et al. 1986; Zhou and Li 1989; Chen 1989; Chen et al. 1997; Zhou 2009).

Adaptation for life in dynamic, sediment-rich, glacial-melt, and monsoon-flooding rivers might have triggered similarities in social behaviors in *Lipotes* and *Platanista*, despite their phylogenetic distance, i.e., being "unrelated" taxa. *Lipotes* appeared to be much more dependent on visual cues unlike blind *Platanista*. Liu et al. (1986, 1994) noted "standing behaviors" in captive *Lipotes*, and Zhou et al. (1998) saw "spy-hopping"-like behaviors in captive and wild *Lipotes*, possibly to detect and evade threats at the surface. Interestingly, Yang et al. (2001) observed *Lipotes* resting without much vocal activity in shaded areas of the tank at night (1800–0600 h), possibly to avoid light (resting = long periods of immobility near the water surface).

Group sizes of wild *Lipotes* typically ranged from 2 to 12 dolphins and displayed consistent group cohesion. Some photo-identified baiji individuals made movements ranging from 200 to 350 km within 1 year, indicating that they (when wild population size went below 50) were highly mobile, possibly moving all the time in response to heavy boat traffic (and constant underwater noise) and destructive fishing practices (rolling hooks, nets, etc.). One baiji travelled almost 100 km in 3 days, based on successive photo-ID locations (Zhou et al. 1998), and one moved 200 km downstream from 14 May 1989 to 24 April 1990. *Lipotes* also swam with foraging groups of Yangtze finless porpoises *Neophocaena asiaeorientalis asiaeorientalis* (Würsig et al. 2000; Zhang et al. 2003).

Unlike *Platanista*, the acoustic repertoire of *Lipotes* was comprised of clicks, cries, creaks, and whistles (Akamatsu et al. 1998; Wang et al. 1999). Hoy (1923) reported a peculiar "rearing noise" heard at night, audible to the human ear, which was apparently from baiji in Dongting Lake. Peak click frequencies were similar in range to *Platanista*, and whistle patterns resembled *Inia* whistles, but differed from marine delphinid whistles. Whistles also showed variable durations and inflection points that could convey specific "messages," which might be a critical need in the very noisy habitats in which the baiji lived out its final years. The absence of

19 Cetacean Sociality in Rivers, Lagoons, and Estuaries

ultrahigh-frequency hearing only in *Lipotes* among all river dolphins (Costeur et al. 2018) might have ostensibly been a factor behind their higher susceptibility to vessel noise, as compared to the sympatric Yangtze River finless porpoise (*Neophocaena asiaeorientalis asiaeorientalis*).

Gao and Zhou (1992) confirmed sexual dimorphism with females larger than males, at 2.5 m and males at 2.16 m (based on Chen et al. 1984). Sexual behaviors occurred during spring and autumn (Liu 1988), i.e., during the low flow or dry season, similar to *Platanista*. Frequent erection, rubbing of penis on tank fixtures, and touching the female Zhen Zhen with it were observed in the captive male Qi Qi in these seasons (Liu et al. 1994). The gestation period is 10–11 months, with one calf born per female in February–March, about every second year (Chen et al. 1984). Captive studies recorded *Lipotes* to play with rings, balls, tubes, and brushes. Rarely, side swimming, tumbling, rolling, and leaping/surface-gliding behaviors were described by Kejie et al. (1985) and Liu et al. (1994). *Inia* dolphins were more playful than *Lipotes* in captivity (Pilleri et al. 1980; Liu et al. 1994). Baiji were generally slow and lethargic swimmers, except when alarmed (Liu et al. 1994). From activity patterns of captive *Lipotes*, and constant evasive movements due to disturbances to wild individuals, inferring much about social behavior will remain difficult (Zhou et al. 1980, 1998; Chen et al. 1997; Turvey et al. 2007).

19.1.5 Neophocaena asiaeorientalis asiaeorientalis

The taxon *Neophocaena* includes the Indo-Pacific finless porpoise, *N. phocaenoides*, and the narrow-ridged finless porpoise, *N. asiaeorientalis*, both inhabiting lower reaches of estuaries and coastal waters up to about 50 m depth. *N. asiaeorientalis* has a subspecies *N. a. asiaeorientalis* (the focus of this section), limited to the middle and lower reaches of the Yangtze River of China (from Yichang to Shanghai) and its adjoining lakes (Poyang and Dongting lakes). It is endangered (Wang and Reeves 2017), with fewer than 1000 individuals left in the wild, and is isolated from coastal and estuarine populations (*N. a. sunameri*). Zhou et al. (2018) reconstructed the demographic history of finless porpoises—they identified genes associated with renal water homeostasis and the urea cycle, likely evolutionary adaptations associated with surviving in freshwater versus saline conditions. Their results suggest that Yangtze finless porpoises are reproductively isolated from coastal and marine porpoise populations and diverged between 50,000 and 100,000 years ago.

In the wild, Yangtze porpoises occur alone or in groups of 2–20 individuals, with most common group size of 2–3. Groups >20 individuals are rare. Genetic analysis of relatedness (Chen et al. 2017) using microsatellite markers proved the presence of matrilineal grouping in these porpoises, with a sign of inbreeding only in the Tian'e'Zhou ex situ reserve population, which is environmentally compromised. Surveys in Dongting Lake showed that socializing was not a predominant behavior (only 1% of 419 sightings were of socializing groups (Zhang et al. 2013b), with this low level of social behavior likely attributable to extreme environmental alterations leading to food resource depletion. In captive situations and ex situ reserves, socializing and resting behaviors were more frequently observed (Jiang 2000; Xian et al. 2010a, b; Zhang et al. 2013b) compared to the extremely altered wild conditions. Predation on Yangtze finless porpoises has not been documented and is probably absent (Wang et al. 2014).

Three Yangtze finless porpoises had been kept in an aquarium for 3–6 years since 2001 (Wei et al. 2004), and 120 observation sessions of these individuals suggested that sociosexual behavior was common. Sexual behavior displays peaked between April–May and September–October. In another captive study of three male porpoises (two adults and one juvenile) with one adult female (Wu et al. 2010), sociosexual behavior peaked in March–July. The monthly mean frequency of sexual behavior of the three males varied, with the youngest male showing highest sexual behavior frequency every month. When the female was absent, these sexual overtures occurred in male-male pairs. The subordinate adult male showed a different behavior—his sexual peak happened from November to January, unlike the dominant adult male, but when the juvenile male porpoise was still intensely social with him. This difference in sexual peaks may have been due to the effects of social factors on hierarchical rank in the male porpoises. In other captive studies, Xian et al. (2010a, b) observed sociosexual behavior in a porpoise calf.

19.1.6 Orcaella brevirostris

Irrawaddy dolphins *Orcaella brevirostris* inhabit rivers, brackish-water lagoons, and coastal waters in South and Southeastern Asia (Beasley 2008). These populations either moved or settled in rivers and lagoons over time given the volume of available prey and reduction of competitive pressure from other coastal dolphin species such as humpback dolphins (Sutaria 2009); or they inhabited these water bodies during the last glacial maximum and got separated from coastal populations as sea levels fell. We compare sociality in a freshwater population of the Mahakam River in Indonesia with one from Chilika Lagoon in India.

Chilika is a bean-shaped, semi-enclosed, muddy, and shallow water body, where the average water depth is 3 m and the connection to the sea is a narrow channel 12–17 km long. The dolphin population in Chilika is less than 150, with socializing the second-most dominant behavior, after foraging. Apparent cooperative foraging for schooling fish is a common strategy. Social structure analyses of the dolphins in Chilika found five primary clusters of associating individuals with association index ≥ 0.5. Of the 48 individuals sighted more than four times, 14 individuals showed strong associations with one or more other individuals. Group size estimates ranged from 1 to 19, with an average group size of 5 while socializing. Behavioral events in intensely socializing groups included touching in some manner, rubbing, apparent mating attempts and diving on top of and underneath other dolphins, chasing, flipper and fluke slaps, somersaults, and back flips. At times, dolphins moving or feeding together touched, rolled over, and swam on the sides of

19 Cetacean Sociality in Rivers, Lagoons, and Estuaries 423

Fig. 19.2 Series of photos of Irrawaddy dolphins, *Orcaella brevirostris*, of events and group structure in a mating chase. A group of dolphins in a tight structure chased one dolphin who eventually turned ventrum up and was mated with by one or more individuals. After a mating bout, the group would often come together head to head as seen in the bottom right picture. Photos by Dipani Sutaria

another individual, or one of the animals leaped out of the water or spy-hopped (often when a calf or juvenile was present in the group). Sutaria (2009) observed that aggressive behavior between conspecifics was rare, unless it was an intensely socializing group, involving mating chases (Fig. 19.2). Mating chases occurred during morning hours from February to May, just before the onset of the monsoons, by 3–7 animals, chasing 1–2 other individuals (probably female with young). A mating chase lasted until a boat disturbed the group or the chases stopped naturally in 25–30 min. It often included a behavioral event lasting 0:35–0.50 min of one dolphin upside down and several individuals on top of and around it, indicating attempts to mate by more than one male. The chasing group often showed synchronized movements with 3 or more animals in front and 2–3 animals behind, with 1–2 satellite individuals moving with this fast-moving mating group. Between mating bouts, chasing males slowed down, logged on the surface, and then converged in a circular formation with their heads facing inward, either preparing for another chase or showing head-to-head aggression (Fig. 19.2). Dolphins in groups of two regularly displayed sideways, lateral swimming positions with pectoral fins on the surface held perpendicular to the body, facing another individual in ventral to ventral position, by the water's edge (Fig. 19.3).

Fig. 19.3 Social behavior between an adult and juvenile Irrawaddy dolphin, *Orcaella brevirostris*, in Chilika, where sideways swimming along with pectoral flipper touching occurs. Photos by Dipani Sutaria

In the Mahakam River and Delta (Kreb 2004), average group size was 5–6 dolphins with socializing groups of larger pod sizes. Home ranges calculated for 53 photo-identified river dolphins sighted 12 times on average showed that dolphins moved freely along a 61 km stretch (individual ranges = 4–181 km^2) and used an average of 10 km^2 of river area (individual home ranges = 0.3–35.5 km^2). Similar to the population in Chilika, female dolphins in the Mahakam had overall smaller home ranges than male home ranges and two center areas of high dolphin density occurred in the river. Site fidelity was high, with residence indices for females significantly higher than for males in Chilika and in Mahakam. Dolphins showed clear preferences for association with certain individuals and had long-term preferred companionships. In 3.5 years, 30 significant long-term dyads were detected, and the associations between females were stronger than for between males. Associations among sexes were fluid, indicating short- or long-term preferred companionships. Mating events occurred in the Mahakam during July–August, March, and December. Mating events took place between 2 and 3 subgroups with total group sizes ranging between 5 and 12 adult individuals per mating event. These interactions were characterized by vocal and behavioral dominance displays: loud blows, fast swimming, rolling along the axis of the dolphin's body, swimming sideward and with belly up, group swimming in small circles and speeding up (chases), jumps, many fin and fluke waves and slaps, and intensive body contact. Interestingly, water spitting (or squirting) that was often associated with feeding also had been observed three times during socializing, where a spitting dolphin targeted another dolphin's body.

Acoustic studies in Chilika and Mahakam showed that vocalizations included click trains, whistles, and pulse calls, both buzzes and creaks, as for coastal dolphins in Australia and Indonesia (Van Parijs et al. 2000; Kreb 2004; Sutaria et al. 2017). Fundamental frequencies of whistles and contour types from Mahakam, Chilika, and Balikpapan Bay are also similar (Sutaria et al. 2017) and comparable to those described for coastal Irrawaddy dolphins by Hoffman et al. (2017) in Malaysia and by Van Parijs et al. (2000) for Australian snubfin dolphins (*Orcaella heinsohni*).

19.1.7 Sotalia *spp.*

The genus *Sotalia* is composed of two species: the tucuxi (*S. fluviatilis*) and the Guiana dolphin (*S. guianensis*) (Caballero et al. 2007; Cunha et al. 2005). The Guiana dolphin inhabits coastal waters, and tucuxi occur in the major freshwater tributaries of the Amazon and Orinoco River basins (Borobia et al. 1991; da Silva and Best 1996; Edwards and Schnell 2001; Flores and Bazzalo 2004; Flores and Silva 2009; Cunha et al. 2005; Caballero et al. 2007; da Silva et al. 2010; Rosas et al. 2010).

Sotalia is a relatively modern delphinid with a fair bit of retention of older traits, so one would expect it to have more complex behavioral repertoires than its counterparts in this chapter (e.g., Santos and Rosso 2007, 2008). Group size ranges are 1–26 (Santos and Rosso 2008; Santos et al. 2010a), often higher than *Inia* sympatric with tucuxi in many areas, where mean group size was 6 (Gómez-Salazár et al. 2010). Similar to other riverine species, seasonal changes in water levels influence movement of tucuxi, with an increase in length and area of usage during monsoon and post-monsoons in the Central Amazon (Faustino and da Silva 2006; Flores and Bazzalo 2004; Gómez-Salazár et al. 2010). Gómez-Salazár et al. (2010) and Trujillo et al. (pers. comm.) observed a higher frequency of intense socializing and mating in low-water periods when the availability of fish is high, and high site fidelity is also noted in marine and riverine populations (Santos et al. 2001; Azevedo et al. 2004; Rossi-Santos et al. 2007; Lopes et al. 2012). Sexual displays and intense socializing followed cooperative foraging encounters (Santos and Rosso 2008), sometimes including foraging associations with birds (Santos et al. 2010b). Key areas where such behavior occurs are shallow stretches of river close to islands and sandbanks. Group sizes were higher in confluence areas, probably owing to higher density of prey. Predation risk on Guiana dolphin may be significant for grouping behaviors and responses (Santos and Gadig 2009), but risk might be higher for marine than for riverine populations. In Gandoca-Manzanillo (Costa Rica), marine tucuxis and bottlenose dolphins often displayed sociosexual behaviors, but these behaviors have not yet been documented in the freshwater tucuxi population.

19.2 Summary and Discussion

The major limitations to documenting sociality in wild river dolphins are the usually turbid waters, near-inability to differentiate sex, and elusive, difficult-to-see behaviors at the surface. For South Asian river dolphins, narrow-ridged Yangtze finless porpoises, and baiji, behavioral descriptions in captivity could be regarded as the only systematic descriptions of sociosexual behaviors. The intense sociosexual activity described in aquaria, e.g., in Yangtze finless porpoises, could be a result of their unnatural life in captivity.

Movements and group sizes of river dolphins tend to be closely related to water flow dynamics and local prey distribution. Sociality is observed primarily in occasional and loose/variable aggregations, and feeding rather than mating opportunities appear to drive most grouping behaviors. River dolphins appear to largely form only-male or only-female groups, with mixed groups mainly associated with socializing or courtship chases, or in mother-calf associations.

If we compare riverine species in terms of their evolutionary history and complexity of social behavior, some consistent patterns emerge. In Fig. 19.4, we tentatively map the species along four visual axes, suggesting that more obligate and basal river dolphin lineages (toward the left) may have less highly evolved sociality and none or low predation risk, and feed mostly on non-schooling benthic prey. An exception in the older lineages is *Pontoporia*, which however also lives in the coastal open ocean and in which grouping behaviors may be in response to predation risk from sharks and killer whales. Among more recent and facultative river/lagoon phocoenid and delphinid species (toward the right), predation risk and tendency to feed on schooling prey appear relatively higher, and social communication also seems to be more complex, than the "true" river dolphins. The complexity of social communication signaling, e.g., whistle complexity, is positively correlated to social structure (May-Collado et al. 2007). Such complexity is lower in *Orcaella* compared to other coastal delphinids such as *Sotalia* or *Sousa*, but might be greater than for more-basal riverine odontocetes such as *Platanista*. This could corroborate our hypothesis of sociality being a function of evolutionary history—possibly driven by combined effects of low predation risk (Morisaka and Connor 2007; Connor 2007) and constraints on acoustic signaling in shallow estuarine-coastal environments (Jensen et al. 2013).

In the phylogenetically distinct *Platanista* and *Pontoporia*, many similarities may exist through evolutionary convergence, such as in their brain/body mass ratios (the lowest of extant cetaceans, Ridgway et al. 2017) and morphology. Yet, unlike *Platanista*, *Pontoporia* show regular grouping behavior in response to higher predation risk. Other curious behavioral similarities might exist between the unrelated *Platanista* and *Lipotes* that live(d) in riverine habitats driven by similar hydroclimatic forcing effects (e.g., from Himalayan-Tibetan seasonality and climate).

Inia, a more recent river dolphin species, exhibits an advanced degree of sociality and overt sociosexual signaling behaviors. Given that *Inia* still retain many "plesiomorphic" traits and with no documented predation risk, a high level of

19 Cetacean Sociality in Rivers, Lagoons, and Estuaries

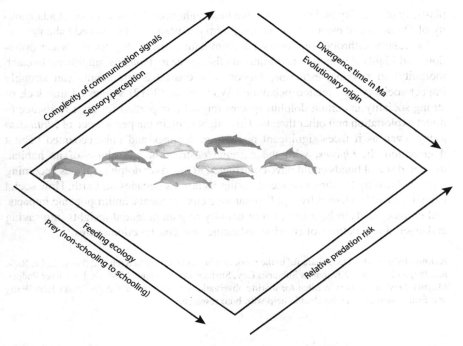

Fig. 19.4 A "map" of the selected species for this chapter, in relation to four axes: (1) evolutionary origin (divergence time: million years), (2) complexity of sensory perception for communication signals, (3) feeding ecology (based on the schooling versus non-schooling prey hypothesis of Connor (2007), and (4) relative predation risk. Species: (from L to R) *Platanista, Lipotes, Inia, Pontoporia, Neophocaena adult and calf, Orcaella, Sotalia*. Illustrations by Uko Gorter, with permission

sociality might be correlated with greater degree of encephalization and habitat-driven selection of visual cue-related behaviors. In contrast, visual limitation in *Platanista* might have constrained the complexity of their social behavior.

In relatively modern species considered as "more recently adapted" to river and lagoon environments than the long-term obligate species, avoidance of resource competition might influence social behaviors, e.g., *Sotalia* sharing habitat with sympatric *Inia* and *Pontoporia*. Such intimate factors, beyond evolutionary effects or selection pressures (Nowak 2003), can translate clear benefits and nontrivial costs of sociality for river and lagoon dolphins. *Orcaella* could be the only dolphin species in some rivers/lagoons (e.g., Chilika, Ayeyarwaddy) that share their range with *Platanista* (in tidal rivers/estuaries, e.g., Hooghly, Sunderbans) or with *Sousa chinensis* and *Neophocaena phocaenoides* (Sunderbans and in coastal habitats, e.g., in Bangladesh, West Bengal, Indonesia, Malaysia, and Thailand). The riverine population of *Orcaella* showed a highly structured society but the population in Chilika showed a more fluid society. This variability in social structure can be explained by differences in productivity, inter-species competition, etc. across habitats. Such plasticity of sociality and grouping behaviors might underlie the ecological adaptability of *Orcaella* and even explain its long-term persistence in threatened habitats.

Increasing anthropogenic pressures from dams, heavy ship traffic, water pollution, and highly destructive and injurious fishing gears (causing significant bycatch mortality) in riverine, estuarine, lagoon, and coastal environments can strongly impact sociality and related behaviors. Wade et al. (2012) suggested that lack of strong sociality in certain dolphin species might be important for their resilience to direct exploitation and other threats. This might explain the persistence of *Platanista* today, even as it faces significant population declines and unprecedented habitat degradation. In *Lipotes*, with the extreme destruction and abuse of its habitat, disrupted social bonds could have contributed to this river dolphin species becoming overwhelmed by the threats it faced during its last six decades on Earth. How social behaviors will be shaped by rapidly changing environments, anthropogenic impacts, and climate needs to be a key area of enquiry to gain potential insights into saving endangered populations of riverine, estuarine, and coastal odontocetes.

Acknowledgments DS is thankful to the work and inputs of Danielle Kreb and thanks Loba, Raja, and Jagga of Chilika. NK thanks Subhasis Dey, Sushant Dey, Tadamichi Morisaka, Oliver Padget, Mayukh Dey, and fishers in Bihar for sharing observations on *Platanista*. CA-W thanks John Wang and Erin Schormans for invaluable help with boto research.

References

Akamatsu T, Wang D, Nakamura K, Wang K (1998) Echolocation range of captive and free-ranging baiji (*Lipotes vexillifer*), finless porpoise (*Neophocaena phocaenoides*), and bottlenose dolphin (*Tursiops truncatus*). J Acoust Soc Am 104:2511–2516

Aliaga-Rossel E (2002) Distribution and abundance of the river dolphin (*Inia geoffrensis*) in the Tijamuchi River, Beni, Bolivia. Aquat Mamm 28(3):312–323

Alves LCPS, Andriolo A, Orams MB, Azevedo AF (2013) Resource defence and dominance hierarchy in the boto (*Inia geoffrensis*) during a provisioning program. Acta Ethol 16:9–19

Andersen S, Pilleri G (1970) Audible sound production in captive *Platanista gangetica*. Invest Cetacea 2:83–86

Araújo CC, da Silva VMF (2014) Spatial distribution of river dolphins, *Inia geoffrensis*, (Iniidae) in the Araguaia River (Central Brazil). Mammalia 78(4):481–486

Araújo CC, Wang JY (2012) Botos (*Inia geoffrensis*) in the Upper Reaches of the Tocantins River (Central Brazil) with observations of unusual behavior, including object carrying. Aquat Mamm 38(4):435–440

Araújo-Wang C (2017) Botos do Cerrado (River dolphins of the Brazilian Savanna). Society of Hong Kong Nature Explorers, 100p. ISBN: 978-988-15468-0-7

Azevedo AF, Lailson-Brito J, Cunha HA, Van Sluys M (2004) A note on site fidelity of marine tucuxis (*Sotalia fluviatilis*) in Guanabara Bay, southeastern Brazil. J Cetacean Res Manag 6:265–268

Beasley I (2008) Conservation of the Irrawaddy dolphin *Orcaella brevirostris* (Owen in Gray 1866) in the Mekong River: biological and social considerations influencing management. School of Earth and Environmental Sciences. James Cook University, Townsville

Best RC, da Silva VMF (1993) *Inia geoffrensis* de Blainville, 1817. Mamm Species 426:1–8

Bordino P, Thompson G, Iniguez M (1999) Ecology and behaviour of the Franciscana dolphin *Pontoporia blainvillei* in Bahía Anegada, Argentina. J Cetacean Res Manag 1(2):213–222

Borobia M, Siciliano S, Lodi L, Hoek W (1991) Distribution of the South-American dolphin *Sotalia fluviatilis*. Can J Zool 69:1025–1039

Bowler MT, Griffiths BM, Gilmore MP, Wingfield A, Recharte M (2018) Potentially infanticidal behavior in the Amazon river dolphin (*Inia geoffrensis*). Acta Ethol 21:141–145. https://doi.org/10.1007/s10211-018-0290-y

Braulik GT, Smith BD (2017) *Platanista gangetica*. The IUCN Red List of Threatened Species 2017: e. T41758A50383612. https://doi.org/10.2305/IUCN.UK.2017-3.RLTS.T41758A50383612.en

Braulik GT, Noureen U, Arshad M, Reeves RR (2015) Review of status, threats, and conservation management options for the endangered Indus river blind dolphin. Biol Conserv 192:30–41

Brownell RL Jr (1975) Progress report on the biology of the Franciscana dolphin, *Pontoporia blainvillei*, in Uruguayan waters. J Fish Res Board Can 32:1073–1078

Caballero S, Trujillo F, Vianna JA, Barrios-Garrido H, Montiel MG, Beltrán-Pedreros S, Marmontel M, Santos MC, Rossi-Santos M, Santos FR, Baker CS (2007) Taxonomic status of the genus *Sotalia*: Species level ranking for "tucuxi" (*Sotalia fluviatilis*) and "costero" (*Sotalia guianensis*) dolphins. Mar Mamm Sci 23(2):358–386

Chen P (1989) Baiji *Lipotes vexillifer* (Miller, 1918). In: Ridgway SH, Harrison RJ (eds) Handbook of marine mammals, vol. 4: River dolphins and the larger toothed whales. Academic Press, London, pp 25–43

Chen PX, Liu R, Kejie L (1984) Reproduction and the reproductive system in the Beiji, *Lipotes vexillifer*. Rep Int Comm Spec Iss 6:445–450

Chen P, Liu R, Wang D, Zhang X (1997) Biology, rearing and conservation of Baiji. Science Press, Beijing

Chen M, Zheng Y, Hao Y, Mei Z, Wang K, Zhao Q et al (2017) Parentage-based group composition and dispersal pattern studies of the Yangtze finless porpoise population in Poyang Lake. Int J Mol Sci 17(8):1268

Choudhary SK, Dey S, Dey S, Sagar V, Nair T, Kelkar N (2012) River dolphin distribution in regulated river systems: implications for dry-season flow regimes in the Gangetic basin. Aquat Conserv Mar Freshwat Ecosyst 22:11–25

Connor RC (2007) Dolphin social intelligence: complex alliance relationships in bottlenose dolphins and a consideration of selective environments for extreme brain size evolution in mammals. Philos Trans R Soc B 362:587–602

Connor RC, Read A, Wrangham R (2000) Male reproductive strategies and social bonds. In: Mann J, Connor RC, Tyack PL, Whitehead H (eds) Cetacean societies: field studies of dolphins and whales. University of Chicago Press, Chicago, IL, pp 247–269

Costeur L, Grohé C, Aguirre-Fernández G, Ekdale E, Schulz G, Müller B, Mennecart B (2018) The bony labyrinth of toothed whales reflects both phylogeny and habitat preferences. Sci Rep 8:78541

Cremer MJ, Simões-Lopes PC (2005) The occurrence of *Pontoporia blainvillei* (Gervais & d'Orbigny) (Cetacea, Pontoporiidae) in an estuarine area in southern Brazil. Rev Bras Zool 22:717–723

Cunha HA, da Silva VMF, Lailson-Brito J Jr, Santos MCO, Flores PAC, Martin AR, Azevedo AF, Fragoso ABL, Zanelatto RC, Solé-Cava AM (2005) Riverine and marine ecotypes of *Sotalia* dolphins are different species. Mar Biol 148:449–457

Curry BE, Ralls K, Brownell RL Jr (2013) Prospects for captive breeding of poorly known small cetacean species. Endanger Species Res 19:223–243

da Silva VMF, Best RC (1996) *Sotalia fluviatilis*. Mamm Species 527:1–7

da Silva VMF, Fettuccia D, Rodrigues ES, Edwards H et al (2010) Report of the working group on distribution, habitat characteristics and preferences, and group size. Lat Am J Aquat Mamm 8 (1–2):31–38

Danilewicz D, Rosas F, Bastida R, Marigo J, Muelbert M, Rodríguez D, Laílson-Brito J, Ruoppolo V, Ramos R, Bassoi M, Ott PH, Caon G, Rocha AM, Catão-Dias JL, Secchi ER (2002) Report of the working group on biology and ecology. Lat Am J Aquat Mamm 1 (1):25–42

Di Beneditto APM (2004) Presence of franciscana dolphin (*Pontoporia blainvillei*) remains in the stomach of a tiger shark (*Galeocerdo cuvieri*) captured in southeastern Brazil. Aquat Mamm 30 (3):311–314

Di Beneditto APM, Ramos RMA (2001) Biology and conservation of the franciscana (*Pontoporia blainvillei*) in the north of Rio de Janeiro State, Brazil. J Cetacean Res Manag 3(2):185–192

Edwards HH, Schnell GD (2001) Status and ecology of *Sotalia fluviatilis* in the Cayos Miskito Reserve, Nicaragua. Mar Mamm Sci 17(3):445–472

Faustino C, da Silva VMF (2006) Seasonal use of Amazon floodplains by the Tucuxi *Sotalia fluviatilis* (Gervais, 1853), in the Central Amazon, Brazil. Lat Am J Aquat Mamm 5(2):95–104

Flores PAC, Bazzalo M (2004) Home ranges and movement patterns of the marine tucuxi dolphin (*Sotalia fluviatilis*) in Baía Norte, Southern Brazil. Lat Am J Aquat Mamm 3:37–52

Flores PAC, Da Silva VMF (2009) Tucuxi and Guiana Dolphin – Sotalia fluviatilis and S. guianensis. In: Perrin WF, Würsig B, Thewissen JGM (eds) Encyclopedia of marine mammals, 2nd edn. Academeic Press, Amsterdam, pp 1188–1191

Fox KCR, Muthukrishna M, Shultz S (2017) The social and cultural roots of whale and dolphin brains. Nat Ecol Evol 1:1699–1705

Gao A, Zhou K (1992) Sexual dimorphism in the baiji, *Lipotes vexillifer*. Can J Zool 70:1484–1493

Gihr M, Kraus C, Pilleri G, Purves PE, Zbinden K (1976) Ethology and bioacoustics of *Platanista indi* in captivity. Invest Cetacea 6:14–69

Gómez-Salazár C, Portocarrero-Aya M, Trujillo F, Caballero S, Bolanos-Jimenez J, Utreras V, McGuire T, Ferrer-Perez A, Pool M, Aliaga-Rossel E (2010) Update on the freshwater distribution of *Sotalia* in Colombia, Ecuador, Peru, Venezuela and Suriname. Lat Am J Aquat Mamm 8(1–2):171–178

Gómez-Salazár C, Trujillo F, Whitehead H (2012) Ecological factors influencing group sizes of river dolphins (*Inia geoffrensis* and *Sotalia fluviatilis*). Mar Mamm Sci 28:E124–E142

Haque AKMA, Nishiwaki M, Kasuya T, Tobayama T (1977) Observations on the behaviour and other biological aspects of the Ganges susu, *Platanista gangetica*. Sci Rep Whales Res Inst 29:87–94

Herald ES (1969) Field and aquarium study of the blind river dolphin *Platanista gangetica*. Steinhart Aquarium, California Academy of Sciences, San Francisco, CA, p 53

Hoffman J, Ponnampalam L, Araújo-Wang C, Kuit SH, Hung SK, Wang JY (2017) Description of whistles of Irrawaddy dolphins (*Orcaella brevirostris*) from the waters of Matang, Peninsular Malaysia. Bioacoustics 26:15–24

Hollatz C, Vilaça ST, Redondo RAF, Marmontel M, Baker CS, Santos FR (2011) The Amazon River system as an ecological barrier driving genetic differentiation of the pink dolphin (*Inia geoffrensis*). Biol J Linn Soc 102:812–827

Hoy CM (1923) The 'white-flag dolphin' of the Tung Ting Lake, China. J Arts Sci 1:154–157

Jensen FH, Rocco A, Mansur RM, Smith BD, Janik VM, Madsen PT (2013) Clicking in shallow rivers: short-range echolocation of Irrawaddy and Ganges river dolphins in a shallow, acoustically complex habitat. PLoS One 8:e59284

Jiang W (2000) Observation on the group of Changjiang finless porpoise conserved in semi-nature conditions. J Anhui Univ (Nat Sci Ed) 24:106–111

Kasuya T (1972) Some information on the growth of the Ganges dolphin with a comment on the Indus dolphin. Sci Rep Whales Res Inst 24:87–108

Kasuya T, Brownell RL (1979) Age determination, reproduction, and growth of the franciscana dolphin, *Pontoporia blainvillei*. Sci Rep Whales Res Inst 31:45–67

Kasuya T, Haque AKMA (1972) Some informations on distribution and seasonal movement of the Ganges dolphin. Sci Rep Whales Res Inst 24:109–115

Kejie L, Pellin L, Peixun C (1985) Observations on the behaviour of *Lipotes* in captivity. Acta Hydrobiol Sin, 01-005. en.cnki.com.cn/Article_en/CJFDTOTAL-SSWX198501005.htm

Kelkar N, Dey S, Deshpande K, Choudhary SK, Dey S, Morisaka T (2018) Foraging and feeding ecology of *Platanista*: an integrative review. Mammal Rev 48(3):194–208

19 Cetacean Sociality in Rivers, Lagoons, and Estuaries

Khan M, Panda S, Pattnaik AK, Guru BC, Kar C, Subudhi M, Samal R (2011) Shark attacks on Irrawaddy dolphin in Chilika Lagoon, India. J Mar Biol Assoc India 53:27–34

Kreb D (2004) Facultative river dolphins: conservation and social ecology of freshwater and coastal Irrawaddy dolphins in Indonesia, PhD dissertation. University of Amsterdam, Amsterdam

Lal Mohan RS, Kelkar N (2015) Ganges river dolphin. In: Johnsingh AJT, Manjrekar N (eds) *Mammals of South Asia*, vol II. Universities Press, Delhi, pp 36–70

Liu R (1988) Study on the regularity of reproduction in *Lipotes*. Aquat Mamm 14(2):63–68

Liu R, Klinowska M, Harrison RJ (1986) The behaviour of *Lipotes vexillifer* and *Neophocaena phocaenoides* in the Changjiang River and in captivity in China. In: Bryden MM, Harrison R (eds) Research on dolphins. University Press, Oxford, pp 433–439

Liu R, Gewalt W, Neurohr B, Winkler A (1994) Comparative studies on the behaviour of *Inia geoffrensis* and *Lipotes vexillifer* in artificial environments. Aquat Mamm 20(1):39–45

Lopes XM, da Silva E, Bassoi M, dos Santos RA, Santos MCO (2012) Feeding habits of Guiana dolphins, *Sotalia guianensis*, from the Brazilian south-eastern: new items and knowledge review. J Mar Biol Assoc UK 8:1723–1733

Martin AR, da Silva VMF (2004) River dolphins and flooded forest: seasonal habitat use and sexual segregation of botos (*Inia geoffrensis*) in an extreme cetacean environment. J Zool 263:295–305

Martin AR, da Silva VMF (2018) Reproductive parameters of the Amazon river dolphin or boto, *Inia geoffrensis* (Cetacea: Iniidae); an evolutionary outlier bucks no trends. Biol J Linn Soc 123:666–676

Martin AR, da Silva VMF, Salmon D (2004) Riverine habitat preferences of botos (*Inia geoffrensis*) and tucuxis (*Sotalia fluviatilis*) in the central Amazon. Mar Mamm Sci 20:189–200

Martin AR, da Silva VMF, Rothery P (2008) Object carrying as a socio-sexual display in an aquatic mammal. Biol Lett 4:243–245

May-Collado LJ, Agnarsson I, Wartzok D (2007) Phylogenetic review of tonal sound production in whales in relation to sociality. BMC Evol Biol 7(1):136–156

McGuire TL, Winemiller KO (1998) Occurrence patterns, habitat associations, and potential prey of the river dolphin, *Inia geoffrensis*, in the Cinaruco River, Venezuela. Biotropica 30(4):625–638

Mintzer VJ, Lorenzen K, Frazer TK, da Silva VMF, Martin AR (2016) Seasonal movements of river dolphins (*Inia geoffrensis*) in a protected Amazonian floodplain. Mar Mamm Sci 32(2):664–681

Moreno P (2004) Ganges and Indus dolphins (Platanistidae). In: Hutchins M, Kleiman D, Geist V, McDade M (eds) Grzimek's animal life encyclopedia mammals IV, vol 15. Thompson Gale, Detroit, pp 13–17

Morisaka T, Connor RC (2007) Predation by killer whales (*Orcinus orca*) and the evolution of whistle loss and narrow-band high frequency clicks in odontocetes. J Evol Biol 20:1439–1458

Nowak RM (2003) Walker's marine mammals of the world, vol 2. John Hopkins University Press, Baltimore, pp 128–130

Nunes ACG (2015) Respostas comportamentais do boto vermelho (*Inia geoffrensis*) ao turismo de interação no baixo rio Negro, Amazonas. Instituto Nacional de Pesquisas da Amazônia, INPA, Brasil. MSc Thesis (In Portuguese)

Ott PH, Danilewicz D (1998) Presence of franciscana dolphins (*Pontoporia blainvillei*) in the stomach of a killer whale (*Orcinus orca*) stranded in southern Brazil. Mammalia 62(4):605–609

Ott PH, Secchi ER, Moreno IB, Danilewicz D, Crespo EA, Bordino P, Ramos RMA, Di Beneditto APM, Bertozzi C, Bastida R, Zanellato R, Perez JE, Kinas PG (2002) Report of the working group on fishery interactions. Lat Am J Aquat Mamm 1(1):55–64

Overstrom NA (1983) Association between burst-pulse sounds and aggressive behavior in captive bottlenose dolphins (*Tursiops truncatus*). Zoo Biol 2:93–103

Parra GJ (2005) Behavioural ecology of Irrawaddy, *Orcaella brevirostris* (Owen in Gray, 1866), and Indo-Pacific humpback dolphins, *Sousa chinensis* (Osbeck, 1765), in northeast Queensland, Australia: a comparative study Page 358. Tropical Environment Studies and Geography. James Cook University, Townsville, Australia

Pilleri G (1970) Observation on the behavior of *Platanista gangetica* in the Indus and Brahmaputra rivers. Invest Cetacea 2:27–60

Pilleri G (1972) Field observations carried out on the Indus dolphin *Platanista indi* in the winter of 1972. Invest Cetacea 4:24–29

Pilleri G (1974) Side-swimming, vision and sense of touch in *Platanista indi* (Cetacea, Platanistidae). Exp Dermatol 30:100–104

Pilleri G, Kraus C, Gihr M (1971a) Further observations on the behaviour of *Platanista indi* in captivity. Invest Cetacea 3(1):34–42

Pilleri G, Kraus C, Gihr M (1971b) Physical analysis of the sounds emitted by *Platanista indi*. Invest Cetacea 3(1):22–30

Pilleri G, Zbinden K, Gihr M, Kraus C (1977) Sonar clicks, directionality of the emission field and echolocating behaviour of the Indus dolphin (*Platanista indi* Blyth 1859). Invest Cetacea 7:13–44

Pilleri G, Gihr M, Kraus C (1980) Play behaviour in the Indus and Orinoco dolphins (*Platanista indi* and *Inia geoffrensis*). Invest Cetacea 11:57–108

Pinedo MC, Praderi R, Brownell Jr RL (1989) Review of the biology and status of the franciscana *Pontoporia blainvillei*. In Perrin WF, Brownell Jr RL, Kaiya Z, Li J (eds) Biology and conservation of the river dolphins. Occasional Papers of the IUCN, Species Survival Commission 3, Gland, Switzerland, pp 46–51

Praderi R (1985) Relaciones entre Pontoporia blainvillei (Mammalia: Cetacea) y tiburones (Selachii) de aguas Uruguayas. Comm Zool Mus Hist Nat Montevideo 11:1–19

Praderi R, Pinedo MC, Crespo EA (1989) Conservation and management of *Pontoporia blainvillei* in Uruguay, Brazil and Argentina. In: Perrin WF, Brownell Jr RL, Zhou K, Li J (eds) Biology and conservation of the river dolphins. Occasional Papers of the IUCN Species Survival Commission 3, Gland, Switzerland, pp 52–55

Reeves RR, Brownell RL Jr (1989) Susu *Platanista gangetica* (Roxburgh, 1801) and *Platanista minor* (Owen, 1853). In: Ridgway SH, Harrison R (eds) Handbook of marine mammals, vol 4. Academic Press, London, pp 69–99

Ridgway SH, Carlin KP, van Alstyne KR, Hanson AC, Tarpley RJ (2017) Comparison of dolphins' body and brain measurements with four other groups of cetaceans reveals great diversity. Brain Behav Evol 88:235–257

Rosas FCW, Marigo J, Laeta M, Rossi-Santos MR (2010) Natural history of dolphins of the genus *Sotalia*. Lat Am J Aquat Mamm 8(1–2):57–68

Rossi-Santos MR, Wedekin LL, Monteiro-Filho ELA (2007) Residence and site fidelity of *Sotalia guianensis* in the Caravelas River Estuary, eastern Brazil. J Mar Biol Assoc UK 87:207–212

Santos MC de O, Ferreira-Neto D (2005) Killer whale (*Orcinus orca*) predation on a franciscana dolphin (*Pontoporia blainvillei*) in Brazilian waters. Lat Am J Aquat Mamm 4(1):62–72

Santos MC de O, Gadig OBF (2009) Evidence of a failed predation attempt on a Guiana dolphin, *Sotalia guianensis*, by a bull shark, *Carcharhinus leucas*, in Brazilian waters. Arquivos de Ciência do Mar 42(2):93–98

Santos MC de O, Rosso S (2007) Ecological aspects of marine tucuxi dolphins (*Sotalia guianensis*) based on group size and composition in the Cananéia estuary, southeastern Brazil. Lat Am J Aquat Mamm 6(1):71–82

Santos MC de O, Rosso S (2008) Social organization of marine tucuxi dolphins, *Sotalia guianensis*, in the Cananéia estuary of southeastern Brazil. J Mammal 89(2):347–355

Santos MC de O, Acuña LB, Rosso S (2001) Insights on site fidelity and calving intervals of the marine tucuxi dolphin (*Sotalia fluviatilis*) in southeastern Brazil. J Mar Biol Assoc UK 81:1049–1052

Santos MC de O, Oshima JEF, da Silva E (2009) Sightings of franciscana dolphins (*Pontoporia blainvillei*): the discovery of a population in the Paranaguá estuarine complex, southern Brazil. Braz J Oceanogr 57:57–63

Santos MC de O, Oshima JEF, Pacífico ES, da Silva E (2010a) Group size and composition of Guiana dolphins (*Sotalia guianensis*) (Van Bénèden, 1864) in the Paranaguá Estuarine Complex, Brazil. Braz J Biol 70(1):111–120

Santos MC de O, Oshima JEF, Pacífico ES, da Silva E (2010b) Feeding associations between Guiana dolphins *Sotalia guianensis* (van Bénèden, 1864) and seabirds in the Lagamar estuary, Brazil. Braz J Biol 70:9–17

Santos GMA, Quaresma AC, Barata RR, Martins BML, Siciliano S, Silva JS Jr, Emin-Lima R (2012) Etho-ecological study of the Amazon River dolphin, *Inia geoffrensis* (Cetacea: Iniidae) and the dolphins of the genus *Sotalia* (Cetacea: Delphinidae) in Guamá River, Amazonia. Mar Biodivers Rec 5:e23

Santos GMA, Rodrigues ALF, Arcoverde DL, Ramos I, Sena L, Silva ML (2014) Unusual records of the behavior of boto *Inia* sp. (Cetartiodactyla, Iniidae) in the lower reaches of the Tocantins and Guamá Rivers, Amazônia. In: Samuels JB (ed) Dolphins: ecology, behavior and conservation strategies. Nova Science, New York

Secchi ER, Ott PH, Crespo EA, Kinas PG, Pedraza SN, Bordino P (2001) A first estimate of franciscana (*Pontoporia blainvillei*) abundance off southern Brazil. J Cetacean Res Manag 3 (2):95–100

Sinha RK, Kannan K (2014) Ganges River dolphin: an overview of biology, ecology, and conservation status in India. Ambio 43:1029–1046

Smith BD, Reeves RR (2012) River cetaceans and habitat change: generalist resilience or specialist vulnerability? J Mar Biol 2012:718935

Stacey PJ, Arnold P (1999) *Orcaella brevirostris*. Mamm Species 616:1–8

Sutaria D (2009) Understanding species conservation in complex socio-ecological systems: case of Irrawaddy dolphins in Chilika Lagoon, India. PhD thesis. James Cook University, Townsville, Australia

Sutaria D, Bopardikar I, Sule M (2017) Irrawaddy dolphins *Orcaella brevirostris* from India. Primary paper SC/67A/SM/08. International Whaling Commission SC Meeting 67, Bled, Slovenia

Taylor B, Chivers SJ, Larese JA, Perrin WF (2007) Generation length and percent mature estimates for IUCN assessments of cetaceans. Southwest Fisheries Science Center, La Jolla, San Diego

Turvey ST, Pitman RL, Taylor BL, Barlow J, Akamatsu T, Barrett LA, Zhao X, Reeves RR, Stewart BS, Wang K, Wei Z, Zhang X, Pusser LT, Richlen M, Brandon JR, Wang D (2007) First human-caused extinction of a cetacean species? Biol Lett 3:573–540

Van Parijs S, Parra G, Corkeron P (2000) Sounds produced by Australian Irrawaddy dolphins, *Orcaella brevirostris*. J Acoust Soc Am 108:1938–1940

Vidal O, Barlow J, Hurtado L, Torre J, Cendon P, Ojeda Z (1997) Distribution and abundance of the Amazon river dolphin (*Inia geoffrensis*) and the tucuxi (*Sotalia fluviatilis*) in the upper Amazon River. Mar Mamm Sci 13:427–445

Wade PR, Reeves RR, Mesnick SL (2012) Social and behavioural factors in cetacean responses to overexploitation: are odontocetes less "resilient" than mysticetes? J Mar Biol 2012:567276

Waller GNH (1983) Is the blind river dolphin sightless? Aquat Mamm 10:106–108

Wang JY, Reeves R (2017) Neophocaena asiaeorientalis. The IUCN Red List of Threatened Species 2017: e.T41754A50381766. doi:https://doi.org/10.2305/IUCN.UK.2017-3.RLTS. T41754A50381766.en. Downloaded on 25 November 2018

Wang D, Wang K, Akamatsu T, Fujita K (1999) Study on whistling of the Chinese river dolphin (*Lipotes vexillifer*). Oceanol Limnol Sin 04. en.cnki.com.cn/Article_en/CJFDTOTAL-HYFZ199904000.htm

Wang Z, Akamatsu T, Wang K, Wang D (2014) The Diel rhythms of biosonar behavior in the Yangtze finless porpoise (*Neophocaena asiaeorientalis asiaeorientalis*) in the Port of the Yangtze River: the correlation between prey availability and boat traffic. PLoS One 9(5): e97907. https://doi.org/10.1371/journal.pone.0097907

Wei Z, Wang D, Zhang X, Wang K, Chen D, Zhao Q, Kuang X, Gong W, Wang X (2004) Observation on some sexual behavior of the Yangtze finless porpoise (*Neophocaena phocaenoides asiaeorientalis*) in captivity. Acta Theriol 24:98–102 [in Chinese]

Wells RS, Bordino P, Douglas DC (2013) Patterns of social association in the franciscana, *Pontoporia blainvillei*. Mar Mamm Sci 29(4):520–528

Wu HP, Hao YJ, Yu XU, Xian YJ, Zhao KJ, Chen DQ, Kuang ZA, Kou ZB, Feng KK, Gong WM, Wang D (2010) Variation in sexual behaviors in a group of captive male Yangtze finless porpoises (*Neophocaena phocaenoides asiaeorientalis*) motivated by physiological changes. Theriogenology 74(8):1467–1475

Würsig B, Beese D, Chen P, Gao A, Tershy B, Liu R, Ding W, Würsig M, Zhang X, Zhou K (2000) Baiji (*Lipotes vexillifer*): travel and respiration behavior in the Yangtze River. In Reeves RR, Smith BD, Kasuya T (eds) Biology and conservation of freshwater cetaceans in Asia. IUCN, Gland, Switzerland and Cambridge, UK. viii + 152 pp

Xian Y, Wang K, Jiang W, Zheng B, Wang D (2010a) Ethogram of Yangtze finless porpoise calves (*Neophocaena phocaenoides asiaeorientalis*). Zool Res 31:523–530

Xian Y, Wang K, Dong L, Hao Y, Wang D (2010b) Some observations on the sociosexual behavior of a captive male Yangtze finless porpoise calf (*Neophocaena phocaenoides asiaeorientalis*). Mar Freshw Behav Physiol 43:221–225

Yan J, Zhou K, Yang G (2005) Molecular phylogenetics of 'river dolphins' and the baiji mitochondrial genome. Mol Phylogenet Evol 37:743–750

Yang J, Xiao W, Kuang X, Wei Z, Liu R (2000) Studies on the distribution, population size and the activity of *Lipotes vexillifer* and *Neophocaena phocaenoides* in Dongting Lake and Boyang Lake. Resour Environ Yangtze Basin 9(4):444–450

Yang J, Wang K, Liu R (2001) Resting behavior of a baiji *Lipotes vexillifer* in captivity. Fish Sci 67:764–766

Zbinden K, Kraus C, Pilleri G (1978) Auditory responses of *Platanista indi*. Invest Cetacea 9:39–64

Zhang X, Wang D, Liu R, Wei Z, Hua Y, Wang Y, Chen Z, Wang L (2003) The Yangtze River dolphin or baiji (*Lipotes vexillifer*): population status and conservation issues in the Yangtze River, China. Aquat Conserv Mar Freshwat Ecosyst 64:51–64

Zhang L, Su F, Yang D, Hao Z, Tong K (2013a) Discharge regime and simulation for the upstream of major rivers over Tibetan Plateau. J Geophys Res Atmos 118:8500–8518

Zhang X, Xian YJ, Wang L, Ding W (2013b) Behaviour and habitat selection of Yangtze finless porpoises in Dongting Lake, China, and the adjacent waters: impact of human activity. Pak J Zool 45:635–642

Zhou K (2009) Baiji *Lipotes vexillifer*. In: Perrin WF, Würsig B, Thewissen JGM (eds) Encyclopedia of marine mammals, 2nd edn. Elsevier, Amsterdam, pp 71–76

Zhou K, Li Y (1989) Status and aspects of the ecology and behaviour of the baiji *Lipotes vexillifer* in the lower Yangtze River. In Perrin WF, Brownell Jr RL, Zhou K, Liu J (eds) Biology and conservation of the river dolphins. IUCN species survival commission occasional paper No. 3, Gland, Switzerland, pp 86–91

Zhou K, Pilleri G, Li Y (1979) Observations on the Baiji (*Lipotes vexillifer*) and the finless porpoise (*Neophocaena asiaeorientalis*) in the Changjiang River between Nanjing and Taiyangzhou, with remarks on some physiological adaptations of the Baiji to its environment. Invest Cetacea 10:109–120

Zhou KY, Pilleri G, Li YM (1980) Observations on Baiji (Lipotes vexillifer) and finless porpoise (Neophocaena asiaeorientalis) in the lower reaches of the Chang Jiang. Sci Sinica 23:785–794

Zhou K, Sun J, Gao A, Würsig B (1998) Baiji (*Lipotes vexillifer*) in the Lower Yangtze River: movements, numbers, threats, and conservation needs. Aquat Mamm 24(2):123–132

Zhou X, Guang X, Sun D, Xu S, Li M, Seim I, Jie W, Yang L, Zhu Q, Xu J, Gao Q, Kaiya A, Dou Q, Chen B, Ren W, Li S, Zhou K, Gladyshev VN, Nielsen R, Fang X, Yang G (2018) Population genomics of finless porpoises reveal an incipient cetacean species adapted to freshwater. Nat Commun 9:1276

Chapter 20
Hector's and Māui Dolphins: Small Shore-Living Delphinids with Disparate Social Structures

Rochelle Constantine

Abstract Hector's dolphins (*Cephalorhynchus hectori hectori*) are a small (~1.5 m long) marine dolphin, primarily inhabiting turbid, coastal waters discontinuously around the South Island of New Zealand. The Māui dolphin (*C. h. maui*) is a critically endangered subspecies of Hector's dolphin, only found along a small part of their original range spanning the west coast of the North Island of New Zealand. Both subspecies have small alongshore home ranges of around 50 km, with high levels of site fidelity and low levels of gene flow. Despite this, some individuals have traveled distances of at least 400 km, interacting with local animals. Hector's dolphins exhibit seasonal movements linked to prey availability and social aggregation behaviors associated with the summer mating and calving period. They typically occur in small groups of 2–10, with high levels of fission-fusion and low levels of association among individuals. Sex segregation occurs in small groups (<5 individuals) of Hector's dolphins throughout the year, but this same pattern does not hold for larger groups. Mother-calf pairs are typically associated with other females, a common pattern for delphinids. Māui dolphins do not show the same pattern, with mixed-sex aggregations of dolphins independent of group size, perhaps an artifact of the extremely small population size. Hector's dolphins largely communicate with ultrasonic clicks, with different vocalizations among social groups and during feeding. Their echolocation clicks are important when foraging in their preferred habitat of low visibility. They forage on a wide range of benthic and demersal fishes and squids, with most prey <10 cm long and some regional differences in species composition, but overall similarities in prey preferences. Despite their distribution around New Zealand and variation in local population sizes, Hector's and Māui dolphins have broad similarities in behavior, association patterns, and habitat use. Where differences exist, the habitat, prey movements, and population size are potential explanatory factors. In New Zealand, a hot spot for cetacean diversity, these dolphins occupy a small and

---

R. Constantine (✉)
School of Biological Sciences and Institute of Marine Science, University of Auckland, Auckland, New Zealand
e-mail: r.constantine@auckland.ac.nz

© Springer Nature Switzerland AG 2019
B. Würsig (ed.), *Ethology and Behavioral Ecology of Odontocetes*, Ethology and Behavioral Ecology of Marine Mammals,
https://doi.org/10.1007/978-3-030-16663-2_20

specific niche that is typical for *Cephalorhynchus* elsewhere in the Southern Hemisphere. Because they occur close to shore in waters affected by humans, they are vulnerable to anthropogenic disturbance. But with recognition of dangers and appropriate protections, the species should flourish in New Zealand's productive coastal waters.

Keywords Hector's dolphin · Māui dolphin · *Cephalorhynchus* · New Zealand · Behavior

20.1 Introduction

The genus *Cephalorhynchus* is represented by four species of small, coastal dolphins in Southern Hemisphere waters, all sharing a common ancestor (Pichler et al. 2001; Dawson 2018). The *Cephalorhynchus* dolphins have similar behavioral characteristics and are typically found in smaller groups, feeding on a wide range of benthic and demersal prey. Their coastal habitat preference exposes them to predation risk from sharks and killer whales (*Orcinus orca*), as well as anthropogenic threats of boat strikes and interactions with fishing gear.

New Zealand has only one species of endemic cetacean, the Hector's and Māui dolphin (*Cephalorhynchus hectori*). They are recognized as subspecies (*C. h. hectori* and *C. h. maui*) based on morphological and genetic differences as a result of around 15,000 years of isolation after the last glacial maxima and low natural dispersal by the species (Baker et al. 2002). Despite this long period of separation, both sister taxa display similar behaviors with variations between populations in different habitats rather than between the two subspecies. For the purposes of this chapter, I refer to them as Hector's dolphins unless there are direct differences between the two subspecies.

20.2 Habitat Preference and Home Range

Hector's dolphins occur discontinuously around the coastal waters of the South Island of New Zealand, with three recognized regional populations along the west, east, and south coasts (Hamner et al. 2012). The North Island is currently populated by the Māui dolphins along only part of their historical range spanning the west coast (Oremus et al. 2012), with occasional reports of Hector's dolphins on the west and east coasts of the North Island, although the provenance of the east coast dolphins remains unknown (Freeman 2003; Hamner et al. 2014).

In almost all areas throughout their range, Hector's dolphins are associated with turbid waters often in association with major watersheds such as river outflows, estuaries, harbors, and/or areas with glacial meltwater (e.g., Rodda and Moore 2013; Derville et al. 2016). These are areas of higher productivity and nutrient flows and therefore are attractive to dolphins as prey hot spots. While adept at swimming in rough coastal waters, often playing in waves just before they crash ashore, Hector's dolphins move away from areas with large ocean swell, perhaps in response to movements in prey toward more turbid waters (Dittmann et al. 2016). As expected with their coastal habitat preference, they usually occur in shallow waters but will use deeper waters if near the coast (Weir and Sagnol 2015). There are suggestions that the turbid waters are preferred habitat because they provide the dolphins protection from predation by sharks and killer whales. While predation by killer whales has not been reported, white sharks (*Carcharodon carcharias*) and seven-gill sharks (*Notorhynchus cepedianus*) have been found with Hector's dolphin remnants in their stomachs (Cawthorn 1988; Hamner et al. 2012). The levels of predation remain largely unknown, but a recent analysis of scars and marks on the bodies of living Māui dolphins suggests that either predation events are successful and result in the death of animals or alternatively that the low level of scars from shark bites indicates low levels of predation (Garg 2017).

Hector's dolphins inhabit cool temperate waters influenced by warm and cool seasonal currents with sea surface temperature (SST) trends around the South Island showing a marked increase over the past 50 years (Shears and Bowen 2017). With SST an important indicator of dolphin distribution (Bräger et al. 2003; Derville et al. 2016), likely a proxy for productivity and prey distribution, it will be interesting to see how dolphins respond as temperatures continue to increase. Hector's have a broad range of thermal tolerance, from approximately 8 to 21 °C, as evidenced by their wide spatial distribution, year-round fidelity to small home ranges, and no long-distance migrations (Rayment et al. 2011a). This, along with their generalist diet (Miller et al. 2013), suggests that changes in SST may not have wide impacts upon this species' distribution as long as there is potential prey. What may influence their behavior is changes in runoff, in particular the west coast South Island regions where glacial meltwater is an important source of turbid water. If there is less turbidity, dolphins need to adapt or move to another habitat that suits their requirements, and this may be challenging given their low dispersal rates. If turbidity is important for evading predators and dolphins remain in less turbid waters, they may form larger groups for predator vigilance, leading to possible shifts in association patterns.

South Island Hector's dolphins along the east and south coasts have seasonal differences in distribution, with smaller, more dispersed groups of dolphins further offshore in winter compared to summer (Dawson and Slooten 1988; Bräger et al. 2003; Turek et al. 2013; Slooten et al. 2006; Rodda 2014; MacKenzie and Clement 2014). Hector's dolphins occur up to ~35 km offshore, often in areas where there is shallow habitat across an extensive continental shelf (MacKenzie and Clement

2014). This pattern is less prevalent for west coast South Island dolphins, where the habitat differs markedly with deeper less turbid water, and the dolphins remain closer to shore across all seasons (Bräger et al. 2003; Rayment et al. 2011a; MacKenzie and Clement 2016). Prey dispersal during winter is likely to be a contributing factor to offshore distribution in the shallower, continental shelf waters, but the summer breeding season is another important driver of clustered social aggregations. Warmer coastal waters provide ideal habitat for females to calve, and thermal demands on newborns may be less than in cooler water.

Despite some seasonal inshore-offshore movement, Hector's dolphins, like other *Cephalorhynchus*, have small home ranges averaging about 50 km alongshore (Rayment et al. 2009a; Oremus et al. 2012). There are some individuals that undertake movements of ~100 km (Bräger et al. 2002), but these are considered unusual for the species. With dorsal fin mark rates typically around 10–20%, the feature most frequently used to identify individual dolphins, we detect only a small proportion of the animals ranging further. There is recent evidence of genetic connectivity between populations, with individuals crossing less optimal habitat (e.g., from the west coast to the south coast South Island) (Hamner et al. 2012). The greatest dispersal distance is a conservative estimate of a $\geq$400 km movement by two female Hector's dolphins from the west coast South Island population to the core range of Māui dolphins, where they were in mixed groups of Māui and Hector's dolphins (Hamner et al. 2014). One of the females remained in the west coast North Island habitat for at least 6 years. There are also other genetically identified South Island dolphins interacting with Māui dolphins (Baker et al. 2016) (Fig. 20.1).

There is no evidence of hybrid offspring between the two subspecies, and with similarities in social structure, size, genetics, and vocalizations, there should not be boundaries to interbreeding (Baker et al. 2002). It is possible that there are more dolphins dispersing from the South Island to the North Island, along the east coast. The origins of these dolphins have yet to be determined, but they are likely to come from the large population of animals from Cloudy and Clifford Bay, east coast South Island (MacKenzie and Clement 2014; Hamner et al. 2017). Although there are few records of longer range dispersals, there is no apparent sex bias in animals ranging further than expected (Bräger et al. 2002; Oremus et al. 2012). Similarly, there appears to be no clear sex bias for dispersal within the normal ranging behavior of Hector's dolphins, with males and females broadly distributed in these coastal subpopulations. With the ability of Hector's dolphins to disperse long distances, as long as they find good-quality prey, they could enhance the gene pool and social structure of the population they move to. This may bode well for the species' future.

Fig. 20.1 A typical short-term aggregation of Māui dolphins, *Cephalorhynchus hectori maui*, swimming in turbid waters off the west coast, North Island. Image credit: Courtesy of Steve Hathaway and the Harbers Family Foundation

20.3 Group Living

Hector's dolphins live in small groups typically numbering between two and ten dolphins, although larger aggregations occur in summer (Dawson 2018). Like most other delphinids, they have a fission-fusion society, but even though they have small home ranges, Hector's have very weak associations between individuals within a population, not different from random (Slooten et al. 1993; Bejder et al. 1998; Bräger 1999). This fluid pattern of association is reflected in field studies where small groups of dolphins occur in close proximity to one another, within a few hundred meters, yet functioning as independent units with regular exchanges of individuals between groups. Males interact with more dolphins than females, suggesting that males might be moving between groups to find females in estrous (Slooten et al. 1993). One of the challenges is that the low mark rate on Hector's dolphin dorsal fins may limit conclusions about the group dynamics of some populations (e.g., Bräger 1999). Nonetheless, the areas with long-term research reveal a persistent pattern of fission-fusion and low individual rates of association, so should be considered typical for the species. In areas with deeper nearshore waters, Hector's have slightly higher levels of association possibly linked to their more limited dispersal patterns compared to east coast dolphins (Bräger 1999).

There are different patterns of age- and sex-class group composition. In the longest-studied population of Hector's dolphins at Banks Peninsula on the east coast South Island, there were clear patterns of sex segregation during spring, summer, and winter. Females were more likely alone and small ($\leq$5 dolphins) groups were typically either all male or all female (Webster et al. 2009). Mother-calf pairs were accompanied by other females, but once groups contained more than five dolphins, they were more likely to be mixed-sex aggregations. This pattern did not hold for Māui dolphins, where mixed-sex groups occurred across all group sizes, noting that there is a female dominant sex bias in this population (Baker et al. 2016). The mixed-sex groups may be due to the very small population size of Māui dolphins (63 dolphins aged 1+, CL 57–75) (Baker et al. 2016) that potentially disrupts typical social structure and breeding aggregations. Whether this variation in sex-biased grouping occurs elsewhere, or if Māui dolphins are an exception, has yet to be determined.

Hector's dolphins often occur in nursery groups, with mother-calf pairs alone, with one or two other female group members, or in loose association with other mother-calf pairs (Bräger 1999; Webster et al. 2009; Oremus et al. 2012). This is a typical pattern for delphinids in a number of other species. Females have a 2- to 3-year inter-birth interval, so the period of calf dependency is short but similar to other small delphinids (Chap. 1). Despite the species' small size, calves grow rapidly within their first years of life, up to adult size at 5 to 6 years old (Webster et al. 2010). It is possible that some of the other females associated with mother-calf pairs may be previous offspring who remain in some association with their mother until they reach sexual maturity.

In summer, high levels of activity occur, indicative of mating competition as part of a multi-mate (often incorrectly termed "promiscuous") breeding system (Slooten et al. 1993). Hector's dolphins regularly engage in jumps clear of the water, repetitive side-slaps, head-slaps, and chases (Fig. 20.2). They use ultrasonic clicks

Fig. 20.2 Hector's dolphins, *Cephalorhynchus hectori hectori*, leaping in Cloudy Bay, South Island. Image credit: Courtesy of Oregon State University and the University of Auckland Collaborative Research Programme

to communicate and produce more complex click types and pulse rates in larger, active groups (Dawson 1991). With only a proportion of the females in estrous, competition by males is likely to be high, and males and females engage in active social behaviors. With their relatively large testes and dynamic larger group aggregations, the summer is an important time for males, when they actively attempt to encounter as many females as possible to increase their chance of fathering offspring (Slooten 1991; Slooten et al. 1993). It is also an important time for females, to produce calves most likely to survive (see Chap. 4).

In winter, groups of Hector's dolphins become more evenly dispersed, and group sizes decrease slightly. Dolphins also move offshore, with the overall winter range generally larger than the summer range as it has a wider offshore and alongshore distribution, most notably along the east coast, South Island (Dawson and Slooten 1988; Bräger et al. 2003; Rodda 2014; MacKenzie and Clement 2014). With social groupings largely driven by breeding behavior rather than the formation of long-term, complex social alliances, once this season has passed, there is no benefit for Hector's dolphins to remain in proximity to other dolphins, hence the change in social structure.

There are several poorly studied small populations dispersed around coastal waters that may reveal a greater disparity in social associations and patterns of group structure than currently understood. With habitat playing an important role in prey availability, water turbidity, and risk of predation, this may influence association rates between isolated populations (e.g., Kaikoura, Hamner et al. 2012; Weir and Sagnol 2015) or the potential for different "inshore" and "offshore" cohorts in areas where the population is widely dispersed (e.g., Cloudy and Clifford Bays, MacKenzie and Clement 2014; Hamner et al. 2017).

20.4 Foraging

Because Hector's dolphins live in turbid waters and only occasionally forage near the sea surface, it is difficult to make direct field observations of foraging events, so our knowledge of diet comes mainly from analysis of gut contents from dead beachcast individuals and as bycatch or entanglement in fishing gear. They are generalist foragers that eat a variety of benthic, demersal, and pelagic fishes and squids throughout the water column (Miller et al. 2013). As they are typically found in shallow, coastal waters, Hector's are able to take advantage of their entire vertical and horizontal habitat within diving range. They are primarily solitary foragers with rare observations of communal foraging behaviors (Dawson 2018). They feed mainly on prey items ranging from <1 to >60 cm, with most prey <10 cm in length. Even though 29 different prey species have been identified, 6 species made up 77% of their total diet (Miller et al. 2013). The patterns of prey types are similar throughout the species' range, although prey species composition varies in particular between the west and east coasts of the South Island (Miller et al. 2013). Observations of deep dives, accompanied by more forceful exhalations upon surfacing, are indications that dolphins are foraging in mid- or benthic waters. Sometimes individuals swim rapidly near the surface, presumably chasing prey, but our behavioral observations of foraging events are limited.

Hector's dolphins have high fidelity to particular coastal or harbor locations, areas of high productivity (Bejder and Dawson 2001; Rayment et al. 2009a, b; Miller et al. 2013; Rodda and Moore 2013). In Akaroa Harbor, Banks Peninsula, the dolphins undertake some diel movements entering the harbor in the morning and leaving at night, most likely in response to prey movements or availability (Stone et al. 1995). Reliable, good-quality prey availability is important (Spitz et al. 2012), and the dolphins move in response to seasonal movements of preferred prey. The summer inshore presence of dolphins is correlated with movements of preferred prey such as red cod (*Pseudophycis bacchus*), which follow their prey into harbors and coastal waters. One of the challenges when determining drivers behind dolphin movements is our poor understanding of marine food webs and dynamics associated with noncommercial fish species.

On the west coast, South Island of New Zealand, the dolphins have a more similar winter and summer distribution than east coast Hector's dolphins and are found considerably closer to shore (<6 nm) than east coast dolphin populations (Rayment et al. 2011a; MacKenzie and Clement 2014, 2016). The exposed west coast has a steep drop-off to deeper waters close to the coast, which may not be a suitable habitat for preferred prey or pose a limitation on the ability of these small dolphins to dive deep enough to capture benthic prey (Schreer and Kovacs 1997). As with the west coast Hector's dolphins, Māui dolphins of the North Island have a limited offshore range (Du Fresne 2010) and similar preferred prey (Miller et al. 2013). One difference is the range of potential preferred habitat on the west coast North Island compared to the South Island, with harbors, turbid waters, and juvenile fish nursery grounds largely underutilized by the population, possibly due to the severely reduced population size and range contraction (Dawson et al. 2001; Rayment et al. 2011b; Oremus et al. 2012; Derville et al. 2016). The current Māui dolphin core range is adjacent to these easily accessible harbors, therefore the low use may reflect a limitation on their current socially transmitted knowledge of these habitats. It is possible that the fisheries closure in core habitat (as part of a marine mammal protected area) has removed prey competition by humans for coastal species, and the small population is able to obtain its nutritional needs from the coastal waters.

Hector's dolphins have not been observed taking fish from gill nets, but some feed behind trawlers (Rayment and Webster 2009). As observed in other species (Chilvers and Corkeron 2001), average group size was significantly larger for trawl-fishing than groups not associated with trawlers (Rayment and Webster 2009). There is an increased availability of prey as fish either are disturbed by the trawl activities or escape from the net, making this an important source of prey for all dolphins, including mothers with calves (as discussed for other dolphins by Fertl and Leatherwood 1997). Hector's dolphins in association with trawlers also increase levels of aerial and sexual behaviors, likely a result of larger numbers of dolphins aggregating (Rayment and Webster 2009). As some Hector's dolphins die from entanglement in trawl nets, it is a risky behavioral strategy, but energetic payoffs must be considerable. Some dolphins feed near trawlers year-round, and it is possible that different communities develop social ("cultural") proclivities for such feeding, as has been observed in other species (e.g., Chilvers and Corkeron 2001; Ansmann et al. 2012; see Chap. 10).

20.5 Interactions

Hector's dolphins are vulnerable to anthropogenic impacts due to their coastal habitat. They are vulnerable to fisheries bycatch and entanglement that has led to significant declines in abundance (Reeves et al. 2013). They are a social dolphin known for boat approaches and are a popular species for dolphin-based tourism operations and/or people swimming out from beaches after they see them from shore (Bejder et al. 1999; Martinez et al. 2010). Hector's dolphins appear to be more interactive with humans during summer, the season when they are most socially active. Overall, Hector's dolphins have similar responses to boats and swimmers as occurs with other small delphinids (Constantine and Bejder 2008). They are attracted to novel stimuli (Martinez et al. 2011) and often play with seaweed, approaching swimmers while engaged in active behaviors, jumping out of waves, and bow-riding boats.

20.6 Concluding Thoughts

Hector's dolphins have social lives typically characterized by small home ranges, weak associations between individuals, an active summer breeding season, and movements largely linked to prey availability. But there are disparities between the east and west coasts of the South Island and the west coast North Island subspecies. The type of social aggregation varies depending on time of year, e.g., short-term male-female associations are linked to breeding, and mother-calf pairs are often associated with other females year-round, perhaps part of younger females' learning associated with calf rearing or as a strategy to minimize risk from predation. In other areas, males and females are equally mixed. Hector's dolphin ranging behavior varies depending on the offshore characteristics with shallow, continental shelf areas leading to greater dispersal and different aggregation behaviors than those living near deeper nearshore waters. Cooperative foraging is rarely observed even in the longest running studies, removing one important driver of delphinid social behavior and affiliations between conspecifics (see Gowans et al. 2007). As far as is known at this time, they do not appear to have complex communication systems characteristic of some larger delphinids, although they may "eavesdrop" on the echolocation signals of other dolphins to locate or secure prey in turbid waters (Gregg et al. 2007), but this is an area requiring more investigation. If the Māui dolphin population recovers, perhaps they will return to the sex-specific social grouping in larger populations of South Island Hector's dolphins, but this is presently unknown. Changes in human land use have resulted in increased runoff, degrading harbor habitats for potential prey. Whether the reduced use of harbors by Māui dolphins is due to small population size influencing social "knowledge" of these habitats or as a result of reduced prey quality remains unknown. Hector's are under threat from anthropogenic impacts, including most recently deaths from the cat-borne disease toxoplasmosis (Roe et al. 2013), but management decisions that act to protect them are having some positive effect on the species and hopefully their future survival (Gormley et al. 2012; MacKenzie and Clement 2014, 2016).

References

Ansmann IC, Parra GJ, Chilvers BL, Lanyon JM (2012) Dolphins restructure social system after reduction of commercial fisheries. Anim Behav 84:575–581

Baker AN, Smith ANH, Pichler FB (2002) Geographical variation in Hector's dolphin: recognition of new subspecies of *Cephalorhynchus hectori*. J R Soc N Z 32:713–727

Baker CS, Steel D, Hamner RM, Hickman G, Boren L, Arlidge W, Constantine R (2016) Estimating the abundance and effective population size of Māui dolphins using microsatellite genotypes in 2015-16, with retrospective matching to 2001-16. Report to Department of Conservation, Auckland, New Zealand

Bejder L, Dawson S (2001) Abundance, residency, and habitat utilisation of Hector's dolphins (*Cephalorhynchus hectori*) in Porpoise Bay, New Zealand. N Z J Mar Freshw Res 35:277–287

20 Hector's and Māui Dolphins: Small Shore-Living Delphinids...

Bejder L, Fletcher D, Bräger S (1998) A method for testing association patterns of social animals. Anim Behav 56:719–725

Bejder L, Dawson SM, Harraway JA (1999) Responses by Hector's dolphins to boats and swimmers in Porpoise Bay, New Zealand. Mar Mamm Sci 15(3):738–750

Bräger S (1999) Association patterns in three populations of Hector's dolphin, *Cephalorhynchus hectori*. Can J Zool 77:13–18

Bräger S, Dawson SM, Slooten E, Smith S, Stone GS, Yoshinaga A (2002) Site fidelity and alongshore range in Hector's dolphin, an endangered marine dolphin from New Zealand. Biol Conserv 108:281–287

Bräger S, Harraway JA, Manly BFJ (2003) Habitat selection in a coastal dolphin species (*Cephalorhynchus hectori*). Mar Biol 143:233–244

Cawthorn MW (1988) Recent observations of Hector's dolphin Cephalorhynchus hectori, in New Zealand. Rep Int Whaling Comm Spec Issue 9:303–314

Chilvers BL, Corkeron PJ (2001) Trawling and bottlenose dolphins' social structure. Proc R Soc B 268:1901–1905

Constantine R, Bejder L (2008) Managing the whale- and dolphin-watching industry: time for a paradigm shift. In: JES H, Lück M (eds) Marine wildlife and tourism management: insights from the natural and social sciences. CABI International, Oxford, pp 321–333

Dawson SM (1991) Clicks and communication: the behavioural and social contexts of Hector's dolphin vocalizations. Ethology 88:265–276

Dawson S (2018) Cephalorhynchus dolphins. In: Würsig B, Thewissen JGM, Kovacs KM (eds) Encyclopedia of marine mammals, 3rd edn. Academic, London, UK, pp 166–172

Dawson SM, Slooten E (1988) Hector's dolphin *Cephalorhynchus hectori*: distribution and abundance. Rep Int Whaling Comm (Sp Iss) 9:315–324

Dawson S, Pichler F, Slooten E, Russell K, Baker CS (2001) The North Island Hector's dolphin is vulnerable to extinction. Mar Mamm Sci 17:366–371

Derville S, Constantine R, Baker CS, Oremus M, Torres LG (2016) Environmental correlates of nearshore habitat distribution by the critically endangered Māui dolphin. Mar Ecol Prog Ser 551:261–275

Dittmann S, Dawson S, Rayment W, Webster T, Slooten E (2016) Hector's dolphin movement patterns in response to height and direction of ocean swell. NZ J Mar Freshw Res 50:228–239

Du Fresne S (2010) Distribution of Maui's dolphin (*Cephalorhynchus hectori maui*) 2000–2009. Department of Conservation Science Research and Development Series 322, Wellington, New Zealand

Fertl D, Leatherwood S (1997) Cetacean interactions with trawls: a preliminary review. J Northwest Atl Fish Sci 22:219–248

Freeman D (2003) A review of records of Hector's dolphin (*Cephalorhynchus hectori*) from the East Coast of the North Island, New Zealand. Technical Support Series Number 11, Department of Conservation, Wellington, New Zealand

Garg R (2017) Photo-identification and demographic assessment of New Zealand's Māui dolphin. BSc (Honours) Thesis, School of Biological Sciences, University of Auckland, New Zealand

Gormley AM, Slooten E, Dawson S, Barker RJ, Rayment W, du Fresne S, Bräger S (2012) First evidence that marine protected areas can work for marine mammals. J Appl Ecol 49:474–480

Gowans S, Würsig B, Karczmarski L (2007) The social structure and strategies of delphinids: predictions based on an ecological framework. Adv Mar Biol 53:195–294

Gregg JD, Dudzinski KM, Smith HV (2007) Do dolphins eavesdrop on the echolocation signals of conspecifics? Int J Comp Psychol 20:65–88

Hamner RM, Pichler FB, Heimeier D, Constantine R, Baker CS (2012) Genetic differentiation and limited gene flow among fragmented population of New Zealand endemic Hector's and Maui's dolphins. Conserv Genet 13:987–1002

Hamner RM, Constantine R, Oremus M, Stanley M, Brown P, Baker CS (2014) Long-range movement by Hector's dolphins provides potential genetic enhancement for critically endangered Maui's dolphin. Mar Mamm Sci 30:139–153

Hamner RM, Constantine R, Mattlin R, Waples R, Baker CS (2017) Genotype-based estimates of local abundance and effective population size for Hector's dolphin. Biol Conserv 211:150–160

MacKenzie DL, Clement DM (2014) Abundance and distribution of ECSI Hector's dolphin. New Zealand Aquatic Environment and Biodiversity Report No. 123, Ministry for Primary Industries, Wellington, New Zealand

MacKenzie DL, Clement DM (2016) Abundance and distribution of WCSI Hector's dolphin. New Zealand Aquatic Environment and Biodiversity Report No. 168. Ministry for Primary Industries, Wellington, New Zealand

Martinez E, Orams MB, Stockin KA (2010) Swimming with an endemic and endangered species: effects of tourism on Hector's dolphins in Akaroa Harbour, New Zealand. Tour Rev Int 14:99–115

Martinez E, Orams MB, Pawley MDM, Stockin KA (2011) The use of auditory stimulants during swim encounters with Hector's dolphins (*Cephalorhynchus hectori hectori*) in Akaroa Harbour, New Zealand. Mar Mamm Sci 28:E295–E315

Miller E, Lalas C, Dawson S, Ratz H, Slooten E (2013) Hector's dolphin diet: the species, sizes and relative importance of prey eaten by *Cephalorhynchus hectori*, investigated using stomach content analysis. Mar Mamm Sci 29:606–628

Oremus M, Hamner RM, Stanley M, Brown P, Baker CS, Constantine R (2012) Distribution, group characteristics and movements of the critically endangered Maui's dolphin *Cephalorhynchus hectori maui*. Endanger Species Res 19:1–10

Pichler FB, Robineau D, Goodall RNP, Meyer MA, Olavarría C, Baker CS (2001) Origin and radiation of Southern Hemisphere coastal dolphins (genus *Cephalorhynchus*). Mol Ecol 10:2215–2223

Rayment W, Webster T (2009) Observations of Hector's dolphins (*Cephalorhynchus hectori*) associating with inshore fishing trawlers at Banks Peninsula, New Zealand. NZ J Mar Freshw Res 43:911–916

Rayment W, Dawson S, Slooten E, Bräger S, Du Fresne S, Webster T (2009a) Kernel density estimates of alongshore home range of Hector's dolphins at Banks Peninsula, New Zealand. Mar Mamm Sci 25:537–556

Rayment W, Dawson S, Slooten E (2009b) Use of T-PODs for acoustic monitoring of *Cephalorhynchus* dolphins: a case study with Hector's dolphins in a marine protected area. Endanger Species Res 10:333–339

Rayment W, Clement D, Dawson S, Slooten E, Secchi E (2011a) Distribution of Hector's dolphin (*Cephalorhynchus hectori*) off the west coast, South Island, New Zealand, with implications for the management of bycatch. Mar Mamm Sci 27:398–420

Rayment W, Dawson S, Scali S, Slooten L (2011b) Listening for a needle in a haystack: passive acoustic detection of dolphins at very low densities. Endanger Species Res 14:149–156

Reeves RR, Dawson SM, Jefferson TA, Karczmarski L, Laidre K, O'Corry-Crowe G, Rojas-Bracho L, Secchi ER, Slooten E, Smith BD, Wang JY, Zhou K (2013) Cephalorhynchus hectori. The IUCN Red List of Threatened Species 2013: e.T4162A44199757

Rodda J (2014) Analysis and geovisualisation of Hector's dolphin abundance and distribution patterns in space and time. PhD Dissertation, University of Otago, New Zealand

Rodda J, Moore A (2013) Hotspots of Hector's dolphins on the south coast. In Proceedings of SIRC NZ Conference, Dunedin, New Zealand

Roe WD, Howe L, Baker EJ, Burrows L, Hunter SA (2013) An atypical genotype of *Toxoplasma gondii* as a cause of mortality in Hector's dolphins (*Cephalorhynchus hectori*). Vet Parasitol 192:67–74

Schreer JF, Kovacs KM (1997) Allometry of diving capacity in air-breathing vertebrates. Can J Zool 75:339–358

Shears NT, Bowen MM (2017) Half a century of coastal temperature records reveal complex warming trends in western boundary currents. Sci Rep 7:14527

Slooten E (1991) Age, growth, and reproduction in Hector's dolphins. Can J Zool 69:1689–1700

Slooten E, Dawson SM, Whitehead H (1993) Associations among photographically identified Hector's dolphins. Can J Zool 71:2311–2318

Slooten E, Rayment W, Dawson S (2006) Offshore distribution of Hector's dolphin at Banks Peninsula: is the Banks Peninsula marine mammal sanctuary large enough? NZ J Mar Freshw Res 40:333–343

Spitz J, Trites AW, Becquet V, Brind'Amour A, Cherel Y, Galois R, Ridoux V (2012) Cost of living dictates what whales, dolphins and porpoises eat: the importance of prey quality on predator foraging strategies. PLoS One 7:e50096

Stone G, Brown J, Yoshinaga A (1995) Diurnal patterns of movement as determined from clifftop observation. Mar Mamm Sci 11:395–402

Turek J, Slooten E, Dawson S, Rayment W, Turek D (2013) Distribution and abundance of Hector's dolphins off Otago, New Zealand. NZ J Mar Freshw Res 47:181–191

Webster TA, Dawson SM, Slooten E (2009) Evidence of sex segregation in Hector's dolphin (*Cephalorhynchus hectori*). Aquat Mamm 35:212–219

Webster T, Dawson S, Slooten E (2010) A simple laser photogrammetry technique for measuring Hector's dolphins (*Cephalorhynchus hectori*) in the field. Mar Mamm Sci 26:296–308

Weir JS, Sagnol O (2015) Distribution and abundance of Hector's dolphins (*Cephalorhynchus hectori*) off Kaikoura, New Zealand. NZ J Mar Freshw Res 49:376–389

Chapter 21
Porpoises the World Over: Diversity in Behavior and Ecology

Jonas Teilmann and Signe Sveegaard

Abstract Unlike the large number of species of true dolphins, there are only seven porpoise or phocoenid species, inhabiting different habitats and climate zones. They range from ice-covered water in the Arctic to the subantarctic islands through tropic waters and even in freshwater river systems. Some species like the harbour porpoise are widely distributed in both the Atlantic and the Pacific, while the vaquita has the most restricted range of any cetacean and only lives in a small northern part of the Gulf of California, or Sea of Cortez, in northwestern Mexico, where it is now close to extinction due to illegal fishery. Little is known about porpoise species compared to many dolphin species as they generally have an elusive behavior or live away from human attention. However, the finless porpoise and the harbour porpoise have been kept in captivity for many years with success, contributing to knowledge on social interactions, breeding behavior, and sensory physiology. In recent years, a growing number of studies of wild porpoises using advanced electronic tagging devices have greatly improved our knowledge of movements, acoustic behavior, feeding ecology, and reactions to noise disturbance. We now know that harbour porpoises are widely distributed over most of the deep north Atlantic during winter and therefore not only a coastal species as previously thought. We also see very similar echolocation behavior between the finless and the harbour porpoises, showing an almost constant hunt for food, but also inactive periods that may be interpreted as sleeping. Ship noise has been shown to have a significant effect on feeding buzzes, and loud sounds may cause pauses in echolocation and fast movements away from the sound source, which again may result in bycatch and in rare cases mass stranding. Tagging data have also shown that harbour porpoises are much more social than previously thought and have acoustic contact to conspecifics up to 58% of the time between mother and calf and 36% between animals tagged alone. With these new findings,

---

The original version of this chapter was revised. A correction to this chapter can be found at https://doi.org/10.1007/978-3-030-16663-2_24

J. Teilmann (✉) · S. Sveegaard
Department of Bioscience, Aarhus University, Roskilde, Denmark
e-mail: jte@bios.au.dk

© Springer Nature Switzerland AG 2019
B. Würsig (ed.), *Ethology and Behavioral Ecology of Odontocetes*, Ethology and Behavioral Ecology of Marine Mammals,
https://doi.org/10.1007/978-3-030-16663-2_21

our growing understanding of porpoise behavior has opened a window to their elusive lives that could help improve management and conservation decisions.

Keywords Dall's porpoise (*Phocoenoides dalli*) · Harbour porpoise (*Phocoena phocoena*) · Vaquita (*Phocoena sinus*) · Burmeister's porpoise (*Phocoena spinipinnis*) · Spectacled porpoise (*Phocoena dioptrica*) · Indo-Pacific finless porpoise (*Neophocaena phocaenoides*) · Narrow-ridged finless porpoise (*Neophocaena asiaeorientalis*) · Social behavior · Distribution · Movements · Anthropogenic disturbance

21.1 Introduction

The small family of porpoises (Phocoenidae) consists of seven species of odontocetes divided into three genera: (1) Phocoenoides, Dall's porpoise (*Phocoenoides dalli*); (2) Phocoena, harbour porpoise (*Phocoena phocoena*), vaquita porpoise (*Phocoena sinus*), Burmeister's porpoise (*Phocoena spinipinnis*), and spectacled porpoise (*Phocoena dioptrica*); and (3) Neophocaena, Indo-Pacific finless porpoise (Neophocaena phocaenoides) and narrow-ridged finless porpoise (*Neophocaena asiaeorientalis*).

The seven species occupy very different habitats and climate zones distributed all over the globe except the poles, and they vary greatly in size of home range. Yet there are many behavioral and ecological similarities. All porpoises are small animals of less than 2.5 m in length; they have robust bodies, blunt rostra, and spade-shaped teeth. These characteristics separate them from dolphins, which are generally larger and tend to have longer tapered rostra and cone-shaped teeth.

The harbour porpoise is by far the most studied of the seven species. Therefore, the information on distribution, movements, social behavior, and acoustic vocalization in this chapter will primarily concern this species but include information on others when available.

21.2 Distribution and Ecological Adaptations

The seven species are widely distributed throughout the seas of the World (Fig. 21.1) and vary immensely with regard to geographical range: the most widely distributed species, the harbour porpoise, lives in the brackish Baltic Sea, in the Black Sea, in shallow shelf waters, and in deep offshore ocean water of the North Atlantic and North Pacific. Dall's porpoise inhabits inshore and offshore waters of the North Pacific, while its range overlaps in coastal areas with the harbour porpoise. The two finless porpoise species are very coastal living in the waters from Japan in the western Pacific Ocean, and eastward from there to the Indian Ocean coastlines, to the Arabian Gulf. *Neophocaena* even inhabit fresh water river deltas, and—as a separate subspecies off the Pacific Ocean—the Yangtze River in China. The Burmeister's porpoise is confined to the coastal shallow water ranging along the

21 Porpoises the World Over: Diversity in Behavior and Ecology

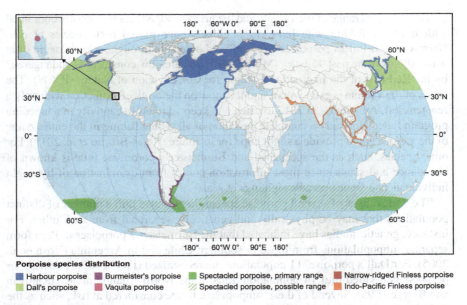

Porpoise species distribution

- Harbour porpoise
- Dall's porpoise
- Burmeister's porpoise
- Vaquita porpoise
- Spectacled porpoise, primary range
- Spectacled porpoise, possible range
- Narrow-ridged Finless porpoise
- Indo-Pacific Finless porpoise

Fig. 21.1 Distribution of the seven porpoise species. Adapted from Encyclopedia of Marine Mammals 3rd Ed. (Würsig et al. 2018; Nielsen et al. 2018). Note overlapping distributions of Dall's, harbour, and finless porpoises around Japan, Dall's and harbour porpoises in Eastern Russia and western North America, Burmeister's and spectacled porpoises along the south-eastern coast of South America, and the two finless porpoise species in the area of Taiwan. See Introduction for genus and species names

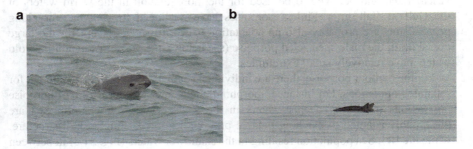

Fig. 21.2 Rare sightings of vaquita, *Phocoena sinus*, in Baja California, Mexico. Photos: Jonas Teilmann

southern part of South America from Peru in the Pacific to Brazil in the Atlantic. The spectacled porpoise has some distributional overlap with Burmeister's porpoise in south-eastern South America, but it primarily roams the southern ocean around Antarctica. The only really range restricted porpoise species is the vaquita (or Gulf of California porpoise), occurring in the upper part of the Gulf of California (or Sea of Cortez) in northern Mexico (Figs. 21.1 and 21.2).

The restricted range of the vaquita and the narrow-ridged finless porpoise combined with intensive fishing with large mesh gillnets have placed both species at risk. Narrow-ridged finless porpoises have suffered a decline throughout their range of at least 50% over the past 45 years (Moore 2015) and are now considered "endangered" by the Red List of the International Union for Conservation of Nature (IUCN). The status of the vaquita is even worse, and it is now on the brink of extinction with only a few tens left. An attempt in 2017 to catch and keep vaquitas in captivity in a breeding program failed, and only a complete cessation of all gillnet fishing in the entire range of the vaquita could provide a slight hope for its future (Rojas-Bracho et al. 2019). For other species, such as the spectacled and Burmeister's porpoises, little is known of abundance and status since most information comes from examination of bycaught individuals and some sporadic sightings at sea.

The wide ranges of several porpoise species are divided into a number of distinct populations that may be physically and genetically separated from each other. For instance, genetic studies have indicated that the Burmeister's porpoises in Peru form separate subpopulations from those in southern Chile and in Argentina (Rosa et al. 2005). For Dall's porpoise, 11 populations are recognized (IWC 2002). The harbour porpoise is divided into 14 population units in the North Atlantic alone (Donovan and Bjørge 1995). Several of these subpopulations are considered at risk, such as the "critically endangered" Baltic Sea harbour porpoise and the "endangered" Black Sea harbour porpoise (as established in the Red List of the IUCN).

These very different ranges also represent most climate zones, from the tropic waters where the vaquita and the finless porpoises occur to Arctic and Antarctic waters inhabited by the harbour and spectacled porpoises, respectively. The vaquita has adapted to the climate by developing large extremities, which are highly vascularized and believed to be used for thermoregulation in the warm waters of the Gulf of California. Finless porpoises have no dorsal fin, but the very long pectoral fins are probably also an adaptation to warm water. Except for the large dorsal fin in the male spectacled porpoise (see below), the porpoises inhabiting cold waters have relatively small extremities.

For the harbour porpoise, there are individual adaptations to their environment for separate populations. For instance, morphometric studies of the rostrum of porpoise populations inhabiting deeper waters in the Baltic Sea and the North Sea are presumably adapted to catching pelagic prey with a more straight skull structure, while the Belt Sea population residing in the shallower waters (Fig. 21.3) in between these two populations are adapted to catching benthic prey with a rostrum pointing downward (Galatius et al. 2012).

Dall's porpoise also seems to be adapted to its deep diving pelagic life by having extremely high blood-oxygen content and a large heart compared to more coastal-dwelling cetacean species (Ridgway and Johnston 1966). This species is divided into two color morphs called the *dalli*-type and the *truei*-type (Kasuya 1978). The *dalli*-type is found throughout the range of the species, while the *truei*-type is found in a smaller area of the northwestern North Pacific. The taxonomic status of the two types remains unclear as their genetic differences are no larger than the differences between neighboring populations of the *dalli*-type (Escorza-Treviño et al. 2004).

Fig. 21.3 A harbour porpoise, *Phocoena phocoena*, in a typical foraging habitat in a coastal narrow strait with strong currents. Photo: Jonas Teilmann

21.3 Movements and Diving Behavior

Movements and migrations have mainly been studied by satellite tracking harbour porpoises in three different areas of the North Atlantic. Porpoises were tagged with satellite transmitters after being either incidentally caught in herring weir (Canada) or pound nets (Denmark) or actively caught (Greenland). The number of tracked animals varied between regions with 9 tagged animals in Canada, 31 in Greenland, and 120 in Denmark, as well as the length of transmission with 1 juvenile male harbour porpoise tagged in Greenland holding the absolute record of 1040 days (yellow track, Fig. 21.4). There are large differences in movements between the three tagging sites: harbour porpoises in Canada and Denmark often remain within a few 100 km from the tagging site and prefer waters with depth <200 m. Quite differently, Greenland porpoises travel up to 2500 km offshore from the tagging site where they swim in waters of 2000–4000 m depth from January to June. Furthermore, they return to the area of tagging the following year, thereby illustrating clear seasonal migrations and strong site fidelity (Nielsen et al. 2018). This difference is probably caused by the cold environment and historic ice conditions off Greenland, forcing porpoises to move into ice-free areas. Maximum dive depth was found to be 410 m, potentially preying on mesopelagic species distributed throughout the North Atlantic.

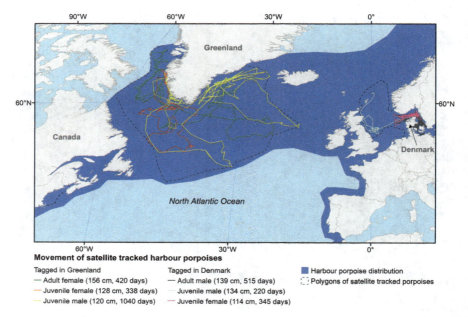

Fig. 21.4 Range of satellite tracked harbour porpoises, *Phocoena phocoena*, in the North Atlantic. This area is the only place where larger satellite tagging programs have been conducted on porpoise species (Read and Westgate 1997; Sveegaard et al. 2011; Nielsen et al. 2018). Polygons around all positions for the three tagging areas in southeast Canada ($N = 9$), west Greenland ($N = 31$), and Denmark ($N = 120$) are shown, as well as three examples of tracks from harbour porpoises tagged in Greenland and Denmark, respectively. Animal length and tag longevity are also given

Similar to Greenland harbour porpoise migrations, a study using visual surveys found that the population of Dall's porpoises of the western North Pacific migrates between wintering grounds off the Pacific coast of northern Japan and summer breeding grounds in the central Okhotsk Sea (IWC 2002; Miyashita and Kaswa 1988). In Dall's porpoises, the timing of calving season in different populations in Japanese waters has been found to correlate with water temperatures (Amano and Kuramochi 1992).

Porpoises often rest motionless at the surface in calm weather. Cetaceans have uni-hemispherical sleep, where only one-half of their brain sleeps at a time, and therefore they can sleep while swimming. While sleep behavior is difficult to observe behaviorally, tagging studies with dive loggers indicate that stereotypic shallow (app. 7 m) parabolic-shaped dives with no or little echolocation activity could be associated with sleeping. These dives were found in 7.5% of the total dive time in all seven harbour porpoises tagged for 2–11 days (Wright et al. 2017).

21.4 Characteristics Related to Social Behavior

The most conspicuous morphological trait within the porpoise family is the large dorsal fin in spectacled porpoise males. This sexual dimorphism is probably a result of sexual selection where males become physically larger and develop very large dorsal fins to secure access to females, perhaps by male-male displays. However, there are only few observations of this species in the wild, with no knowledge of their social behavior. Sexual dimorphism is also found in Dall's porpoises, where males become larger and more muscular than females, males having a post-anal hump, larger skull structures, and dorsal fins canted forward (Frandsen and Galatius 2013; Jefferson 2018). Spectacled and Dall's porpoises are the largest porpoises, with males growing up to 224 and 239 cm, respectively (Goodall and Brownell 2018; Jefferson 2018). This suggests a polygynous mating system, which is supported in the case of Dall's porpoises as they have relatively small testes and mate guarding has been observed (Willis and Dill 2007). Almost all newly calved females in estrus were accompanied by an adult male, and the males made significantly shorter dives when together with other animals, to maintain contact with the female. In contrast, both sexes of harbour, finless, Burmeister's, and vaquita porpoises are similar in size, or females grow slightly larger than males, suggesting no physical pre-copulatory competition between males. In addition, these latter species have unusually large testes (4–5% of body mass) suggesting a promiscuous mating system with sperm competition, where several males copulate with each female during estrus (Gao and Zhou 1993; Sørensen and Kinze 1994; Hohn et al. 1996; Keener et al. 2018).

All porpoise species have small hard bumps emerging from the leading edge of the dorsal fin or the dorsal ridge in finless porpoises, termed epidermal tubercles. These may serve as an organ for mutual contact as they contain many nerve endings (Kasuya 1999) and seem to be used during social contact when captive harbour porpoises engage in rubbing, e.g., the rostrum against the dorsal fin. Amundin and Amundin (1971) suggested one function of the tubercles in two captive porpoises that had been separated, with much loose skin covering their bodies, but when they were moved together, they immediately began to rub each other with their dorsal fins, and the loose skin disappeared by the following day.

21.5 Social Behavior

Most porpoises are notoriously shy animals that seldomly approach boats and people, except for Dall's porpoise that regularly bowrides fast-moving boats in a similar manner to some dolphin species. Furthermore, their elusive appearance with little visible activity, except breathing at the surface, provides limited options for behavioral studies. Their similar appearance and skin color have until recently prevented extensive long-term photo ID studies. In Washington State, USA, 53 harbour porpoises were identified on photos based on fin shape, skin coloration, and scars, and 35% were resighted over a 3-year study period (Elliser et al. 2018). This study revealed some site fidelity and group size between 1 and 9 individuals, being highest in spring (mean = 3.0 individuals, SD = 1.7, $n = 70$) and lowest in summer (mean = 1.8, SD = 0.9, $n = 79$) and highest during feeding events (mean = 2.9, SD = 1.6, $n = 69$) and lower during travel (mean = 2.0, SD = 1.4, $n = 76$). Porpoises tend to occur in small fluid groups of 1–10 individuals, but occasionally large apparently loosely structured feeding or traveling groups up to 100 animals have been reported (e.g., Reyes 2018). The only strong bond between individuals seems to be mother-calf pairs that possibly stay together for 1 year until the next calf is born or possibly 2 years if the females become pregnant every other year (Teilmann et al. 2007; Read 2018; Fig. 21.5). Two mother and calf pairs were satellite tagged in Denmark in the beginning of the calving season in spring. Their relationship was genetically confirmed, and the ages of the calves were estimated to be 1 and 2 years old, respectively (Teilmann et al. 2004). The first pair swam together for the duration of the track (41 days), while the other pair—mother and calf—went separate routes after 6 days (Fig. 21.6). The 1-year-old calf showed a gradual independent diving behavior over the course of the tracking period, which happened during the beginning of the calving season when the older calf is believed to have separated from the mother (Teilmann et al. 2007). This could also be the reason for the divided track of the 2-year-old calf if the mother was giving birth to a new calf.

Studies of captive porpoises and observations from bridges and drones have provided some insight to their social interactions. Some of the more detailed descriptions of social behavior in harbour porpoises were made by Andersen and Dziedzic (1964) and Amundin and Amundin (1971) studying some of the first captive harbour porpoises as part of a NATO research program on underwater noise and animal vocalization in Denmark. Although porpoises—with rare exceptions (Fig. 21.7)—are shy in the wild, harbour porpoises in captivity are curious and easy to train. They seem to learn faster when another porpoise carries out an exercise, suggesting that porpoises often learn from each other by visual and acoustic contact. They have been found to invent games, playing with objects such as sea stars or seaweed, either alone or with others.

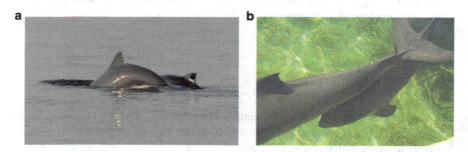

Fig. 21.5 A calf harbour porpoise, *Phocoena phocoena*, often swims directly beneath the mother with its dorsal fin touching the mother, or directly behind the mother, in both cases making it difficult to see the calf when observing porpoises in the wild. Photo: Jeppe Dalgaard Balle (left) and Jonas Teilmann (right)

21 Porpoises the World Over: Diversity in Behavior and Ecology

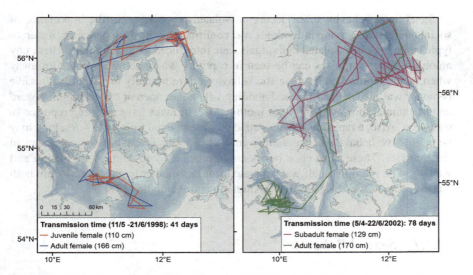

Fig. 21.6 Maps of satellite tracked mother-calf pair of harbour porpoises, *Phocoena phocoena*, displaying tracks from the time when both animals were tracked. Tracks are illustrated based on one position per day. The adult female was accompanied by an approximately 1-year-old calf (left panel) and an approximately 2-year-old calf (right panel), respectively. The tracking period, duration of track, and length of animals are given

Fig. 21.7 A rare occasion of a curious harbour porpoise, *Phocoena phocoena*, approaching to and apparently investigating a vessel. Photo: Jonas Teilmann

Stereotypic behavior often occurs in isolated porpoises, and the importance of social interactions among individuals was confirmed by the disappearance of stereotypic behavior when isolated individuals were introduced to other animals. Hierarchy among harbour porpoises can be seen in captivity where some animals turn away when being approached by agonistic behavior or are sexually harassed. Aggressive behavior was also defined by head nodding with either open or closed mouth toward another animal; however, biting or tooth mark skin rakes have not been reported in porpoises. Also, a rapid click series called buzzes or social calls up to 4.5 s. long may be aggressive behavior or an alarm signal when animals are stressed and swim in small circles. These short click bursts are also heard when animals are hungry and waiting to be fed or if a calf was separated from its mother [concatenated from both Andersen and Dziedzic (1964) and Amundin and Amundin (1971)].

21.5.1 Sexual Behavior

Harbour and finless porpoises exhibit active sexual behavior as an important part of social interactions. Sociosexual behavior is seen between and among females and males. It may involve many forms of genital contact and has been observed from the age of 1 month with up to 5 events per hour within the first year of life in finless porpoises (Xian et al. 2010). While sexual behavior has mostly been studied in captive finless and harbour porpoises, recent bridge observations by the Golden Gate Cetacean Research group in San Francisco have shown regular mating attempts year-round, where a male swims rapidly from below, always approaching the left flank of the presumed female and leaping out of the water in 69% of the times when on top of the female, who may quickly expose her belly toward the male (Fig. 21.8, Keener et al. 2018). A male can erect his penis (~50 cm) within less than one second when approaching the female from behind and then turns his body abruptly to attempt a brief intercourse; whereafter the penis is rapidly retracted again. In captivity, male harbour porpoises regularly rub the penis against the dorsal fin of a conspecific. Behavior interpreted as masturbation of the erect penis being rubbed against objects is characterized by loud social calls, alternating fluke positions, with the body in an s-posture (Amundin and Amundin 1971).

21.5.2 Acoustic Behavior

Unlike most dolphins that produce broadband echolocation clicks for foraging and whistles for communication, porpoises produce narrowband high-frequency (NBHF) clicks as their only vocalization (Silber 1991). The main energy of the clicks is from 120 to 140 kHz, and the characteristic narrow bandwidth and an almost constant sound production day and night make the signals ideal to detect in acoustic recordings. These high-frequency clicks are not audible to the human ear, but they produce a weak audible sound probably as an artifact from the sound

Fig. 21.8 Harbor porpoise mating at the surface. The female in front lifts her flukes high out of the water in response to the male's approach. The penis is barely visible near the genital area. Note that the male approach on the left side of the female is similar to all observations in Keener et al. (2018). Photo credit: Kurt Videbæk Jensen

production system, at around 2 kHz. Møhl and Andersen (1973) originally documented the existence of the much louder high-frequency signal based on blindfolded porpoise's ability to detect very thin metal wires, which would only be possible based on wave length if their primary echolocation sense was at much higher frequency than the 2 kHz. There are few other sounds at 120–140 kHz, so it is an acoustically "clean" environment, and the short wavelength also makes it possible for porpoises to detect very small fish. It is speculated that in narrowband high-frequency echolocation evolved for predator avoidance, as killer whales cannot hear such high frequencies (Kyhn et al. 2013; Galatius et al. 2019).

Besides studies in captivity, bio-logging studies have opened a window into understanding behavior of porpoises. Echolocation clicks are primarily used for orientation, prey detection, and prey capture (Wisniewska et al. 2016, 2018a). Click intervals can indicate at what range the porpoises are searching. A comparative study between finless porpoises from the Yangtze River and a harbour porpoise in Danish waters found that they use echolocation in a similar manner. For porpoises to interpret the information based on their own echolocation, they would need to receive the echo and understand its meaning before emitting a new click. Therefore, inter-click intervals represent the "active space" in which the animal is searching. Finless porpoises used an average inter-click interval of 60 ms, while the harbour porpoise used 80 ms. This represents an active space of 30 m and 45 m, respectively. This difference may reflect the narrower and shallower waters of the river habitat of these finless porpoises, compared to the oceanic habitat of harbour porpoises (Akamatsu et al. 2007).

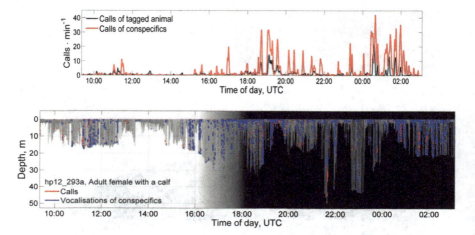

Fig. 21.9 Dive depth, feeding buzzes, and social calls of a tagged harbour porpoise over a 15-hour period. The tagged animal was an adult female accompanied by its calf. The dive profile in (**a**) also indicates when the animal is making social calls (red) or feeding buzzes (blue). Time of day is illustrated by daylight (white), twilight (gray), and night (black). (**b**) Social call rate for each minute for the tagged animal (black line) and what is believed to be the calf or other conspecifics (red line). Note that social calls take place at any depth and that there is a general correlation between call rate from the mother and other animals, presumably the calf (Sørensen et al. 2018)

Social calls, defined as a rapid click sequence (<15 ms between clicks) without any previous sequence of normal echolocation clicks (>15 ms between clicks), have previously been recorded only in captive animals (Clausen et al. 2011). As higher frequencies travel shorter distances than lower frequencies, porpoises are physically restricted to less than 1000 m in the range they can communicate. Porpoises usually occur alone or in small fluid groups that are not stable over time. Therefore, it was believed that porpoises had limited social relationships with other conspecifics. However, recent tagging studies of six harbour porpoises reveal that social calls are often heard from another animal than the one carrying the tag (Fig. 21.9). For mother-calf pairs, call rate was 54–58% of the time, while for single animals call rate from conspecifics was 10–36 % of the time (Sørensen et al. 2018). This is a surprisingly high level of social contact given group size and limited acoustic active space and suggests that we should consider porpoises much more social than previously believed.

21.5.3 Disturbance Behavior

When captive porpoises are scared, they swim faster, stay underwater longer, and stop vocalizing (Teilmann et al. 2006). This is also seen in harbour porpoises outfitted with acoustic tags (DTAG, Fig. 21.10), where intense ship noises significantly reduce

Fig. 21.10 Harbour porpoise after having been caught in a Danish pound net, released with a suction cup tag recording sound, 3D movements, and GPS positions (DTAG4). Such bio-logging tags provide new insight to the social behavior, physiology, and effects of noise on these shy and difficult-to-study animals. Photo: Florian Graner

foraging echolocation and in some cases make animals cease echolocation and turn to passive hearing to potentially avoid collision (Wisniewska et al. 2018b). This silent flight behavior may pose a tremendous risk of bycatch in areas with gillnet fishery, and it has been suggested as the main contributing factor to an unusual large mass stranding of 85 mainly bycaught harbour porpoises. The strandings occurred before a large naval exercise in the North Sea in 2005, while ships in the area were testing equipment at high volume prior to the main exercise (Wright et al. 2013).

Harbour porpoises also react strongly to very intense sounds from pile driving of wind turbine foundations into the seabed. An avoidance distance of up to 25 km has been shown around pile driving construction sites (Dähne et al. 2013). Based on acoustic monitoring, operation of wind farms has positive and negative long-term effects on harbour porpoises, probably depending on importance and motivation of porpoises to use the habitat after it has been changed relative to fishery, background noise, and introduction of artificial reef habitats around the foundations (Scheidat et al. 2011; Teilmann and Carstensen 2012).

21.6 Conclusion

Although the seven species of porpoises have similar physical appearance, they occur in very different habitats and climate zones. The difference in habitats and water temperature has a significant impact on their behavior, ecology, and biology. We know that spectacled and Dall's porpoise and the population of harbour porpoises in Greenland spend months in the polar or temperate deep part of the oceans, while the two finless porpoise species live in very shallow coastal or riverine tropical or subtropical environments year-round. The vaquita has the most restricted distribution, only living in the subtropical upper Gulf of California. Our knowledge on movements, diving, and social behavior is limited to the well-studied harbour porpoise and the less studied finless and Dall's porpoises, while we know virtually nothing about how vaquita, spectacled, and Burmeister's porpoises socially behave. Limited knowledge on behavior, ecology, and abundance is part of the reason for lack of timely management intervention to conserve the poor or unknown status of porpoises as established in the IUCN Red List. Vaquitas and Baltic Sea harbour porpoises are "critically endangered." Narrow-ridged finless porpoises and Black Sea harbour porpoises are "endangered." Indo-Pacific finless porpoises are "vulnerable." For Burmeister's and spectacled porpoises, the status is "data deficient," which means that we have no relevant data to provide a reliable status of the species. Only for Dall's porpoise and the overall world status of harbour porpoises is the status "least concern." In many areas porpoise habitats overlap with intense human activities, such as oil and mineral exploitation, dredging, shipping, wind farming, pollution, and fishery. We are only beginning to understand how noise affects their behavior, but initial results show that it may disturb their foraging by ceasing echolocation. Without echolocation, porpoises are virtually "blind," which may increase the risk of swimming into fishing nets. Bycatch in set net fishery is known to cause the death of thousands of porpoises every year and is the main reason for the near-extinction of the vaquita, where only a few tens remain of the entire species. Knowledge of most porpoise species is limited, and it is a challenge for the future to manage and mitigate anthropogenic impacts to ensure protection of this widespread cetacean family.

References

Akamatsu T, Teilmann J, Miller LA, Tougaard J, Dietz R, Wang D, Wang K, Siebert U, Naito Y (2007) Comparison of echolocation behaviour between coastal and riverine porpoises. Deep Sea Res Part II 54(3–4):290–297

Amano M, Kuramochi T (1992) Segregative migration of Dall's porpoise (*Phocoenoides dalli*) in the Sea of Japan and Sea of Okhotsk. Mar Mamm Sci 8:143–151

Amundin B, Amundin M (1971) Några etologiska iagttagelser över tumlaren, *Phocoena phocoena* (L.), I fångenskap. Zool Revy 3–4:51–59

Andersen S, Dziedzic A (1964) Behaviour patterns of captive harbour porpoise, *Phocoena phocoena* (L.). Bull Int Oceanogr 63(1316):1–20

Clausen KT, Wahlberg M, Beedholm K, Deruiter S, Madsen PT (2011) Click communication in harbour porpoises. Bioacoustics 20(1):1–28

Dähne M, Gilles A, Lucke K, Peschko V, Adler S, Krügel K, Sundermeyer J, Siebert U (2013) Effects of pile-driving on harbour porpoises (*Phocoena phocoena*) at the first offshore wind farm in Germany. Environ Res Lett 8:025002

Donovan GP, Bjørge A (1995) Harbour porpoises in the North Atlantic: edited extract from the report of the IWC scientific committee, Dublin 1995. Rep Int Whal Commn 16:3–25

21 Porpoises the World Over: Diversity in Behavior and Ecology

Elliser CR, MacIver KH, Green K (2018) Group characteristics, site fidelity, and photo-identification of harbor porpoises, *Phocoena phocoena*, in Burrows Pass, Fidalgo Island, Washington. Mar Mamm Sci 34(2):365–384

Escorza-Treviño S, Dizon AE, Pastene LA (2004) Molecular analyses of the truei and dalli morphotypes of Dall's porpoise (*Phocoenoides dalli*). J Mamm 85:347–355

Frandsen MMS, Galatius A (2013) Sexual dimorphism of Dall's porpoise and harbor porpoise skulls. Mamm Biol 78:153–156

Galatius A, Kinze CC, Teilmann J (2012) Population structure of harbour porpoises in the greater Baltic region: evidence of separation based on geometric morphometric comparisons. J Mar Biol Asoc UK 92:1669–1676

Galatius A, Olsen MT, Steeman ME, Racicot RA, Bradshaw CD, Kyhn LA, Miller LA (2019) Raising your voice: evolution of narrow-band high-frequency signals in toothed whales (Odontoceti). Biol J Linn Soc 126(2):213–224

Gao A, Zhou K (1993) Growth and reproduction of three populations of finless porpoises *Neophocaena phocaenoides* in Chinese waters. Aquat Mamm 19(1):3–12

Goodall RNP, Brownell RL Jr (2018) Spectacled porpoise, *Phocoena dioptrica*. In: Würsig B, Thewissen JGM, Kovacs K (eds) Encyclopedia of marine mammals. Academic, San Diego, CA

Hohn A, Read AJ, Fernandez S, Vidal O, Findley LT (1996) Life history of the vaquita, *Phocoena sinus* (Phocoenidae, Cetacea). J Zool 239:235–251

International Whaling Commission (2002) Report of the standing sub-committee on small cetaceans. J Cetacean Res Manag 4:325–338

Jefferson TA (2018) Dall's porpoise, *Phocoenoides dalli*. In: Würsig B, Thewissen JGM, Kovacs K (eds) Encyclopedia of marine mammals. Academic, San Diego, CA

Kasuya T (1978) The life history of Dall's porpoise with special reference to the stock off the pacific coast of Japan. Sci Rep Whales Res Inst 30:1–63

Kasuya T (1999) Finless porpoise, *Neophocaena phocaenoides* (G. Cuvier, 1829). In: Handbook of marine mammals, the second book of dolphins and the porpoises, vol 6. Academic, London, pp 411–442

Keener W, Webber MA, Szczepaniak ID, Markowitz TM, Orbach DN (2018) The sex life of Harbor porpoises (*Phocoena phocoena*): lateralized and aerial behavior. Aquat Mamm 44(6):620–632

Kyhn LA, Tougaard J, Beedholm K, Jensen FH, Ashe E, Williams R, Madsen PT (2013) Clicking in a killer whale habitat: narrow-band, high-frequency biosonar clicks of Harbour porpoise (*Phocoena phocoena*) and Dall's porpoise (*Phocoenoides dalli*). PLoS One 8(5):e63763

Miyashita T, Kaswa T (1988) Distribution and abundance of Dall's porpoise off Japan. Sci Rep Whales Res Inst 39:121–150

Møhl B, Andersen S (1973) Echolocation: high-frequency component in the click of the harbour porpoise (*Phocoena ph. L.*). J Acoust Soc Am 54(5):1368–1372

Moore J (2015) Intrinsic growth (rmax) and generation time (T) estimates for odontocetes of the genus *Sousa, Orcaella,* and *Neophocaena*, in support of IUCN Red List Assessments. NOAA Technical Memorandum NMFS SWFSC

Nielsen NH, Teilmann J, Sveegaard S, Hansen RG, Sinding MHS, Dietz R, Heide-Jørgensen MP (2018) Oceanic movements, site fidelity and deep diving in harbour porpoises from Greenland show limited similarities to animals from the North Sea. Mar Ecol Prog Ser 597:259–272

Read AJ (2018) Porpoise, overview. In: Würsig B, Thewissen JGM, Kovacs K (eds) Encyclopedia of marine mammals. Academic, San Diego, CA

Read AJ, Westgate AJ (1997) Monitoring the movements of harbour porpoises (*Phocoena phocoena*) with satellite telemetry. Mar Biol 130(2):315–322

Reyes JC (2018) Burmeister's porpoise, *Phocoena spinipinnis*. In: Würsig B, Thewissen JGM, Kovacs K (eds) Encyclopedia of marine mammals. Academic, San Diego, CA

Ridgway SH, Johnston DG (1966) Blood oxygen and ecology of porpoises of three genera. Science 151(3709):456–458

Rojas-Bracho L, Gulland FMD, Smith CR, Taylor B, Wells RS, Thomas PO, Bauer B, Heide-Jørgensen MP, Teilmann J, Dietz R, Balle JD, Jensen MV, Sinding MHS, Jaramillo-Legorreta A, Abel G, Read AJ, Westgate AJ, Colegrove K, Gomez F, Martz K, Rebolledo R, Ridgway S, Rowles T, van Elk CE, Boehm J, Cardenas-Hinojosa G, Constandse R, Nieto-Garcia E, Phillips W, Sabio D, Sanchez R, Sweeney J, Townsend F, Vivanco J, Vivanco JC, Walker S (2019) A field effort to capture critically endangered vaquitas (*Phocoena sinus*) for protection from entanglement in illegal gillnets. Endanger Species Res 38:11–27

Rosa S, Milinkovitch MC, Van Waerebeek K, Berck J, Oporto JA, Alfaro-Shigueto J, Van Bressem MF, Goodall R, Cassens I (2005) Population structure of nuclear and mitochondrial DNA variation among South American Burmeister's porpoises (*Phocoena spinipinnis*). Conserv Genet 6:431–443

Scheidat M, Tougaard J, Brasseur S, Carstensen J, Petel TVP, Teilmann J, Reijnders P (2011) Harbour porpoises (*Phocoena phocoena*) and wind farms: a case study in the Dutch North Sea. Environ Res Lett 6:025102

Silber GK (1991) Acoustic signals of the vaquita (*Phocoena sinus*). Aquat Mamm 17(3):130–133

Sørensen TB, Kinze CC (1994) Reproduction and reproductive seasonality in Danish harbour porpoises, *Phocoena Phocoena*. Ophelia 39(3):159–176

Sørensen PM, Wisniewska DM, Jensen FH, Johnson M, Teilmann J, Madsen PT (2018) Click communication in wild harbour porpoises (*Phocoena phocoena*). Sci Rep 8:9702

Sveegaard S, Teilmann J, Tougaard J, Dietz R, Mouritsen KN, Desportes G, Siebert U (2011) High density areas for harbor porpoises (*Phocoena phocoena*) identified by satellite tracking. Mar Mamm Sci 27(1):230–246

Teilmann J, Carstensen J (2012) Negative long term effects on harbour porpoises from a large scale offshore wind farm in the Baltic - evidence of slow recovery. Environ Res Lett 7:045101

Teilmann J, Dietz R, Larsen F, Desportes G, Geertsen BM, Andersen LW, Aastrup PJ, Hansen JR, Buholzer L (2004) Satellitsporing af marsvin i danske og tilstødende farvande. Danmarks Miljøundersøgelser. Sci Rep 484: 86. http://www2.dmu.dk/1_viden/2_Publikationer/3_fagrapporter/rapporter/FR484_samlet.PDF

Teilmann J, Tougaard J, Kirketerp T, Anderson K, Labberté S, Miller L (2006) Reactions of captive harbour porpoises (*Phocoena phocoena*) to pinger-like sounds. Mar Mamm Sci 22(2):240–260

Teilmann J, Larsen F, Desportes G (2007) Time allocation and diving behaviour of harbour porpoises (*Phocoena phocoena*) in Danish waters. J Cetacean Res Manag 9(3):35–44

Willis PM, Dill LM (2007) Mate guarding in male Dall's porpoises (*Phocoenoides dalli*). Ethology 113:587–597

Wisniewska DM, Johnson M, Teilmann J, Rojano-Donate L, Shearer J, Sveegaard S, Miller LA, Siebert U, Madsen PT (2016) Ultra-high foraging rates of harbor porpoises make them vulnerable to anthropogenic disturbance. Curr Biol 26(11):1441–1446

Wisniewska DM, Johnson M, Teilmann J, Rojano-Donate L, Shearer J, Sveegaard S, Miller LA, Siebert U, Madsen PT (2018a) Response to "resilience of harbor porpoises to anthropogenic disturbance: must they really feed continuously?". Mar Mamm Sci 34(1):265–270

Wisniewska DM, Johnson M, Teilmann J, Siebert U, Galatius A, Dietz R, Madsen PT (2018b) High rates of vessel noise disrupt foraging in wild harbour porpoises (*Phocoena phocoena*). Proc R Soc B 285:20172314

Wright AJ, Maar M, Mohn C, Nabe-Nielsen J, Siebert U, Jensen LF, Baagøe HJ, Teilmann J (2013) Possible causes of a harbour porpoise mass stranding in Danish waters in 2005. PLoS One 8(2): e55553

Wright AJ, Akamatsu T, Nørgaard KM, Sveegaard S, Dietz R, Teilmann J (2017) Silent porpoise: potential sleeping behaviour identified in wild harbour porpoises. Anim Behav 133:211–222

Würsig B, Thewissen JGM, Kovacs K (eds) (2018) Encyclopedia of marine mammals, 3rd edn. Elsevier Press, San Diego, CA

Xian Y, Wang K, Dong L, Hao Y, Wang D (2010) Some observations on the sociosexual behavior of a captive male Yangtze finless porpoise calf (*Neophocaena phocaenoides asiaeorientalis*). Mar Freshw Behav Physiol 43(3):221–225

Chapter 22
Endangered Odontocetes and the Social Connection: Selected Examples of Species at Risk

Thomas A. Jefferson

Abstract Despite centuries of whaling focused mostly on mysticete species, eight of the ten most endangered species of cetaceans in the world today are odontocetes. These species have certain features of their ecology in common, such as coastal habitats and usually ranges in developing countries, but also have some shared behavioral and social traits, such as strong susceptibility to entanglement in fishing nets and acoustic disturbance. I use four species of small cetaceans as case studies to examine the elements that have caused their predicaments. It is likely that the vaquita (*Phocoena sinus*) will soon become the second species of cetacean to go extinct in modern times, and the Atlantic humpback dolphin (*Sousa teuszii*) appears to be the next most endangered species. Several other cetacean species are facing similar levels of risk—despite some having misleading status assessments. There is a need to learn from our past mistakes to provide better protection to those species at risk and thereby avoid future extinctions.

Keywords Cetaceans · Dolphins · Endangered species · Extinction · Management · Porpoises · Small cetaceans

22.1 Introduction[1]

Cetaceans have long been prized by people looking for a readily available source of food, oil, and a whole host of other products. Because they are large, they were particularly attractive subjects for human exploitation, but their relatively inaccessible habitats made them hard to hunt until the past several hundred years (Ellis 2018;

---

[1]This introduction is mostly paraphrased from Jefferson (2019).

T. A. Jefferson (✉)
Clymene Enterprises, Lakeside, CA, USA

© Springer Nature Switzerland AG 2019
B. Würsig (ed.), *Ethology and Behavioral Ecology of Odontocetes*, Ethology and Behavioral Ecology of Marine Mammals,
https://doi.org/10.1007/978-3-030-16663-2_22

Reeves 2018a). Although there is evidence that prehistoric humans took advantage of fortuitous strandings of fresh whales or dolphins, most cetaceans were safe from large-scale human exploitation until relatively recently (see Moore et al. 2018).

The first known large-scale hunting of whales was by the Basques (who lived in what is presently part of northern Spain), starting in the first millennium AD. They mainly targeted the North Atlantic right whale (*Eubalaena glacialis*) and were so effective in killing such numbers that the species' recovery is still in doubt (Ellis 2018; Kraus 2018). Norse and Icelandic whalers also hunted in the North Atlantic, and the Japanese began their culture of whale hunting in the 1600s (Kasuya 2018). In the 1700s, the "Yankee whaling" era began, focusing largely on sperm whales (*Physeter macrocephalus*), and this led to the United States becoming a major player in the commercial whaling game (Townsend 1935; Ellis 2018). In the late 1800s, development of steam-powered vessels and the exploding harpoon ushered in the modern era of commercial whaling (Brownell et al. 2018; Clapham and Baker 2018). Fast-swimming species, such as the rorquals (blue, fin, sei, Bryde's, Omura's, and minke whales, *Balaenoptera* species), were now within the realm of commercial whalers. It did not take long for them to decimate multiple species, starting with the largest and working their way down the list. This often extended to the point of "commercial extinction," which is defined as the stage at which it is no longer financially viable to continue the hunt.

However, the public perception that all large whales are endangered is not correct. Most large whales are no longer commercially hunted, and many are recovering from past exploitation—major exceptions being the North Atlantic and North Pacific right whale (*E. japonica*) species (Kraus 2018). Most of the truly serious conservation problems now lie with several of the smaller cetacean species and subspecies (Brownell et al. 1989; Reeves et al. 2003). The vaquita (*Phocoena sinus*), Indus susu (*Platanista minor*)[2], North Island Hector's dolphin (*Cephalorhynchus hectori maui*), Atlantic humpback dolphin (*Sousa teuszii*), and Taiwanese white dolphin (*Sousa chinensis taiwanensis*) are among those in worst shape.

In recent decades, direct killing of whales and dolphins has become less significant, and the indirect deaths of small cetaceans (especially dolphins and porpoises) have increased (Jefferson and Curry 1994; Reeves et al. 2013a, b; Northridge 2018a; Reeves 2018b). More cetaceans now die incidentally in fishing nets each year than from any other threat, including whale and dolphin hunting. Other major threats include habitat degradation, environmental contamination, climate change, noise pollution (including from naval sonar and seismic surveys), and even live captures for oceanaria displays and research (Evans 2018; Northridge 2018b; Reijnders et al. 2018; Southall 2018). Despite a number of populations of cetaceans in specific regions having been wiped out by humans (e.g., the Atlantic gray whale by commercial whaling), it is only recently that an entire cetacean species (see below) has

---

[2]This geographic form is currently recognized as a subspecies, *Platanista gangetica minor,* but there is strong evidence for its species-level distinctiveness (Braulik et al. 2015), and I expect it to be split out some time soon.

gone extinct at the hands of humans. Unfortunately, several other species are now on the verge of that very fate.

The baiji (*Lipotes vexillifer*) was assessed to be probably extinct after an extensive survey of their entire known range in 2006 resulted in no sightings or acoustic detections (Turvey et al. 2007). The baiji was found only in the Yangtze River (and some connected lakes) in China. Besides incidental deaths in fishing gear and problems of severe pollution, baiji suffered from overall habitat loss and degradation, due to rapid modification of the river for human use, this done with little or no concern for impacts on its original inhabitants (Turvey 2008, 2010). The Chinese government largely ignored calls of scientists and conservationists and thus allowed their only endemic cetacean species to go extinct. They must bear the primary responsibility for this environmental catastrophe.

Twenty years ago, Perrin (1999) provided an insightful summary of small cetaceans at risk of extinction. He pointed out the special problems that make many small cetaceans highly susceptible to extinction risk: (1) ease of capture, (2) vulnerable habitats, (3) development of new markets, (4) difficulties of monitoring and regulation, and (5) lack of international means of management. Since that time, one species that he reviewed has become extinct (the baiji), and a second one (the vaquita) is very close. It is hoped that the current review, which in many ways is an extension of Perrin (1999), will further help to clarify views and promote effective plans for preventing future extinctions among odontocetes.

22.2 The Most Endangered: What Puts Them at Risk?

The five factors that Perrin (1999) highlighted are important issues in what makes a species prone to extinction. Internationally, the IUCN Red List (http://www.iucnredlist.org/) reviews information related to species status and provides an empirical assessment of extinction risk, based on a set of predetermined criteria, many of which are quantitative. The Red List categories are (from least at risk to most):

Least Concern (LC)
Near Threatened (NT)
Vulnerable (VU)
Endangered (EN)
Critically Endangered (CR)
Extinct in the Wild (EW)
Extinct (EX)

Species in the categories of VU, EN, and CR are considered to be threatened. Species for which there is not enough information to place them into one of the above categories are listed as Data Deficient (DD).

Although the IUCN Red List designations—as well as national endangered species listings that are produced by many countries—are helpful in determining which species are most at risk of extinction, listings can be misleading (see Merrick et al. 2018). There are political and technical considerations that can mask a species' true potential for extinction. Examples of this are the sperm whale (*Physeter macrocephalus*), which is globally distributed, with a population of at least hundreds of thousands, and for which the main threat (commercial whaling) has largely been removed. Yet this species is listed on the Red List as Vulnerable and on the US Endangered Species List as Endangered. But most biologists agree that sperm whales are in no immediate danger of extinction. At the opposite end of the spectrum are species such as the Chilean dolphin (*Cephalorhynchus eutropia*), which is found in a small range in South America, numbers at most in the low thousands, and is facing continuing (and very possibly escalating) threats to its survival. However, the species is listed on the Red List at Near Threatened, and is not even listed on the US Endangered Species List.

Although there are bound to be disagreements, I generated a list of what I consider the ten most endangered cetacean species on Earth (Table 22.1). For this review, I attempted to consider factors such as how much is known about the species, whether there is at least one area of the range where the species is quite "safe," how much potential there is for ex situ conservation methods, the attention paid to the species by NGOs, and what the political and management-related situations are within their range countries. Eight of the ten species are odontocetes (Table 22.1).

To my thinking the Atlantic humpback dolphin is at higher risk of extinction than the North Atlantic right whale, despite the former species having a population that may be almost one order of magnitude higher. But, there are almost no effective management or conservation measures in place in any of the range countries for the Atlantic humpback dolphin, and no major NGOs are focusing effort on this species at present. Some will disagree with my decision to put *S. teuszii* above the right whales (a few perhaps quite strongly), but I attempted to minimize political and personal biases.

22.3 Selected Species at Risk

Among the eight species of odontocetes that have the dubious distinction of having made my top ten most endangered list (Table 22.1), I discuss four of them in detail here; I hope these examples demonstrate the varied influences of the wide range of factors that make all of them particularly vulnerable to extinction.

22.3.1 Vaquita (Phocoena sinus): Critically Endangered

The vaquita (Fig. 22.1) is number one on the most endangered cetacean species list. It has long been listed as Critically Endangered on the IUCN Red List (Rojas-Bracho and Taylor 2017), one of only two cetacean species in this most threatened category. This porpoise has probably been declining in numbers since before it was

Table 22.1 The ten most endangered cetacean species

Rank	Common name	Scientific name	IUCN status	Population size	Trend	Population fragmented	Management/ Conservation	Threats	References
1	Vaquita	*Phocoena sinus*	CR	<20	↓	No	Yes	Continuing	Rojas-Bracho and Taylor (2017)
2	Atlantic humpback dolphin	*Sousa teuszii*	CR	<3000?	↓	Yes	No	Continuing	Collins et al. (2017)
3	North Atlantic right whale	*Eubalaena glacialis*	EN	300–400	↓?	No	Yes	New	Reilly et al. (2012)
4	North Pacific right whale	*Eubalaena japonica*	EN	ca. 500	?	No	Yes	New?	Reilly et al. (2008)
5	Indus river dolphin	*Platanista minor*[a]	EN	1200–1800	?	Yes	Yes	Continuing	Braulik and Smith (2017)
6	Ganges river dolphin	*Platanista gangetica*[a]	EN	<4000	↓?	Yes	Yes	Continuing	Braulik and Smith (2017)
7	Chilean dolphin	*Cephalorhynchus eutropia*	NT	Low 1000s	↓	Yes?	Some	Continuing	Heinrich and Reeves (2017)
8	Irrawaddy dolphin	*Orcaella brevirostris*	EN	<10,000?	↓	Yes	Some	Continuing	Minton et al. (2017)
9	Indian Ocean humpback dolphin	*Sousa plumbea*	EN	Low 10,000s	↓	Yes	Some	Continuing	Braulik et al. (2017)
10	Hector's dolphin	*Cephalorhynchus hectori*	EN	<7400	↓	Yes	Yes	Continuing	Reeves et al. (2013a, b)

[a]Indus and Ganges river dolphins are currently classified by IUCN as a single species, but will likely be split soon

Fig. 22.1 A vaquita, *Phocoena sinus*, mother and calf surfacing in the northern Gulf of California, Mexico. The vaquita is the most endangered marine mammal species in the world, with <30 individuals left on the planet. Photo by the author

"discovered" by science in 1958 (Norris and McFarland 1958), and it is now decreasing at a rate of at least 50%/year (CIRVA 2018). The entire population was estimated to be approximately 30 individuals in November 2016 (Thomas et al. 2017), and the current number may be in the single digits by the time this paper is published in 2019. It has at most a year or two left, unless the threat of illegal gillnet fishing is removed or dramatically reduced.

The early history of vaquita conservation actions was reviewed by Perrin (1999). Vaquitas have been threatened by a series of gillnet fisheries within their small range in the northern Gulf of California, Mexico (Rojas-Bracho and Reeves 2013). Conservation efforts are largely through a series of recommendations made by CIRVA (known in English as the International Committee for the Recovery of the Vaquita), a recovery team composed of government, academic, and NGO scientists. At first, the relevant fisheries were legal, though poorly monitored and managed, and they caused the vaquita's initial decline to about 100 individuals by 2014 (CIRVA 2014). Since 2015, a gillnet ban has been in place within the vaquita's range, and now most gillnet fishing in the vaquita's range is illegal, but it has continued, and probably even increased, due to the Chinese demand for swim bladders from a large fish (the

22 Endangered Odontocetes and the Social Connection: Selected Examples... 471

totoaba, *Totoaba macdonaldi*), which shares much of its range with the vaquita and becomes entangled along with vaquitas (Robles et al. 1987).

The gillnet ban followed a plan by the Mexican government to eliminate the gillnet threat through voluntary means (buyouts and "rent-outs") to reduce fishing gear causing vaquita mortality (Rojas-Bracho and Reeves 2013). This did not work, and a more stringent gillnet ban was then seen to provide renewed hope, but also was ineffective. A desperate program to capture several of the remaining vaquitas and remove them from the dangers of life at sea (called VaquitaCPR) was attempted in late 2017 (CIRVA 2017a, b). After two individuals were captured, both showing signs of major stress, and the second animal died in human care, this effort was canceled (CIRVA 2018). There is now little chance that the vaquita can be saved. It will take an unlikely radical change that results in quick elimination of the demand for totoaba bladders or a massive increase in the efficiency of fishery enforcement efforts through the donation of tens/hundreds of millions of dollars from a rich donor. Most people involved are not giving up hope, but there is also the realization that the clock has nearly run out (Thomas et al. 2017).

The vaquita situation demonstrates the tragic case of a species that occurs only within an area where conservation efforts (necessarily expensive and disruptive to some human activities) are viewed by much of the local population as in conflict with the welfare of people (see Cantu-Guzman et al. (2015) for a review of Mexico's history of vaquita 'mismanagement' efforts). The "its us or them" argument, when made in a situation like this, although not necessarily valid, is often effective in hampering or slowing conservation efforts by those who oppose them.

22.3.2 Atlantic Humpback Dolphin (Sousa teuszii): Critically Endangered

The Atlantic humpback dolphin (Fig. 22.2), an obligate shallow-water species of West Africa, is listed as Critically Endangered on the IUCN Red List, largely based on inferred evidence of a decline in population size over the past three generations (about 75 years, Collins et al. 2017). Only the vaquita, with less than 30 individuals remaining, shares that dire listing among cetaceans. Yet, the Atlantic humpback dolphin was not even viewed as in great danger of extinction until recently. It had traditionally been in one of the "data-poor" categories (data deficient or insufficiently known). It was listed as Vulnerable (the least severe of the threatened categories) when previously assessed in 2012, and it was not until its reassessment in 2015 that it was recommended for a higher level of extinction risk (Critically Endangered, Collins 2015). There have been some recent calls for attention to the plight of this species (Van Waerebeek and Jefferson 2004; Weir et al. 2011), but there has been no new concerted action, with no recovery team or recovery plan in place.

Part of the reason for little or no action on Atlantic humpback dolphins is that the species ranks among the least known of all small cetaceans, and even its taxonomic

Fig. 22.2 An Atlantic humpback dolphin, *Sousa teuszii*, breaks the surface near shore off Angola, West Africa. Although the population biology of this species is very poorly known, there is little doubt that it is seriously threatened by human activities and is in danger of extinction in the next decade or two. Photo by C. Weir/Ketos Ecology

status as a distinct species has been uncertain until recently (see Jefferson and Van Waerebeek 2004; Jefferson and Rosenbaum 2014). There are likely fewer than 3000 of these animals remaining, and the separate populations appear to be small and declining (Collins et al. 2017). There is apparent isolation of many populations, and much of this may have been caused by human impacts resulting in severe fragmentation (e.g., Van Waerebeek et al. 2004). However, it should be noted that this species has been virtually unstudied in most of its range, many of the conclusions about its status are based on old or incomplete data, and its current IUCN status makes liberal use of inference and assumption.

The species faces multiple threats, including fishery bycatch; directed killing (and use in the "marine bush meat" trade); prey reduction from overfishing; coastal development and associated habitat loss, degradation, and disturbance; and environmental contamination (Collins et al. 2017). Fishery impacts (both direct and indirect) are especially concerning and probably are the major factors causing the species to decline. With the exception of the creation of some marine protected areas (most of which are ecosystem-oriented, without much specific focus on dolphins), conservation measures designed to help the species are nearly nonexistent. Most or all of the countries in the species' range are poor, with a shortage of available sources of protein for humans, and the political situations in many of them are unstable. Even basic assessment needs are unmet, such as population structure and monitoring, and specifics on exact range. Where there has been work, it is quite recent and not well funded (see Collins 2015; Collins et al. 2017). Because of fragmentation and population declines, it is almost certain that genetic diversity of the species has declined, although empirical data are lacking.

Conservation measures that are effective in stopping and reversing population declines will have to be bold and will need to bridge international borders (e.g., there are 19 countries within the species' moderate ribbon-like range of <10,000 linear km, although only 11–12 of these currently have known records—see Collins 2015; Weir and Collins 2015). Extinction risk is considered to be very real for this species in the next decade or two (Collins et al. 2017). There is a very real need for robust, quantitative population assessments in major portions of its range (such as that recently undertaken in Guinea by Weir 2015), and presently conservation action to benefit the species is virtually nonexistent. For these reasons, I consider the Atlantic humpback dolphin to be the second-most endangered cetacean species on the planet, after the vaquita.

22.3.3 *Indus River Dolphin* (Platanista minor)*: Endangered (Under* P. gangetica)

The Indus River dolphin (Fig. 22.3) is currently recognized as a subspecies of the South Asian river dolphin (*Platanista gangetica*—classified by IUCN as Endangered—Braulik and Smith 2017), though it has vacillated several times in the past few hundred years between being viewed as a distinct species and a rank somewhere below the species level (Braulik et al. 2015). Recent molecular and morphometric work (primarily by G. Braulik and her colleagues) has shown strong evidence from multiple markers of species-level distinctness from the form in the Ganges/Brahmaputra/Meghna and Karnaphuli/Sangu river systems, and I here tentatively recognize the Indus form as a full species (*Platanista minor*), in anticipation of an impending split by the Society for Marine Mammalogy's Taxonomy Committee.

The Indus "species" occurs only in the lower portions of the Indus River system of Pakistan and India (Reeves and Brownell 1989). This group has been fragmented by a series of barrages, built primarily to divert water for agricultural irrigation, into 17 relatively distinct segments, essentially making it a meta-population. Of these, at least ten groupings have already been extirpated, and only six to seven still contain dolphins, though most individuals are found in just three of these (Braulik and Smith 2017). Overall, there has been a nearly 80% reduction in the size of the range of this species since the 1870s (Anderson 1879; Reeves et al. 1991). Abundance and density decrease as one moves upstream, away from the mouth of the river.

The major threat to the Indus River dolphin is severe fragmentation caused by the creation of barrages (Reeves and Brownell 1989; Braulik 2012), as dolphin movement through the barrage gates is almost impossible. Another serious threat is water pollution, from industrial wastes, agricultural (including DDT and other pesticides) effluent, and poorly treated human sewage. Though largely unstudied, impacts of

Fig. 22.3 An Indus River dolphin (bhulan), *Platanista* sp.; mother and calf swim next to a ferry boat in the Beas River of India. It is likely that the Indus form is distinct from the Ganges form at the species level, and therefore the bhulan moves up the list of most endangered cetacean species. Photo by H. Aisha, courtesy of G. Braulik

massive pollution on these animals must be significant. Bycatch in fishing gear is another significant problem (Braulik and Smith 2017). Despite the fact that much fishing occurs in side channels of the river (where dolphins are less frequently found), incidental kills are still likely unsustainable. Between 1993 and 2012, at least 95 dolphins were killed in fishing gear in the main sections of the river (Waqas et al. 2012). Finally, hunting was a significant threat in the past, but since the early 1970s, it has been largely curbed (Pilleri and Zbinden 1973/74).

The Indus River dolphin numbers fewer than 2000 individuals. Four surveys of their range between 2001 and 2017 resulted in counts of 1200–1800 individuals (Braulik 2006; Braulik et al. 2012a, b; Noureen 2013). However, there is some good news. Although overall trends in the entire population are not known, there is good evidence of an increase in the size of the subpopulation between Guddu and Sukkur barrages. The best evidence suggests a 5.65% annual increase in size of this subpopulation since the 1970s, though methodological differences may explain some portion of this (Braulik et al. 2012a, b). Although its future is by no means assured, the ability of this subpopulation to increase in numbers in the face of multiple anthropogenic problems is somewhat encouraging.

Fig. 22.4 Two Chilean dolphins, *Cephalorhynchus eutropia*, cavort in the fjords of Chile. Much uncertainty surrounds the Chilean dolphin, as it has only recently been studied in a few locations. But there are concerns about heavy mortality and habitat issues, and the species is likely declining. Photo by S. Heinrich

22.3.4 Chilean Dolphin (Cephalorhynchus eutropia): Near Threatened

The Chilean dolphin (Fig. 22.4) is listed on the Red List as Near Threatened (since 2008—Heinrich and Reeves 2017), and before that was in one of the "data-poor" categories. It remains a poorly known species, with detailed studies only conducted in a handful of small areas within the species' range in Chile and Argentina (Goodall 1994; Heinrich 2006). The species is thought to number in the "low thousands," and there are currently no quantitative data on the decline rate (though it is assumed to be declining), resulting in the IUCN assessment not leading to one of the threatened categories (Heinrich and Reeves 2017). Although it may not be as rare as once thought, due to it shyness and evasive behavior, there is little doubt that the species is at risk. Dawson (2018) cautioned that the species numbers may only be in the hundreds and suggested "urgent consideration."

There are a number of threats to Chilean dolphins. These include hunting for food and crab bait, which was much more serious in the past (late 1900s) than it is thought to be at present (Lescrauwaet and Gibbons 1994). Fishery bycatch is a serious threat and probably affects the species throughout its range, with gillnets a particularly serious problem. The magnitude of the take is largely unknown. Negative effects of aquaculture farming (mostly for salmon and various invertebrates) have also been implicated, though impact levels are undetermined (Heinrich and Reeves 2017).

There has been little directed conservation or management work on Chilean dolphins (Viddi et al. 2016), and it seems likely to me that the species should qualify for one of the threatened categories (perhaps Endangered), but lack of quantitative data on population levels and decline rates has prevented it from being placed into the appropriate category. However, the situation here is not much different than for the Atlantic humpback dolphin (which similarly has virtually no data on population or trends, but for which inference has been used to move it into the Critically Endangered category), so it seems that a more precautionary approach using more liberal use of informed assumption and inference would place the Chilean dolphin into a category that more accurately reflects its extinction risk. I hope that will be done soon.

22.4   The Behavioral and Social Connections

There are several factors related to behavioral ecology of these threatened species, which make them especially vulnerable to human impacts. These are divided into two categories below: behavioral and social.

Behavioral Factors With the possible exception of the North Pacific right whale (for which the range/habitat is still poorly known), the most endangered species are characterized by coastal or inshore/riverine habitats, and those that are endemic to particularly small areas, such as the vaquita and Chilean dolphin, are obviously at even greater risk. Nearshore areas have dramatically increased levels of human activities and thus expose coastal and riverine species to elevated levels of anthropogenic threats. There is even a tendency for some species to be attracted to dangerous or noxious human activities. Examples are the dolphins that ride bow waves of ships and thus become exposed to capture by harpooning (though none of these are on my list) and Indo-Pacific humpback dolphins (*Sousa chinensis*) in Hong Kong, which at times are attracted to dredging activities, putting them at greater risk of being struck by equipment and of acoustic trauma (see Jefferson 2000). In fact, vulnerability to acoustic disturbance and injuries is an issue of particular concern for all odontocete cetaceans, which are sensitive to a wide range of acoustic frequencies and are dependent on their echolocation abilities to navigate, find food and mates, and avoid predators and other threats (Southall 2018).

Last but not least, susceptibility to fishing net entanglement is a major issue for these cetaceans (Northridge 2018a). All eight of the most endangered odontocetes, plus the North Atlantic right whale, have fishing gear entanglement as a major (if not, *the* major) threatening factor. In fact, for the North Atlantic right whale, this issue has now replaced whaling as the main cause of the species' decline.

Social Factors Odontocetes live in groups, and many species have strong and pervasive social bonds. The increased levels of social cohesion of many toothed whale species may help them deal with many external threats in their environment, but they can also be detrimental. This happens when entire cetacean schools perish if intact herds can be driven ashore in drive fisheries (such as those that occur in Japan and the Faroe Islands—Kasuya 2018; Reeves 2018b) or when a single sick or injured individual leads the entire group into a mass stranding event (Odell et al. 1989). The very large school sizes of some species of dolphins and pilot whales (which can be in the hundreds or even thousands of individuals) only exacerbate the problem.

Finally, for odontocetes there is recent recognition that even survival and reproductive success may be influenced by social factors. This particular issue has been discussed by Wade et al. (2012), who argued that odontocetes are more vulnerable to exploitation than mysticetes, due to unique issues related to their social systems. Examples are the importance of ecological or cultural knowledge and the leadership or other specialized roles of certain individuals in social groups (see also Whitehead and Rendell 2015). This is an issue that requires greater consideration in dolphin, porpoise, and toothed whale management.

22.5 Conclusions and Some Lessons Learned

Some "endangered species lists" include species mainly for political or historical reasons, and their legal status listing may not be accurate. Thus, there is a difference between Endangered (the official status listing, with a capital E) and endangered (the true status of a species, with a lowercase e). For example, the sperm whale is listed as Vulnerable by the International Union for the Conservation of Nature (IUCN) and Endangered by the US Government, yet it is globally distributed, numbers at least hundreds of thousands, and many populations are quite healthy (Whitehead 2003). Thus, the sperm whale, while technically and legally an Endangered species, is not really endangered.

All species of small cetaceans that are in serious danger of extinction are coastal or inshore/riverine species. Despite the incidental kills of millions of pelagic dolphins of several species in offshore drift net and purse seine fisheries (Reeves et al. 2013a, b), none of the truly oceanic species are endangered. This should not surprise us, as the negative effects of human activities on seas and freshwater bodies of the world are much more intense close to land, where most of our activities occur and where the vast majority of people live. One overriding exception may be large-scale climate change, which could affect even deep ocean basins and the cetaceans living in them (Moore 2005).

The social and behavioral issues that have been identified as potential factors in how resilient or vulnerable odontocete species are to exploitation (and, by analogy, any kind of "removals" from populations) should be more fully considered in future assessments. In the past, most management of these species has focused on numbers of individuals removed by outright mortality (see Barlow 2018) and has not considered ages, sexes, or social roles that those individuals play in a cohesive functioning society. In light of evidence that cetacean culture relies on specialized roles of herd defense, alloparenting, assisted reproduction, and other aspects of tight-knit societies, this issue needs to be taken into consideration in future comprehensive management schemes.

There is no doubt that several cetacean species (and many other populations) are in danger of extinction in the next decade or two. That this is the case in a world of great human wealth and enormous technical achievement is a sad statement on our lack of concern for the natural environment. I hope this paper will help especially our new generations of scientists and environmental biologists to better appreciate the diversity and fragility of the world's odontocetes, to better understand which species are truly most at risk from our actions, and to inspire them to change their behavior and that of others and work toward the long-term preservation of these fascinating large-brained social mammals. The chair of the IUCN Cetacean Specialist Group, Randall R. Reeves, recently gave a sobering prediction: "Only with broad-scale, deeply felt changes in how we value and care for what remains of the world's natural variety and abundance can we hope to head off, or even just slow down, a cascade of marine mammal extinctions in the coming decades" (Reeves 2018a, p. 229). The baiji and the vaquita, as well as the rest of these endangered species, remind us that we need to listen and heed his warning and soon.

Acknowledgments The ideas presented in this paper, though my own, were formulated through discussions with many valued colleagues over the years, and I thank them for their inspiration and wisdom.

References

Anderson J (1878/1879) Anatomical and zoological researches: comprising an account of the zoological results of the two expeditions to western Yunnan in 1868 and 1875; and a monograph of the two cetacean genera, *Platanista* and *Orcella*. Bernard Quartich, London, UK

Barlow J (2018) Management. In: Würsig B, Thewissen JGM, Kovacs KM (eds) Encyclopedia of marine mammals, 3rd edn. Academic Press, San Diego, CA, pp 555–558

Braulik G (2006) Status assessment of the Indus River dolphin, *Platanista gangetica minor*, March–April 2001. Biol Conserv 129:579–590

Braulik GT (2012) Conservation ecology and phylogenetics of the Indus River dolphin (*Platanista minor*). Ph.D. Thesis, University of St. Andrews, p 259

Braulik GT, Smith BD (2017) *Platanista gangetica*. The IUCN Red List of Threatened Species 2017: e.T41758A50383612. https://doi.org/10.2305/IUCN.UK.2017-3.RLTS.T41758A50383612.en. Accessed 16 Apr 2018

Braulik GT, Reichert AP, Ehsan T, Khan S, Northridge SP, Alexander JS, Garstang R (2012a) Habitat use by a freshwater dolphin in the low-water season. Aquat Conserv Mar Freshwat Ecosyst 22:533–546

Braulik GT, Bhatti ZI, Ehsan T, Hussain B, Khan AR, Khan A, Khan U, Kundi K, Rajput R, Reichert AP, Northridge SP, Bhaagat HB, Garstang R (2012b) Robust abundance estimate for endangered river dolphin subspecies in South Asia. Endanger Species Res 17:201–215

Braulik GT, Barnett RV, Islas-Villanueva V, Hoelzel AR, Graves JA (2015) One species or two? Vicariance, lineage divergence and low mtDNA diversity in geographically isolated populations of South Asian river dolphin. J Mamm Evol 22:111–120

Braulik GT, Findlay K, Cerchio S, Baldwin R, Perrin W (2017) *Sousa plumbea*. The IUCN Red List of Threatened Species 2017:e.T82031633A82031644. https://doi.org/10.2305/IUCN.UK.2017-3.RLTS.T82031633A82031644.en. Accessed 16 Apr 2018

Brownell RL, Ralls K, Perrin WF (1989) The plight of the 'forgotten' whales. Oceanus 32:5–11

22 Endangered Odontocetes and the Social Connection: Selected Examples... 479

Brownell RL Jr, Yablokov AV, Ivaschenko YV (2018) Whaling, illegal and pirate. In: Würsig B, Thewissen JGM, Kovacs KM (eds) Encyclopedia of marine mammals, 3rd edn. Academic Press, San Diego, CA, pp 1063–1066

Cantu-Guzman JC, Oliviera-Bonavilla A, Sanchez-Saldana ME (2015) A history (1990-2015) of mismanaging the vaquita into extinction - A Mexican NGO's perspective. J Mar Anim Ecol 8:15–25

CIRVA (International Committee for the Recovery of the Vaquita) (2014) Report of the fifth meeting of the international committee for the recovery of the vaquita (CIRVA), Ensenada, Baja California, México, 8–10 July 2014, p 43

CIRVA (International Committee for the Recovery of the Vaquita) (2017a) Report of the eight meeting of the international committee for the recovery of the vaquita (CIRVA), La Jolla, CA, 29–30 Nov 2016, p 69

CIRVA (International Committee for the Recovery of the Vaquita) (2017b) Report of the ninth meeting of the international committee for the recovery of the vaquita (CIRVA), La Jolla, CA, 25–26 Apr 2017, p 32

CIRVA (International Committee for the Recovery of the Vaquita) (2018) Report of the tenth meeting of the international committee for the recovery of the vaquita (CIRVA), La Jolla, CA, 11–12 Dec 2017, p 65

Clapham PJ, Baker CS (2018) Whaling, modern. In: Würsig B, Thewissen JGM, Kovacs KM (eds) Encyclopedia of marine mammals, 3rd edn. Academic Press, San Diego, CA, pp 1070–1074

Collins T (2015) Re-assessment of the conservation status of the atlantic humpback dolphin, *Sousa teuszii* (Kükenthal, 1892) using the IUCN red list criteria. In: Jefferson TA, Curry BE (eds) Humpback dolphins (*Sousa* spp.): current status and conservation, Part 1: Advances in marine biology. Elsevier, London, UK, pp 47–78

Collins T, Braulik GT, Perrin W (2017) *Sousa teuszii*. The IUCN Red List of Threatened Species 2017: e.T20425A50372734. https://doi.org/10.2305/IUCN.UK.2017-3.RLTS.T20425A50372734.en. Accessed 16 Apr 2018

Dawson SM (2018) *Cephalorhynchus* dolphins. In: Würsig B, Thewissen JGM, Kovacs KM (eds) Encyclopedia of marine mammals, 3rd edn. Academic Press, San Diego, CA, pp 166–172

Ellis R (2018) Whaling, aboriginal and western traditional. In: Würsig B, Thewissen JGM, Kovacs KM (eds) Encyclopedia of marine mammals, 3rd edn. Academic Press, San Diego, CA, pp 1054–1063

Evans PGH (2018) Habitat pressures. In: Würsig B, Thewissen JGM, Kovacs KM (eds) Encyclopedia of marine mammals, 3rd edn. Academic Press, San Diego, CA, pp 441–447

Goodall RNP (1994) Chilean dolphin *Cephalorhynchus eutropia* (Gray, 1846). In: Ridgway SH, Harrison R (eds) Handbook of marine mammals, volume 5: the first book of dolphins. Academic Press, London, UK, pp 269–287

Heinrich S (2006) Ecology of Chilean dolphins and Peale's dolphins at Isla Chiloe, southern Chile. Ph.D. Thesis, University of St. Andrews, p 258

Heinrich S, Reeves R (2017) *Cephalorhynchus eutropia*. The IUCN Red List of Threatened Species 2017:e.T4160A50351955. https://doi.org/10.2305/IUCN.UK.2017-3.RLTS.T4160A50351955.en. Accessed 16 Apr 2018

Jefferson TA (2000) Population biology of the Indo-Pacific hump-backed dolphin in Hong Kong waters. Wildl Monogr 144:1–65

Jefferson TA (2019) Save the whales website. https://savethewhales.org/threatened-and-endagered/. Accessed 24 Jan 2019

Jefferson TA, Curry BE (1994) A global review of porpoise (Cetacea, Phocoenidae) mortality in gillnets. Biol Conserv 67:167–183

Jefferson TA, Rosenbaum HR (2014) Taxonomic revision of the humpback dolphins (*Sousa* spp.), and description of a new species from Australia. Mar Mamm Sci 30:1494–1541

Jefferson TA, Van Waerebeek K (2004) Geographic variation in skull morphology of humpback dolphins (*Sousa* spp.). Aquat Mamm 30:3–17

Kasuya T (2018) Whaling, Japanese. In: Würsig B, Thewissen JGM, Kovacs KM (eds) Encyclopedia of marine mammals, 3rd edn. Academic Press, San Diego, CA, pp 1066–1070

480   T. A. Jefferson

Kraus SD (2018) Entanglement of whales in fishing gear. In: Würsig B, Thewissen JGM, Kovacs KM (eds) Encyclopedia of marine mammals, 3rd edn. Academic Press, San Diego, CA, p 336

Lescrauwaet AC, Gibbons J (1994) Mortality of small cetaceans and the crab bait fishery in the Magallanes area of Chile since 1980. Rep Int Whal Commn (Spec Iss) 15:485–494

Merrick RL, Silber GK, DeMaster DP (2018) Endangered species and populations. In: Würsig B, Thewissen JGM, Kovacs KM (eds) Encyclopedia of marine mammals, 3rd edn. Academic Press, San Diego, CA, pp 313–318, 1157

Minton G, Smith BD, Braulik GT, Kreb D, Sutaria D, Reeves R (2017) *Orcaella brevirostris*. The IUCN Red List of Threatened Species 2017:e.T15419A50367860. https://doi.org/10.2305/IUCN.UK.2017-3.RLTS.T15419A50367860.en. Accessed 16 Apr 2018

Moore SE (2005) Long-term environmental change and marine mammals. In: Reynolds JE, Perrin WF, Reeves RR, Montgomery S, Ragen TJ (eds) Marine mammal research: conservation beyond crisis. Johns Hopkins University Press, Baltimore, MD, pp 137–148

Moore KM, Simeone CA, Brownell RL Jr (2018) Strandings. In: Würsig B, Thewissen JGM, Kovacs KM (eds) Encyclopedia of marine mammals, 3rd edn. Academic Press, San Diego, CA, pp 945–951

Norris KS, McFarland WN (1958) A new harbor porpoise of the genus *Phocoena* from the Gulf of California. J Mammal 39:22–39

Northridge S (2018a) Bycatch. In: Würsig B, Thewissen JGM, Kovacs KM (eds) Encyclopedia of marine mammals, 3rd edn. Academic Press, San Diego, CA, pp 149–152

Northridge S (2018b) Fisheries interactions. In: Würsig B, Thewissen JGM, Kovacs KM (eds) Encyclopedia of marine mammals, 3rd edn. Academic Press, San Diego, CA, pp 375–384

Noureen U (2013) Indus River dolphin (*Platanista gangetica minor*) abundance estimations between Chashma and Sukkur barrages, in the Indus River, Pakistan. M. Phil. Thesis, Quaid-e-Azam University

Odell DK, Walsh MT, Asper ED (1989) Cetacean mass strandings: healthy vs. sick animals. Whalewatcher 23:9–10

Perrin WF (1999) Selected examples of small cetaceans at risk. In: Twiss JR, Reeves RR (eds) Conservation and management of marine mammals. Smithsonian University Press, Washington, DC, pp 296–310

Pilleri G, Zbinden K (1973-74) Size and ecology of the dolphin population (*Platanista indi*) between Sukkur and Guddu Barrages, Indus River. Invest Cetacea 5:59–70

Reeves RR (2018a) Conservation. In: Würsig B, Thewissen JGM, Kovacs KM (eds) Encyclopedia of marine mammals, 3rd edn. Academic Press, San Diego, CA, pp 215–229, 1157

Reeves RR (2018b) Hunting. In: Würsig B, Thewissen JGM, Kovacs KM (eds) Encyclopedia of marine mammals, 3rd edn. Academic Press, San Diego, CA, pp 492–496, 1157

Reeves RR, Brownell RL (1989) Susu *Platanista gangetica* (Roxburgh, 1801) and *Platanista minor* Owen, 1853. In: Ridgway SH, Harrison R (eds) Handbook of marine mammals, volume 4: river dolphins and the larger toothed whales. Academic Press, London, UK, pp 69–99

Reeves RR, Chaudhry AA, Umeed K (1991) Competing for water on the Indus plain: is there a future for Pakistan's river dolphins? Environ Conserv 18:341–350

Reeves RR, Smith BD, Crespo EA, Notarbartolo Di Sciara G (2003) Dolphins, whales and porpoises: 2002–2010 conservation action plan for the world's cetaceans. IUCN - The World Conservation Union, Gland, Switzerland

Reeves RR, McClellan K, Werner TB (2013a) Marine mammal bycatch in gillnet and other entangling net fisheries, 1990 to 2011. Endanger Species Res 20:71–97

Reeves RR, Dawson SM, Jefferson TA, Karczmarski L, Laidre K, O'Corry-Crowe G, Rojas-Bracho L, Secchi ER, Slooten E, Smith BD, Wang JY, Zhou K (2013b) *Cephalorhynchus hectori*. The IUCN Red List of Threatened Species 2013:e.T4162A44199757. https://doi.org/10.2305/IUCN.UK.2013-1.RLTS.T4162A44199757.en. Accessed 16 Apr 2018

Reijnders PJH, Borrell A, Van Franeker JA, Aguilar A (2018) Pollution. In: Würsig B, Thewissen JGM, Kovacs KM (eds) Encyclopedia of marine mammals, 3rd edn. Academic Press, San Diego, CA, pp 746–753

Reilly SB, Bannister JL, Best PB, Brown M, Brownell RL Jr, Butterworth DS, Clapham PJ, Cooke J, Donovan GP, Urbán J, Zerbini AN (2008) *Eubalaena japonica*. The IUCN Red List

of Threatened Species 2008:e.T41711A10540463. https://doi.org/10.2305/IUCN.UK.2008.
RLTS.T41711A10540463.en. Accessed 16 Apr 2018

Reilly SB, Bannister JL, Best PB, Brown M, Brownell RL Jr, Butterworth DS, Clapham PJ, Cooke J, Donovan G, Urbán J, Zerbini AN (2012) *Eubalaena glacialis*. The IUCN Red List of Threatened Species 2012:e.T41712A17084065. https://doi.org/10.2305/IUCN.UK.2012. RLTS.T41712A17084065.en. Accessed 16 Apr 2018

Robles A, Vidal O, Findley LT (1987) La totoaba y la vaquita: Mexicanas en peligro de extincion. Informacion Cientifica y Tecnologica (CONACYT) 9:4–6

Rojas-Bracho L, Reeves RR (2013) Vaquitas and gillnets: Mexico's ultimate cetacean conservation challenge. Endanger Species Res 21:77–87

Rojas-Bracho L, Taylor BL (2017) *Phocoena sinus*. The IUCN Red List of Threatened Species 2017: e.T17028A50370296. https://doi.org/10.2305/IUCN.UK.2017-2.RLTS.T17028A50370296.en. Accessed 16 Apr 2018

Southall BL (2018) Noise. In: Würsig B, Thewissen JGM, Kovacs KM (eds) Encyclopedia of marine mammals, 3rd edn. Academic Press, San Diego, CA, pp 637–645

Thomas L, Jaramillo-Legorreta A, Cardenas-Hinejosa GA, Nieto-Garcia E, Rojas-Bracho L, Ver Hoef J, Moore J, Taylor BL, Barlow J, Tregenza N (2017) Last call: passive acoustic monitoring shows continued rapid decline of critically endangered vaquita. J Acoust Soc Am 142:EL512

Townsend CH (1935) The distribution of certain whales as shown by logbook records of American whaleships. Zoologica 19:3–50

Turvey S (2008) Witness to extinction: how we failed to save the Yangtze river dolphin. Oxford University Press, Oxford, UK

Turvey ST (2010) Failure of the baiji recovery program: conservation lessons for other freshwater cetaceans. In: Ruiz-Garcia M, Shostell J (eds) Biology, evolution and conservation of river dolphins within South America and Asia. Nova Science, Enfield, NH, pp 377–394

Turvey ST, Pitman RL, Taylor BL, Barlow J, Akamatsu T, Barrett LA, Zhao X, Reeves RR, Stewart BS, Wang K, Wei Z, Zhang Z, Pusser LT, Richlen M, Bandon JR, Wang D (2007) First human-caused extinction of a cetacean species. Biol Lett 3:537–540

Van Waerebeek K, Jefferson TA (2004) Dolphins under threat: conservation of humpback dolphins. Species 41:6

Van Waerebeek K, Barnett L, Camara A, Cham A, Diallo M, Djiba A, Jallow A, Ndiaye E, Ould-Bilal AOS, Bamy IL (2004) Distribution, status, and biology of the Atlantic humpback dolphin, *Sousa teuszii* (Kükenthal, 1892). Aquat Mamm 30:56–83

Viddi FA, Harcourt RG, Hucke-Gaete R (2016) Identifying key habitats for the conservation of Chilean dolphins in the fjords of southern Chile. Aquat Conserv Mar Freshwat Ecosyst 26 (3):506–516

Wade PR, Reeves RR, Mesnick SL (2012) Social and behavioural factors in cetacean responses to overexploitation: are odontocetes less "resilient" than mysticetes? J Mar Biol 2012:567276. https://doi.org/10.1155/2012/567276

Waqas U, Malik MI, Khokhar LA (2012) Conservation of Indus River dolphin (*Platanista gangetica minor*) in the Indus River system, Pakistan: an overview. Rec Zool Surv Pak 21:82–85

Weir CR (2015) Photo-identification and habitat use of Atlantic humpback dolphins *Sousa teuszii* around the Río Nuñez Estuary in Guinea, West Africa. Afr J Mar Sci 37:325–334

Weir CR, Collins T (2015) A review of the geographical distribution and habitat of the Atlantic humpback dolphin (*Sousa teuszii*). In: Jefferson TA, Curry BE (eds) Humpback dolphins (*Sousa* spp.): current status and conservation, Part 1: Advances in marine biology. Elsevier, London, UK, pp 79–118

Weir CR, Van Waerebeek K, Jefferson TA, Collins T (2011) West Africa's Atlantic humpback dolphin (*Sousa teuszii*): endemic, enigmatic, and soon endangered? Afr Zool 46:1–17

Whitehead H (2003) Sperm whales: social evolution in the ocean. University of Chicago Press, Chicago, IL, p 431

Whitehead H, Rendell L (2015) The cultural lives of whales and dolphins. University of Chicago Press, Chicago, IL, p 417

Chapter 23
Ethology and Behavioral Ecology of Odontocetes: Concluding Remarks

Bernd Würsig

> *The Huzza Porpoise. This is the common porpoise found almost all over the globe. The name is of my own bestowal; for there are more than one sort of porpoises, and something must be done to distinguish them. I call him thus, because he always swims in hilarious shoals, which upon the broad sea keep tossing themselves to heaven like caps in a Fourth-of-July crowd. Their appearance is generally hailed with delight by the mariner. Full of fine spirits, they invariably come from the breezy billows to windward. They are the lads that always live before the wind. They are accounted a lucky omen. If you yourself can withstand three cheers at beholding these vivacious fish, then heaven help ye; the spirit of godly gamesomeness is not in ye.*
>
> (Melville 1851)

Abstract The odontocetes—especially the sperm whale (*Physeter macrocephalus*) and delphinids (family Delphinidae)—have been the subjects of much attention by ancient to modern cultures, exemplified well by Herman Melville's writings of the 1850s. Most odontocetes have multilayered sophisticated societies, probably relying much on living together for several decades, knowing each other well, and remembering the past to unknown degree. Odontocete schooling has similarities to moving terrestrial ungulate herds, and perhaps even more similarities to three-dimensional flocking of birds and schooling of fishes. As mammals, they have the disadvantage of needing to stop feeding and other activities to regularly come to the surface to breathe, and the advantages of echolocation and large brains. It is possible but unproven that dolphins can learn about each other to some degree by echolocating into each other. Large brains and long lives make cultural ways particularly possible,

---

B. Würsig (✉)
Department of Marine Biology, Ocean and Coastal Sciences, Texas A&M University at Galveston, Galveston, TX, USA

© Springer Nature Switzerland AG 2019
B. Würsig (ed.), *Ethology and Behavioral Ecology of Odontocetes*, Ethology and Behavioral Ecology of Marine Mammals,
https://doi.org/10.1007/978-3-030-16663-2_23

but culture can be or become maladaptive if, for example, a particular way of feeding is no longer efficient but is not abandoned. There is evidence especially from captivity that individuals of a species—just as in humans—have vastly different capabilities, but this aspect of individuality has not been explored in detail in nature. Odontocetes are being impacted by humans, often but not always in detrimental ways. We strive for a greater understanding of them, our impacts on them, and their relationships and impacts on us.

Keywords Societies · Schools · Aggregations · Intelligence · Large brains · Matriarchies · Echolocation · Conservation · Culture

One of the finest books about cetaceans in the English literature is that of Herman Melville's *Moby Dick* (1851 in original three-book form as *The Whale*), a fantasy but with plenty of truths, written over 150 years ago. In it, Melville describes the behavior of ocean-going dolphins (such as those in Chap. 9). I am imagining Melville's whaling and US Navy voyages through the eastern tropical Pacific in the 1840s, encountering ancestors of the same multi-species shoals described in more modern terms in the present book (Fig. 23.1).

Fig. 23.1 Short-beaked common dolphins, *Delphinus delphis delphis*, "running" off the Channel Islands, Southern California Bight. While this kind of hurried movement is probably what Melville was describing, when movement is this rapid, it tends to signal that the animals are not joyful but more likely fearful and in fleeing mode. This could be due to killer whales, *Orcinus orcas*, a shark attack, speedboats associated with tuna fishing, etc. In this photo, no obvious cause for alarm was seen. Photo by Sophie Webb

Another fine book about cetaceans is Karen Pryor's *Lads Before the Wind* (1975), taking Melville's quote as an inspiration for her own description of a scientific voyage with dolphins and other toothed whales in nature and as subjects of training. Both books have been inspirations to my life. While dolphins habitually approach ships and smaller vessels to ride the bow and stern waves "for fun" (Fig. 23.2), it is unlikely that the "lads" (and lasses and their offspring) do so more often from windward than any other direction. What Melville may have meant by his oft-quoted expression is that dolphins approaching the vessel are "making haste", are "in a hurry", as "before the wind" usage of that time may have meant. Never mind our post-analyses, poetry does not need to ascribe to strict science, and "dolphins before the wind" as they approach a ship's bow they will always be for me.

The present book opens with a poem from one of my former graduate students, but the rest sticks close to science (see also Chap. 3). I had asked the authors to speculate—if they wish to, even strongly—about the ethology and behavioral ecology of the animals and systems they are describing, perhaps as conclusion sections to their chapters. Speculation is accepted more easily in a compendium such as this book, unlike for most mainstream science journals. Wisely, none took my suggestion, and all stayed remarkably close to facts of science, with solid cautious interpretations based solely on the data. We chose good scientists.

While largely sticking to science as we know at present, I take this opportunity to also discuss digressions from science, as there has been much written especially on topics of supposed dolphin intelligence. I take issue with some (most?) of this

Fig. 23.2 Long-beaked common dolphins, *Delphinus delphis bairdii*, riding the bow pressure wave of the 44 m barquentine research vessel *Regina Maris*, off Panama. Bow-riding was a common sight for sailors in Melville's days. Photo by B. Würsig

discussion, in part because we do not know much about intelligence (not even in our own species, Gould 1981); in part because there are almost 40 species of dolphins and about 30 species of other odontocetes, with vastly different capabilities among them; and in part because discussion of purported intelligence almost perfectly ignores the individuality of animals, points to get back to later.

Shannon Gowans (Chap. 1) discusses the social structural plasticity of most delphinids, as well as the strong matriarchies of several odontocete species. While there appear to be general rules—such as flexibility in social grouping per behaviors in inshore societies with predictable food resources and low predation risk vs. enhanced capabilities of large (and "more stable"?) societies in open waters with dispersed prey and high danger of predation—there are exceptions. Thus, rough-toothed dolphins (*Steno bredanensis*) may occur in generally small gregarious societies with temporary aggregations of such small groups in near-island oceanic waters (Baird et al. 2008; Baird 2016), and nearshore societies of killer whales (*Orcinus orca*) may be virtually closed (Chap. 11).

Serengeti and other migrating herds of ungulates are aggregations that follow simple (seeming) rules of (1) avoiding crowding their nearest neighbors, (2) steering toward an average heading of neighbors, and (3) positioning themselves at some average distance, give or take, from their nearest neighbors. There is no central leadership, no central control (Delgado-Mata et al. 2007). Within that aggregation there is almost constant feeding (part of the reason for migrating) and constant vigilance for predators (another part of the reason for migrating), with the need of each individual to pay at least some attention to others nearby, by sight, sound, small intention movements, and perhaps taste, which has been termed a sensory integration system, SIS (Norris and Schilt 1988; Whitehead and Rendell 2015). Within that aggregation there is also much sociality going on: juvenile play as probably parts of learning to be a sexual adult, attempted mating, avoidance of such attempts, mother-calf interactions including giving birth in the moving herd (!), and nursing and other care-giving behaviors. At times, such individual activities result in sub-structuring of the herd, so that (for example) mating groups may form (probably largely because others avoid the rambunctious activities of mating); mothers and calves may have some tendency to stay together as well, possibly to combat boisterous activities by males while needing to be especially vigilant for predators of their calves (Boinski and Garber 2000 provide several excellent reviews).

Bird flocks and fish schools are similar to the above terrestrial mammal summary, but their herding goes on in three-dimensional space, so there is that extra complication of "depth" while staying close to but not too close to nearest neighbors. As well, they move much faster than herding terrestrial mammals. In three dimensions, the concept of a three-dimensional "chorus line," of sensing nearest neighbors but also those beyond them, is apparently of critical need to keep the ever-moving, ever-gyrating flock or school aggregation in synchrony (Potts 1984). Otherwise, overall movements, including avoidance of predators, could not be as rapid as they are, and animals would more often than they do (very rarely) crash into each other, or the school split apart. Just like herding creatures on land, flocking and schooling individuals can carry out personal functions—including breathing—without concern for the others.

23 Ethology and Behavioral Ecology of Odontocetes: Concluding Remarks 487

Dolphins in a school, especially a large one, are a (social) aggregation, with movement apparently without leadership, almost-synchronous turns and occasional bursts of speed, etc. (but see Markowitz 2004, for possible indications of "leadership" in determining directions of movement). Norris and Dohl (1980) pointed out that a resting group of spinner dolphins (*Stenella longirostris*) behaviorally appears similar to a fish school—when humans approach with a kayak, for example, the resting group may deviate left or right, or dive, or split apart only to reform behind the kayak's intrusion, almost in ameboid-shape fashion. A school of fish and a flock of birds confronted by an object do the same, but more rapidly.

However, for odontocetes, there are several major added aspects to those of three-dimensional bird flocks and fish schools. (1) As mammals, odontocetes need to come to the surface to breathe. They need to stop most other activities to ascend from at times prodigious depths to gain that life-giving set of breaths. Dusky dolphins (*Lagenorhynchus obscurus*) can feed down to about 130 m depths, but that's it—they can take a few bites of aggregations of myctophids or squid and then need to ascend those 130 m, take several breaths, and go back down again to 130 m while the prey, accessibility time-limited as prey are only at these shallow depths for a few hours at night, is available. Whether dolphins do this as a cohesive subgroup of animals, perhaps for safety in sensory integration and for potential coordinated herding at depth, or alone, depends on the occasion and whims we do not yet fully understand (Benoit-Bird et al. 2004; Benoit-Bird and Au 2009). Sperm whale (*Physeter macrocephalus*) females with young occur in tight matriarchies. They need to dive deep and for long, and there is evidence that they do so synchronously when without small calves but asynchronously when newborns are present, presumably for alloparenting calves at and near the surface while mother is diving (Whitehead 1996). (2) All odontocetes have "that extra" sense of discerning each other, prey, predators, and the outside world in general, by echolocation. Only one other mammalian group, the bats, have this capability in similarly sophisticated fashion, and bats also feed and avoid predators in a three dimensional environment. (3) Dolphins are social mammals that have evolved large brains, larger than the terrestrial ungulates to whom they are related and rivaling the carnivores that they also resemble, for after all, they are carnivorous creatures that prey on fish and squid and—for several species such as killer whales, *Orcinus orca*—on other marine mammals as well.

The mention of echolocation above leads me to one aspect of "seeing" by sound that has received little attention by researchers, and probably merits more. The logic goes thusly: Dolphins can detect fish prey some distance below a sandy bottom, leading to "crater feeding" by Bahamian spotted dolphins (*Stenella frontalis*), apparently by echolocation alone (Rossbach and Herzing 1997). It has long been known that bottlenose dolphins and several other species of odontocetes can be trained to detect human explosive mines and other metal objects buried below the substrate, and details of this capability were recently published (Ridgway et al. 2018a). Human-made sonar within the frequency range of echolocation of many toothed whales can "see" the lungs of dolphins even many meters below the surface (Benoit-Bird et al. 2004). While Norris and Harvey (1974) demonstrated that blubber and muscle tissue are not very good sound propagators, it is nevertheless possible that dolphins can "see" into each other by echolocation. At least, they may be able to see lungs, trachea, and other cavities with gasses and perhaps bones of fetuses and some particularly hard-bodied tumors, etc. I am surprised that apparently very little work has been carried out in this realm (confirmed by Sam Ridgway, pers. comm, 22 Dec 2018), as even partial positive answers would open up a wealth of new understanding of odontocete capabilities, including how much information (if any) they may have on aspects of physiological state of conspecifics.

Dolphins while resting/sleeping appear to operate at the level of a sensorally integrated fish school (Norris and Schilt 1988) but can quickly "wake up" and be representatives of a sophisticated social network of cognizant higher-level mammals, with complex patterns of communication (Chap. 2), divisions of synchronous feeding (Chap. 3), social/sexual strategies (Chap. 4), mother-calf interactions and long-term teaching and care (Chap. 5), most efficient movements for life support (Chap. 6), fearing and attempting to avoid predators (Chap. 7), and conducting mammalian-type interactions while constrained by the need to gain the surface every few minutes but empowered by the capabilities of echolocation and large brains (all other chapters of this book).

Large brains and sophisticated social groupings and behaviors take us to the oft-repeated idea that toothed whales and dolphins have reached a pinnacle of "intelligence" unrivalled on this Earth, save perhaps humans. The literature is replete with such assertions, as one of the earliest by Lilly (1962), and there has been much else written on the subject. One of the first to do so is the edited book of *Mind in the Waters: A Book to Celebrate the Consciousness of Whales and Dolphins* (McIntyre 1974), and others have lent considerable intellectual discussion. Donald R. Griffin's *The Question of Animal Awareness* (1976) asked us to open a window on the potential minds of all creatures; Rachel Smolker described her experiences with and thoughts about dolphins in *To Touch a Wild Dolphin* (2001); Toni Frohoff and Brenda Peterson edited a compendium *Between Species: Celebrating the Dolphin-Human Bond* (2003); Denise Herzing and Christine Johnson edited a book *Dolphin Communication and Cognition: Past, Present, and Future* (2015); Hal Whitehead and Luke Rendell wrote the finest book available on *The Cultural Lives of Whales and Dolphins* (2015); Janet Mann edited a book on *Deep Thinkers: Inside the Minds of Whales, Dolphins, and Porpoises* (2017), and Richard Connor wrote a compelling description of his life with dolphins in *Dolphin Politics in Shark Bay: A Journey of Discovery* (2018). I especially like the book edited by Philippa Brakes and Mark Peter Simmonds on *Whales and Dolphins: Cognition, Culture, Conservation and Human Perceptions* (2011) and Justin Gregg's summary *Are Dolphins Really Smart?* (2013). There are more, and please (you editors and authors) do not chastise me too strongly for not mentioning all. Such contributions represent a wealth of knowledge and speculation, with at times bold assertions about possible or believed capabilities of odontocete lives.

There are missteps, and they should not be minimized. John Cunningham Lilly was one of the earliest authors, but much of what he had to say about dolphin behavior and supposed intelligence turned out to be incorrect (Lilly 1962, 1967, 1975), as are

23 Ethology and Behavioral Ecology of Odontocetes: Concluding Remarks   489

the writings of Joan Ocean, *Dolphin Connection: Interdimensional Ways of Living* (1989), who seems to have no concept of the real lives of dolphins—they with amazing capabilities, constraints, and dangers. There are more such "inspirational" writers, and while assertions from "feelings" or beliefs can be good, the inspiration so often presented could stick more to the realisms of what we know, do not know, and might imagine.

Some delphinids do indeed have, on average, very large brains as compared to carnivores and even great apes (Marino et al. 2004). It is likely that these brains make possible sophisticated processing abilities related to communicating and interacting in complex societies, and with detailed spatiotemporal memories of conspecifics and events, such as in humans, other great apes, some terrestrial carnivores (de Waal and Tyack 2003), and elephants (Douglas-Hamilton et al. 2006). While there have been numerous speculations on how such large brains evolved in the sea, Ridgway et al. (2018b) point out that carnivorous cetaceans (instead of their related vegetarian terrestrial ungulates) (1) have a high caloric intake per time spent feeding, providing much energy for the energy-expensive brain; (2) live in a buoyant environment, freeing a large brain from the constraints of gravity; and (3) gestate calves that already have well-developed hearing capabilities in utero in later stages of pregnancy, can likely hear external sounds including echolocation and whistle communication signals of conspecifics, and that this capability may have further driven large brain size from an early age. By the way, it stands to reason that if the fetus can hear echolocation clicks, that enough echoes bounce back for the outside emitter to (potentially) detect aspects of the fetus as well, as discussed for "seeing into bodies" above.

The largest odontocete on Earth, the largest toothed creature on Earth, the sperm whale, also has the largest brain on Earth. It has a matriarchal society and complicated interweaving sets of cultures (Whitehead 2003) but as a part of cultural beings (such as humans!) also engages in some apparently maladaptive behaviors, such as mass stranding to an entire groups' final detriment (Whitehead and Rendell 2015). Killer whales also have intricate tight matriarchies and sophisticated communication and cooperative behaviors but due to cultural ways of feeding on salmon, for example, may be so linked to that feeding style that as salmon are depleted, certain pods do not change behaviors and may not survive (Whitehead and Ford 2018). Rough-toothed dolphins have particularly large brain to body size ratios ("encephalization quotient," Jerison 1973), as well as amazingly quick and complicated learning abilities in captivity (Pryor et al. 1969), Fig. 23.3. They along with bottlenose dolphins and killer and sperm whales may possibly be the "brightest" creatures in the seas, with the understanding that we humans do not have present ways to rigorously define or compare "intelligences" (Würsig 2018).

There are small societies of dolphins living in particular near-shore (inshore) bays and bayous (Wells 2018 and Chap. 15, provide an excellent long-term study example), and we have huge "societies" of oceanic dolphins traveling over hundreds to thousands of kilometers, quite certainly with subgroupings of animals that know each other reasonably well. But, how much can one intelligent mammal really take in? How many animals can be accepted as a part of the closest associations, with a concept of a large complex social system that requires every member to keep track of

Fig. 23.3 Subgroup of rough-toothed dolphins, *Steno bredanensis*, off Kona, Hawai'i. They often seem unhurried and unbothered, as seen here. From a distance, "steno" could be confused with bottlenose dolphins (*Tursiops* sp.), but their forehead to upper jaw is not demarcated by a surface crease as in many other delphinids and a sidewise glance reminds of some dinosaur heads. They are often termed "lizard dolphin" in the earlier literature. They have very large brain to body size ratios and sophisticated rapid learning. Photo by Deron S. Verbeck

(i.e., to remember) a large number of individuals "personally," as well as inter-relationships and histories (and how far back?), as well as spatiotemporal memory? Such complex societies with needs for long-term memory may also be drivers for large brain development, as in carnivores, elephants, and great apes including humans (Marino et al. 2004). In humans, there may well be a practical limit of what "we" can efficiently remember and how many colleagues with whom we can knowledgeably associate, within a framework of stable relationships. This limit has been termed Dunbar's Number (Hill and Dunbar 2003) and seems to be around 150 well-known individuals. It is possible that coastal bottlenose dolphin society sizes (e.g., Vermeulen and Bräger 2015) may not have developed simply by chance but what "our" (human and perhaps delphinid) societies are capable of handling. Such potential limits of the closest of social networks need to be explored further and will be with time.

An important conclusion that Gowans (Chap. 1) brings to the discussion is that it would be advantageous for us to progress to a more "agent based model" (Axelrod 1997; Sun 2006) of appreciating odontocete social organization, ethology, and aspects of behavioral ecology. I interpret this as meaning we should be concerned with the group, of course, but should also progress to the stage where we can obtain data on—and discuss—the individual. To put this in human terms, we have a collective of university students in (say) math class. Some are math prodigies and understand and appreciate everything the professor writes on the board; some others do so after long thought while doing problems at night; some others do not, even after concerted study, understand the ways or the concepts. In a different class such as (say) history, the roles may be at least partially reversed—some of the brightest math students may have problems in keeping up with historical (and philosophical) discussions, while some of the worst math students may excel here. One is not more capable than the other, one is not more "intelligent," but their capabilities go in very different directions. Of course, we also have the students (and, by extension, the dolphins) who may be very good at everything and those who may not be good at much of anything. It is this flavor that we generally tend to omit in aspects of animal associations and behavior, in part due to our lack of depth of having or understanding the data, in part due to our present sophistication of thinking about animal capabilities. This has been put well for members of the great apes, where we know more about societies and individuals than for dolphins (Byrne 1995). Do dusky dolphins efficiently cooperating to herd fish into a tight ball have those members who are good at it and those who are not? The bridge to individuality of better understanding odontocete societies will not be easy.

Many of the "herding-type" cetaceans—*Stenella*, *Lagenorhynchus*, *Delphinus*, and *Cephalorhynchus*—are perhaps not as variably capable ("intelligent") as *Physeter*, *Orcinus*, *Tursiops*, and *Steno* (for the latter, Pryor et al. 1969). And, it is also possible (likely?) that among each of these species, among each of each species' populations, among each group, there are those individuals who are the run of the mill of them as capabilities of mind and action are concerned, those that are the "Einsteins" among them, and those that are merely biding time and not all that helpful to themselves or their society. The latter may be quirks, and they may be tolerated and perhaps even respected as quirks within society, as was the European medieval "village idiot" (or "village savant") of human society (Oliver 1989). It is up to a new set of data and paradigms (and science history) to ascertain whether my assertion here is "mere twaddle," perhaps similarly as history has judged the writings of John C. Lilly.

It would be wrong to write about odontocetes and not mention our human interactions with them and the amazing, often distressing, changes in habitats we have created and continue to create. These go from as little as a near-shore factory putting chemicals into the environment to the huge global conundrum of climate change and ocean acidification (Reeves 2018). Quite a few populations and several species of odontocetes—especially in human-degraded areas near shores—are in danger due to human activities. While "only" one species of odontocete cetacean, the Chinese baiji (*Lipotes vexillifer*; Turvey et al. 2007), has gone extinct in modern times, the Gulf of California harbor porpoise, vaquita (*Phocoena sinus*), is hanging on by a tenuous thread, and several others are also faring poorly (Chap. 22). Bearzi et al. (Chap. 10) explore the multifaceted ways that many odontocetes have adapted to live with humans and the ways that we humans have done so as well. Some adaptation, perhaps much of it, comes with costs such as disruption of human fisheries, and Bearzi et al. enjoin us—as human individuals and societies—to more adequately adapt to, respect, and enjoy the odontocetes so ubiquitous in our human-altered oceans, seas, and several mighty rivers (as, e.g., Bezamat et al. 2018). An important part of this appreciation is for us to pro-actively understand and appreciate the rich social lives of odontocetes, and to thereby aid local and international efforts to save cultural entities and diversity through conservation/management actions (Brakes et al. 2019). A giant biologist of our time, Edward O. Wilson, asks us to more actively and more fully engage in a love of all life and living systems (the concept of biophilia, Wilson 1984, 2010), even as we encroach evermore on the nature that is an intricate part of us.

References

Axelrod R (1997) The complexity of cooperation: agent-based models of competition and collaboration. Princeton University Press, Princeton, NJ. isbn: 978-0-691-01567-5

Baird RW (2016) The lives of Hawai'i's dolphins and whales. University of Hawai'i Press, Honolulu, HI

Baird RW, Webster DL, Mahaffy SD, McSweeney DJ, Schorr GD, Ligon AD (2008) Site fidelity and association patterns in a deep-water dolphin: rough-toothed dolphins (*Steno bredanensis*) in the Hawaiian Archipelago. Mar Mamm Sci 24:535–553

Benoit-Bird KJ, Au WWL (2009) Cooperative prey herding by the pelagic dolphin, *Stenella longirostris*. J Acoust Soc Am 125:125. https://doi.org/10.1121/1.2967480

Benoit-Bird KJ, Würsig B, McFadden CJ (2004) Dusky dolphin (*Lagenorhynchus obscurus*) foraging in two different habitats: active acoustic detection of dolphins and their prey. Mar Mamm Sci 20:215–231

Bezamat C, Simões-Lopes PC, Castilho PV, Daura-Jorge FG (2018) The influence of cooperative foraging with fishermen on the dynamics of a bottlenose dolphin population. Mar Mamm Sci, Online Issue. https://doi.org/10.1111/ms.12565

Boinski S, Garber PA (2000) On the move: how and why animals travel in groups. University of Chicago Press, Chicago, IL

Brakes P, Dall SRX, Aplin LM, Bearhop S, Carroll EL, Ciucci P, Fishlock V, Ford JKB, Garland EC, Keith SA, McGregor PK, Mesnick SL, Noad MJ, di Sciara GN, Robbins MM, Simmonds MP, Spina F, Thornton A, Wade PR, Whiting MJ, Williams J, Rendell L, Whitehead H, Whiten A, Rutz C (2019) Animal cultures matter for conservation: understanding the rich social lives of animals benefits international conservation efforts. Science 363:1032–1034. http://science.sciencemag.org/content/363/6431/1032

Brakes P, Simmonds MP (2011) Whales and dolphins: cognition, culture, conservation, and human perceptions. Earthscan, London, UK

Byrne R (1995) The thinking ape: evolutionary origins of intelligence. Oxford University Press, New York, NY

Connor RC (2018) Dolphin politics in Shark Bay: a journey of discovery. The Dolphin Alliance Project, New Bedford, MA

de Waal FBM, Tyack PL (2003) Animal social complexity: intelligence, culture, and individualized societies. Harvard University Press, Cambridge, MA

Delgado-Mata C, Ibanez J, Bee S, Ruiz R, Aylett R (2007) On the use of virtual animals with artificial fear in virtual environments. New Gener Comput 25:145–169. https://doi.org/10.1007/s00354-007-0009-5

Douglas-Hamilton I, Bhalla S, Wittemyer G, Vollrath F (2006) Behavioural reactions of elephants towards a dying and deceased matriarch. Appl Anim Behav Sci 100:87–102

Frohoff T, Peterson B (2003) Between species: celebrating the dolphin-human bond. Sierra Club Books, San Francisco, CA

Gould SJ (1981) The mismeasure of man. W. W. Norton, New York, NY

Gregg J (2013) Are dolphins really smart? The mammal behind the myth. Oxford University Press, Oxford, UK

Griffin DR (1976) The question of animal awareness: evolutionary continuity of mental experiences. Rockefeller University Press, New York, NY

Herzing DL, Johnson CM (2015) Dolphin communication and cognition: past, present, and future. MIT Press, Cambridge, MA

Hill RA, Dunbar RIM (2003) Social network size in humans. Hum Nat 14:53–72. https://doi.org/10.1007/s12110-003-1016-y

Jerison HJ (1973) Evolution of the brain and intelligence. Academic Press, New York, NY

Lilly JC (1962) Man and dolphin. Victor Gollancz Press, London, UK

Lilly JC (1967) The mind of the dolphin: a non-human intelligence. Doubleday, Garden City, NY

Lilly JC (1975) Lilly on dolphins. Doubleday, Garden City, NY

Mann J (2017) Deep thinkers: inside the minds of whales, dolphins, and porpoises. University of Chicago Press, Chicago, IL

Marino L, McShea DW, Uhen MD (2004) Origin and evolution of large brains in toothed whales. Anat Rec A 281A:1247–1255

Markowitz TM (2004) Social organization of the New Zealand dusky dolphin. Dissertation, Texas A & M University, Galveston, TX, p 278

McIntyre J (1974) Mind in the waters. Charles Scribner, New York, NY

Melville H (1851) The Whale. Book 3, Chapter 1. Page 1. B. Clay Printer, London, UK. (See also: The Project Gutenberg EBook of Moby Dick; or The Whale, by Herman Melville: http://www.gutenberg.org/files/2701/2701-h/2701-h.htm)

Norris KS, Dohl TP (1980) The structure and functions of cetacean schools. In: Herman LM (ed) Cetacean behavior: mechanisms and functions. Wiley-Interscience, New York, NY

Norris KS, Harvey GW (1974) Sound transmission in a porpoise head. J Acoust Soc Am 56:659–664

Norris KS, Schilt CR (1988) Cooperative societies in three-dimensional space: on the origins of aggregations, flocks, and schools, with special reference to dolphins and fish. Ethol Sociobiol 9:149–179

Ocean J (1989) Dolphin connection. Dolphin Connection, Kailua, HI

Oliver M (1989) Disability and dependency: a creation of industrial societies? In: Barton L (ed) Disability and dependency. Routledge; Taylor and Francis, New York, NY. isbn:978-1-85000-616-9

Potts WK (1984) The chorus-line hypothesis of manoeuvre coordination in avian flocks. Nature 309:344–345

Pryor K (1975) Lads before the wind: adventures in porpoise training. Harper and Rowe Press, New York, NY

Pryor KW, Haag R, O'Reilly J (1969) The creative porpoise: training for novel behavior. J Exp Anal Behav 12:653–661

Reeves RR (2018) Conservation. In: Würsig B, Thewissen JGM, Kovacs K (eds) Encyclopedia of marine mammals, 3rd edn. Academic/Elsevier Press, San Diego, CA

Ridgway SH, Dibble DS, Kennemer JA (2018a) Timing and context of dolphin clicks during and after mine simulator detection and marking in the open ocean. Biol Open 7:bio031625. https://doi.org/10.1242/bio.031625

Ridgway SH, Carlin KP, Van Alstyne KR (2018b) Delphinid brain development from neonate to adulthood with comparisons to other cetaceans and artiodactyls. Mar Mamm Sci 34:420–439

Rossbach KA, Herzing DL (1997) Underwater observations of benthicfeeding bottlenose dolphins (*Tursiops truncatus*) near Grand Bahama Island, Bahamas. Mar Mam Sci 13:498–504

Smolker R (2001) To touch a wild dolphin: a journey of discovery with the sea's most intelligent creatures. Anchor Books, Random House, New York, NY

Sun R (2006) Cognition and multi-agent interaction: from cognitive modeling to social simulation. Cambridge University Press, Cambridge. isbn:978-0-521-83964-8

Turvey ST, Pitman RL, Taylor BL, Barlow J, Akamatsu T, Barrett LA, Zhao X, Reeves RR, Stewart BS, Wang K, Wei Z, Zhang Z, Pusser LT, Richlen M, Bandon JR, Wang D (2007) First human-caused extinction of a cetacean species. Biol Lett 3:537–540

Vermeulen E, Bräger S (2015) Demographics of the disappearing bottlenose dolphin in Argentina: a common species on its way out? Plos One 10(3):e0119182. https://doi.org/10.1371/journal.pone.0119182

Wells RW (2018) Bottlenose dolphin, *Tursiops truncatus*, common bottlenose dolphin. In: Würsig B, Thewissen JGM, Kovacs K (eds) Encyclopedia of marine mammals, 3rd edn. Academic/Elsevier Press, San Diego, CA

Whitehead H (1996) Babysitting, dive synchrony, and indications of alloparental care in sperm whales. Behav Ecol Sociobiol 38(4):237–244

Whitehead H (2003) Sperm whales: social evolution in the ocean. University of Chicago Press, Chicago, IL

Whitehead H, Ford JKB (2018) Consequences of culturally-driven ecological specialization: killer whales and beyond. J Theor Biol 456:279–294

Whitehead H, Rendell L (2015) The cultural lives of whales and dolphins. University of Chicago Press, Chicago, IL

Wilson EO (1984) Biophilia. Harvard University Press, Cambridge, MA. isbn:0-674-07442-4

Wilson EO (2010) The creation: an appeal to save life on earth. W.W. Norton, New York, NY

Würsig B (2018) Intelligence. In: Würsig B, Thewissen JGM, Kovacs K (eds) Encyclopedia of marine mammals, 3rd edn. Academic/Elsevier, San Diego, CA

Correction to: Ethology and Behavioral Ecology of Odontocetes

Bernd Würsig

Correction to:
B. Würsig (ed.), *Ethology and Behavioral Ecology*
*of Odontocetes***, Ethology and Behavioral Ecology**
of Marine Mammals,
https://doi.org/10.1007/978-3-030-16663-2

The book was inadvertently published with the incorrect captions for Fig. 9.1 in Chap. 9 and for Fig. 21.8 in Chap. 21, which has been updated as below:

Fig. 9.1:
  "Dolphins of the eastern tropical Pacific (ETP). Top: Central American spinner dolphin, "whitebelly" spinner dolphin; Middle: striped dolphin, coastal pantropical spotted dolphin; Bottom: offshore pantropical spotted dolphin, short-beaked common dolphin. Latin names are in Appendix 1. The species have overlapping ranges but distinct habitat preferences and different school sizes, social structure, and behavior (see text). Photo credit: Southwest Fisheries Science Center, NOAA Fisheries"

Fig. 21.8:
  "Harbor porpoise mating at the surface. The female in front lifts her flukes high out of the water in response to the male's approach. The penis is barely visible near the genital area. Note that the male approach on the left side of the female is similar to all observations in Keener et al. (2018). Photo credit: Kurt Videbæk Jensen"

The corrections have been carried out in the chapters and the updated chapters have been approved by the authors.

The updated online versions of the chapters can be found at
https://doi.org/10.1007/978-3-030-16663-2_9
https://doi.org/10.1007/978-3-030-16663-2_21

© Springer Nature Switzerland AG 2020
B. Würsig (ed.), *Ethology and Behavioral Ecology of Odontocetes*, Ethology and
Behavioral Ecology of Marine Mammals,
https://doi.org/10.1007/978-3-030-16663-2_24

Index[1]

A

Abundance, 148, 190, 191, 199, 202, 254, 297, 306, 309, 310, 320, 325, 373, 404, 443, 452, 462, 473, 478

Acoustics*, 169, 243, 308, 332, 416, 450, 467, 476
  behavior, 244, 246, 247, 253, 266, 283, 308, 458, 460
  communication, 6, 7, 29–31, 33, 34, 36, 37, 39–41, 43, 44, 80, 171, 271, 376, 378
  repertoire, 272, 274
  signaling, 26, 32, 99, 187, 270, 376, 378, 379, 426

Acrobatic behavior, 379

Activity budgets, 351, 389, 390

Adaptations, 28, 42, 79, 88, 118, 212–226, 241, 253, 262, 264, 285, 293, 354, 420, 421, 450, 452, 491

Admiralty Bay, 11, 52, 60, 62, 63, 173, 390, 392, 396–400, 404, 406

Adoption, 359

Affiliative behavior, 26, 196, 349, 358, 415

Age of first birth, 105

Alliances*, 166, 198, 347, 441

Allocare, 170, 174

Allomaternal care, 100, 190, 194, 263, 269, 275

Allonursing, 100, 170, 174

Alloparental care, 184, 187, 202, 276, 359

Altruism, 55

Amazon, 28, 79, 106, 135, 168, 215, 417–419, 425

Amazon river dolphin (*Inia geoffrensis*), 28, 79, 106, 135, 168, 215, 220, 221, 417, 418

Animal communication, 27

Antarctica, 62, 63, 154, 220, 221, 240, 242, 248, 252, 282, 451

Anthropogenic disturbance, 18, 177, 186, 189

Anthropogenic interaction, 174, 175

Anthropogenic sounds, 325

Anti-predator behaviors, 146, 149, 151, 153, 157, 158

Arctic, 107, 130, 152, 155, 173, 221, 452

Argentina, 52, 53, 60, 63, 135, 217, 251, 388, 390–394, 396, 402, 403, 416, 417, 452, 475

Associations*, 168, 194, 216, 254, 309, 340, 347, 372, 415, 437, 489
  index, 321–324, 395, 397, 422
  patterns, 11, 12, 15, 45, 66, 193, 244, 245, 290, 309, 319–325, 393, 398, 437, 441

Atlantic, xi, 12, 43, 54, 122, 149, 217, 243, 269, 282, 308, 339, 370, 401, 450, 466
  herring (*Clupea harengus*), 62, 129
  humpback dolphins, 54, 339, 466, 468, 471–473, 476
  spotted dolphin (*Stenella frontalis*), 12, 54, 60, 65, 104–106, 149, 174, 218, 400, 402, 403, 487

Azores, 128, 131, 132, 272, 285, 291–293

---

[1]An entry with an asterisk (*) after the index word is represented >100 times in the text, and is indexed only upon first appearance per chapter

© Springer Nature Switzerland AG 2019
B. Würsig (ed.), *Ethology and Behavioral Ecology of Odontocetes*, Ethology and Behavioral Ecology of Marine Mammals,
https://doi.org/10.1007/978-3-030-16663-2

B

Baby position, 85, 88
Babysitting, 100, 101, 190, 359
Behavior*, 172, 175, 186, 220, 251, 306, 331, 349, 370, 414, 436, 450, 475, 484
Behavioral ecology, ix, 4, 118, 146, 147, 178, 185, 263, 264, 270, 275, 285, 297, 306, 308, 325, 352, 356, 380, 476, 484, 485, 490
Behavioral flexibility, 12, 171, 172, 174–176, 178, 223, 398, 399
Beluga (*Delphinapterus leucas*), ix, 77, 98, 99, 105, 107, 130, 152, 153, 168, 170, 173, 221
Bigg's killer whale, 241, 252
Biophilia, 225, 492
Biosonar, 29, 263, 265
Birth, 16, 30, 76, 96, 97, 102, 103, 107, 177, 245, 268, 283, 319, 342, 347, 357, 361, 440, 456, 486
Blainville's beaked whale (*Mesoplodon densirostris*), 53, 58, 59, 64, 129, 306–326
Bluefin tuna (*Thunnus thynnus*), 129, 243, 246, 250
Bottlenose dolphins (*Tursiops truncatus* and *Tursiops aduncus*), 4–6, 9–13, 15, 16, 18, 26–33, 35, 37, 52, 54, 55, 57–65, 174, 213, 322, 331, 345, 375, 425, 487
Brain sizes, 171, 353, 389, 398, 489
Breaches, 58, 270, 307, 415
Breaching, 270, 307, 415, 418
Breeding, 75–77, 79, 85, 96, 98, 102, 103, 128, 167, 170, 240, 250, 268, 271, 296, 438, 440, 441, 444, 454
Bull sharks, 101
Burmeister's porpoise (*Phocoena spinipinnis*), 450–452, 455, 462
Buzzes, 61, 265, 266, 270, 316, 333, 425, 458, 460
Bycatch, 104, 199, 201, 216, 298, 339, 340, 403, 428, 442, 443, 461, 462

C

Calf mortality, 103, 104, 155
Call subunits, 39, 40
Canary Islands, 129, 282–289, 291, 293, 296, 308, 320, 321
Captivity, 30, 31, 52, 77, 100, 194, 196, 356, 359, 420, 421, 426, 452, 456, 458, 459, 489
Carousel feeding, 36, 249

Cephalopods, 53, 54, 80, 156, 265, 316, 346, 350, 357, 416
Cephalorhynchus
 C. eutropia, 134, 468, 475, 476
 C. hectori hectori, 441
 C. hectori maui, 439, 466
Cetaceans, x, xii, xiii, 4, 11, 13–19, 27, 28, 52, 64, 65, 75–77, 79–83, 85, 87, 88, 96–98, 100, 102, 103, 108, 109, 118, 129, 137, 147, 158, 177, 184, 187, 190, 197, 202, 212, 215, 216, 221–225, 241–243, 247, 262, 296, 306, 317, 325, 341, 379, 398, 404, 405, 414, 436, 452, 454, 458, 462, 465–468, 471, 473, 474, 476–478, 484, 489, 491
Chemical sensing, 26–45
Chilean and Hector's dolphins (*Cephalorhynchus eutropia* and *C. hectori*), 134
Chilean dolphins, 404, 468, 475, 476
Chilika, 218, 424, 425, 427
Chondrichthyes (sharks and rays), 129, 152, 241
Clans, 27, 37, 38, 41–45, 85, 172, 245, 247, 269, 272–275, 290
Clicks*, 246, 319, 333, 350, 416, 440, 458, 489
Climate change, 137, 152, 158, 298, 403, 404, 477, 491
Codas, 7, 40–44, 85, 109, 266, 269–274
Coexistence, 223–225
Cognition, xii, 171, 353–354, 359, 398, 488
Commerson's dolphins (*Cephalorhynchus commersonii*), 27, 54, 135
Common bottlenose dolphin, xii, 5, 6, 9–11, 13, 15, 16, 18, 26, 27, 31–33, 35, 37, 87, 104, 148, 166, 169, 190, 191, 204, 213–215, 220, 223, 331–342, 375, 393, 404
Common dolphin (*Delphinus delphis*), 17, 54, 58, 60, 83, 131, 132, 185, 190, 192, 197, 198, 203, 214, 218, 222, 484, 485
Communal defense, 189
Communication*, 171, 187, 246, 320, 350, 376, 414, 444, 458, 488
Community*, 171, 186, 245, 341, 353, 374
Competition, 4, 8, 11, 65, 76, 77, 79–84, 86–88, 167, 192, 197, 198, 212, 216, 224, 248, 264, 268, 271, 276, 298, 322, 351, 353, 361, 373, 391, 393, 395, 397, 417, 427, 441, 443, 455
Conditioning, 340
Conflicts, 6, 27, 87, 170, 202, 213, 216, 223–226, 255, 349, 352, 353, 404, 471
Conformity, 274

Index 497

Conservation, xiii, 19, 45, 66, 137, 158, 176–178, 184–202, 215, 224, 275, 325, 342, 381, 382, 389, 403–405, 452, 466, 468, 470–473, 476, 488
Consortships, 10, 86, 347, 348, 350, 354, 361
Consumptive predation, 145
Contact calls, 30, 31, 34, 36, 99, 416
Contact swimming, 349, 358
Contain prey, 59
Contest competitions, 77, 80, 81
Cookie-cutter shark, 101, 308, 375
Cooperation, 6, 8, 9, 11, 52, 154, 156, 167, 169, 171, 172, 189, 217, 248, 349, 391–392
Cooperative behaviors, 169, 170, 172, 177, 294, 339
Coordinated foraging, 52, 64, 332, 340, 393, 394, 397, 398, 403
Coordination, xii, 36, 52, 59, 62, 63, 65, 99, 156, 171, 176, 251, 372, 378, 391–392, 399
Copulations, 75, 77, 81, 84–86, 88, 201, 347, 393, 395, 399
Courtship, 77, 418, 426
Crater feeding, 335, 487
Creaks, 61, 265, 420, 425
Crittercams, 51
Crustaceans, 53, 60, 346, 350, 361
Crypsis, 30, 155
Cultural fad, 356
Cultural traditions, 173, 241, 254, 274, 275, 397, 406
Cultural transmission, 64, 65, 85, 108, 109, 176, 178, 223, 241, 247, 248, 253, 338, 340, 397
Cultures, 18, 40, 44, 96, 108, 174, 175, 262, 269, 275, 277, 283, 341, 398, 466, 477, 488, 489
Cuvier's beaked whale (*Ziphius cavirostris*), 29, 129, 306–326

D
Dall's porpoise (*Phocoenoides dalli*), 79, 81, 135, 242, 251, 450–452, 454, 455, 461
Data tags (Dtags), 51, 263, 334, 460
Decompression, 317
Deep diving, 15, 52, 58, 59, 61, 63, 66, 99, 100, 129, 264, 317, 452
Deep oceans, 29, 262, 264, 297, 477
Defend, 10, 76, 77, 81, 85, 268, 276, 353, 376
Delphinapterus leucas, 77, 130, 153, 170

Delphinids, xii, 7, 8, 13, 30, 52, 85, 99, 118, 172, 186, 198, 203, 222, 243, 246, 282–298, 306, 359, 362, 372, 376, 378, 388, 389, 400–403, 420, 425, 426, 436–444, 486, 489, 490
Depredations, 178, 212, 214, 216, 217, 220, 222, 224, 250, 251, 255, 339
Development, 8, 14, 18, 31, 33, 44, 45, 77, 83, 96–109, 137, 158, 170, 171, 198, 214, 215, 267, 285, 298, 338, 341, 352, 359, 405, 466, 467, 472, 490
Dialects, 34, 37, 108, 153, 244–247, 253, 254, 269, 272–274
Diel, 317
cycle, 371
pattern, 135, 173, 317
Diets, 15, 80, 149, 187, 213, 241, 242, 250, 254, 275, 333, 352, 437, 442
Discovery curves, 310, 311
Dispersal, 8, 118, 136, 171, 172, 195, 241, 244, 246–248, 253, 275, 436–438, 440, 444
Displacements, 118, 121, 129, 130, 133, 267, 273, 336
Display competitions, 77, 79, 80
Distances, 7, 17, 26, 31, 41, 42, 57, 58, 87, 100, 121, 128, 130, 131, 133, 135, 136, 151, 186, 187, 194, 196, 267, 291, 313, 318, 319, 358, 359, 362, 371, 373, 395, 420, 438, 460, 461, 486, 487, 490
Distribution*, 166, 185, 220, 240, 352, 414, 437, 451
Diversity, 9, 31, 36, 58, 63, 65, 88, 166, 171–175, 178, 184–202, 240, 273–275, 378, 414, 450–462, 473, 478
Dives, 15, 29, 32, 40, 43, 53, 57, 59, 63, 66, 82, 85, 99–101, 128, 152, 153, 216, 249, 263–266, 268, 271, 274, 275, 283, 306, 308, 315–319, 325, 333, 371, 374, 399, 442, 443, 453–455, 460, 487
Diving
behavior, 109, 118, 267, 309, 315–319, 453, 454, 456, 462
synchrony, 319
tradeoffs, 101
DNA, 37, 128, 274
Dolphins, xi, 3, 26, 52, 77, 96, 118, 148, 166, 184, 212, 240, 282, 308, 332, 347, 369, 388–389, 414, 436, 450, 466
Dusky dolphins (*Lagenorhynchus obscurus*), 11, 12, 52, 54, 57, 58, 60–64, 82–84, 108, 150–152, 172, 173, 379, 381, 388–406, 487, 491

498   Index

E

Eastern tropical Pacific Ocean (ETP), xii, 8, 17, 58, 185, 187–189, 192, 272, 369, 484

Eavesdropping, 30–32, 267, 268

Echelon, 97, 98, 358

Echolocation, 6, 29–32, 40, 44, 52, 57, 61, 246, 247, 264–266, 270, 271, 316, 319, 332, 333, 335, 336, 359, 375–379, 416, 444, 454, 458, 459, 461, 462, 487–489

Echolocation clicks, 6, 7, 29–32, 40, 43, 44, 57, 61, 246, 247, 265, 266, 270, 271, 319, 333, 375, 377, 379, 458, 489

Echosounders, 51, 375

Ecological adaptations, 450, 452

Ecological specializations, 172, 176, 241, 253

Ecology*, ix, xiii, 4, 44, 66, 118, 146, 147, 171, 185, 243, 264, 283, 306, 350–353, 370, 389, 427, 462, 485

Ecology of fear, xii, 145–158, 393

Ecotypes, 12, 27, 34, 42, 129, 132, 133, 152, 171, 241, 243, 246–248, 251, 253–255, 362

Endangered, 176, 177, 225, 254, 414, 421, 428, 452, 462, 465–478

Endangered species, 225, 467, 477, 478

Endurance competition, 81, 82

Entanglements, 178, 200, 212, 277, 339, 443, 476

Environmental degradations, 66

Environmental disturbance, 174

Environments*, 3, 28, 53, 76, 99, 118, 146, 167, 184, 212, 262, 283, 354, 373, 397, 414, 453, 476, 487

Epipelagic, 53, 54, 63

Estuaries, 9, 133, 134, 346, 355, 356, 414, 437

Eulachon (*Thaleichthys pacificus*), 130

Evasive behaviors, 84, 85, 152, 192, 475

Evolution*, xiii, 7–8, 29, 80, 151, 170, 197, 265, 354, 389

Exogamy, 285, 293, 296

Extinction, 212, 452, 466–468, 471–473, 476–478

Extra-group mating, 285, 290, 293, 295, 296

F

Fad, 356

Feeding

    behaviors, 63, 391, 395, 396, 400

    buzzes, 333, 460

Finless porpoise (*Neophocaena phocaenoides*), 29, 57, 135, 361, 420–422, 426, 450–452, 455, 458, 459, 462

Fish*, 8, 36, 53, 99, 130, 148, 173, 213, 241, 265, 332, 340, 346, 374, 391, 416, 443, 470

Fisheries*, 12, 158, 189, 214, 243, 264, 403, 404, 443, 470, 491

Fishery interactions, 175

Fishing, 58, 103, 137, 176, 178, 186, 188, 199, 200, 203, 212, 214, 216, 217, 222, 250, 277, 336, 339–341, 355, 420, 428, 436, 442, 452, 462, 466, 467, 470, 474, 476, 484

Fishwhacking, 336

Fission-fusion, 4, 9, 11, 32, 34, 52, 58, 82, 101, 108, 167, 169, 172–174, 193, 347, 351, 373, 393, 440

Fission-fusion dynamics, 169, 170, 173, 372, 373, 388, 394, 398, 399, 403

Flipper-rubbing, 26, 357, 399

Food safety trade-offs, 148, 151

Foraging*, xi, 8, 33, 52, 79, 99, 117, 148, 169, 184, 212, 241, 264, 309, 331, 346, 371, 391, 416, 417, 420, 422, 425, 442, 443, 458

    behaviors, 15, 42, 52, 59, 61, 63–66, 129, 152, 176, 253, 265, 276, 331–342, 352, 362, 404, 442

    specializations, 12, 65, 175, 187, 241, 248, 255

    tactics, 55, 64, 65, 109, 175, 178, 222, 243, 247, 249, 354, 362

G

Ganges river dolphin, 414, 416

Genealogy, 244, 247

Genitalia, 84, 86

Geographic isolation, 172

Globicephala, 7, 52, 54, 56, 63, 77, 106, 107, 170, 217, 282, 286, 291–293, 298, 375

Golfo Nuevo, 390, 391, 393, 405

Golfo San José, 390, 391, 393, 396–398

Goosing, 349, 359

Gray whale (*Eschrichtius robustus*), 62, 136, 155, 242, 466

Great white sharks, 395

Group

    decisions, 267

    identification, 18, 324

    living, 4, 52, 146, 151, 167, 184, 190, 201, 440, 441

    sizes, 4, 5, 7, 11–13, 101, 102, 149, 152, 156, 168, 171–173, 175, 190, 191, 243,

Index

246, 248, 251, 253, 269, 275, 290, 295, 310, 312, 320, 321, 323, 324, 332, 360, 361, 391, 393–397, 400, 401, 405, 417, 420–422, 424–426, 440, 441, 443, 456, 460

Grouping, xi, 4, 5, 8, 9, 11, 14, 15, 17–19, 32, 101, 149, 166, 167, 172–174, 194, 244, 246, 268, 282, 309, 320, 321, 347, 351, 352, 389, 390, 394, 396, 397, 405, 414, 417, 421, 425, 426, 428, 440, 441, 444, 473, 486, 488

Grouping patterns, xi, 4–6, 8, 9, 14, 18, 19, 166, 167, 173, 309, 346, 347, 351, 388–390, 394, 396, 397, 405

Group-specific calls, 248

Group-specific dialects, 34, 37, 247, 253

Guiana dolphin (*Sotalia guianensis*), 134, 414, 425

Gulf of California porpoise, 451

H

Habitat
degradation, 177, 428, 466
features, 331–342
loss, 137, 176, 403, 404, 467, 472
preference, 154, 175, 185, 436–438

Harbor porpoise (*Phocoena phocoena*), 7, 45, 81, 87, 135, 136, 222, 242, 456, 491

Heaviside's dolphin (*Cephalorhynchus heavisidii*), 135

Hector's dolphins, 7, 13, 134, 436–438, 440–444, 466

Herding, 12, 60, 61, 64, 249, 336, 337, 347, 348, 352, 375, 392, 396, 397, 486, 487, 491

Home ranges, 11, 107, 109, 118, 129–137, 153, 267, 332, 333, 351, 355, 362, 417, 424, 436–438, 440, 444, 450

Homophily, 352

Human encroachment, 212–214, 221–225

Human impacts, 104, 176, 212–226, 472, 476

Human interactions, 340, 341

Humpback dolphins (*Sousa chinensis; Sousa sahulensis; Sousa plumbea*), xii, 54, 61, 79, 102, 106, 133, 177, 214, 221, 339, 466, 468, 471–473, 476

Hunting development, 298

I

Inbreeding, 37, 86, 247, 248, 294, 296, 421

Indian Ocean, 33, 61, 149, 191, 217, 221, 283, 370, 376, 450

Individual foraging techniques, 351

Individual identifications, 16, 32, 308

Indo-Pacific, xii, 4, 6, 9, 10, 12, 16, 26, 30, 33, 81, 82, 85, 96, 97, 103, 104, 148, 170, 176, 336, 345–362, 462, 476

Indo-Pacific finless porpoise (*Neophocaena phocaenoides*), 421, 450, 462

Indo-Pacific humpback dolphins, 102, 172, 214, 215, 476

Indus River dolphin, 414, 473, 474

Infant safety hypothesis, 395

Infanticide, 85, 86, 100, 102, 103, 276, 348, 419

Inia, 79, 106, 135, 167, 215, 221, 414, 417, 418, 420, 421, 425–427

Insular population, 313

Intelligence, xii, 171, 222, 276, 398, 485, 488, 489

Intragroup communication, 44

Intrapopulation differences, 172

Irrawaddy dolphins, 172, 177, 414, 416, 422–425

Island-associated, 129, 172, 310, 313–315, 370, 371, 373, 376

IUCN Red List, 168, 176, 414, 462, 467, 468, 471

J

Juvenile period, 105, 107, 169

K

Kaikoura, 11, 61, 134, 150–152, 173, 388, 390, 394–400, 402, 403, 405, 406, 442

Kerplunking, 335, 340, 352

Killer whale (*Orcinus orca*)*, 7, 27, 52–56, 58, 60–65, 77, 96, 119, 146, 168, 187, 221, 240, 282, 317, 362, 375, 416, 426, 436, 486

Kin selection, 55, 82, 248, 359

Kinship, 5, 11, 15, 18, 174, 246, 268, 274, 276

Knowledge, ix, xii, 52, 55, 56, 65, 66, 109, 119, 130, 136, 137, 153, 167, 170, 174, 176, 186, 190, 202, 241, 245, 267, 268, 270, 274, 275, 277, 294, 297, 306, 341, 342, 354, 370, 389, 403, 405, 406, 442–444, 455, 462, 477, 488

L

Lactation, 7, 80, 83, 96, 98, 100, 102, 108, 268, 359, 388, 395

Lagenorhynchus obscurus, 52, 54, 60, 79, 86, 108, 150, 172, 379, 388–390, 392, 402, 487

Lagoons, 10, 218, 220, 373, 414
Large-bodied, 85, 188, 262, 264
Lateral side swimming, 336
Leadership, 170, 174, 176, 190, 202, 283, 294, 296, 477, 486, 487
Leaping behavior, 421
Leaps, 58, 82, 196, 337, 338, 351, 379–381, 388, 389, 391, 394, 395, 399, 400, 459
Learning*, xi, 33, 53, 96, 174, 190, 223, 241, 265, 283, 336, 356, 418, 486
Lipotes, 7, 167, 414, 420–421, 426–428, 467, 491
Long-term associations, 52, 53, 82, 172, 322, 324
Lunar cycles, 316

M
Macaronesia, 291–293, 308
Macrorhynchus, 56, 106, 203, 217, 282, 286, 291, 292, 298, 375
Madeira, 285, 289–293
Maladaptive, 77, 177, 178, 213, 489
Male alliances, 12, 148, 170, 174, 350, 352, 353, 361
Male coalitions, 33
Management, 14, 19, 186, 199, 224, 225, 275, 381, 444, 462, 467, 468, 476, 477
Marine Mammal Protection Act, 189
Mate choice, 77, 80, 84, 117, 118, 395
Mate choice copying, 84, 85
Maternal, 5, 9, 11, 15, 37, 41, 96–109, 128, 186, 245, 248, 263, 269, 273, 341, 347, 359
  bonds, 5, 9, 246
  care, xi, 18, 41, 96–109, 186
Mating
  behavior, 75, 76, 82, 85, 88, 354, 393, 416
  chase, 82, 84, 85, 423
  strategy, 395
  system, 76, 87, 189, 197, 200, 201, 319, 347–349, 373, 393, 415, 417, 455
  tactics, 75–88
Matriarchal, xii, 9, 76, 177, 285, 294, 489
Matriarchies, 55, 170, 174, 244, 245, 267, 282–298, 317, 486, 487, 489
Matrifocal, 245
Matriline, 171, 244, 245, 247–249, 253
Matrilineal societies, 52, 245, 285, 293–296, 298
Māui dolphins, 436–444
Melas, 63, 76, 107, 218, 282
Menopause, 174, 245
Mesopelagic, 54, 61, 63, 129, 151, 186, 264, 265, 267, 374, 453

Mesoplodon densirostris, 53, 129
Migrations, 11, 108, 118, 119, 128, 130, 135, 137, 173, 174, 221, 240, 242, 250, 264, 267, 297, 317, 374, 375, 437, 453
Milk composition, 98, 99
Mill, 65, 491
Mitochondrial DNA, 273, 290
Mixed species schools, 190, 191
Monodon monoceros, 80, 130, 153, 170
Mounting, 348, 349, 359
Mourning (responses to calf death), 177
Movements*, 15, 36, 57, 82, 117, 148, 187, 214, 241, 264, 288, 308, 332, 350, 391, 414, 437, 450, 486
Mud plume feeding, 335
Mud ring feeding, 337, 338
Multilevel societies, 297–298
Multiple male mating, 77, 81
Mussel farms, 220, 390, 404

N
Narrow-ridged finless porpoise (*Neophocaena asiaeorientalis*), 421, 450, 452, 462
Narwhal (*Monodon monoceros*), xiii, 54, 80, 98, 99, 130, 153, 155, 170, 173, 221
Natal philopatry, 174, 244, 246, 285, 290, 293–296, 347
Neophocaena, 57, 135, 361, 414, 420–422, 427, 450
Networks, 9, 10, 16, 18, 34, 105, 108, 109, 152, 167, 168, 170, 172, 174, 175, 269, 288, 290, 292, 310, 312–314, 353, 399, 403, 404, 488, 490
Newborn, 32, 96, 97, 99, 102, 106, 244, 247, 355, 374, 393, 395, 438, 487
New Zealand*, xiii, 9, 52, 58, 60–63, 104, 131, 150, 172, 249, 388, 402, 436
Niche partitioning, 313–316
Noises, 15, 29, 45, 58, 137, 158, 176, 283, 335, 361, 389, 416, 420, 456, 460–462, 466
Nomadic, 152, 267, 273
Non-consumptive predation, 157
Nursery groups, 85, 152, 173, 395, 396, 405, 440

O
Observational learning, 65, 336, 341
Oceanic dolphins, 80, 119, 130, 184–202, 489
Odontocetes*, ix, 28, 52, 96, 118, 147, 165, 184, 212, 264, 308, 369, 414, 450, 467
Offshore killer whales, 222, 242
Olfaction, 28, 30

Index 501

Ontogeny of diving, 265
Opportunistic foraging, 152, 213, 217
*Orca**, 7, 27, 52, 54, 56, 119, 146, 168, 217, 240, 264, 282, 317, 375, 393, 416, 426, 436
Orcaella brevirostris, 172, 217, 218, 422–425
*Orcinus**, 7, 27, 52, 54, 56, 101, 118, 168, 187, 240, 317, 375, 393, 416, 426, 436
*Orcinus orca**, 7, 27, 52, 54, 56, 101, 119, 146, 168, 187, 217, 240, 282, 317, 375, 393, 416, 426
Orinoco, 417, 425
Outbreeding, 248

P

Pacific*, 7, 34, 55, 77, 98, 128, 149, 185, 217, 241, 267, 283, 345, 370, 395, 450, 466, 484
  halibut (*Hippoglossus stenolepis*), 129
  herring (*Clupea pallasii*), 135
Pantropical spotted dolphins, 106, 185, 186, 188, 191, 193, 194, 198, 376
Passive listening, 247, 251, 317, 332, 335
Patagonia, 251, 333, 390, 405, 406
Patagonian toothfish (*Dissostichus eleginoides*), 129, 216, 243, 250
Paternity, 15, 76, 77, 81–84, 86, 102, 103, 198, 348
Penguins, 129, 243, 246
Península Valdés, 390–394, 396, 397, 402, 406
Petting, 97, 349, 357
Philopatry, 8, 15, 167, 170, 290, 293, 294, 296, 347, 348, 354, 417
Phocoena sinus, 450, 451, 466, 468, 470, 471, 491
Photo-identification (Photo-ID), 14, 15, 128, 284–286, 289, 291, 294–296, 308–310, 325, 356, 360, 395, 397, 406, 420
Phylogenetics, 197, 420
Physeter, 4, 29, 42, 52, 54, 105, 118, 156, 170, 216, 219, 222, 263, 266, 269, 284, 317, 466, 468, 487, 491
Physeter macrocephalus, 4, 29, 42, 52, 54, 79, 105, 128, 156, 170, 216, 217, 222, 263, 266, 269, 284, 317, 466, 468
Pigmentation patterns, 27, 308, 309
Pinnipeds, 109, 118, 157, 241–243, 247
Pinwheeling, 336
Plasticity, 4, 8–13, 76, 178, 213, 427, 486
Platanista, 135, 167, 221, 414–416, 420, 421, 426–428, 466, 473, 474
  P. gangetica, 136, 214, 220, 221, 473

  P. minor, 466, 473, 474
Play, 27, 34, 75, 100, 137, 157, 174, 176, 187, 194, 196, 270, 326, 352, 359, 372, 376, 378, 417, 421, 443, 477, 486
Pod, 34, 37, 102, 153, 175, 177, 245, 247, 286, 289, 290, 295, 372, 374–376, 378–381, 424
Polar bear, xi
Polyestry, 84, 85
Polyandry, 76
Polygynandry, 76, 197, 198, 373, 393
Pontoporia, 107, 149, 169, 414, 416, 426, 427
Porpoises*, xiii, 7, 29, 57, 75, 77, 79, 81, 87, 98, 135, 157, 167, 222, 242, 361, 420, 450, 466
Post-anal keel, 187, 192, 194, 197, 198
Postcopulatory, 76, 79, 83, 86, 88
Post-reproductive, 102, 170, 174, 202, 245, 248, 253, 283, 285, 286, 293, 296, 297
Post-reproductive females, 186, 194, 245, 283, 294
Precopulatory, 76, 79
Predation, 4, 30, 76, 79, 96, 173, 185, 214, 241, 264, 309, 350, 375–376, 389, 414, 436, 486
Predation risk, 4, 7, 11, 101, 102, 135, 145, 146, 149, 150, 152–154, 158, 173, 192, 309, 320, 323, 325, 350, 352, 360, 389, 394–397, 400, 405, 414, 417, 425–427, 436, 486
Predator avoidance, xii, 117, 118, 137, 149, 184, 189, 315–319, 459
Prey balls, 60, 62, 64, 391, 392, 396, 397, 399, 400
Prey capture, 59–64, 216, 283, 316, 332, 336, 338, 392, 393, 397, 459
Prey containment, 61
Prey searching, 57
Prey sharing, 108, 248
Provisioning, 18, 88, 103, 214, 248, 339, 340
Pygmy killer whale (*Feresa attenuata*), 129, 170, 282, 375

R

Range of communication, 6, 26, 27
Rate of interaction model, 353, 361
Reciprocal altruism, 55
Red List, 414, 452, 462, 467, 468, 471, 475
Relatedness, 6, 11, 37, 85, 103, 137, 170, 174, 276, 290, 294, 295, 323, 348, 352, 354, 421
Relationship complexity, 4, 6

Reproduction, 5, 18, 79, 83, 105, 117, 170, 176, 195, 200, 254, 325, 373–374, 381, 395, 406, 477

Reproductive senescence, 104, 170, 296

Reproductive success, 4, 5, 11, 15, 76, 77, 81, 85, 87, 102, 148, 170, 176–178, 184, 189, 198, 347, 352, 354, 477

Residency, 131, 286, 287, 289–292, 297, 313–315, 332, 417

Resident killer whales, 9, 55, 57, 170, 171, 177, 241, 244, 245, 247–249, 253, 254

Resilience, 178, 189, 200, 202, 213, 214, 223, 255, 275, 428

Resting, 12, 151, 172, 173, 191, 263, 371–374, 376, 378, 380–382, 388, 390, 393, 394, 397, 400, 403, 405, 420, 422, 487, 488

Risk effects, 146, 147, 149, 152, 155, 157

Risso's dolphin (*Grampus griseus*), xii, 81, 101, 106, 131, 190, 203, 218, 222, 282

Rivers*, 7, 28, 79, 106, 119, 149, 167, 214, 356, 414, 437, 450, 467, 492

Roles, 5, 7, 8, 27, 28, 52, 53, 55, 57, 59, 61, 63–65, 102, 109, 137, 149, 170, 174, 186, 187, 194, 195, 201, 202, 245, 247, 265, 267, 270, 283, 295, 296, 336–338, 348–350, 352, 354, 357, 359, 372, 375, 376, 378, 441, 477, 491

Rove, 77, 88

S

Salmon sp. (*Salmonidae*), 53–55, 57, 61, 65, 129, 130, 171, 241, 242, 245, 248, 249, 253, 254, 297, 404, 475, 489

Satellite tagging, 308, 325, 454

Scavenging, 214, 216, 217, 222, 339, 341

Schools*, 8, 36, 58, 173, 186, 249, 332, 350, 375, 476, 486

School structure, 189, 193, 202

Scramble competition, 82, 391, 395, 397

Seal predation, 149

Seasonal, 55, 65, 79, 83, 96, 108, 118, 128, 130–133, 135, 136, 148, 149, 151, 152, 171–174, 188, 198, 240, 243, 253, 267, 288, 290, 293, 295, 297, 353, 371, 393–396, 398, 400, 414, 425, 437, 442, 453

Sensory integration, 189, 202, 486, 487

Sex, xii, 5, 8, 10, 11, 14, 15, 17, 34, 76, 88, 103, 149, 151, 156, 167, 172, 174, 187, 191, 193, 201, 248, 253, 265, 288, 290, 294, 297, 308, 310, 312, 319, 320, 322–324,

347, 349, 351, 357, 358, 388, 394, 399, 402, 406, 426, 438, 440, 444

Sexual
behavior, 75, 202, 349, 372, 373, 393, 421, 422, 426, 443, 458

dimorphism, 77, 79, 82, 88, 197, 198, 262, 271, 276, 283, 373, 415, 421, 455

selection, 76, 77, 87, 102, 197, 271, 455

strategies, xi, 76, 373, 488

Shark predation, 101, 149

Sharks, 11, 96, 101, 102, 104, 129, 137, 146–149, 152, 157, 167, 170, 187, 224, 241, 242, 308, 317, 346, 360, 371, 375, 376, 378, 393, 395, 397, 403, 404, 414, 416, 426, 436, 437, 484

Short-finned pilot whale (*Globicephala macrorhynchus*), 56, 106, 107, 129, 170, 203, 282–285, 289, 291, 293, 296–298, 322, 375

Signal discrimination, 84

Signature whistles, 15, 30–33, 35, 99, 171, 350

Silverfish, 243

Site fidelity, 129, 131–134, 370

Small cetaceans, 242, 404, 466, 467, 471, 477

Social
associations, 172, 173, 244, 441

behavior, 7, 9, 15, 17, 100, 108, 172–174, 268, 270, 271, 274, 276, 316, 319–325, 414, 415, 417, 420, 421, 424, 426–428, 441, 444, 450, 455–462

bonds, 11, 13, 30, 33, 75, 108, 165, 176, 177, 189, 196, 197, 202, 393, 396, 398, 400, 428

cohesion, 12, 41, 59, 176, 189, 245, 381, 476

complexity, 5, 168, 170, 171, 178, 202, 270, 362, 398

disruption, 176, 189

intelligence, 171

interactions, 18, 178, 187, 192, 202, 349–350, 357–360, 372, 418, 419, 456, 458

learning, 44, 75, 96, 109, 175, 178, 190, 202, 241, 253, 273–275, 283, 337–341, 356, 398

network, 18, 64, 152, 167, 170, 172, 174, 269, 288, 290, 292, 310, 312–314, 353, 399, 403, 488, 490

organization, 4, 14, 16, 19, 167, 170, 171, 189, 240–254, 285, 309, 319–325, 417, 490

Index

separation, 192, 200

structure*, 3, 31, 82, 101, 118, 148, 170, 186, 241, 269, 285, 348, 372, 393, 422, 438

units, 41–43, 129, 174, 243–245, 263, 268, 269, 272–275, 295

vulnerability, 176–178

Sociality, xii, xiii, 165, 172, 173, 175, 176, 178, 184, 186, 202, 246, 268, 372, 398, 414, 428

Socializing, 8, 11, 40, 43, 53, 81, 173, 263, 288, 351, 371, 372, 378, 380, 397, 400, 418, 421–423, 425, 426

Sociosexual behavior, 169, 359, 360, 414, 458

Soniferous fishes, 335

Sotalia, 98, 102, 106–108, 133, 414, 425–428

Sound playback experiment, 45

Sousa chinensis, 102, 133, 172, 214, 215, 219, 427, 466, 476

Sousa teuszii, 217, 339, 466, 471, 473

South America, 135, 187, 282, 403, 414, 451, 468

South Asia, 414

South Asian river dolphin (*Platanista gangetica*), 136, 214, 215, 220, 221, 426, 466, 473

Southeast Asia, 370, 414

Southern Ocean, 240, 242, 250, 451

Specializations, 12, 156, 172, 175, 176, 187, 241, 248, 253–255

Speciation, 12–13, 362

Spectacled porpoise (*Phocoena dioptrica*), 451, 452, 455, 462

Sperm competition, 76, 77, 83, 87, 88, 197, 198, 322, 373, 393, 417, 455

Sperm whale (*Physeter macrocephalus*), 119, 468

Sperm whale codas, 273

Spinner dolphins, 13, 17, 54, 61, 105, 149, 172, 187, 188, 191, 192, 197–199, 201, 204, 369–382, 392, 403, 487

Sponging (sponge carrying), 65, 109, 351, 352

Spotted dolphins, 12, 54, 60, 65, 99, 104, 106, 109, 131, 149, 174, 185, 186, 188, 191–193, 198–200, 203, 375, 376, 403, 487

Squid, 63, 99, 101, 173, 187, 241, 242, 263, 265, 283, 294, 297, 374, 375, 391, 394, 399, 442, 487

Stenella attenuata, 185, 203, 375, 376

Stenella frontalis, xiii, 12, 60, 104, 131, 149, 174, 400, 472, 487

Stenella longirostris, 17, 52, 54, 105, 149, 172, 204, 369, 379, 380, 392, 400, 487

Striped dolphin (*Stenella coeruleoalba*), 185, 190, 197, 203

Subantarctic, 243, 246, 251

Sublethal effects, 150

Sympatric, 13, 37, 41, 42, 129, 174, 241, 247, 254, 269, 272, 275, 316, 322, 414, 421, 425, 427

Sympatry, 241, 275

Synchronous, 101, 273, 274, 350, 358, 359, 391, 392, 399, 487, 488

Synchrony, 6, 97, 130, 169, 195, 201, 319, 350, 358, 372, 486

T

Tactics, xii, 55, 57, 58, 60, 62, 63, 65, 75–88, 109, 146, 151, 165–178, 243, 247–249, 251, 253, 354, 362

Tactile communication, 26–45, 187

Tagging, 65, 130, 193, 308, 313, 314, 325, 453, 454, 460

Terminal investment, 336

Testis, 197, 393, 417

Threats, 150, 158, 166, 340, 349, 350, 393, 403, 404, 420, 428, 436, 466, 468, 472, 475, 476

Tiger sharks, 96, 101, 102, 148, 149

Toothed whales, ix, xi, xii, 3–19, 26–30, 36, 44, 45, 58, 75, 77, 105, 107, 118, 147, 148, 150, 156, 166, 178, 476, 477, 485, 487, 488

Toothfish, 129, 216, 243, 250

Touch, 26, 169, 488

Tourism, 104, 137, 403, 443

Trade-offs, 76, 79, 146, 148

Traditions, 44, 137, 175, 254, 264, 274, 275, 277, 397, 406

Trait-mediated indirect interactions (TMII), 146, 157, 158

Transient killer whales, 7, 65, 154, 171, 242, 245, 248, 251, 254

Transit speeds, 130, 132, 135

Travelling, 187, 263, 267, 287, 291, 372, 456

Trophic cascades, 157, 158

Tuna, 17, 60, 129, 132, 175, 185, 188, 189, 193, 198, 202, 203, 246, 250, 484

Tuna-dolphin issue, 202

Tuna purse seine fishery, 106, 186, 198

Tursiops aduncus, 4, 26, 97, 107, 148, 214, 219, 336, 345–362

Tursiops truncatus, 5, 26, 35, 102, 105, 132, 148, 166, 169, 190, 217, 219, 393

V
Vaquita (*Phocoena sinus*), 466, 468, 471, 491
Vertical transmission, 65, 108, 351
Visual communication, 187, 270, 378
Visual signals, 26, 63
Vocal communication, 169, 270
Vocal cultures, 44
Vocalizations, 26, 33, 37, 44, 45, 58, 59, 63, 82, 251, 270, 274, 350, 357, 425
Vocal learning, 33, 37, 39, 44, 253, 273, 274
Vocal repertoires, 269

W
Wave wash, 60, 62, 252
Weaning, 98, 104, 106–108, 200, 340, 346, 347
Whale-watching, 298, 405

Y
Yangtze River, 7, 168, 420, 421, 450, 459, 467

Z
Ziphius cavirostris, 29, 306, 307, 311, 312, 314, 321

9783030166656